I0049038

Agribusiness Management in Sustainable Agricultural Enterprises

Richard Skiba

AFTER MIDNIGHT PUBLISHING

Copyright © 2024 by Richard Skiba

All rights reserved.

No portion of this book may be reproduced in any form without written permission from the publisher or author, except as permitted by copyright law.

This publication is designed to provide accurate and authoritative information in regard to the subject matter covered. While the publisher and author have used their best efforts in preparing this book, they make no representations or warranties with respect to the accuracy or completeness of the contents of this book and specifically disclaim any implied warranties of merchantability or fitness for a particular purpose. No warranty may be created or extended by sales representatives or written sales materials. The advice and strategies contained herein may not be suitable for your situation. You should consult with a professional when appropriate. Neither the publisher nor the author shall be liable for any loss of profit or any other commercial damages, including but not limited to special, incidental, consequential, personal, or other damages.

Skiba, Richard (author)

Agribusiness Management in Sustainable Agricultural Enterprises

ISBN 978-0-9756552-2-1 (paperback) 978-0-9756552-3-8 (eBook)

Non-fiction

Contents

Preface

The industrialization of farming was a boon to food production. It allowed for the cheap and fast production and transportation of food and made life easier for many across the world. The problem with the current industrial farming sector is that it is not sustainable. That is, it uses up more resources than can be naturally replenished, and it causes various types of pollution.

Sustainable agriculture takes a more natural approach. It consists of three main goals: a better, cleaner environment, profitability for all in the system from growers to sellers to waste managers, and promotion of social and economic equity. The environmentally friendly aspect includes using methods that enhance soil health through rotating crops and decreasing tillage. Healthier soil produces more and better plants. The use of natural fertilizers leads to less pollution and ancillary damage to wildlife. Water is used more efficiently to ensure that vital water resources are not drained.

Sustainable agriculture practices aim to minimize the negative impact on the environment by reducing pollution, conserving water and soil resources, and preserving biodiversity. By adopting techniques such as crop rotation, organic farming, and integrated pest management, sustainable agriculture helps protect ecosystems and maintain the health of natural habitats.

Sustainable agriculture prioritizes soil health and fertility through practices like crop rotation, cover cropping, and minimal tillage. Healthy soils are crucial for maintaining high agricultural productivity, as they provide essential nutrients to plants, regulate water retention, and support diverse microbial communities.

Sustainable agriculture practices help conserve water resources by promoting efficient irrigation methods, water recycling, and soil moisture management techniques. By reducing water usage and minimizing runoff, sustainable farming contributes to water conservation and mitigates the risk of water scarcity. Further, sustainable agriculture plays a role in mitigating climate change by sequestering carbon in soils, reducing greenhouse gas emissions, and promoting carbon-neutral farming practices. Agroforestry, conservation tillage, and the use of renewable energy sources are examples of sustainable practices that contribute to climate change mitigation.

Sustainable agriculture is economically viable in the long term, as it promotes resilience to environmental challenges, reduces input costs, and enhances market access for farmers. By diversifying income streams, improving soil health, and promoting local food systems, sustainable farming practices can improve the economic stability of farming communities.

Sustainable agriculture emphasizes fair labour practices, community engagement, and equitable access to resources, ensuring that farming benefits both present and future generations. By promoting social responsibility and fostering partnerships between farmers, consumers, and policymakers, sustainable agriculture contributes to building resilient and inclusive food systems.

Sustainable agriculture is essential for ensuring food security, protecting the environment, and promoting the well-being of both people and the planet. By adopting sustainable practices, farmers can meet the needs of current generations without compromising the ability of future generations to meet their own needs.

By adopting practices that prioritize ecological balance, social equity, and long-term profitability, sustainable agriculture offers a holistic approach to meeting the needs of the present generation without compromising the ability of future generations to meet their own needs.

Several management skills are essential for implementing sustainable agriculture practices effectively. Strategic planning plays a crucial role in sustainable agriculture, requiring long-term planning to integrate environmentally friendly practices while maintaining productivity and profitability. Managers must develop strategic plans that outline goals, priorities, and action steps for implementing sustainable farming techniques.

Resource management is another critical aspect of sustainable agriculture. Effective resource management is essential for optimizing the use of land, water, energy, and other resources while minimizing waste and environmental impact. This includes implementing practices like crop rotation, efficient irrigation, and soil conservation measures.

Risk management is integral to sustainable agriculture as well. Managers must handle various risks, including weather-related events, pest and disease outbreaks, and market fluctuations. They need to assess potential risks, develop contingency plans, and implement resilience-building strategies to mitigate adverse impacts on farm operations.

Financial management plays a crucial role in ensuring the economic viability of sustainable agriculture. Managers need to develop budgets, monitor expenses, and analyze financial performance to ensure profitability while investing in sustainable practices. This includes identifying cost-effective solutions, securing financing for sustainable investments, and optimizing revenue streams.

Adaptability is another key skill required for sustainable agriculture. Managers must be adaptable to changing environmental conditions, market trends, and regulatory requirements. They need to stay informed about emerging technologies, research findings, and industry developments to continually improve farm practices and remain competitive in the marketplace.

Effective communication and collaboration are essential for implementing sustainable agriculture practices across farm operations. Managers must communicate goals and expectations to employees, collaborate with stakeholders, and engage with the community to build support for sustainable farming initiatives.

Continuous monitoring and evaluation are critical for assessing the effectiveness of sustainable agriculture practices and identifying areas for improvement. Managers need to collect data, analyse performance metrics, and adjust strategies based on feedback to optimize outcomes and achieve sustainability goals.

By developing and applying these management skills, farmers and agricultural professionals can successfully implement sustainable agriculture practices that promote environmental stewardship, economic viability, and social responsibility. The focus of this book is to facilitate the development of the skills and knowledge required to manage sustainable agricultural practices.

Part 1 - Approaches to Sustainable Agriculture

Part 1 of this book covers various topics related to sustainable agriculture, including organic farming, agroforestry, precision agriculture, vertical farming, integrated pest management (IPM), soil conservation and regenerative agriculture, community-supported agriculture (CSA) and local food systems, challenges, future directions, and sustainable agriculture and farm management. It provides an overview of different approaches and practices aimed at promoting environmental sustainability, biodiversity conservation, and food security in agricultural systems.

Chapter One

Introduction to Sustainable Agriculture

Agriculture is undeniably the world's largest industry, providing livelihoods for over one billion individuals and contributing over $1.3 trillion annually to global food production (Holzschuh et al., 2012). It occupies nearly 50 percent of habitable land on Earth, serving as vital habitats and food sources for countless species. However, the expansion of agricultural land at the cost of semi-natural and natural habitats has led to negative effects on biodiversity and ecosystem functioning (Holzschuh et al., 2012). The percentage of agricultural land has increased globally, leading to the intensification of agriculture and the decline of traditional agricultural practices (Blount et al., 2021). This expansion has contributed to habitat and biodiversity loss, particularly in the United States (Melstrom, 2020).

Agricultural landscapes have historically provided habitats for a vast number of bird species, but agricultural intensification has negatively affected habitat composition, leading to a decrease in habitat quality for certain species (Schöll et al., 2023). However, there is evidence that annually cropped agriculture may provide suitable winter habitat for some species of migratory songbirds in certain regions (Valdez-Juarez et al., 2019). Additionally, extensive agriculture has been shown to benefit certain bird species, as seen in the case of the little bustard in southern France (Wolff et al., 2001).

The impact of agriculture on wildlife habitat extends beyond birds. For instance, the avifauna of primary forest edge and primary forest was found to be more species-rich and diverse than that of agricultural habitats (Dawson et al., 2011). Furthermore, the presence of agricultural land cover has been associated with a decrease in habitat quality for stream macroinvertebrates (Genito et al., 2002). In contrast, agricultural field margins in Southwestern Ontario, Canada, have been found to provide food and nesting resources to bumble bees, highlighting the potential for agricultural landscapes to support certain pollinator species (Purvis et al., 2019).

While agricultural intensification has led to the loss and fragmentation of natural and semi-natural habitats, there is also evidence that certain agricultural practices, such as the presence of uncropped agricultural field margins, can increase the value of agricultural landscapes for providing resources to certain species (Diekötter & Crist, 2013). However, monoculture systems and pesticide application within agricultural systems have been identified as detrimental to pollinators and insect communities (Buchori et al., 2019).

When agricultural practices are managed sustainably, they play a pivotal role in preserving and rejuvenating critical ecosystems, safeguarding watersheds, and enhancing soil fertility and water purity. However, unsustainable methods wield significant repercussions on both human populations and the natural environment.

The urgency for sustainable resource management intensifies as the demand for agricultural commodities escalates in tandem with the world's burgeoning population. Given agriculture's profound interconnectedness with the global economy, human societies, and biodiversity, it emerges as a paramount realm for conservation efforts worldwide.

Farming and fishing permeate every facet of daily life, implicating individuals across the entire supply chain. From producers and agrichemical manufacturers to retailers and consumers, all are intricately linked to the sustenance that graces dining tables, whether it stems from the land or the sea.

As organic and sustainable food options proliferate in mainstream markets, consumers rightfully exhibit growing apprehension about the origins of their sustenance. Queries regarding chemical usage, labour conditions, and environmental impacts underscore a broader concern: the sustainability of our agricultural and fishing practices, and the essence of sustainability itself.

Figure 1: Members of an organic community supported agriculture farm near Rostock, Germany, support the farmer by plugging weeds from the beet root field. Smaack, CC BY-SA 4.0, via Wikimedia Commons.

The initial hurdle lies in clarifying that 'sustainable' does not equate to 'organic', nor vice versa, though there can be overlap:

- Organic agriculture entails adhering to specific guidelines concerning genetic modification and production methods, emphasizing limited fertilizer and pesticide usage. However, organic production may overlook considerations such as land utilization or the proximity of food sourcing. Fishing, by its nature, cannot consistently align with organic standards due to the uncontrollable dietary habits of wild fish.

- Conversely, sustainable agriculture and fishing strive to produce food while conserving natural resources and ensuring long-term viability. The objective is to augment natural assets, operate with fiscal prudence, and enhance the well-being of both producers and consumers.

Practitioners of sustainable agriculture pursue the integration of three primary objectives: environmental health, economic viability, and social equity. Every entity involved in the food chain, from growers and processors to retailers and consumers, bears a role in fostering a sustainable agricultural ecosystem.

Numerous practices commonly employed in sustainable agriculture and food systems underscore this endeavour. Growers implement methods to bolster soil health, curtail water consumption, and diminish farm-based pollution. Similarly, consumers and retailers attuned to sustainability seek out "values-based" foods that promote farmworker welfare, environmental stewardship, or local economic resilience. Meanwhile, researchers traverse interdisciplinary boundaries, melding biology, economics, engineering, chemistry, and community development to advance sustainable agriculture.

Yet, sustainable agriculture transcends mere adoption of practices; it embodies a dynamic process of negotiation—a tug-of-war between the sometimes conflicting interests of individual farmers or communities as they navigate the intricacies of food and fibre production.

Sustainable agriculture epitomizes farming practices geared toward meeting current societal needs for food and textiles without compromising the ability of present or future generations to meet their own. It stems from an appreciation of ecosystem services and embraces a multitude of methods to bolster agricultural sustainability. Within sustainable food systems, fostering flexible business processes and farming techniques is crucial.

Agriculture wields a colossal environmental footprint, significantly contributing to climate change, water scarcity, pollution, deforestation, and other ecological upheavals. Nevertheless, sustainable agriculture champions environmentally benign farming methods that nurture crops and livestock without imperilling human or natural systems. Key elements include permaculture, agroforestry, mixed farming, multiple cropping, and crop rotation.

Through decades of scientific inquiry and practical application, several farming practices have emerged as effective means of achieving sustainability, particularly when deployed in concert:

- Rotating crops and embracing diversity: Cultivating a diverse array of crops yields numerous benefits, including improved soil health and enhanced pest control. Intercropping and complex multiyear crop rotations exemplify strategies to enhance crop diversity.

- Planting cover crops and perennials: Cover crops like clover and rye, as well as perennial crops, safeguard soil integrity, replenish soil nutrients, and deter weed proliferation. These crops mitigate erosion, bolster soil health, and curtail the need for fertilizers and herbicides.

- Reducing or eliminating tillage: No-till or reduced-till methods circumvent soil loss inherent to traditional ploughing, fostering erosion mitigation and soil health enhancement.

- Employing integrated pest management (IPM): IPM encompasses a range of techniques, from mechanical to biological controls, aimed at managing pest populations while minimizing reliance on chemical pesticides.

- Integrating livestock and crops: Introducing livestock into crop operations can optimize farm efficiency and profitability by capitalizing on synergies between plant and animal production.

- Embracing agroforestry practices: Incorporating trees or shrubs into farming operations provides shade, shelter, and potential additional income streams while fortifying ecosystem services.

- Managing whole systems and landscapes: Sustainable farms treat uncultivated or less intensively cultivated areas as integral components, leveraging their ecological contributions to control erosion, reduce nutrient runoff, and foster biodiversity.

Figure 2: Agroforestry in Burkina Faso with Borassus akeassii and Faidherbia albida. Marco Schmidt [1], CC BY-SA 2.5, via Wikimedia Commons.

Central to many of these practices is their emphasis on soil health. Preserving fertile, biologically active soil is pivotal to resolving myriad issues associated with industrial agriculture. Healthy soil nurtures robust crops, retains moisture, prevents pollution, and sustains farming communities.

Moreover, diversity emerges as a common thread linking many of these practices. The most sustainable and productive agricultural systems mimic the complexity and diversity of natural ecosystems.

Sustainable agriculture epitomizes a form of farming that seeks to safeguard the environment while fulfilling present consumption needs and preserving resources for future generations. It embodies stewardship on the part of farmers and encompasses individual practices as much as state policies.

Farmers, whether individual or industrial, hold the agency to adopt practices that render their operations more environmentally sound. Alternatively, governments can enact regulations and policies to uplift ecological standards across a nation's agricultural sector.

A less intrusive approach entails the establishment of a governmental regulatory body tasked with certifying farms as 'sustainable' upon adherence to specified guidelines or the adoption of sustainable practices. Through such mechanisms, states and individuals can opt to collaborate to varying degrees in the pursuit of agricultural sustainability.

Methods of sustainable farming span a spectrum:

- Rotational crop farming and grazing optimize nutrient usage and combat pest pressures.

- Ecological pest management reduces reliance on chemical pesticides through preventive measures.

- Reduced soil disturbance curtails erosion and enhances soil health.

- On-site energy production promotes self-sufficiency and mitigates dependence on non-renewable energy sources.

Figure 3: Oregon vineyard in the Willamette Valley wine region utilizing solar power. eyeliam, CC BY 2.0, via Wikimedia Commons.

Sustainable agriculture embodies a holistic approach to farming that balances environmental, economic, and social imperatives. It represents a conscientious commitment to the well-being of both present and future generations, underscoring the critical importance of agricultural sustainability in safeguarding our planet's resources and nourishing its inhabitants.

Since the end of World War II, agriculture has undergone profound transformations, characterized by heightened productivity in food and fibre production facilitated by technological advancements, mechanization, increased chemical usage, specialization, and government policies prioritizing output maximization and price reduction (Brodt et al., 2011; Dubey, 2024). While these changes have yielded numerous benefits, including increased efficiency and reduced farming risks, they have also incurred significant costs. Challenges such as topsoil depletion, groundwater contamination, air pollution, greenhouse gas emissions, the decline of family farms, labour rights neglect, and rural community disintegration have surfaced.

In response to these concerns, a sustainable agriculture movement has emerged over the past four decades, advocating for innovative alternatives (Brodt et al., 2011). Sustainable agriculture revolves around three core objectives: environmental health, economic profitability, and social equity. It emphasizes the imperative of meeting present needs without compromising the ability of future generations to meet their own, necessitating the stewardship of both natural and human resources. This stewardship extends to considerations of land and natural resource preservation, labour conditions, animal welfare, and consumer health and safety.

An agroecosystems and food systems perspective underpins the understanding of sustainability, emphasizing the interconnectedness of agricultural and food production, distribution, and consumption. A systems approach facilitates a comprehensive examination of the impacts of agricultural enterprises on human communities and the environment while enabling assessments of societal influences on farming and environmental sustainability (Brodt et al., 2011).

Resilience, adaptability, and diversity emerge as pivotal attributes of sustainable agroecosystems and food systems. These systems endure over time due to their ability to withstand disturbances, adapt to changing conditions, and exhibit high levels of biodiversity (Brodt et al., 2011). Additionally, fostering sustainability necessitates collaborative efforts across diverse stakeholders, including researchers, farmers, labourers, retailers, consumers, and policymakers (Brodt et al., 2011).

It is important to recognize that sustainable agriculture is a dynamic, evolving concept influenced by contemporary issues, perspectives, and values. Definitions of sustainability are continually evolving, with scientific understanding adapting to changing circumstances and cultural contexts. Hence, it is more practical to view agricultural systems along a continuum of sustainability rather than adhering to a binary sustainable/unsustainable dichotomy.

Dubey (2024) outlines that Sustainable agriculture prioritizes the cultivation of diverse crops, including heirloom varieties tailored to the specific climate of a region. Instead of relying on a single crop within an industrial monoculture system, sustainable agriculture promotes polyculture, where multiple crops are cultivated together. Although polyculture typically requires more labour compared to industrial monoculture, it offers benefits such as reduced reliance on chemical pesticides and fertilizers, as well as overall improvements in soil quality. Additionally, crop rotation can help maintain soil productivity and minimize the need for agricultural chemicals for fertilization and pest management. Incorporating nitrogen-fixing cover crops, smother crops, and green manures aids in soil restoration and erosion reduction. The practice of composting crop residues and other agricultural wastes serves to recycle nutrients back into the farmland.

Further, Dubey (2024) outlines that sustainable agriculture endeavours to reform animal agriculture, recognizing its significant contribution to greenhouse gas emissions, which fuel anthropogenic global warming. Strategies such as managing manure and incorporating animal feed additives can mitigate methane emissions, a potent greenhouse gas.

Alternative approaches to enhancing the sustainability of animal agriculture focus on maintaining the health of livestock. Intensive animal farming practices can precipitate health crises, with diseases like Nipah virus and swine flu arising from overcrowded factory farms. Avian influenza poses a threat, easily transmitted between wild birds and poultry farms, resulting in widespread culling of chickens and other poultry worldwide. Sustainable agriculture initiatives aim to mitigate planetary health risks associated with livestock farming by reducing animal density and elevating farm hygiene standards. On smaller-scale farms, integrating animal and crop production fosters interconnected systems that minimize waste.

Preserving water resources stands as a crucial component of sustainable agriculture. Globally, agriculture consumes approximately 70 percent of available freshwater reserves (Dubey, 2024). Strategies to curb water wastage encompass enhancing water storage methods to mitigate losses from evaporation and seepage, as well as cultivating drought-resistant or climate-appropriate crops. While many agricultural regions rely on conventional flooding or surface irrigation, this approach often inundates fields with excess water, leading to significant evaporation and transportation losses. In response, some sustainable farmers advocate for reduced-volume irrigation, delivering controlled streams of water tailored to the precise needs of crops, thus minimizing water waste.

Moreover, sustainable agriculture endeavours to tackle surface and groundwater contamination. Industrial-scale agriculture frequently generates pollutants such as agrochemical runoff and pathogen-laden animal waste, which infiltrate water bodies, endangering surrounding ecosystems and human health. Soil erosion exacerbates water quality degradation, diminishing crop yields and available agricultural land. To combat these issues, farmers can adopt practices such as reduced tillage or no-till methods to limit soil disturbance. Careful application of fertilizers and pesticides, preferably during dry weather conditions, can mitigate runoff and air pollution resulting from chemical drift. Additionally, some farmers deploy buffer plants along waterways to absorb nutrient pollutants before they reach aquatic environments.

Sustainable agriculture advocates for a transition away from non-renewable fossil fuels towards clean and renewable energy sources, such as solar, wind, nuclear, and hydroelectric power. Numerous sustainable farms integrate on-site wind turbines or solar panels to fulfill their electricity requirements and may incorporate electric vehicles for various farm operations. Innovations like energy-efficient farm machinery and enhanced insulation of farm structures contribute to minimizing energy consumption in agriculture. The use of fossil fuels is linked to air pollution, acid rain, and the emission of carbon dioxide, a primary contributor to global warming.

From an agroecosystems standpoint, sustainability involves considering how agricultural practices impact natural resources, biodiversity, and ecosystem health. By viewing agriculture as part of a larger ecological system, we can better understand the interplay between farming practices and environmental outcomes. For example, agricultural activities such as pesticide use or monocropping can have far-reaching consequences on soil fertility, water quality, and wildlife habitats. Adopting an agroecosystems perspective allows us to assess the ecological implications of different farming methods and identify strategies for minimizing environmental harm while promoting sustainable land stewardship.

Similarly, a food systems approach broadens our understanding of sustainability by examining the entire lifecycle of food, from production to consumption. This perspective recognizes that agricultural sustainability is not solely determined by on-farm practices but is also influenced by factors such as food processing, transportation, and waste management. By analysing the interconnectedness of these elements, we can identify opportunities to reduce resource consumption, minimize food waste, and enhance the resilience of the food system.

Moreover, a systems approach enables us to consider the social dimensions of sustainability, including the impacts of agriculture on human communities and livelihoods. Sustainable agriculture should not only prioritize environmental conservation but also address issues of equity, social justice, and community well-being. By examining the socioeconomic dynamics of food production and distribution, we can identify strategies for promoting fair labour practices, supporting small-scale farmers, and ensuring access to healthy and affordable food for all members of society.

In essence, an agroecosystems and food systems perspective offers a comprehensive lens through which to evaluate sustainability in agriculture. By recognizing the interconnectedness of agricultural activities with ecological, economic, and social systems, we can develop holistic solutions that promote the long-term health and resilience of food systems while safeguarding the well-being of both people and the planet.

Organic Farming

Organic farming encompasses a set of principles and practices aimed at cultivating crops and raising livestock in a manner that promotes environmental sustainability, soil health, and biodiversity while minimizing the use of synthetic inputs such as pesticides and fertilizers. Key components of organic farming include soil health management, crop rotation, and natural pest control methods.

Organic farming is a sustainable agricultural system that prioritizes ecologically based pest control methods and utilizes biological fertilizers primarily derived from animal and plant wastes, as well as nitrogen-fixing cover crops. This approach to farming emerged as a response to the detrimental environmental impacts associated with the use of chemical pesticides and synthetic fertilizers in conventional agriculture. Organic farming offers numerous ecological benefits compared to conventional methods (Adamchak, 2023).

In contrast to conventional agriculture, organic farming employs fewer pesticides, mitigates soil erosion, reduces nitrate leaching into groundwater and surface water, and incorporates animal wastes back into the farm. However, these advantages are accompanied by higher food costs for consumers and generally lower yields. Organic crop yields have been observed to be approximately 25 percent lower overall than conventionally grown crops, although this can vary significantly depending on the crop type (Adamchak, 2023). Moving forward, the challenge for organic agriculture lies in maintaining its environmental benefits, increasing yields, and reducing prices while addressing the challenges posed by climate change and a growing global population.

The history of organic agriculture traces back to the early 1900s, with pioneers such as Sir Albert Howard, F.H. King, and Rudolf Steiner advocating for farming practices that emphasized the use of animal manures, cover crops, crop rotation, and biologically based pest controls (Adamchak, 2023). These principles gained traction in the mid-20th century, spurred by publications like "Organic Gardening and Farming" magazine and Rachel Carson's "Silent Spring," which highlighted the environmental harms of chemical pesticides.

The demand for organic food surged in the late 20th century, driven by increased environmental awareness and concerns over pesticide residues and genetically modified crops. Retail sales of organic food experienced significant growth, with the United States and Europe leading the market. However, organic food typically commands higher prices compared to conventionally grown produce, reflecting the costs associated with organic certification and production methods.

Organic agriculture is formally defined and regulated by governments, with strict standards governing the certification and labelling of organic products. These standards prohibit the use of synthetic pesticides, fertilizers, ionizing radiation, sewage sludge, and genetically engineered plants or products (Adamchak, 2023). Certification is carried out by approved organic control bodies, ensuring compliance with organic standards.

In terms of farming methods, organic agriculture emphasizes soil health management, crop rotation, and natural pest control strategies. Organic farmers prioritize building and maintaining fertile soil through the addition of organic matter, such as compost and manure. Crop rotation and cover cropping are employed to enhance soil fertility, prevent erosion, and suppress weeds. Pest control in organic farming relies on biological pesticides derived from natural sources, as well as biological, cultural, and genetic controls to minimize pest damage.

Overall, organic farming represents a holistic approach to agriculture that seeks to balance environmental sustainability, economic viability, and social responsibility (Adamchak, 2023). By promoting regenerative farming practices and reducing reliance on synthetic inputs, organic agriculture offers a promising path towards a more resilient and sustainable food system.

Soil health management is a critical aspect of organic farming, emphasizing the significance of soil in sustainable agriculture. Healthy soil, enriched with organic matter, nutrients, and beneficial microorganisms, plays a pivotal role in promoting plant growth and resilience (Díaz-Pérez & Batal, 2002). Organic soil management practices such as composting, cover cropping, and mulching are integral to maintaining soil health. Composting involves the decomposition of organic materials to create nutrient-rich compost, enhancing soil structure and fertility (Díaz-Pérez & Batal, 2002). Cover cropping, on the other hand, involves planting non-cash crops during fallow periods to protect the soil from erosion, suppress weeds, and add organic matter, thereby improving soil structure and fixing nitrogen in the soil (Díaz-Pérez & Batal, 2002). Mulching, the application of organic materials to the soil surface, conserves moisture, suppresses weeds, and moderates soil temperature, further contributing to soil health (Díaz-Pérez & Batal, 2002).

Research has shown that the use of mulches, such as coloured plastic film, paper, and organic materials, can significantly impact soil health and plant growth. For instance, studies have demonstrated that coloured plastic film mulches affect root-zone temperature, which in turn influences tomato growth and yield (Díaz-Pérez & Batal, 2002). Additionally, the application of plastic mulches has been found to improve soil microbial diversity and community, contributing to the sustainable development of crops (Díaz-Pérez & Batal, 2002). Furthermore, the performance of different mulching materials has been evaluated in various regions, indicating their influence on soil moisture content, weed infestation, and growth of crops such as maize and groundnut (Díaz-Pérez & Batal, 2002).

Figure 4: Soybean Field with Healthy Soil. USDA NRCS South Dakota, Public domain, via Wikimedia Commons.

Moreover, the type and colour of mulches have been found to affect soil temperature, with implications for plant growth and yield. For example, red plastic mulch has been shown to increase soil temperature, while yellow plastic mulch has been associated with lower soil temperatures (Díaz-Pérez & Batal, 2002). Additionally, the use of black polyethylene mulch has been found to significantly impact soil properties, leaf nutrient status, and weed growth in pomegranate cultivation under rainfed conditions (Díaz-Pérez & Batal, 2002).

Crop rotation is a fundamental practice in organic farming that involves alternating the types of crops grown in a particular field over time. This practice has been shown to have numerous benefits for soil health, pest and disease management, and overall agricultural sustainability. emphasize the economic value of grain legume pre-crop benefits, highlighting the competitiveness and positive effects of grain legumes in crop rotations (Preißel et al., 2015). Additionally, Gaudin et al. (2013) stress the importance of improving the resilience of field crop systems through the use of inter-seeded red clover, which contributes to nitrogen inputs and efficient use of resources. Furthermore, Liu et al. (2022) discuss the mapping of complex crop rotation systems, providing insights into the seasonal dynamics and diversity of crop rotations, which are essential for understanding and implementing effective crop rotation strategies.

In their study, Liu et al. (2022) delve into the intricate mapping of complex crop rotation systems, shedding light on the seasonal dynamics and diversity inherent within these systems. Crop rotations, a fundamental aspect of sustainable agriculture, involve the sequential planting of different crops in the same field over a period of time. This practice offers numerous benefits, including improved soil health, enhanced nutrient cycling, and reduced pest and disease pressure.

One of the key insights offered by Liu et al. (2022) is the recognition of the nuanced patterns and transitions that characterize crop rotations throughout the year. By comprehensively mapping these rotations, researchers can gain a deeper understanding of the temporal dynamics involved, including the timing of planting and harvesting, as well as the sequence and variety of crops grown. This information is crucial for developing and implementing effective crop rotation strategies that optimize agricultural productivity while minimizing environmental impacts.

Moreover, the study highlights the importance of diversity within crop rotations, both in terms of the types of crops grown and their spatial distribution across agricultural landscapes. Diverse crop rotations not only promote soil health and fertility but also contribute to pest and weed management by disrupting pest life cycles and reducing monoculture pressures. By elucidating the diversity of crop rotations, Liu et al. (2022) provide valuable insights into how farmers can design more resilient and sustainable cropping systems.

Furthermore, the findings of the study have practical implications for agricultural decision-making and land management. By understanding the seasonal dynamics and diversity of crop rotations, farmers and land managers can make informed choices regarding crop selection, rotation sequences, and agronomic practices. This knowledge enables them to optimize resource use, minimize risks, and maximize yields while promoting long-term sustainability.

The benefits of crop rotation are not limited to soil health and resource efficiency. (Garrison et al., 2014) illustrate the weed suppression benefits of diversifying crop rotations, emphasizing the role of crop diversity in sustainable weed management. Moreover, Stevenson and Kessel (1996) highlight the positive impact of pulse crops in rotation systems, leading to greater seed yields in succeeding cereal crops, demonstrating the weed and pest management benefits of crop rotation. Additionally, (Baumhardt, 2015) discusses the importance of crop choices and rotation principles, emphasizing the need for profitable crop selection and rotation implementation to protect natural resources.

Furthermore, the environmental and climate resilience aspects of crop rotation are addressed by Yu et al. (2022) and He et al. (2021). Yu et al. (2022) emphasize the importance of crop rotation in improving climate resilience and addressing the shortcomings of continuous crop methodologies. He et al. (2021) highlight the role of crop rotation in enhancing agricultural sustainability and supporting a synergistic service to both society and nature by increasing resource efficiency and reducing pest epidemics.

Natural pest control methods are crucial in organic farming to manage pests and diseases without relying on synthetic chemicals. These methods include biological control, crop diversity, and habitat manipulation (Gajda et al., 2016). Biological control involves the use of beneficial organisms such as predators, parasites, and pathogens to control pest populations. For example, introducing ladybugs to control aphids or using Bacillus thuringiensis (Bt) bacteria to control caterpillars (Tiffin & Balcombe, 2011). Crop diversity, achieved through intercropping or polyculture, can disrupt pest cycles and reduce the likelihood of pest outbreaks. Planting a variety of crops in close proximity can confuse pests and provide habitat for beneficial insects (Nowak et al., 2013). Habitat manipulation involves creating favourable habitats for natural enemies of pests, such as flowering plants to attract pollinators or predator-friendly environments for insect predators (Suwanmaneepong et al., 2020).

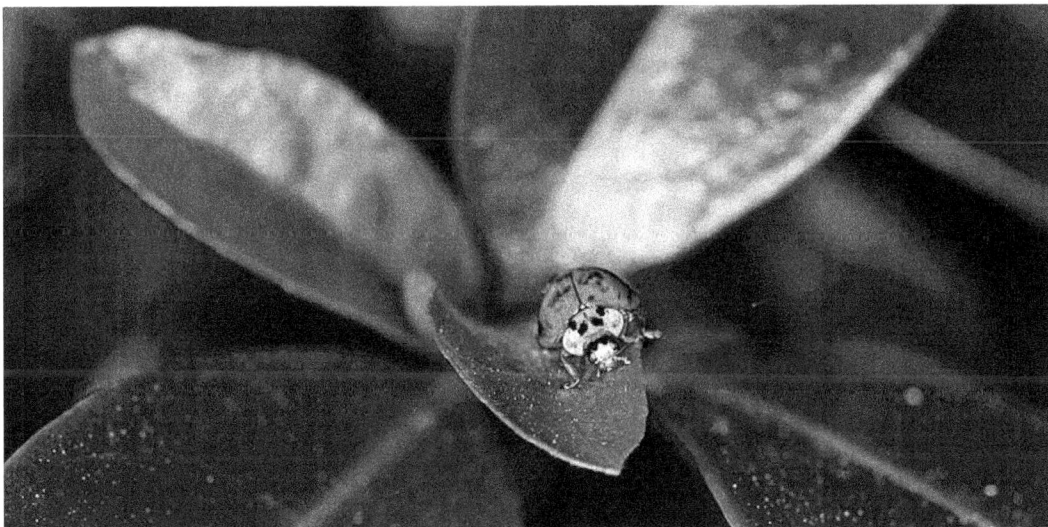

Figure 5: Lady bug eating aphids. DAVID S. FERRY III from PLANT CITY, FL., U.S.A., CC0, via Wikimedia Commons.

The use of natural pest control methods in organic farming has been shown to have positive effects on soil quality. Research has indicated that organic farming leads to a greater amount of organic matter in arable soils, influencing the readily-dispersible clay content (Lee & Eom, 2009). Additionally, the glomalin content in soils from organic farms was significantly higher than in soils from conventional farms, indicating the positive impact of organic farming on soil health (Pawlewicz et al., 2020). Furthermore, lower soil bulk densities and high soil microbial activity were found in areas where organic farming was practiced, highlighting the potential benefits of organic farming systems on soil microbial communities and soil structure (Ma et al., 2017).

In addition to the environmental benefits, organic farming has also been found to have economic advantages. Studies have shown that organic farming can offer business opportunities and better income for farmers, especially if marketing linkages are established, allowing them to fetch a better price for organic products than conventional products (Hrabalova & Zander, 2006). Moreover, the economic analysis of commercial organic and conventional vegetable farming revealed a higher benefit to cost ratio in organic farms compared to conventional farms (McGee & Alvarez, 2016).

Furthermore, the adoption of organic farming is influenced by various factors. Better educated farmers with larger farms are more likely to adopt organic farming, while farm specialization appears to have a negative impact on the willingness to adopt (Shrestha, 2015). Additionally, the majority of farmers started organic production immediately after hearing about the concept of organic farming, indicating the importance of awareness and education in promoting organic farming practices (Makaju & Kurunju, 2021). Moreover, the environmental attitudes and the social embedding of the farmers within organic agriculture played a decisive role in those enterprises that chose to continue farming organically (Vaish et al., 2020).

Organic farming presents a comprehensive approach to agriculture, rooted in its holistic and ecologically conscious principles. This approach yields various environmental benefits by minimizing reliance on synthetic inputs, emphasizing soil health, and fostering biodiversity. The discussion below elaborates on some of the key environmental advantages associated with organic farming.

Reduced Chemical Inputs - Organic farming significantly reduces reliance on synthetic pesticides and fertilizers, opting instead for natural and organic alternatives like compost, crop rotation, and biological pest control methods. By eschewing synthetic chemicals, organic farming mitigates negative impacts such as chemical runoff and pesticide residues in soil, water, and ecosystems. Consequently, this practice helps preserve water quality and safeguard wildlife habitats.

Improved Soil Fertility - The focus of organic farming on building and maintaining soil health is evident through practices like incorporating organic matter, utilizing cover crops, and implementing crop rotations. These strategies enhance soil structure, improve water retention, and facilitate nutrient cycling. Through composting, organic farming enriches the soil with vital nutrients and beneficial microorganisms, fostering a diverse and healthy soil ecosystem. Healthy soil is better equipped to support plant growth, sequester carbon, and resist erosion, thereby contributing to long-term soil fertility and productivity.

Enhanced Biodiversity - Organic farming promotes biodiversity by establishing diverse and balanced agroecosystems that emulate natural ecosystems. Practices such as crop diversity, intercropping, and agroforestry support a wide array of plant and animal species. By providing habitat and food sources for beneficial insects, birds, and other wildlife, organic farms contribute to biodiversity conservation and ecosystem resilience. Beneficial organisms, including pollinators and natural predators, play crucial roles in pest control and crop pollination, thereby reducing the need for synthetic pesticides and fostering ecosystem stability.

Preservation of Ecosystem Services - Organic farming practices aid in preserving essential ecosystem services such as soil formation, nutrient cycling, and water purification. These services are vital for sustaining agricultural productivity and supporting human well-being. By maintaining healthy soil, minimizing chemical pollution, and promoting biodiversity, organic farming enhances the capacity of agroecosystems to provide ecosystem services that benefit agricultural production and the broader environment.

Overall, organic farming embodies a multifaceted approach to sustainable agriculture, prioritizing environmental stewardship and ecosystem health. Through strategies such as reducing chemical inputs, improving soil fertility, enhancing biodiversity, and preserving ecosystem services, organic farming contributes to the conservation of natural resources, the protection of wildlife habitats, and the promotion of a resilient and sustainable food system.

Agroforestry

Agroforestry encompasses a diverse array of land-use systems and practices that integrate trees and shrubs with agricultural crops and/or livestock. These systems are designed with deliberate spatial arrangements or temporal sequences to foster ecological and economical interactions among their components. Essentially, agroforestry seeks to harmonize the coexistence of agriculture and trees within the same management unit, aiming to maximize both productivity and sustainability (Gold, 2023).

Defined as a dynamic and ecologically based natural resource management system, agroforestry plays a crucial role in diversifying and sustaining agricultural production while generating social, economic, and environmental benefits. Particularly vital for smallholder farmers and rural communities, agroforestry enhances food security, income generation, and overall health. These multifunctional systems offer a wide range of economic, sociocultural, and environmental advantages.

There are three main types of agroforestry systems:

1. Agrisilvicultural systems combine crops and trees, exemplified by practices like alley cropping or homegardens.

2. Silvopastoral systems integrate forestry with animal grazing on pastures or rangelands.

3. Agrosylvopastoral systems incorporate trees, crops, and animals in integrated agricultural landscapes, such as homegardens featuring both livestock and scattered trees on croplands.

Agroforestry systems aim to balance various objectives, including timber production, food provision, and environmental conservation (Gold, 2023). By harnessing the positive interactions between trees, crops, and livestock, agroforestry promotes diverse and sustainable land use practices. These systems yield a myriad of benefits, ranging from improved livelihoods and economic growth to enhanced ecosystem resilience and environmental sustainability.

The classification of these systems is drawn from P.K. Ramachandran Nair's seminal work "An Introduction to Agroforestry," reflecting his foundational role in modern agroforestry and as a co-founder of the World Agroforestry Centre (ICRAF) in 1993 (Food and Agriculture Organisation if the United Nations, 2022).

Agrisilvicultural systems primarily involve the combination of trees with crops, offering several distinct arrangements (Food and Agriculture Organisation if the United Nations, 2022):

- Improved fallows entail planting woody species during fallow periods, allowing them to grow.

- Taungya represents a combined stand of woody and agricultural species during the early stages of plantation establishment.

- Alley cropping, or hedgerow intercropping, features woody species planted in hedges with agricultural species cultivated in alleys between the hedges in a microzonal or strip arrangement.

- Multilayer tree gardens comprise multispecies, multilayer dense plant associations with no organized planting arrangements.

- Multipurpose trees on crop lands involve trees scattered haphazardly or according to some systematic patterns on bunds, terraces, or plot/field boundaries.

- Plantation crop combinations include integrated multistorey mixtures of plantation crops, plantation crops intercropped with shade trees, and various other configurations.

Figure 6: Alley cropping walnut and soybeans. East to west is best to maintain sunlight in alleys. Missouri, USA. National Agroforestry Center, CC BY 2.0, via Wikimedia Commons.

Additionally, home gardens exemplify a multistorey combination of various trees, crops, and sometimes animals, surrounding homesteads. Trees are also employed in soil conservation and reclamation efforts, such as planting them on bunds, terraces, raisers, etc., with or without grass strips, for soil reclamation. Shelterbelts, windbreaks, live hedges, and fuelwood production schemes are further examples of agrisilvicultural systems.

Silvopastoral systems integrate trees with pastures or animals and include:

- Trees on rangeland or pastures, where trees are scattered irregularly or arranged systematically.

- Protein banks, which focus on producing protein-rich tree fodder on farm/rangelands for cut-and-carry fodder production.

- Plantation crops combined with pastures and animals, such as cattle under coconut trees in South-East Asia and the South Pacific.

Agrosilvopastoral systems incorporate animals, trees, and crops and encompass:

- Homegardens involving animals, which feature an intimate, multistorey combination of various trees, crops, and animals around homesteads.

- Multipurpose woody hedgerows, which are utilized for browse, mulch, green manure, soil conservation, and other purposes.

- Apiculture with trees, where trees are cultivated for honey production.

- Aquaforestry, which involves trees lining fish ponds and using tree leaves as 'forage' for fish, representing a unique integration of aquaculture and agroforestry practices.

Figure 7: Silvopasture system in Georgia. National Agroforestry Center, CC BY 2.0, via Wikimedia Commons.

With its roots dating back to ancient times, agroforestry has gained renewed attention in modern agricultural practices. It offers practical and cost-effective solutions for integrated land management, promoting long-term sustainability and resilience, especially for small-scale producers (Gold, 2023). Through careful management and strategic planning, agroforestry contributes to a green economy by fostering sustainable forest management and enhancing ecosystem services.

Despite variations in terminology and practices across different regions and ecosystems, agroforestry remains a globally recognized approach to sustainable agriculture. Its widespread adoption and continued research efforts underscore its potential to address contemporary challenges while preserving natural resources for future generations.

Precision Agriculture

Precision agriculture, also known as precision farming or smart farming, is an approach to farm management that utilizes advanced technology to optimize various aspects of agricultural production. The goal of precision agriculture is to increase the efficiency, profitability, and sustainability of farming operations by precisely managing inputs such as water, fertilizer, pesticides, and seeds, while minimizing waste and environmental impact.

At the heart of precision agriculture is the collection and analysis of data from various sources, including satellite imagery, drones, sensors, GPS technology, and farm machinery equipped with advanced sensors and monitoring systems. These technologies enable farmers to gather detailed information about soil conditions, crop health, weather patterns, and other relevant factors, allowing for more informed decision-making.

Figure 8: Avocado plant monitoring precision agriculture. Simple loquat, CC0, via Wikimedia Commons.

At the core of precision agriculture lies the utilization of various technologies for the collection and analysis of data from multiple sources. These sources include satellite imagery, drones, sensors, GPS technology, and advanced monitoring systems integrated into farm machinery. The amalgamation of these technologies enables farmers to gather detailed information about soil conditions, crop health, weather patterns, and other relevant factors, thereby facilitating more informed decision-making (Ashworth et al., 2018; Danbaki et al., 2020; Nie & Yang, 2021; Shafi et al., 2019; Sharma & Srushtideep, 2022; Безносов et al., 2019).

Precision agriculture encompasses a suite of technologies and practices aimed at optimizing farm management through data-driven decision-making and targeted resource application. Key technologies in precision agriculture include Global Positioning System (GPS), remote sensing, and variable rate application (VRA), each offering unique capabilities to improve resource efficiency, minimize inputs, and maximize yields.

1. **Global Positioning System (GPS)**: GPS technology allows farmers to precisely determine the geographic coordinates of field locations, enabling accurate mapping, monitoring, and management of agricultural activities. GPS receivers installed on farm equipment, such as tractors and harvesters, provide real-time positioning information, allowing farmers to precisely navigate fields and perform tasks with high accuracy. This capability is particularly valuable for tasks such as planting, spraying, and harvesting, where precise equipment control is essential for optimal crop management.

2. **Remote Sensing**: Remote sensing technologies, including satellites, drones, and aerial imaging systems, capture detailed information about crop health, soil characteristics, and environmental conditions over large geographic areas. These data can be used to generate high-resolution maps and imagery, providing valuable insights into crop growth patterns, pest infestations, nutrient deficiencies, and other factors affecting agricultural productivity. By analysing remote sensing data, farmers can identify areas of the field requiring attention and take targeted actions to address specific issues, such as adjusting irrigation schedules, applying fertilizers or pesticides, or implementing soil conservation practices.

3. **Variable Rate Application (VRA)**: VRA technology enables farmers to apply inputs, such as fertilizers, pesticides, and seeds, at variable rates across a field based on spatial variations in soil properties, crop requirements, and environmental conditions. By

utilizing data from GPS, remote sensing, soil maps, and on-the-go sensors, VRA systems automatically adjust input application rates in real-time as farm equipment moves through the field. This precision approach to input application allows farmers to optimize resource use, minimize input costs, and maximize yields by tailoring inputs to the specific needs of different areas within the field.

Applications of these precision agriculture technologies include:

- **Optimizing Resource Efficiency**: By precisely mapping field variability and monitoring crop performance, farmers can optimize resource allocation and minimize waste. For example, GPS-guided equipment can ensure accurate placement of seeds and fertilizers, reducing overlap and avoiding under- or over-application.

- **Minimizing Inputs**: Remote sensing data can identify areas of the field with healthy crop growth and areas with stress or pest damage, allowing farmers to target inputs only where needed. This targeted approach minimizes input usage while maximizing crop health and productivity.

- **Maximizing Yields**: By leveraging data-driven insights and precision application technologies, farmers can make informed decisions to enhance crop yields. For instance, VRA systems can adjust seed planting densities based on soil fertility maps, optimizing plant populations for maximum yield potential.

Overall, precision agriculture technologies offer powerful tools for improving farm management practices, enhancing productivity, and promoting sustainability in modern agriculture. By harnessing the capabilities of GPS, remote sensing, and variable rate application, farmers can achieve greater precision, efficiency, and profitability while minimizing environmental impacts.

Precision agriculture, often referred to as site-specific farming, is a modern approach to farm and resource management, guiding farmers towards more efficient and environmentally sensible utilization of farm inputs (Buick, 1997). The adoption of precision agriculture technologies has been observed to vary across different regions and farm types, with an increasing trend in the United States for various field crops (Mitchell et al., 2018; Schimmelpfennig & Lowenberg-DeBoer, 2020). The technology has also been studied and found to have great potential in countries like China and Bangladesh, where it can aid in preventing agricultural product damage caused by natural calamities (Afroj et al., 2016; Hashim et al., 2018).

The implementation of precision agriculture is underpinned by the integration of intelligent systems for decision-making in agricultural operations, as well as the use of GPS and GIS technologies for precise field operations and data collection (Koleshko et al., 2012; Liu et al., 2012). Furthermore, precision agriculture aims to optimize the management of agricultural inputs to meet site-specific needs, with the ultimate goal of achieving sustainable and profitable agricultural practices (Agüera et al., 2013; Dokin & Aletdinova, 2021).

While precision agriculture holds promise for enhancing productivity and reducing environmental impacts, its widespread adoption faces barriers such as socio-economic, agronomic, and technological challenges (Dunn et al., 2006; Gusev et al., 2020). However, successful adoption and implementation of precision agriculture have the potential to increase production, reduce costs, and minimize environmental impacts.

One key aspect of precision agriculture is variable rate technology (VRT), which enables farmers to apply inputs at variable rates across a field based on the specific needs of different areas. For example, rather than applying a uniform amount of fertilizer or pesticides to an entire field, VRT allows farmers to adjust application rates based on soil nutrient levels, crop growth stage, and other factors, thereby optimizing resource use and minimizing waste.

Another important component of precision agriculture is the use of automated machinery and robotics to perform tasks such as planting, spraying, and harvesting with greater accuracy and efficiency. These technologies can be equipped with sensors and GPS guidance systems to precisely control their movements and ensure uniform coverage of fields.

In addition to improving productivity and efficiency, precision agriculture can also have environmental benefits by reducing the use of inputs such as water, fertilizer, and pesticides, and minimizing soil erosion and runoff. By optimizing farm management practices based on real-time data and analysis, precision agriculture enables farmers to achieve higher yields, lower costs, and greater sustainability in their operations.

Variable rate technology (VRT) is a pivotal component of precision agriculture that empowers farmers to apply inputs such as fertilizers, pesticides, and seeds at varying rates across a field, tailored to the specific requirements of different areas within that field. Traditionally, farmers would apply these inputs uniformly across the entire field, regardless of variations in soil conditions, crop health, or other factors. However, VRT enables a more precise and targeted approach to input application, thereby optimizing resource utilization and minimizing waste.

Instead of treating the entire field as homogeneous, VRT leverages data from various sources, such as soil maps, satellite imagery, and on-the-ground sensors, to create detailed maps of soil properties, nutrient levels, moisture content, and other relevant parameters. These maps allow farmers to identify areas of the field with differing needs, such as areas with nutrient deficiencies, weed infestations, or pest pressures.

With VRT, farmers can then program their equipment, such as fertilizer spreaders or sprayers, to adjust application rates in real-time as they traverse the field. For instance, in areas with low soil fertility, fertilizer application rates can be increased to meet the crop's nutrient requirements, while in areas with adequate fertility, application rates can be reduced to avoid overapplication and minimize environmental impact.

By tailoring input application to the specific needs of different areas within the field, VRT enables farmers to optimize resource use, improve crop yields, and reduce costs. Moreover, by minimizing overapplication of inputs, VRT helps mitigate environmental concerns associated with excessive fertilizer or pesticide use, such as nutrient runoff and groundwater contamination. Overall, VRT plays a crucial role in enhancing the efficiency, profitability, and sustainability of modern agricultural practices.

Vertical Farming

Vertical farming is an innovative agricultural approach that involves cultivating crops in vertically stacked layers within controlled indoor environments, integrating various soilless farming techniques such as hydroponics, aquaponics, and aeroponics with controlled-environment agriculture to optimize plant growth (Kumar et al., 2021). This method is gaining traction globally, with approximately 30 hectares (74 acres) of operational vertical farmland worldwide as of 2020 (Kumar et al., 2021). Structures commonly used for vertical farming systems include buildings, shipping containers, tunnels, and even abandoned mine shafts (Kumar et al., 2021).

Unlike traditional horizontal farming, which relies on expansive land areas, vertical farming maximizes space efficiency by utilizing vertical space. By stacking plants in multiple layers, this innovative approach offers numerous advantages over conventional agriculture. Vertical farming eliminates the need for large amounts of arable land, making it particularly suitable for urban areas with limited land availability. Furthermore, vertical farms can operate year-round, unaffected by seasonal changes or adverse weather conditions, ensuring a consistent food supply.

Figure 9: Vertical farm Hydroponics, Finland (iFarm.fi). ifarm.fi, CC BY-SA 4.0, via Wikimedia Commons.

Precise control over environmental factors such as temperature, humidity, and lighting is a key feature of vertical farming. This control enables farmers to optimize crop growth and quality while minimizing resource inputs such as water and fertilizer. Additionally, the closed indoor environment of vertical farms reduces reliance on pesticides and herbicides, thus helping prevent pest infestations and disease outbreaks.

Despite its potential benefits, vertical farming also faces economic challenges. Start-up costs are significantly higher compared to traditional farms, and large energy demands, particularly for supplementary lighting like LEDs, can lead to increased pollution if non-renewable energy sources are used. However, despite these challenges, vertical farming technologies have attracted substantial investment from venture capitalists, governments, financial institutions, and private investors, highlighting their growing importance in the agricultural sector.

Hydroponics, a key component of vertical farming, involves growing crops under soil-less conditions by immersing the roots in a water solution composed of chemical nutrients to support plant growth (Margaret et al., 2022). This method has been shown to produce high-quality produce, such as hydroponic spinach with high vitamin C and low NO3 contents, by transferring the plants to N-free medium prior to harvest (Kimura & Rodriguez–Amaya, 2003). Additionally, hydroponic farming has been demonstrated to suppress bacterial wilt disease and control pathogens through the application of electrochemically activated solutions (Hải et al., 2023). Furthermore, the use of hydroponics in vertical farming has been associated with unique rhizosphere characteristics, disease control, and the development of root hairs, contributing to the overall success of the system (Fujiwara et al., 2012). The potential for hydroponics to produce a wide range of marketable and regional crops, including tomatoes, peppers, cucumbers, strawberries, and green vegetables, underscores its versatility and commercial viability (Kumar et al., 2021). However, the feasibility of hydroponics as a home industry is still being explored, with challenges related to the relatively high investment value for new entrepreneurs (Kholis et al., 2022).

These are various techniques and types of vertical farming:

Hydroponics: Growing plants without soil, using liquid solutions containing essential nutrients. Plant roots are submerged in these solutions or supported by inert mediums like gravel. Hydroponics offers increased yield per area and reduced water usage compared to conventional farming.

Figure 10: Nutrient film technique hydroponics. Oregon State University, CC BY-SA 2.0, via Wikimedia Commons.

Aquaponics: Integrating aquaculture (fish farming) with hydroponics. Nutrient-rich wastewater from fish tanks is filtered and converted into nutritious nitrate by bio-filters. Plants absorb these nutrients and purify the wastewater, which is recycled back to the fish tanks. Aquaponics helps maintain greenhouse temperatures and is a closed-loop system.

Figure 11: The raft tank at the Crop Diversification Centre (CDC) South Aquaponics greenhouse in Brooks, Alberta. Bryghtknyght, CC BY 3.0, via Wikimedia Commons.

Aeroponics: Growing plants in air chambers without soil or liquid mediums. Nutrient solutions are misted onto the plant roots. Aeroponics is highly sustainable, using minimal water and energy. Its vertical design saves space and is suitable for vertical farming.

Figure 12: Tomato and Lettuce growing aeroponicly. MyAeroponics, CC BY-SA 3.0, via Wikimedia Commons.

Controlled-Environment Agriculture (CEA): Modifying the natural environment to optimize crop yield and extend the growing season. Implemented in enclosed structures like greenhouses, CEA allows control over environmental factors such as temperature, light, and humidity. It is often combined with soilless farming techniques in vertical farming systems.

Types:

1. Building-based farms: Repurposing abandoned buildings or constructing new ones for vertical farming. Examples include "The Plant" in Chicago, which was transformed from an old meatpacking plant. These structures provide ample space for vertical farming operations.

2. Shipping-container vertical farms: Using recycled shipping containers to create modular chambers for growing plants. Equipped with LED lighting, hydroponic systems, and climate controls, these containers offer a convenient and space-efficient solution for vertical farming.

3. Deep farms: Building vertical farms in refurbished underground tunnels or abandoned mine shafts. Underground environments provide stable temperature and humidity, reducing the need for heating. Deep farms can produce significantly more food than conventional farms and can be fully self-sufficient with automated harvesting systems.

4. Floating farms: Deploying floating platforms and barges for vertical farming in urban areas with limited land availability. Projects like the Science Barge in New York City have demonstrated urban hydroponic agriculture. Floating farms utilize open water spaces to capture sunlight and offer innovative solutions for urban agriculture.

Figure 13: Floating farm on Inle Lake in Maing Thauk. Christophe95, CC BY-SA 4.0, via Wikimedia Commons.

In essence, vertical farming is a modern agricultural technique that involves growing crops in vertically stacked layers, often in controlled environments such as skyscrapers, warehouses, or shipping containers. This method of farming has gained attention due to its potential to address various challenges faced by traditional agriculture, and it offers several advantages including year-round production, reduced water usage, and the potential for urban food production.

One of the primary advantages of vertical farming is its capacity for year-round production. Unlike traditional farming methods which are often limited by seasonal changes and weather conditions, vertical farming can operate continuously throughout the year. This is made possible by the controlled environment in which crops are grown, allowing for the manipulation of factors such as temperature, humidity, and light to create optimal growing conditions regardless of the external climate. As a result, vertical farms can produce multiple harvests in a single year, increasing overall crop yield and reducing the impact of seasonal fluctuations on food supply.

In addition to year-round production, vertical farming offers significant advantages in terms of water usage. Traditional agriculture is heavily reliant on water for irrigation, and in many regions, water scarcity is a critical concern. Vertical farming addresses this issue through the implementation of hydroponic or aeroponic systems, which use significantly less water compared to conventional soil-based farming. Hydroponic systems deliver nutrients directly to the plant roots in a recirculating system, minimizing water wastage, while aeroponic systems mist the plant roots with nutrient-rich water, further reducing overall water consumption. The efficient use of water in vertical farming not only helps to conserve this precious resource but also mitigates the environmental impact associated with excessive water usage in traditional agriculture.

Furthermore, vertical farming has the potential to revolutionize urban food production. As the global population continues to urbanize, the demand for fresh produce in urban areas has increased. Vertical farming offers a solution to this challenge by enabling the cultivation of crops within or in close proximity to urban centres. By utilizing vertical space in buildings, abandoned warehouses, or repurposed shipping containers, vertical farms can bring food production closer to the consumers, reducing the need for long-distance transportation and minimizing the carbon footprint associated with food distribution. This localized approach to food production also enhances food security in urban areas, as it reduces reliance on distant rural farms and minimizes the vulnerability of food supply chains to disruptions.

Vertical farming offers numerous advantages over traditional farming methods. One of the key benefits is its efficiency in land use. Traditional farming requires large amounts of arable land, which is becoming increasingly unsustainable due to rapid population growth and urbanization (Zhang et al., 2018). With the expected decrease in arable land per person by 66% in 2050 compared to 1970, vertical farming presents a solution by allowing for over ten times the crop yield per acre than traditional methods (Mok et al., 2013). This increased productivity is further amplified by the ability to produce crops year-round, multiplying the productivity of the farmed surface by a factor of four to six, depending on the crop, and up to 30 times for crops like strawberries . Additionally, vertical farming allows for the production of a larger variety of harvestable crops due to its usage of isolated crop sectors, enabling the simultaneous growth and harvest of multiple crops .

Furthermore, vertical farming mitigates climate change by lowering emissions and reducing water usage . The reduced need for chemical pesticides in vertical farms also makes it easier to grow organic crops compared to traditional farming, contributing to environmental conservation . This method of farming also reduces the invasive impact on native flora and fauna, as it causes nominal harm to wildlife due to its limited space usage (Eldridge et al., 2020). Additionally, vertical farming reduces the need for deforestation and desertification caused by agricultural encroachment on natural biomes, as it requires less farmland, thus saving natural resources (Wood et al., 2021).

Vertical farming also addresses the issue of resistance to adverse weather conditions, which is a significant challenge in traditional outdoor farming. Crops grown in traditional outdoor farming are heavily dependent on supportive weather and are susceptible to various natural disasters such as floods, wildfires, and droughts . Vertical farming provides a solution to food insecurity in remote northern communities and areas where traditional farming is largely impossible, by offering fresh produce year-round at a lower cost and with less susceptibility to disruption.

However, despite these advantages, vertical farming faces economic challenges. The financial burden of large startup costs and high urban occupancy costs results in a longer break-even time compared to traditional rural farms . Additionally, the potential profitability of vertical farming is questioned, as high-value crops must be grown to be financially successful, and the costs of operating these farms must decrease or the price of traditional farming must increase (Zhang et al., 2018). The energy requirements of vertical farming also pose a challenge, as it demands significant energy consumption for lighting, temperature, humidity control, and other factors, making it potentially economically uncompetitive with current market prices.

Integrated Pest Management (IPM)

Integrated Pest Management (IPM) is a comprehensive approach to managing pests in agriculture that aims to minimize the use of chemical pesticides and reduce the environmental impact of pest control strategies (Khanal et al., 2021). It involves the integration of various pest control methods such as biological control, habitat manipulation, modification of cultural practices, and use of resistant varieties, with the judicious use of pesticides when necessary. The adoption of IPM practices has been shown to have economic benefits for farmers, as it reduces the reliance on and misuse of chemical pest control methods, leading to improved profitability (Bulkis et al., 2020). Additionally, IPM has been found to be effective in reducing damage caused by pests without harming the environment (Rejesus & Jones, 2020). It involves understanding the nature of pests and implementing a combination of strategies to manage them effectively. By incorporating biological, cultural, and chemical practices, IPM aims to control pest populations while minimizing environmental impact.

Pests, which include insects, mites, diseases, weeds, and certain animals, can cause damage to crops and pose risks to agricultural production. However, not all insects and mites are pests, and it's important to distinguish between harmful organisms and beneficial ones that aid in pest management.

Key principles of IPM include knowing the history and nature of pests, being proactive in prevention, prioritizing sanitation, and regularly monitoring crops and growing environments. It involves recording observations, gaining confidence through education and training, and utilizing available resources to implement integrated pest management strategies effectively.

IPM practices encompass forward planning, regular monitoring, and timely decision-making. Control methods include cultural practices that modify growing conditions, physical methods to prevent pest entry or remove them, genetic methods utilizing pest-resistant crop varieties, biological control using natural predators or parasites, and chemical control as a last resort, employing selective pesticides only when necessary.

IPM can be applied across various settings, including farms, homes, gardens, workplaces, and natural spaces such as national parks and schools. It combines ecological and sustainable practices to manage pests effectively while minimizing environmental impact.

IPM offers a holistic approach to pest management that promotes sustainability and long-term solutions. By integrating multiple control methods and prioritizing ecological practices, IPM aims to maintain pest populations below economic thresholds without relying heavily on synthetic pesticides.

The impact of IPM adoption has been studied extensively, with research focusing on its effects on pesticide use, yields, and farm profits (Simončič et al., 2015). Studies have also compared the effectiveness of IPM production with conventional pest control methods, demonstrating the superiority of IPM in managing insect pests and reducing the number of chemical sprays required (Halder et al., 2020; Maurya et al., 2017; Stephens et al., 2017). Furthermore, the use of IPM has been associated with improved knowledge, pest control, and practices in various agricultural settings (Akter et al., 2016).

The adoption of IPM has been assessed in different agricultural contexts, including rice farming in Indonesia and vegetable cultivation in Nepal and Bangladesh (Blake et al., 2007; Brodt et al., 2007; Khanal et al., 2021). These studies have highlighted the benefits of IPM in terms of reducing pesticide use, improving yields, and enhancing profitability. Additionally, surveys have been conducted to identify barriers to the adoption of IPM practices, with the aim of addressing biological, educational, social, and political factors that may hinder the widespread implementation of IPM.

The United States Environmental Protection Agency (US EPA) has devised a comprehensive four-tiered strategy for implementing Integrated Pest Management (IPM), aimed at promoting effective pest control while minimizing environmental impact (NSW EPA, 2024):

1. **Setting Action Thresholds**: The first step involves establishing action thresholds, which are specific points indicating when pest populations or environmental conditions require intervention to prevent pests from becoming economic or environmental threats. Merely observing a single pest does not automatically necessitate control measures.

2. **Monitoring and Identifying Pests**: Accurate identification and ongoing monitoring of pest populations and behaviour are crucial in determining when action thresholds have been reached and selecting appropriate control methods. Many organisms classified as pests may actually be harmless or beneficial. Monitoring and identification efforts help mitigate the risk of employing incorrect pesticides or using pesticides unnecessarily when alternative strategies may be more effective.

3. **Preventing Pests from Becoming a Threat**: Effective prevention measures can often mitigate pest threats with minimal risk to human health or the environment. Prevention strategies include selecting pest-resistant plant varieties and implementing crop rotation in agriculture, as well as reducing clutter and maintaining hygiene in buildings.

4. **Control**: When prevention methods prove insufficient and monitoring indicates the need for pest control, evaluating control options becomes necessary. IPM prioritizes methods with minimal environmental and human health risks. These methods include physical controls such as trapping or weeding, as well as highly targeted chemical controls like pheromones to disrupt pest reproduction. If less risky methods are ineffective, targeted pesticide spraying may be employed, while broad-spectrum pesticide application is considered a last resort.

Figure 14: Pheromone Dispenser in a vineyard, Hagnau, Lake Constance, Germany. Dietrich Krieger, CC BY-SA 3.0, via Wikimedia Commons.

IPM programs operate as a series of pest management assessments, decisions, and interventions, rather than relying on a single pest control method. Through this approach, growers can proactively manage potential pest infestations by adhering to the following steps (NSW EPA, 2024):

- **Setting Action Thresholds**: Establishing thresholds that indicate when pest control action is warranted based on pest populations or environmental conditions.

- **Monitoring and Identifying Pests**: Conducting ongoing monitoring to accurately identify pests and make informed decisions aligned with action thresholds.

- **Prevention**: Implementing measures to manage crops, lawns, or indoor spaces to prevent pests from posing a threat, utilizing techniques such as cultural methods and selecting pest-resistant varieties.

- **Control**: Assessing and implementing control methods based on effectiveness and risk, prioritizing less risky options like targeted chemicals and mechanical controls, with broad-spectrum pesticide application reserved as a last resort.

Soil Conservation and Regenerative Agriculture

Soil conservation and regenerative agriculture practices are indispensable for safeguarding soil health and fertility, which are fundamental for sustainable agriculture and ecosystem stability.

Soil Conservation

Soil conservation encompasses the implementation of strategies aimed at preventing soil erosion, degradation, and nutrient loss. It is imperative because soil, being a finite and non-renewable resource, sustains plant growth, water filtration, and carbon sequestration. By conserving soil, we can uphold its structure, fertility, and capacity to support plant life.

1. Preventing Erosion: Soil erosion, induced by factors like wind and water, can deplete valuable topsoil, diminishing fertility and disturbing ecosystems. Conservation practices such as contour ploughing, terracing, and employing cover crops aid in stabilizing soil and curtailing erosion.

2. Maintaining Soil Structure: Healthy soil exhibits a well-balanced structure characterized by good aggregation, aeration, and water-holding capacity. Conservation practices like reduced tillage and mulching assist in preserving soil structure by mitigating disruption and compaction.

3. Protecting Nutrient Content: Soil conservation endeavours target to prevent nutrient loss via runoff and leaching. Practices such as nutrient management planning, crop rotation, and the utilization of organic amendments serve to sustain soil fertility by retaining essential nutrients within the soil profile.

Regenerative Agriculture

Regenerative agriculture transcends soil conservation by focusing on revitalizing and enhancing ecosystem functions, biodiversity, and resilience. It seeks to ameliorate soil health while fostering carbon sequestration, water retention, and ecosystem services.

1. Building Soil Organic Matter: Regenerative practices such as no-till farming, cover cropping, and composting bolster soil organic matter, thereby enhancing soil structure, water retention, and nutrient cycling. Optimal levels of soil organic matter contribute to heightened fertility and productivity.

2. Enhancing Biodiversity: Agriculturally biodiverse systems bolster a plethora of plant and animal species, including beneficial microbes and insects. Practices like agroforestry, crop diversity, and integrated pest management foster biodiversity, culminating in more resilient and productive ecosystems.

3. Carbon Sequestration: Robust soils serve as substantial carbon sinks, sequestering carbon dioxide from the atmosphere and mitigating climate change. Regenerative practices bolster carbon storage in soil through the incorporation of organic matter and the advocacy of perennial vegetation, thus offsetting greenhouse gas emissions.

4. Improving Water Management: Regenerative agriculture endeavours revolve around refining water infiltration, retention, and efficiency. Techniques such as conservation tillage, agroforestry, and the restoration of wetlands and riparian buffers curtail water runoff, mitigate erosion, and elevate water quality.

In conclusion, soil conservation and regenerative agriculture practices are pivotal for nurturing soil health and fertility, thereby ensuring the enduring sustainability of agricultural systems and the well-being of the planet. By embracing these practices, farmers can safeguard invaluable soil resources, fortify ecosystem resilience, and contribute to global initiatives aimed at combating climate change and biodiversity loss.

Regenerative agriculture encompasses various practices aimed at promoting soil health and ecosystem resilience. Among these techniques are cover cropping, no-till farming, and rotational grazing, each playing a vital role in soil regeneration, carbon sequestration, and water conservation.

Cover Cropping: Cover cropping involves planting crops specifically to cover the soil rather than for harvesting. These cover crops protect the soil from erosion, suppress weed growth, and enhance nutrient cycling. Leguminous cover crops, such as clover and vetch, contribute to nitrogen fixation, enriching the soil. By keeping the soil covered, cover cropping minimizes erosion, promotes soil aggregation, and fosters beneficial microbial activity. Overall, cover cropping rejuvenates soil health, boosts carbon sequestration, and conserves water.

No-Till Farming: No-till farming avoids soil disturbance through tillage. Instead of plowing fields, farmers directly sow seeds into untilled soil or residues of previous crops. This practice preserves soil structure, minimizes erosion, and retains moisture. No-till farming reduces carbon loss, promotes biodiversity, and enhances nutrient cycling. It also lowers fuel consumption and greenhouse gas emissions, contributing to agricultural sustainability.

Rotational Grazing: Rotational grazing involves moving livestock through pastures, allowing areas to rest and regenerate. This mimics natural grazing patterns, promotes plant diversity, and prevents overgrazing. Rotational grazing reduces soil compaction, increases organic matter input, and enhances plant growth. It also minimizes erosion, conserves soil moisture, and preserves structure. By integrating livestock into cropping systems, rotational grazing fosters resilient agricultural ecosystems.

Figure 15: No-till farming system in Brookings, Co., SD. USDA NRCS South Dakota, Public domain, via Wikimedia Commons.

Cover cropping, no-till farming, and rotational grazing are effective techniques for promoting soil health, carbon sequestration, and water conservation. Implementing these practices enhances soil fertility, mitigates climate change, and ensures sustainable food production while preserving ecosystem integrity.

Community Supported Agriculture (CSA) and Local Food Systems

Community Shared Agriculture, or CSA, presents a distinctive approach to food production and distribution that establishes a direct connection between farmers and consumers. Under the CSA model, members of the community invest in shares or subscriptions from a local farm at the start of the growing season, ensuring financial stability for the farmer upfront. In return, these members receive regular deliveries of fresh produce throughout the season, often available for pickup or delivery at designated locations (Stratton, 2022).

The essence of CSA lies in its ability to promote local food systems by linking consumers directly with nearby farms. By patronizing local producers, CSA members contribute to the vitality of the local economy, diminish their carbon footprint associated with food transportation, and reinforce community bonds. Typically, CSA farms prioritize sustainable and organic farming practices, thereby championing environmental conservation and the preservation of agricultural diversity (Stratton, 2022).

Moreover, CSA facilitates direct relationships between farmers and consumers, fostering transparency and trust in the food supply chain. Members have the opportunity to engage with the farm, meet the farmers, and gain insights into the cultivation methods employed. This direct engagement fosters accountability and empowers consumers to make informed decisions regarding their food choices, considering factors such as production methods, environmental impact, and nutritional value.

Additionally, CSA plays a pivotal role in advancing food sovereignty, allowing communities to determine their own food systems based on local knowledge, resources, and preferences. By supporting local farmers and engaging in CSA initiatives, communities assert greater control over their food supply chain, reducing reliance on global food markets dominated by large corporations. CSA encourages diversified farming practices, promotes crop diversity, and celebrates cultural food traditions, empowering communities to prioritize food security, health, and resilience.

Community Shared Agriculture serves as a catalyst for promoting local food systems, nurturing direct farmer-consumer relationships, and championing food sovereignty. By fostering connections between consumers and nearby farms, CSA contributes to sustainable agriculture, fosters community cohesion, and empowers communities to shape their food future.

Supporting local agriculture offers a myriad of benefits that extend beyond just the quality of the food. Buying locally-produced food significantly reduces the carbon footprint associated with transportation. Instead of traveling long distances from farm to table, locally grown food travels shorter distances, thus reducing greenhouse gas emissions from transportation vehicles. This reduction in carbon emissions helps mitigate climate change and promotes environmental sustainability.

Locally grown food tends to be fresher compared to produce that has travelled long distances. Since local farmers can harvest their crops at peak ripeness and deliver them directly to consumers, the produce retains more nutrients and flavour. This freshness enhances the taste and nutritional value of the food, providing consumers with healthier and more enjoyable eating experiences.

Buying from local farmers provides crucial support to small-scale agricultural operations. These farmers often face numerous challenges, including competition from large-scale industrial farms, fluctuating market prices, and limited access to resources. By purchasing their products, consumers help sustain local farming communities, preserve agricultural traditions, and promote economic diversity in rural areas. Supporting small-scale farmers also fosters local employment opportunities and strengthens the resilience of regional food systems.

Local farms often prioritize diverse crop varieties and sustainable farming practices, contributing to greater biodiversity and ecosystem health. Unlike monoculture industrial farms, which rely heavily on a single crop variety, local farms cultivate a wide range of fruits, vegetables, and grains. This diversity helps protect against crop failures, pests, and diseases, while also preserving genetic diversity within plant populations. Additionally, sustainable farming practices such as crop rotation, cover cropping, and agroforestry promote soil health, water conservation, and wildlife habitat preservation.

Purchasing locally grown food fosters a sense of connection and belonging within the community. Consumers have the opportunity to meet and interact with local farmers at farmers' markets, community-supported agriculture (CSA) programs, and farm-to-table events. These connections create meaningful relationships between producers and consumers, fostering trust, transparency, and mutual support. Additionally, local food initiatives often serve as hubs for community gatherings, cultural exchange, and education about sustainable agriculture and food systems.

Supporting local agriculture offers a multitude of benefits, including reduced carbon emissions, fresher produce, support for small-scale farmers, promotion of biodiversity, and connection to the community. By choosing to buy locally grown food, consumers contribute to a more sustainable, resilient, and vibrant food system that benefits both people and the planet.

Challenges and Future Directions

Implementing sustainable agriculture practices faces several challenges and barriers that can hinder progress.

Policy Frameworks: Inadequate or outdated policy frameworks can pose significant barriers to the adoption of sustainable agriculture practices. Government policies may prioritize conventional farming methods or provide subsidies that Favor industrial agriculture over sustainable alternatives. Additionally, regulatory frameworks may lack incentives or enforcement mechanisms to encourage farmers to adopt environmentally friendly practices. Without supportive policies that promote sustainable agriculture, farmers may be reluctant to invest time and resources into implementing new techniques.

To address the issue of inadequate or outdated policy frameworks posing barriers to the adoption of sustainable agriculture practices, it is crucial to understand the impact of government policies and regulatory frameworks on the promotion of sustainable agriculture. The lack of supportive policies may deter farmers from investing in environmentally friendly practices. Piñeiro et al. (2020) emphasize that policy instruments are more effective when designed considering the characteristics of the target population and the trade-offs between economic, environmental, and social outcomes. This highlights the need for policy frameworks to be tailored to the specific needs and challenges faced by farmers in adopting sustainable practices.

Economic incentives play a crucial role in the adoption of sustainable agricultural practices, as highlighted in various studies. Piñeiro et al. (2020) emphasized that adoption depends on various factors, including the conditions of the program, incentives offered, and economic characteristics of farmers. Aznar-Sánchez et al. (2020) identified income level, capital, and access to labour as factors influencing the adoption of sustainable practices. Ntakirutimana et al. (2019) stressed the importance of implementing economic incentive policies to increase the adoption of sustainable agriculture practices. Additionally, Zhao (1991) highlighted the significance of economic incentives, such as price policies, in promoting agricultural growth in developing countries.

Technological limitations can indeed hinder the adoption of sustainable agriculture practices, particularly in the case of sustainable farming techniques that rely on advanced technologies such as precision agriculture, remote sensing, and biological pest control methods. These limitations can stem from factors such as limited access to advanced technologies in certain regions, prohibitively expensive costs for small-scale farmers, and a lack of research and development focused on adapting sustainable practices to local conditions ((Adão et al., 2017; Aslan et al., 2022; Degieter et al., 2023; Lu et al., 2020; Segarra et al., 2020; Sun et al., 2022).

For instance, the use of hyperspectral imaging in precision agriculture is limited by factors such as the influence of climate on data acquired by traditional platforms like satellites and aircraft, which can hamper its use, especially in precision agriculture where information detail matters (Adão et al., 2017). Similarly, the accessibility of hyperspectral images outside of the scientific community is limited, thus hindering their widespread use in precision agriculture (Lu et al., 2020). Furthermore, the deployment of unmanned aerial vehicles (UAVs) for precision agriculture is affected by limitations such as battery life, load capacity, and weather conditions, which can impact their efficiency and application (Aslan et al., 2022).

In addition, the limitations of remote sensing techniques for precision agriculture have been addressed in recent years with the launch of Sentinel-2 A + B, which has helped overcome many of the previous limitations in the application of remote sensing techniques for precision agriculture (Segarra et al., 2020). However, challenges and future directions in the use of high-resolution satellite sensors and other types of remote sensors for precision agriculture are still being discussed, indicating ongoing limitations in this area (Yang, 2018).

Sentinel-2A and Sentinel-2B are Earth observation satellites operated by the European Space Agency (ESA) as part of the Copernicus program. These satellites are equipped with advanced multispectral imaging sensors capable of capturing high-resolution optical imagery of the Earth's surface.

The Sentinel-2 mission aims to monitor various aspects of the Earth's environment, including land cover, land use, vegetation health, agricultural practices, water quality, and natural disasters. By providing frequent and systematic coverage of the Earth's surface, Sentinel-2A and Sentinel-2B contribute valuable data for environmental monitoring, resource management, urban planning, and disaster response efforts.

Both Sentinel-2A and Sentinel-2B carry a MultiSpectral Instrument (MSI), which consists of 13 spectral bands ranging from the visible to the shortwave infrared region of the electromagnetic spectrum. These bands capture imagery at different wavelengths, allowing scientists to analyse various properties of the Earth's surface, such as vegetation health, soil moisture, and water quality.

Sentinel-2A was launched on June 23, 2015, followed by Sentinel-2B on March 7, 2017. Together, these satellites provide global coverage of the Earth's land surface with a revisit time of five days at the equator and shorter intervals at higher latitudes. The data collected by Sentinel-2A and Sentinel-2B are freely available to the public and can be accessed through the Copernicus Open Access Hub, making them valuable resources for researchers, policymakers, and other stakeholders involved in Earth observation and environmental monitoring initiatives.

Moreover, the adoption of sustainable agriculture practices is also limited by challenges in biological pest control methods. For example, the current designs for precision-guided sterile insect techniques are limited by the rapid evolution of resistance, necessitating future research to develop drives that can overcome evolved resistance (Kandul et al., 2019).

Furthermore, the limitations of advanced technologies in precision agriculture are not only technical but also socio-economic. Farmers' acceptance of robotics and UAVs, for instance, is influenced by factors such as the price of the technology, compatibility with other software and equipment, and labour scarcity, which can impact their adoption and use in sustainable agriculture practices (Degieter et al., 2023).

Market Demand and Consumer Awareness: The demand for sustainably produced food is growing, but many consumers may still prioritize factors such as price and convenience over environmental considerations. Additionally, there may be a lack of awareness or understanding among consumers about the benefits of sustainable agriculture and the importance of supporting farmers who adopt these practices. Without strong market demand for sustainably produced food, farmers may struggle to justify the transition to sustainable farming methods.

Knowledge and Training: Implementing sustainable agriculture practices requires specialized knowledge and training, but many farmers may lack access to educational resources and technical assistance. Training programs and extension services focused on sustainable agriculture may be limited or unavailable in certain regions, leaving farmers without the necessary skills and support to make the transition. Additionally, cultural and language barriers may further impede access to information and training opportunities for farmers from marginalized or underserved communities.

Environmental and Climatic Factors: Environmental and climatic factors, such as soil quality, water availability, and climate variability, can also influence the feasibility and effectiveness of sustainable agriculture practices. Certain practices may be better suited to specific agroecological zones or climates, making it challenging to implement a one-size-fits-all approach to sustainable agriculture. Moreover, extreme weather events, such as droughts, floods, and heatwaves, can pose significant challenges for farmers implementing sustainable practices, highlighting the importance of climate resilience in agricultural systems.

Addressing these challenges and barriers to implementing sustainable agriculture practices requires a multifaceted approach that involves collaboration between policymakers, researchers, farmers, consumers, and other stakeholders. This approach may include developing supportive policy frameworks, providing financial incentives and technical assistance to farmers, investing in research and development of sustainable farming technologies, raising consumer awareness about the benefits of sustainable agriculture, and promoting knowledge sharing and capacity building among farmers. By addressing these challenges collectively, stakeholders can work towards building a more sustainable and resilient food system that benefits both people and the planet.

Sustainable Agriculture and Farm Management

Implementing sustainable agriculture from a farm management perspective involves several key components and considerations.

Assessment of Current Practices is the initial step, requiring a thorough evaluation of the farm's existing methods to pinpoint areas for improvement in sustainability. This assessment encompasses various aspects such as soil health, water management practices, pesticide and fertilizer usage, crop rotation strategies, and overall resource efficiency.

Goal Setting is paramount to guide decision-making and prioritize actions towards sustainability. Clear sustainability goals are established, which may include reducing environmental impact, enhancing soil health, conserving water resources, promoting biodiversity, and bolstering the resilience of the farm's operations.

Education and Training play a crucial role in effectively implementing sustainable practices. Farm managers and workers receive necessary education and training through workshops, seminars, field demonstrations, and access to resources and information on sustainable farming techniques.

Adoption of Sustainable Practices involves integrating techniques that minimize negative environmental impacts while maintaining or improving productivity. This includes practices like conservation tillage, cover cropping, crop rotation, integrated pest management (IPM), precision agriculture, agroforestry, and organic farming methods.

Resource Management is fundamental to sustainable agriculture, requiring efficient management of resources such as soil, water, energy, and nutrients. This involves implementing soil conservation measures, water-saving irrigation technologies, energy-efficient equipment, and nutrient management plans to minimize runoff and leaching.

Monitoring and Evaluation are essential for assessing the effectiveness of implemented practices and tracking progress towards sustainability goals. Regular monitoring includes measuring soil health parameters, water quality, biodiversity indices, greenhouse gas emissions, and crop yields over time.

Continuous Improvement is integral to sustainable agriculture, necessitating ongoing learning, adaptation, and improvement of practices. Farms should regularly review their methods, gather feedback from stakeholders, stay informed about advancements in sustainable farming techniques, and adjust strategies accordingly to optimize sustainability outcomes.

Stakeholder Engagement is crucial for building support, fostering collaboration, and collectively addressing sustainability challenges. Engaging with employees, neighbouring communities, customers, government agencies, and environmental organizations through open communication and transparency about sustainability efforts cultivates trust and goodwill.

By integrating these components into farm management practices, agricultural operations can transition towards greater sustainability, benefiting both the environment and the long-term viability of the farm.

Part 2 - Developing and Implementing Sustainable Land Use Strategies

This part of the book outlines procedures for formulating and executing sustainable land use strategies aimed at enhancing the ecological sustainability of cultivated land. These strategies are tailored for individuals overseeing agricultural and horticultural operations, requiring them to analyse information, exercise judgment, and undertake various advanced tasks showcasing a profound understanding of land management practices. These individuals are tasked with the development and implementation of sustainable land use strategies.

Chapter Two

Assessing Requirements for Improved Land Use

Assessing requirements for enhanced land utilization demands a comprehensive approach that delves into various critical facets. Firstly, it entails conducting a meticulous audit of natural resources present at the site to uncover existing vulnerabilities and potential threats to sustainability. This encompassing evaluation involves scrutinizing factors such as soil quality, water availability, biodiversity, and overall ecosystem health to gain a holistic understanding of the site's environmental condition.

Moreover, reviewing existing production and management plans is essential to ensure their alignment with sustainable land use principles. This review process involves evaluating current practices related to soil conservation, water management, agrochemical use, and biodiversity conservation to identify areas for improvement and optimization.

Additionally, the assessment involves evaluating water resources comprehensively, considering aspects like availability, quality, and their influence on erosion processes. This entails analyzing various factors such as water supply sources, hydrological patterns, water quality parameters, and the risk factors associated with erosion to develop effective mitigation strategies.

Furthermore, assessing soil health and fertility in relation to production plan requirements is crucial. This step entails conducting detailed soil tests, analyzing nutrient levels, evaluating soil structure and texture, and identifying any deficiencies or imbalances that could potentially impact productivity and sustainability.

Analyzing vegetation types and coverage across the site is another integral aspect of the assessment process. This involves identifying dominant plant species, assessing the health of vegetation, and determining the extent of vegetation cover. Additionally, it entails identifying necessary changes to optimize vegetation management practices for enhanced land utilization.

Finally, the assessment culminates in planning and prioritizing measures and structures for land and water protection based on the evaluation findings. This involves identifying areas vulnerable to erosion, waterlogging, or pollution and implementing appropriate protective measures such as erosion control structures, vegetative buffers, and water retention features. Prioritization of these measures is guided by the severity of threats, ecological significance, and potential impact on land productivity and sustainability.

Audit Site Natural Resources and Identify Threats to Sustainability

From the perspective of individuals managing agricultural and production horticulture enterprises, developing and implementing sustainable land use strategies involves a comprehensive approach to auditing site natural resources and identifying threats to sustainability. This includes:

1. Assessment Planning: Begin by planning the assessment process, outlining objectives, scope, and methodology for auditing site natural resources. Consider factors such as the scale of the operation, the diversity of natural resources present, and any specific

sustainability goals or regulatory requirements.

2. Data Collection: Gather relevant data and information about the site's natural resources, including soil characteristics, water availability and quality, vegetation cover, biodiversity, and land use history. Utilize tools such as soil tests, water quality assessments, biodiversity surveys, and aerial imagery to collect data.

3. On-Site Inspection: Conduct a thorough on-site inspection to visually assess the condition of natural resources and identify any observable threats to sustainability. This may involve examining soil erosion, water runoff patterns, vegetation health, presence of invasive species, and signs of habitat degradation.

4. Stakeholder Consultation: Engage with stakeholders, including farm managers, workers, environmental experts, and local communities, to gather insights and perspectives on the state of natural resources and potential sustainability challenges. Encourage open dialogue and collaboration to gain a comprehensive understanding of the site's dynamics.

5. Threat Identification: Analyse the collected data and observations to identify potential threats to the sustainability of land under production. This may include soil degradation, water pollution, habitat loss, nutrient depletion, pest infestations, climate-related risks, and regulatory compliance issues.

6. Risk Assessment: Assess the likelihood and potential impact of each identified threat on ecological sustainability, considering factors such as severity, frequency, duration, and spatial extent. Prioritize threats based on their significance and urgency for mitigation.

7. Root Cause Analysis: Investigate the underlying causes of identified threats, examining factors such as land management practices, external influences, socio-economic factors, and climate variability. Identify key drivers contributing to sustainability challenges to inform targeted interventions.

8. Sustainability Gap Analysis: Compare current land management practices and sustainability outcomes against desired sustainability targets and benchmarks. Identify gaps and areas for improvement to guide the development of sustainable land use strategies.

9. Solution Design: Develop tailored solutions and management interventions to address identified threats and enhance ecological sustainability. Consider a range of sustainable land management practices, technologies, and innovations suited to the specific context and needs of the operation.

10. Communication and Documentation: Communicate audit findings, risk assessments, and proposed solutions to relevant stakeholders in a clear, concise, and accessible manner. Document the audit process, including data sources, methodologies, findings, and recommendations, to facilitate transparency, accountability, and continuous improvement.

11. Implementation Planning: Develop an action plan outlining specific tasks, responsibilities, timelines, and resources required to implement sustainable land use strategies effectively. Collaborate with stakeholders to ensure buy-in and support for implementation efforts.

12. Monitoring and Evaluation: Establish monitoring protocols and performance indicators to track the effectiveness of implemented strategies and measure progress towards sustainability goals. Regularly review and evaluate outcomes, adjusting strategies as needed to optimize results and adapt to changing conditions.

Assessment planning is a crucial first step in the process of auditing site natural resources to develop sustainable land use strategies. This phase requires meticulous consideration of several factors to ensure a thorough and effective assessment.

The assessment process begins with the clear definition of its objectives. These objectives should align with broader sustainability goals and address the specific needs of the agricultural or horticultural enterprise. They may include identifying threats to ecological sustainability, evaluating the health of natural resources, assessing compliance with regulatory standards, or informing decision-making for land management practices.

Defining the scope of the assessment involves specifying its boundaries, focus areas, and the resources to be included. Factors such as the size of the farm or horticultural site, as well as the spatial extent of the assessment area, must be considered. Decisions need to be made regarding whether the assessment will cover all natural resources present on the site or focus on specific aspects such as soil, water, biodiversity, or land use practices.

Selecting appropriate methodologies and data collection techniques is crucial for conducting a robust assessment. The diversity of natural resources present on the site should be taken into account when choosing methods that are scientifically sound, cost-effective, and suitable for the scale of the operation. Common methodologies may include field surveys, remote sensing, laboratory analysis, interviews, and data modelling.

Various factors that may influence the assessment process and outcomes must be considered. These include the diversity of natural resources such as soil types, water bodies, vegetation cover, and wildlife habitats. Additionally, specific sustainability goals or performance indicators established for the enterprise, as well as regulatory requirements related to environmental protection, land use planning, water management, and biodiversity conservation, should be taken into account.

Allocating resources, including personnel, equipment, time, and budget, is essential to support assessment activities adequately. Sufficient resources must be made available to conduct a comprehensive evaluation of site natural resources while adhering to project timelines and budget constraints.

Identifying potential risks and challenges that may impact the assessment process or outcomes is necessary. Logistical, technical, or regulatory obstacles that could arise should be anticipated, and contingency plans should be developed to address them effectively. Factors such as weather conditions, access to the site, data quality issues, and stakeholder engagement should be considered.

Engaging with relevant stakeholders, including farm managers, landowners, workers, local communities, government agencies, and environmental organizations, is crucial. Gathering input and perspectives from diverse stakeholders ensures that the assessment process is comprehensive and that all relevant knowledge is considered.

By carefully planning the assessment process and considering these factors, individuals managing agricultural and production horticulture enterprises can lay the groundwork for a systematic, rigorous, and informative evaluation of site natural resources. This sets the stage for informed decision-making and the development of effective strategies to promote ecological sustainability and responsible land management practices. The requirements for assessment planning are summarised in an Assessment Planning Checklist shown as Figure 16.

Assessment Planning Checklist

Objective Definition
- ☐ Clearly define the objectives of the assessment process.
- ☐ Ensure alignment with broader sustainability goals and specific enterprise needs.
- ☐ Identify objectives, such as identifying threats to ecological sustainability, evaluating natural resource health, assessing compliance with regulations, or informing land management decisions.

Scope Determination
- ☐ Define the boundaries, focus areas, and resources to be included in the assessment.
- ☐ Consider the scale of the operation (e.g., farm or horticultural site).
- ☐ Determine the spatial extent of the assessment area.
- ☐ Decide whether the assessment will cover all natural resources or specific aspects (e.g., soil, water, biodiversity).

Methodology Selection
- ☐ Select appropriate methodologies and data collection techniques.
- ☐ Ensure methods are scientifically robust, cost-effective, and suitable for the operation's scale.
- ☐ Consider methods such as field surveys, remote sensing, laboratory analysis, interviews, and data modelling.

Consideration of Factors
- ☐ Take into account the diversity of natural resources present on the site.
- ☐ Consider specific sustainability goals and regulatory requirements.
- ☐ Identify factors such as soil types, water bodies, vegetation cover, and wildlife habitats.

Scale and Diversity
- ☐ Assess the scale of the operation.
- ☐ Evaluate the diversity of natural resources present.
- ☐ Consider how the scale and diversity may impact the assessment process and outcomes.

Sustainability Goals and Regulatory Requirements
- ☐ Review existing sustainability goals for the enterprise.
- ☐ Identify relevant regulatory requirements related to environmental protection, land use planning, water management, and biodiversity conservation.

Resource Allocation
- ☐ Allocate resources, including personnel, equipment, time, and budget.
- ☐ Ensure sufficient resources are available to conduct a comprehensive assessment.
- ☐ Consider project timelines and budget constraints.

Review and Adjust
- ☐ Review the assessment plan periodically.
- ☐ Adjust objectives, scope, and methodologies as needed.
- ☐ Incorporate feedback from stakeholders and lessons learned from previous assessments.

Stakeholder Engagement
- ☐ Engage relevant stakeholders in the planning process.
- ☐ Gather input and perspectives from diverse stakeholders.
- ☐ Ensure transparency and collaboration throughout the assessment planning phase.

This checklist serves as a guide for planning the assessment process, outlining objectives, scope, and methodology for auditing site natural resources. It considers factors such as the scale of the operation, diversity of natural resources present, and specific sustainability goals or regulatory requirements. Adjustments can be made based on the unique characteristics and needs of each assessment.

Figure 16: Assessment Planning Checklist.

The checklist can be used as a tool to systematically guide the assessment planning process for auditing site natural resources. Here's how it can be utilized:

1. **Preparation**: Before starting the assessment planning process, gather relevant stakeholders and resources, including information about the operation, regulatory requirements, and sustainability goals.

2. **Completion**: Use the checklist to work through each criterion, ensuring that all aspects of assessment planning are considered thoroughly. Discuss each item with the team and document responses accordingly.

3. **Discussion and Decision-Making**: As you go through the checklist, engage in discussions with stakeholders to gather input and perspectives. Use the checklist as a framework for making decisions about objectives, scope, methodologies, and resource allocation.

4. **Adjustment**: If needed, adjust the assessment plan based on feedback and discussions. Consider revising objectives, expanding or narrowing the scope, or modifying methodologies to better align with project goals and constraints.

5. **Documentation**: Keep a record of the completed checklist, including responses, decisions made, and any adjustments to the assessment plan. This documentation serves as a reference for future planning and ensures accountability and transparency in the process.

6. **Implementation**: Once the assessment plan is finalized, implement the planned activities according to the established objectives, scope, and methodologies.

Records that should emerge from the process include:

- Completed checklist with responses documented for each criterion.

- Documentation of discussions and decisions made during the planning process.

- Revised assessment plan, if adjustments were made based on feedback or changes in circumstances.

- Resource allocation plan detailing personnel, equipment, time, and budget allocated for assessment activities.

- Stakeholder engagement records, including meeting minutes, feedback received, and communication logs.

- Any additional documentation or reports generated during the assessment planning process, such as maps, data collection protocols, or risk assessments.

By using the checklist and maintaining thorough records throughout the assessment planning process, you can ensure that the resulting plan is comprehensive, well-informed, and aligned with the goals and requirements of the agricultural or horticultural enterprise.

Expanding on the second step, data collection is an important phase in the assessment process, as it provides the foundation for understanding the current state of the site's natural resources. The data collection process involves gathering a wide range of information pertaining to the site's natural resources. This includes detailed assessments of soil characteristics, such as texture, structure, pH levels, nutrient content, and organic matter content. Water availability and quality should also be assessed, considering factors such as sources of water, water storage capacity, water table levels, and potential contaminants. Additionally, data on vegetation cover, including plant species diversity, biomass, and canopy coverage, should be collected to evaluate habitat suitability and ecosystem health. Biodiversity surveys can further contribute by identifying species present on the site and assessing their abundance and distribution patterns. Lastly, collecting information on the site's land use history, including past farming practices, land management techniques, and any previous environmental impacts, provides valuable context for understanding current conditions and potential challenges.

Various tools and techniques can be employed to collect data effectively. Soil tests, conducted using soil sampling kits and laboratory analysis, provide quantitative measurements of soil properties and fertility levels. Water quality assessments involve collecting water samples from different sources, such as wells, rivers, or irrigation ponds, and analysing them for parameters like pH, turbidity, dissolved oxygen, and nutrient concentrations. Biodiversity surveys may involve field observations, species identification guides, camera traps, or acoustic monitoring devices to document the presence and abundance of flora and fauna. Aerial imagery, obtained through satellite imagery or drone technology, offers valuable insights into landscape features, vegetation patterns, and land use changes over time. By utilizing these tools and techniques, data collection can be conducted efficiently and accurately, providing a comprehensive understanding of the site's natural resources.

It's essential to integrate data collected from various sources to develop a holistic picture of the site's ecological characteristics and sustainability status. Combining information from soil tests, water quality assessments, biodiversity surveys, and land use history records allows for a multidimensional analysis of environmental conditions and trends. Geographic Information System (GIS) software can be used to organize, analyse, and visualize spatial data, facilitating the integration of diverse datasets and the identification of spatial patterns and relationships. By integrating data from multiple sources, a more robust understanding of the site's natural resources can be achieved, enabling informed decision-making and the development of targeted land use strategies that promote ecological sustainability. Figure 17 shows a sample template for recording the observations related to the site's natural resources assessment.

Data Collection Template: Site Natural Resources Assessment

Date: [Date of Assessment]
Location: [Site Name or Address]
Assessment Team: [Names of Team Members]

1. Comprehensive Data Gathering:
Soil Characteristics:

- Texture:
- Structure:
- pH Levels:
- Nutrient Content:
- Organic Matter Content:

Water Availability and Quality:

- Sources of Water:
- Water Storage Capacity:
- Water Table Levels:
- Potential Contaminants:

Vegetation Cover:

- Plant Species Diversity:
- Biomass:
- Canopy Coverage:

Biodiversity Surveys:

- Species Identification:
- Abundance:
- Distribution Patterns:

Land Use History:

- Past Farming Practices:
- Land Management Techniques:
- Previous Environmental Impacts:

2. Utilization of Tools and Techniques:
Soil Tests:

- Soil Sampling Kits:

Notes and Observations:
[Space for Additional Notes, Observations, or Comments]

- Laboratory Analysis:

Water Quality Assessments:

- Water Sampling Locations:
- Parameters Measured (pH, Turbidity, Dissolved Oxygen, etc.):

Biodiversity Surveys:

- Field Observations:
- Species Identification Guides:
- Camera Traps:
- Acoustic Monitoring Devices:

Aerial Imagery:

- Source (Satellite Imagery/Drone Technology):
- Observations:

3. Integration of Data Sources:
Data Sources:

- Soil Tests:
- Water Quality Assessments:
- Biodiversity Surveys:
- Land Use History Records:

Analysis:

- Geographic Information System (GIS) Software:
- Spatial Data Visualization:
- Identification of Patterns and Relationships:

Recommendations:
[Space for Recommendations Based on Data Analysis]

Next Steps:
[Outline Next Steps for Developing Sustainable Land Use Strategies]

Approval:
[Signature and Date of Approval]

Record Keeping:
All data collected during the assessment process should be documented and stored for future reference. This includes raw data, analysis reports, and any recommendations or action plans developed based on the assessment findings. Ensure that records are organized, easily accessible, and securely maintained to support ongoing monitoring and decision-making processes.

Figure 17: Data collection template for recording site natural resources.

Engaging stakeholders is paramount in the assessment process, as it offers invaluable insights into the site's natural resources and sustainability challenges. Here's a detailed explanation of this step:

Identifying Stakeholders: It's essential to identify key stakeholders who hold a vested interest in the site's natural resources and sustainability outcomes. These stakeholders may include farm managers, workers, environmental experts, local communities, government agencies, non-governmental organizations (NGOs), and other relevant parties. Each stakeholder brings diverse perspectives and expertise to the discussion.

Establishing positive and collaborative relationships with stakeholders is crucial to fostering open communication and trust. This involves reaching out to stakeholders through various channels such as meetings, workshops, focus groups, interviews, surveys, and social media platforms. Providing opportunities for stakeholders to express their concerns, share knowledge, and contribute to the assessment process is essential.

Encouraging open dialogue and the exchange of ideas among stakeholders is key to gaining insights into the site's dynamics and sustainability challenges. Creating a supportive and inclusive environment where stakeholders feel comfortable expressing their views and engaging in constructive discussions is vital. Facilitating brainstorming sessions, group discussions, and interactive activities can help generate innovative solutions and identify common goals.

Actively listening to the perspectives and insights shared by stakeholders is crucial, acknowledging the diversity of opinions and experiences represented. Paying attention to key issues, concerns, and priorities raised by stakeholders related to natural resource management, environmental conservation, community well-being, and sustainable development is essential. Capturing valuable information and data shared during stakeholder consultations for further analysis and consideration is necessary.

Collaborating with stakeholders to identify practical solutions and strategies for addressing sustainability challenges identified during the assessment process is essential. Fostering a sense of ownership and shared responsibility among stakeholders by involving them in the decision-making process and empowering them to contribute to the development of actionable plans and initiatives is critical. Leveraging the collective expertise and resources of stakeholders to co-create sustainable land use strategies that are contextually relevant and socially inclusive is paramount.

Providing feedback to stakeholders on the outcomes of the assessment process, including key findings, recommendations, and proposed actions, is essential. Maintaining ongoing communication channels with stakeholders to keep them informed about progress, updates, and opportunities for involvement in implementation efforts is vital. Fostering transparency, accountability, and mutual respect in all interactions with stakeholders to build long-term partnerships and support for sustainable land management initiatives is necessary.

By engaging stakeholders in meaningful dialogue and collaboration, individuals managing agricultural and horticultural enterprises can harness collective wisdom and expertise to develop effective strategies for promoting ecological sustainability and responsible land use practices. These factors are summarised in Figure 18 as a checklist and Figure 19 as a template for recording stakeholder engagement.

Stakeholder Consultation Checklist: Enhancing Collaboration for Sustainable Land Use

1. **Identifying Stakeholders:**
   - ❐ Identify key stakeholders who have a vested interest in the site's natural resources and sustainability outcomes.
   - ❐ Consider including farm managers, workers, environmental experts, local communities, government agencies, non-governmental organizations (NGOs), and other relevant parties.
   - ❐ Assess the diverse range of perspectives and expertise each stakeholder brings to the table.

2. **Building Relationships:**
   - ❐ Establish positive and collaborative relationships with stakeholders to foster open communication and trust.
   - ❐ Reach out to stakeholders through various channels, such as meetings, workshops, focus groups, interviews, surveys, and social media platforms.
   - ❐ Create opportunities for stakeholders to express their concerns, share knowledge, and contribute to the assessment process.

3. **Facilitating Dialogue:**
   - ❐ Encourage open dialogue and exchange of ideas among stakeholders to gain insights into the site's dynamics and sustainability challenges.
   - ❐ Provide a supportive and inclusive environment where stakeholders feel comfortable expressing their views and engaging in constructive discussions.
   - ❐ Facilitate brainstorming sessions, group discussions, and interactive activities to generate innovative solutions and identify common goals.

4. **Gathering Insights:**
   - ❐ Actively listen to the perspectives and insights shared by stakeholders, acknowledging the diversity of opinions and experiences represented.
   - ❐ Pay attention to key issues, concerns, and priorities raised by stakeholders related to natural resource management, environmental conservation, community well-being, and sustainable development.
   - ❐ Capture valuable information and data shared during stakeholder consultations for further analysis and consideration.

5. **Collaborative Problem-Solving:**
   - ❐ Collaborate with stakeholders to identify practical solutions and strategies for addressing sustainability challenges identified during the assessment process.
   - ❐ Foster a sense of ownership and shared responsibility among stakeholders by involving them in the decision-making process and empowering them to contribute to the development of actionable plans and initiatives.
   - ❐ Leverage the collective expertise and resources of stakeholders to co-create sustainable land use strategies that are contextually relevant and socially inclusive.

6. **Feedback and Communication:**
   - ❐ Provide feedback to stakeholders on the outcomes of the assessment process, including key findings, recommendations, and proposed actions.
   - ❐ Maintain ongoing communication channels with stakeholders to keep them informed about progress, updates, and opportunities for involvement in implementation efforts.
   - ❐ Foster transparency, accountability, and mutual respect in all interactions with stakeholders to build long-term partnerships and support for sustainable land management initiatives.

Figure 18: Stakeholder consultation checklist.

Stakeholder Consultation Recording Template: Enhancing Collaboration for Sustainable Land Use
1. **Identifying Stakeholders:**
   - Stakeholder Name/Group:
   - Vested Interest in Site's Natural Resources and Sustainability Outcomes:
   - Relevant Perspective and Expertise:
   - Additional Notes:
2. **Building Relationships:**
   - Communication Channel Used (e.g., Meeting, Workshop, Interview):
   - Date of Interaction:
   - Key Discussion Points and Concerns Raised:
   - Opportunities Created for Stakeholder Engagement:
   - Follow-up Actions Needed:
3. **Facilitating Dialogue:**
   - Type of Dialogue Facilitated (e.g., Brainstorming Session, Focus Group):
   - Date and Duration of Session:
   - Key Insights and Ideas Generated:
   - Level of Participation and Engagement:
   - Identified Common Goals:
4. **Gathering Insights:**
   - Date of Stakeholder Consultation:
   - Perspectives and Insights Shared:
   - Key Issues, Concerns, and Priorities Identified:
   - Data and Information Captured for Further Analysis:
   - Recommendations for Consideration:
5. **Collaborative Problem-Solving:**
   - Solutions and Strategies Identified:
   - Stakeholder Involvement in Decision-Making Process:
   - Actions Taken to Foster Ownership and Shared Responsibility:
   - Utilization of Stakeholder Expertise and Resources:
   - Contextual Relevance and Social Inclusivity of Strategies:
6. **Feedback and Communication:**
   - Date of Feedback Session:
   - Key Findings Communicated:
   - Recommendations and Proposed Actions Presented:
   - Ongoing Communication Channels Established:
   - Measures Taken to Ensure Transparency and Accountability:

This template serves as a structured framework for recording stakeholder consultations and documenting the outcomes of collaborative efforts towards sustainable land use.

Figure 19: Stakeholder engagement recording template.

In the process of sustainable land use management, threat identification plays a pivotal role in safeguarding the ecological integrity and productivity of the land under production. Here's an expansion on the fifth step:

After gathering comprehensive data and observations about the site's natural resources, the next crucial step is to analyse this information to identify potential threats to the sustainability of the land under production. This analytical process involves a systematic examination of various factors that could compromise the ecological balance, productivity, and long-term viability of the land. This includes:

a. Soil Degradation: Assess the health and quality of the soil by examining factors such as erosion, compaction, salinization, acidity, and loss of organic matter. Soil degradation can impair fertility, water retention capacity, and nutrient cycling processes, leading to reduced agricultural productivity and increased vulnerability to environmental stresses.

b. Water Pollution: Evaluate the quality and availability of water resources, including surface water bodies, groundwater aquifers, and irrigation sources. Identify sources of contamination such as agricultural runoff, industrial discharge, and improper waste disposal practices that may contribute to water pollution. Addressing water pollution is critical for maintaining ecosystem health, supporting aquatic life, and ensuring safe drinking water supplies.

c. Habitat Loss: Identify changes in land use patterns, habitat fragmentation, and encroachment on natural ecosystems that may result in habitat loss and biodiversity decline. Loss of habitat can disrupt ecological processes, diminish species diversity, and increase the risk of species extinction. Assess the potential impacts of habitat loss on ecosystem services, wildlife populations, and ecosystem resilience.

d. Nutrient Depletion: Examine nutrient cycling dynamics and nutrient management practices to identify potential imbalances or deficiencies in soil nutrients. Nutrient depletion can result from excessive fertilizer use, poor soil management practices, and loss of organic matter. Addressing nutrient depletion is essential for maintaining soil fertility, supporting plant growth, and minimizing nutrient runoff into water bodies.

e. Pest Infestations: Monitor pest populations and pest damage levels to assess the risk of pest outbreaks and crop losses. Identify factors contributing to pest proliferation, such as monoculture cropping, lack of crop rotation, and pesticide resistance. Implement integrated pest management (IPM) strategies to minimize reliance on chemical pesticides and promote biological control methods.

f. Climate-Related Risks: Evaluate the vulnerability of the land to climate change impacts, including extreme weather events, temperature fluctuations, droughts, floods, and shifting precipitation patterns. Assess the adaptive capacity of the agricultural system to cope with climate variability and identify measures to enhance resilience, such as improved water management, diversified cropping systems, and soil conservation practices.

g. Regulatory Compliance Issues: Ensure compliance with relevant environmental regulations, land use policies, and sustainability standards governing agricultural practices. Identify areas of non-compliance or potential regulatory risks that may result in legal penalties, fines, or restrictions on land use activities. Take proactive measures to address compliance issues and promote adherence to environmental best practices.

By conducting a thorough analysis of these potential threats to sustainable land use, stakeholders can develop targeted mitigation strategies and adaptive management approaches to minimize risks, enhance resilience, and promote the long-term sustainability of agricultural production systems. A sample recording form to record identified threats is shown as Figure 20.

Environmental Assessment Record Template

1. Soil Degradation:
- Factors to Examine:
  - Erosion
  - Compaction
  - Salinization
  - Acidity
  - Loss of organic matter
- Potential Impacts:
  - Impaired fertility
  - Reduced water retention capacity
  - Disrupted nutrient cycling processes
- Mitigation Measures:
  - Soil conservation practices
  - Organic matter management
  - Erosion control techniques

2. Water Pollution:
- Quality and Availability Assessment:
  - Surface water bodies
  - Groundwater aquifers
  - Irrigation sources
- Sources of Contamination:
  - Agricultural runoff
  - Industrial discharge
  - Improper waste disposal practices
- Remedial Actions:
  - Pollution prevention measures
  - Water quality monitoring
  - Implementation of best management practices

3. Habitat Loss:
- Indicators of Habitat Change:
  - Land use patterns
  - Habitat fragmentation
  - Encroachment on natural ecosystems
- Ecological Implications:
  - Disruption of ecological processes
  - Diminished species diversity
  - Increased risk of species extinction
- Conservation Strategies:
  - Habitat restoration efforts
  - Biodiversity conservation initiatives
  - Sustainable land use planning

4. Nutrient Depletion:
- Nutrient Cycling Dynamics:
  - Imbalances or deficiencies in soil nutrients
- Contributing Factors:
  - Excessive fertilizer use
  - Poor soil management practices
  - Loss of organic matter
- Soil Health Improvement:
  - Nutrient management plans
  - Organic farming methods
  - Cover cropping and crop rotation

5. Pest Infestations:
- Pest Population Monitoring:
  - Pest outbreaks and crop losses
- Factors Contributing to Proliferation:
  - Monoculture cropping
  - Lack of crop rotation
  - Pesticide resistance
- Pest Management Strategies:
  - Integrated pest management (IPM) practices
  - Biological control methods
  - Pest-resistant crop varieties

6. Climate-Related Risks:
- Vulnerability Assessment:
  - Extreme weather events
  - Temperature fluctuations
  - Droughts, floods, and precipitation changes
- Adaptation Measures:
  - Water management improvements
  - Crop diversification
  - Soil conservation and erosion control

7. Regulatory Compliance Issues:
- Regulatory Requirements:
  - Environmental regulations
  - Land use policies
  - Sustainability standards
- Compliance Monitoring:
  - Identification of non-compliance areas
  - Documentation of regulatory adherence
  - Implementation of corrective action

Figure 20: Environmental Assessment Record.

Risk assessment plays a pivotal role in the process of evaluating potential threats to ecological sustainability within a given context. This step involves a systematic examination of the likelihood and potential impact of each identified threat on the ecological balance of the environment. Various factors are taken into account during this assessment, including the severity of the threat, the frequency of occurrence, the duration of its effects, and the spatial extent over which it may manifest.

The severity of a threat refers to the degree of harm or damage it can inflict on the natural resources and ecosystems under consideration. This could range from minor disturbances to catastrophic events with far-reaching consequences. Assessing the severity helps in understanding the extent of the potential impact and determining the appropriate level of response required to mitigate the threat effectively.

Frequency and duration are key considerations in assessing the likelihood of a threat occurring and its persistence over time. Threats that occur frequently or persist for extended periods pose greater risks to ecological sustainability as they have a more significant cumulative impact on the environment. Understanding the temporal dynamics of threats allows for the development of strategies to manage and mitigate their effects proactively.

Spatial extent refers to the geographical area or range over which a threat may spread and exert its influence on the environment. Some threats may be localized, affecting only specific areas or ecosystems, while others may have widespread implications across larger landscapes or regions. Evaluating the spatial extent helps in delineating the geographic scope of the threat and identifying areas that are most vulnerable to its effects.

Based on the outcomes of the risk assessment, threats are prioritized according to their significance and urgency for mitigation. This involves weighing the severity, likelihood, and spatial extent of each threat to determine its relative importance in terms of ecological sustainability. Prioritizing threats allows for the allocation of resources and efforts towards addressing those with the highest potential impact or those that require immediate attention to prevent further degradation of natural resources and ecosystems.

Risk assessment provides a structured approach to understanding and prioritizing threats to ecological sustainability based on their severity, likelihood, duration, and spatial extent. By systematically evaluating these factors, decision-makers can identify key areas of concern and develop targeted strategies for mitigating risks and promoting the long-term health and resilience of the environment. A checklist for inclusions to the risk assessment is shown as Figure 21.

Risk Assessment Checklist

1. **Threat Identification**
   - ❏ Identify potential threats to ecological sustainability within the given context.
   - ❏ Consider various factors such as soil degradation, water pollution, habitat loss, nutrient depletion, pest infestations, climate-related risks, and regulatory compliance issues.

2. **Severity Assessment**
   - ❏ Evaluate the degree of harm or damage each identified threat can inflict on natural resources and ecosystems.
   - ❏ Rank the severity of threats on a scale from minor disturbances to catastrophic events with far-reaching consequences.

3. **Likelihood and Duration**
   - ❏ Assess the likelihood of each threat occurring and its persistence over time.
   - ❏ Determine the frequency of occurrence and duration of effects for each threat to understand their temporal dynamics.

4. **Spatial Extent**
   - ❏ Determine the geographical area or range over which each threat may spread and exert its influence on the environment.
   - ❏ Evaluate whether threats are localized to specific areas or ecosystems or have widespread implications across larger landscapes or regions.

5. **Prioritization**
   - ❏ Prioritize threats based on their severity, likelihood, and spatial extent.
   - ❏ Identify threats with the highest potential impact or those requiring immediate attention for mitigation efforts.

6. **Mitigation Strategies**
   - ❏ Develop targeted strategies for mitigating identified threats to ecological sustainability.
   - ❏ Allocate resources and efforts towards addressing prioritized threats effectively.

7. **Monitoring and Review**
   - ❏ Establish monitoring mechanisms to track the effectiveness of mitigation strategies.
   - ❏ Regularly review and update the risk assessment to reflect changes in threat dynamics and environmental conditions.

8. **Documentation**
   - ❏ Document all aspects of the risk assessment process, including identified threats, severity assessments, prioritization criteria, mitigation strategies, and monitoring outcomes.
   - ❏ Maintain records for future reference and decision-making purposes.

By utilizing this structured risk assessment template, decision-makers can systematically evaluate and prioritize threats to ecological sustainability, leading to the development of targeted strategies for promoting the long-term health and resilience of the environment.

Figure 21: Risk Assessment Checklist.

Root cause analysis is a pivotal step in understanding the fundamental factors driving sustainability challenges within a given context. It involves delving deeper into the underlying causes of identified threats to ecological sustainability, aiming to uncover the root drivers that perpetuate these challenges. By examining various factors such as land management practices, external influences, socio-economic factors, and climate variability, this analysis seeks to identify the primary drivers shaping the current state of the environment.

One crucial aspect of root cause analysis is evaluating land management practices employed within the area under assessment. This entails scrutinizing agricultural techniques, irrigation methods, soil management strategies, and other land-use practices to ascertain their impact on natural resources. For instance, unsustainable farming practices such as excessive pesticide use, monoculture cropping, or improper soil tillage may contribute to soil degradation, water pollution, and loss of biodiversity. By identifying these practices as root causes, targeted interventions can be devised to promote more sustainable land management approaches.

External influences, including market dynamics, policy frameworks, and global trends, also play a significant role in shaping sustainability challenges. Economic pressures, market demands, and regulatory frameworks can influence land-use decisions and agricultural practices, impacting the health of ecosystems. Additionally, globalization and trade policies may drive deforestation, habitat destruction, and resource exploitation in pursuit of economic gains. Understanding these external factors is essential for developing interventions that address systemic issues and promote sustainable land use.

Socio-economic factors, such as population growth, urbanization, poverty levels, and access to resources, can exacerbate ecological sustainability challenges. Rapid population growth and urban expansion may lead to land conversion, encroachment on natural habitats, and increased demand for agricultural products. Moreover, socio-economic disparities and limited access to education and resources can hinder the adoption of sustainable practices among certain communities. Addressing these socio-economic drivers requires holistic approaches that integrate environmental conservation with social equity and economic development.

Furthermore, climate variability and environmental change are significant drivers of sustainability challenges, exacerbating threats such as droughts, floods, wildfires, and ecosystem disruptions. Climate change-induced shifts in temperature, precipitation patterns, and extreme weather events can amplify existing vulnerabilities and disrupt ecosystems' ability to adapt. Understanding the linkages between climate variability and ecological sustainability is crucial for implementing adaptive strategies that enhance resilience and mitigate climate-related risks.

By conducting a comprehensive root cause analysis that considers land management practices, external influences, socio-economic factors, and climate variability, decision-makers can gain insights into the underlying drivers of sustainability challenges. This understanding forms the basis for developing targeted interventions and policy measures aimed at addressing root causes and fostering long-term ecological resilience and sustainability.

Root cause analysis (RCA) is a problem-solving technique used to identify the underlying causes of an issue or problem. It involves systematically investigating the factors that contributed to a particular event or outcome, with the goal of understanding the root causes rather than just addressing symptoms. The purpose of RCA is to prevent recurrence of problems by addressing their fundamental causes.

In various fields such as engineering, healthcare, manufacturing, and business management, root cause analysis is employed to identify why an undesired outcome occurred and to develop effective solutions. The process typically involves several steps, including:

1. Problem Identification: Clearly defining the problem or issue that needs to be addressed. This could be a safety incident, equipment failure, product defect, customer complaint, or any other undesirable outcome.

2. Data Collection: Gathering relevant data and information related to the problem, including facts, observations, and documentation. This may involve reviewing records, conducting interviews, analysing data, and examining the affected processes or systems.

3. Cause Analysis: Analysing the collected data to identify the factors that contributed to the problem. This may involve using techniques such as brainstorming, fishbone diagrams (Ishikawa diagrams), fault tree analysis, or 5 Whys analysis to systematically explore potential causes.

4. Identifying Root Causes: Distinguishing between the immediate causes (symptoms) of the problem and the deeper underlying root causes. Root causes are the fundamental factors or systemic issues that, if addressed, could prevent recurrence of the problem.

5. Developing Solutions: Once the root causes are identified, developing and implementing corrective actions or solutions to address them. These solutions aim to eliminate or mitigate the root causes, thereby preventing similar problems from occurring in the future.

6. Monitoring and Verification: Monitoring the effectiveness of the implemented solutions and verifying that they have successfully addressed the root causes. This may involve tracking key performance indicators, conducting follow-up assessments, and making adjustments as needed.

Root cause analysis, which can be conducted following the procedure shown in Figure 22, is a proactive approach to problem-solving that emphasizes learning from past mistakes, improving processes, and preventing future incidents. It helps organizations identify opportunities for continuous improvement and fosters a culture of accountability and problem-solving.

Root Cause Analysis Procedure

To conduct a root cause analysis (RCA) effectively:

1. **Define the Problem**: Clearly articulate the sustainability challenge or issue that requires investigation. This could include threats to ecological sustainability such as soil degradation, water pollution, habitat loss, or climate-related risks.
2. **Gather Relevant Data**: Collect data and information related to the identified problem. This may involve reviewing records, conducting interviews with stakeholders, analysing environmental reports, and examining historical data on land management practices and environmental conditions.
3. **Identify Potential Root Causes**: Analyse the collected data to identify potential factors contributing to the sustainability challenge. Consider factors such as land management practices, external influences, socio-economic factors, and climate variability.
4. **Evaluate Land Management Practices**: Scrutinize agricultural techniques, irrigation methods, soil management strategies, and other land-use practices to assess their impact on natural resources. Identify unsustainable practices that may be contributing to the problem.
5. **Assess External Influences**: Consider market dynamics, policy frameworks, global trends, and other external factors that may influence land-use decisions and agricultural practices. Evaluate how these influences contribute to sustainability challenges.
6. **Examine Socio-Economic Factors**: Investigate population growth, urbanization, poverty levels, access to resources, and other socio-economic factors that may exacerbate ecological sustainability challenges. Identify disparities and barriers hindering the adoption of sustainable practices.
7. **Analyse Climate Variability**: Assess the impact of climate change and environmental change on sustainability challenges. Consider shifts in temperature, precipitation patterns, extreme weather events, and their effects on ecosystems and natural resources.
8. **Prioritize Root Causes**: Rank the identified root causes based on their significance and urgency for mitigation. Focus on addressing the most critical factors that have the greatest impact on ecological sustainability.
9. **Develop Solutions**: Once root causes are identified, develop and implement targeted interventions to address them. These solutions should aim to eliminate or mitigate the root causes, thereby preventing recurrence of the sustainability challenge.
10. **Monitor and Adjust**: Continuously monitor the effectiveness of implemented solutions and make adjustments as needed. Track key performance indicators, conduct follow-up assessments, and refine interventions to ensure sustained progress towards ecological resilience and sustainability.

Figure 22: Procedure for conducting root cause analysis.

Sustainability gap analysis serves as a critical tool in assessing the effectiveness of current land management practices and their alignment with desired sustainability goals and benchmarks. This step involves a systematic comparison between the existing state of land management and sustainability outcomes against predetermined targets or standards. By conducting this analysis, stakeholders can identify disparities or gaps that exist between current practices and desired sustainability outcomes, thereby guiding the development of strategies to enhance ecological resilience and promote sustainable land use.

To begin the sustainability gap analysis, it is essential to establish clear and measurable sustainability targets or benchmarks that reflect the desired ecological, social, and economic outcomes. These targets may encompass various aspects of sustainability, including soil health, water quality, biodiversity conservation, carbon sequestration, and resilience to climate change. By defining specific targets, stakeholders can establish a framework for assessing progress and identifying areas requiring improvement.

Once sustainability targets are defined, the next step is to evaluate the current state of land management practices and their associated sustainability outcomes. This involves collecting data and information on various indicators related to land use, resource management, environmental quality, and socio-economic factors. Common indicators may include soil erosion rates, nutrient levels, water usage efficiency, habitat quality, crop diversity, income distribution, and community well-being.

After gathering relevant data, stakeholders can conduct a comparative analysis to identify gaps between current practices and desired sustainability targets. This analysis may reveal areas where land management practices fall short in achieving sustainability objectives or where negative environmental impacts exceed acceptable thresholds. By quantifying these gaps, stakeholders gain insights into priority areas for intervention and improvement.

Based on the findings of the sustainability gap analysis, stakeholders can then develop targeted strategies and action plans, see Figure 23 and Figure 24 for sample templates, to address identified gaps and enhance sustainability outcomes. These strategies may involve implementing best management practices, adopting innovative technologies, promoting conservation measures, strengthening regulatory frameworks, and fostering stakeholder engagement and capacity-building initiatives. By aligning interventions with specific areas of need identified through the gap analysis, stakeholders can maximize the effectiveness and impact of their sustainability efforts.

Furthermore, sustainability gap analysis is an iterative process that should be periodically reassessed and updated to track progress over time and adapt strategies as needed. By monitoring changes in sustainability indicators and regularly revisiting sustainability targets, stakeholders can ensure that their land management practices remain responsive to evolving environmental, social, and economic conditions. Ultimately, sustainability gap analysis provides a structured approach for bridging the divide between current practices and desired sustainability outcomes, thereby promoting the long-term health and resilience of ecosystems and communities.

Sustainability Action Plan

1. Identified Gaps:
   o [List of gaps identified through sustainability gap analysis]
2. Targeted Strategies: a. Implementing Best Management Practices:
   o [Specific best management practices to address identified gaps]
   o [Responsible parties for implementation]
   o [Timeline for implementation]

b. Adopting Innovative Technologies:
   o [Innovative technologies to improve sustainability outcomes]
   o [Implementation process and timeline]
   o [Training or resource needs]

c. Promoting Conservation Measures:
   o [Conservation initiatives to protect natural resources]
   o [Communication and outreach strategies]
   o [Expected outcomes and metrics for success]

d. Strengthening Regulatory Frameworks:
   o [Proposed changes or enhancements to regulations]
   o [Engagement with relevant stakeholders and policymakers]
   o [Timeline for advocacy and implementation]

e. Fostering Stakeholder Engagement:
   o [Strategies to involve stakeholders in decision-making processes]
   o [Methods for gathering feedback and input]
   o [Capacity-building activities to empower stakeholders]

3. Implementation Plan:
   o [Detailed action plan for each targeted strategy]
   o [Allocation of resources, budget, and personnel]
   o [Monitoring and evaluation mechanisms]
4. Timeline:
   o [Timeline for implementing each strategy and achieving milestones]
   o [Regular review and adjustment schedule]
5. Expected Outcomes:
   o [Anticipated improvements in sustainability outcomes]
   o [Key performance indicators to measure success]
6. Responsible Parties:
   o [Assigned responsibilities for each action item]
   o [Clear lines of communication and coordination]
7. Resources Needed:
   o [Required resources, including funding, equipment, expertise, etc.]
   o [Plans for securing necessary resources]
8. Communication Plan:
   o [Strategies for communicating the action plan to stakeholders]
   o [Channels for sharing progress updates and soliciting feedback]
9. Contingency Plans:
   o [Plans for addressing unforeseen challenges or setbacks]
   o [Strategies for adapting the action plan as needed]
10. Review and Evaluation:
    o [Process for monitoring progress and evaluating outcomes]
    o [Criteria for assessing the effectiveness of implemented strategies]
    o [Opportunities for continuous improvement]

Figure 23: Sustainability Action Plan.

Sustainability Action Plan Summary

1. Identified Gaps:
   - [Brief overview of gaps identified through sustainability gap analysis]
2. Targeted Strategies:
   - [Summary of key strategies to address identified gaps]
3. Implementation Plan:
   - [Overview of the planned actions and initiatives]
4. Timeline:
   - [High-level timeline for implementing strategies]
5. Expected Outcomes:
   - [Anticipated improvements in sustainability outcomes]
6. Responsible Parties:
   - [Key stakeholders responsible for implementing actions]
7. Resources Needed:
   - [Summary of required resources and potential sources]
8. Communication Plan:
   - [Brief outline of communication strategies]
9. Review and Evaluation:
   - [Overview of the monitoring and evaluation process]

This template can be customized according to the specific needs and context of the sustainability action plan.

Figure 24: Sustainability Action Plan Summary.

Solution design is a critical phase in the process of enhancing ecological sustainability within a given context. This step involves developing customized solutions and management interventions to effectively address the threats identified through the assessment process and bridge the sustainability gap. By tailoring solutions to the specific context and needs of the operation, stakeholders can implement targeted strategies that promote long-term ecological resilience and sustainability.

One key aspect of solution design is to consider a range of sustainable land management practices that are suitable for the unique characteristics of the landscape and ecosystem. This may involve implementing practices such as conservation tillage, cover cropping, crop rotation, agroforestry, and integrated pest management (IPM) to enhance soil health, biodiversity, and ecosystem services. By adopting these practices, land managers can minimize environmental impacts, improve resource efficiency, and promote the overall health and productivity of the land.

In addition to traditional land management practices, solution design should also explore the use of innovative technologies and advancements in agricultural science. Technologies such as precision agriculture, remote sensing, soil moisture sensors, and data analytics offer valuable tools for optimizing resource use, improving decision-making, and monitoring environmental conditions. Incorporating these technologies into land management strategies can enhance efficiency, productivity, and environmental sustainability while reducing reliance on inputs such as water, fertilizers, and pesticides.

Furthermore, solution design should take into account the socio-economic context and the needs of local communities and stakeholders. It's essential to engage with stakeholders throughout the design process to ensure that solutions are socially acceptable, culturally appropriate, and economically viable. This may involve conducting participatory workshops, stakeholder consultations, and community engagement activities to gather input, build consensus, and foster ownership of the proposed solutions.

By combining sustainable land management practices, innovative technologies, and stakeholder engagement strategies, solution design aims to develop holistic and integrated approaches to addressing ecological sustainability challenges. These solutions should be adaptable, scalable, and contextually relevant, taking into account the dynamic nature of ecosystems and the evolving needs of the community. Ultimately, the goal of solution design is to create resilient and sustainable land management systems that support the well-being of both people and the planet over the long term.

Recording the solution design process involves documenting the key elements and decisions made during the development of customized strategies and interventions to address sustainability challenges. The following outlines how this process can be recorded:

1. Introduction: Provide an overview of the importance of solution design in enhancing ecological sustainability and bridging the sustainability gap within a specific context.

2. Identified Threats: List the threats identified through the assessment process, highlighting their significance and potential impact on ecological sustainability.

3. Tailored Solutions: Describe the customized solutions and management interventions proposed to address each identified threat. This may include a range of sustainable land management practices, innovative technologies, and stakeholder engagement strategies.

4. Sustainable Land Management Practices: Outline the sustainable land management practices considered for implementation, such as conservation tillage, cover cropping, crop rotation, agroforestry, and integrated pest management (IPM). Explain how these practices contribute to enhancing soil health, biodiversity, and ecosystem services.

5. Innovative Technologies: Discuss the innovative technologies and advancements in agricultural science explored for integration into land management strategies. This may include precision agriculture, remote sensing, soil moisture sensors, and data analytics, emphasizing their potential to optimize resource use and improve environmental sustainability.

6. Socio-Economic Considerations: Address the socio-economic context and the needs of local communities and stakeholders in the solution design process. Describe how stakeholder engagement activities, participatory workshops, and consultations were conducted to ensure that solutions are socially acceptable, culturally appropriate, and economically viable.

7. Holistic Approach: Emphasize the importance of developing holistic and integrated approaches to addressing sustainability challenges. Explain how the combination of sustainable land management practices, innovative technologies, and stakeholder engagement strategies aims to create resilient and sustainable land management systems.

8. Adaptability and Scalability: Discuss the adaptability and scalability of the proposed solutions, considering the dynamic nature of ecosystems and the evolving needs of the community. Highlight how solutions can be adjusted and expanded over time to address emerging challenges and opportunities.

9. Long-Term Goals: Conclude by reiterating the long-term goals of solution design, which are to create resilient and sustainable land management systems that support the well-being of both people and the planet. Emphasize the importance of ongoing monitoring, evaluation, and adaptation to ensure the effectiveness and sustainability of implemented strategies.

Effective communication and documentation are crucial components of the sustainability assessment process and the development of solutions. Clear and concise communication is essential for conveying audit findings, risk assessments, and proposed solutions to relevant stakeholders in an understandable manner. Utilizing plain language, visual aids, and engaging formats helps ensure that diverse audiences, including farm managers, workers, community members, policymakers, and others, can comprehend key messages easily.

Stakeholder engagement is integral to the communication process, ensuring that stakeholders' input, feedback, and perspectives are considered. This may involve hosting meetings, workshops, or forums to discuss audit findings, share information, and gather insights. By encouraging two-way communication and active participation, stakeholders become invested in the decision-making process, fostering collaboration and ownership of proposed solutions.

Documentation of the audit process is essential for transparency, accountability, and continuous improvement. Detailed records of data sources, methodologies, observations, and analysis procedures provide a clear record of the assessment process. This documentation facilitates replication, verification, and review by internal and external stakeholders, serving as a valuable resource for future reference and evaluation of sustainability practices.

Comprehensive reporting is key to summarizing audit findings, conclusions, and recommendations. Reports should offer a comprehensive overview of the current state of natural resources, sustainability challenges, and proposed solutions. Using a structured format, including executive summaries, background information, methodology descriptions, and actionable recommendations, ensures that reports are informative, relevant, and actionable for different stakeholders.

Dissemination of results is crucial for reaching a wide audience and raising awareness of sustainability issues. Reports should be distributed electronically and in print to relevant stakeholders and presented at meetings, conferences, and workshops. Leveraging social media, press releases, and other communication channels helps amplify the reach and impact of messages, engaging broader audiences and catalysing action towards sustainable land use practices.

By prioritizing clear communication and comprehensive documentation throughout the sustainability assessment process, stakeholders can promote transparency, accountability, and continuous improvement in their efforts to enhance ecological sustainability and responsible land management practices.

Implementation planning is a pivotal phase in the journey towards sustainable land use, as it translates the identified strategies and solutions into actionable steps. This step involves developing a detailed action plan that outlines specific tasks, responsibilities, timelines, and required resources to effectively implement sustainable land use strategies. By establishing a clear roadmap for implementation, stakeholders can streamline efforts, allocate resources efficiently, and ensure accountability throughout the process.

Central to implementation planning is the development of an action plan that delineates the sequence of tasks and activities required to achieve sustainability goals. This involves breaking down overarching strategies into smaller, actionable steps, each with clear objectives and timelines. By defining specific tasks and milestones, stakeholders can track progress, identify potential bottlenecks, and adjust plans as needed to stay on course towards achieving sustainability objectives.

Assigning responsibilities is another critical aspect of implementation planning, ensuring that individuals or teams are accountable for executing specific tasks within the action plan. Clearly defining roles and responsibilities helps avoid confusion, duplication of efforts, and gaps in implementation. Stakeholders should collaborate to assign tasks based on expertise, capacity, and availability, ensuring that responsibilities are distributed effectively and equitably among all involved parties.

Timelines are essential for setting realistic deadlines and milestones for each task or activity outlined in the action plan. By establishing clear timelines, stakeholders can prioritize tasks, monitor progress, and track adherence to project schedules. It's crucial to consider factors such as seasonality, resource availability, and external dependencies when setting timelines to ensure feasibility and avoid potential delays.

Moreover, implementation planning requires careful consideration of the resources needed to execute the action plan effectively. This includes human resources, financial resources, technical expertise, equipment, and materials. Stakeholders should assess resource requirements and allocate resources judiciously to support implementation efforts. Collaboration with stakeholders is key to identifying resource needs, leveraging existing capacities, and exploring opportunities for resource mobilization or partnerships.

Collaboration with stakeholders is fundamental to ensuring buy-in and support for implementation efforts. Engaging stakeholders throughout the planning process fosters a sense of ownership, shared responsibility, and commitment to achieving sustainability goals. By soliciting input, addressing concerns, and building consensus, stakeholders can align their interests and priorities, enhancing the likelihood of successful implementation.

Implementation planning is a dynamic and collaborative process that transforms sustainability strategies into actionable plans, for example using the template shown as Figure 25. By developing a clear action plan, assigning responsibilities, setting realistic timelines, allocating resources, and fostering collaboration with stakeholders, organizations can effectively implement sustainable land use strategies and drive positive environmental outcomes.

Implementation Planning Template

1. **Objective:** Describe the overarching goal or objective of the implementation plan.
2. **Action Plan Overview:**
   o **Tasks:** Break down the strategies into specific tasks and activities required for implementation.
   o **Responsibilities:** Clearly define who is responsible for each task or activity.
   o **Timelines:** Set realistic deadlines and milestones for each task to track progress.
   o **Resources:** Identify the necessary resources, including human, financial, technical, and material resources.
3. **Task Breakdown:**
   o **Task ID:** Assign a unique identifier to each task for easy reference.
   o **Task Description:** Provide a brief description of the task or activity.
   o **Responsible Party:** Specify the individual or team responsible for executing the task.
   o **Start Date:** Set the planned start date for the task.
   o **End Date:** Establish the deadline for completing the task.
   o **Dependencies:** Identify any dependencies or prerequisites for the task.
   o **Progress:** Track the progress of each task throughout the implementation process.
4. **Resource Allocation:**
   o **Human Resources:** Specify the personnel needed for each task and their roles.
   o **Financial Resources:** Estimate the budget required for implementation, including costs for materials, equipment, and other expenses.
   o **Technical Expertise:** Identify any specialized knowledge or skills needed for certain tasks.
   o **Materials and Equipment:** List the materials, tools, or equipment required for implementation.
5. **Stakeholder Collaboration:**
   o **Engagement Strategy:** Outline how stakeholders will be involved in the implementation process.
   o **Communication Plan:** Describe the communication channels and methods for keeping stakeholders informed and engaged.
   o **Feedback Mechanism:** Establish a process for soliciting feedback and addressing concerns from stakeholders.
6. **Monitoring and Evaluation:**
   o **Key Performance Indicators (KPIs):** Define measurable indicators to track progress and evaluate the effectiveness of implementation.
   o **Evaluation Criteria:** Specify the criteria for assessing the success of the implementation plan.
   o **Monitoring Schedule:** Establish a schedule for monitoring progress and conducting periodic evaluations.
7. **Contingency Planning:**
   o **Risk Assessment:** Identify potential risks or obstacles that may arise during implementation.
   o **Mitigation Strategies:** Develop strategies to mitigate risks and address unforeseen challenges.
   o **Adaptation Plan:** Outline how the implementation plan will be adjusted in response to changing circumstances or new information.
8. **Conclusion:**
   o Summarize the key points of the implementation plan.
   o Reinforce the importance of collaboration and stakeholder engagement for successful implementation.
   o Emphasize the commitment to driving positive environmental outcomes through sustainable land use practices.

Figure 25: Implementation Planning Template.

Monitoring and evaluation are integral components of sustainable land management practices, ensuring that implemented strategies effectively contribute to ecological sustainability goals. This step involves establishing robust monitoring protocols and performance indicators to systematically track the implementation of strategies and measure progress towards sustainability objectives. Additionally, regular review and evaluation of outcomes are essential to assess the effectiveness of implemented measures, identify areas for improvement, and adapt strategies to changing environmental conditions.

To begin, it's crucial to develop monitoring protocols that outline the procedures and methodologies for collecting data and assessing the performance of implemented strategies. These protocols should specify the frequency of monitoring activities, the parameters to be measured, and the tools or techniques to be used for data collection. For example, monitoring protocols may include methods for assessing soil health, water quality, biodiversity, crop yields, or greenhouse gas emissions, depending on the objectives of the sustainability initiatives.

Performance indicators play a central role in monitoring and evaluation, providing quantifiable measures to assess progress towards sustainability goals. These indicators should be carefully selected to reflect key aspects of ecological health, resource use efficiency, and socio-economic well-being. Common performance indicators may include metrics such as soil organic matter content, water retention capacity, species diversity indices, carbon sequestration rates, or economic returns on investment in sustainable practices. By tracking these indicators over time, stakeholders can gauge the effectiveness of implemented strategies and make informed decisions about future actions.

Regular review and evaluation of monitoring data are essential to determine the success of implemented strategies and identify areas where adjustments may be needed. This involves analyzing monitoring results against established performance indicators, comparing actual outcomes to predefined targets or benchmarks, and identifying trends or patterns that may indicate areas of concern or success. Stakeholders should engage in periodic reviews to discuss monitoring findings, assess progress towards sustainability goals, and collaboratively identify strategies for improvement.

Based on the findings of monitoring and evaluation, stakeholders should be prepared to adjust strategies and interventions as needed to optimize results and adapt to changing environmental conditions. This may involve refining management practices, reallocating resources, scaling up successful initiatives, or discontinuing ineffective measures. Flexibility and responsiveness are essential qualities of effective monitoring and evaluation processes, allowing stakeholders to iteratively improve their approach to sustainable land management over time.

In conclusion, monitoring and evaluation are critical components of the sustainable land management cycle, providing feedback mechanisms to assess the effectiveness of implemented strategies and guide decision-making processes. By establishing robust monitoring protocols, selecting appropriate performance indicators, regularly reviewing outcomes, and adapting strategies as needed, stakeholders can enhance their ability to achieve ecological sustainability goals and promote resilient land management practices.

Production and Management Plans for Sustainable Land Use Principles

In the realm of sustainable land use management, individuals tasked with developing and implementing strategies must possess a nuanced understanding of both production and management plans. Specifically tailored to agricultural and horticultural enterprises, this review process involves a comprehensive examination of existing plans through the lens of sustainable land use principles. As stewards of the land, these individuals must demonstrate a deep knowledge of ecological systems, agricultural practices, and sustainability frameworks to effectively evaluate and enhance existing plans.

The review process begins with a meticulous analysis of production plans, which outline the strategies and practices employed to cultivate crops or manage horticultural production. These plans typically encompass a range of activities, including crop selection, cultivation techniques, irrigation methods, pest management strategies, and harvesting practices. Individuals with accountability for such plans must scrutinize them with a critical eye, assessing their alignment with sustainable land use principles such as soil conservation, water efficiency, biodiversity preservation, and ecosystem resilience.

Simultaneously, the review extends to management plans, which govern the overall operation and administration of agricultural or horticultural enterprises. Management plans encompass a broader spectrum of activities, including resource allocation, budgeting, staffing, compliance with regulations, and risk management strategies. When evaluating these plans through the lens of sustainability, individuals must consider not only the immediate economic outcomes but also the long-term environmental impacts and social implications of management decisions.

To effectively review production and management plans for sustainable land use principles, individuals must possess a diverse skill set that encompasses analytical thinking, technical expertise, and a deep understanding of ecological systems. They must be proficient in analysing data related to soil health, water usage, biodiversity, and climate patterns to identify areas where current practices may fall short of sustainability objectives. Moreover, they must exercise judgment to discern trade-offs and prioritize interventions that maximize ecological sustainability without compromising productivity or profitability.

Communication skills are also paramount in this process, as individuals must effectively communicate their findings and recommendations to stakeholders, including farm managers, owners, workers, regulatory agencies, and community members. Clear and persuasive communication is essential for building consensus, garnering support, and catalysing action towards the adoption of sustainable land use practices. Through well-articulated presentations, reports, and discussions, individuals can leverage their expertise to influence decision-making processes and drive positive change within agricultural and horticultural enterprises.

In summary, reviewing production and management plans for sustainable land use principles requires individuals with advanced skills and knowledge in land management practices, as well as the ability to analyse information, exercise judgment, and communicate solutions to complex problems. By conducting thorough reviews and advocating for sustainable practices, these individuals play a crucial role in advancing ecological sustainability within agricultural and production horticulture enterprises.

The meticulous analysis of production plans serves as the cornerstone of the review process, offering a detailed examination of the strategies and practices that underpin agricultural and horticultural production. These plans serve as blueprints, guiding farmers and horticulturists through the intricacies of crop cultivation and management. Within production plans, a diverse array of activities is encapsulated, ranging from initial crop selection to the final stages of harvesting. Each phase of production is meticulously outlined, including the techniques employed for cultivation, irrigation, pest management, and harvesting practices.

Individuals entrusted with the accountability of these plans must approach their analysis with a discerning eye, recognizing the pivotal role they play in shaping the environmental footprint of agricultural and horticultural operations. The scrutiny applied to these plans extends beyond mere compliance with regulatory standards; it necessitates a thorough evaluation of their alignment with sustainable land use principles. These principles encompass a multifaceted approach to environmental stewardship, including considerations for soil conservation, water efficiency, biodiversity preservation, and ecosystem resilience.

Within the realm of soil conservation, the review delves into the techniques utilized to maintain or enhance soil health and productivity. This entails an assessment of practices such as conservation tillage, cover cropping, and organic soil amendments, which serve to minimize soil erosion, enhance soil structure, and promote nutrient cycling. Similarly, the evaluation of water efficiency focuses on the utilization of irrigation methods that minimize water wastage and optimize water use efficiency, thereby reducing the strain on water resources and mitigating the risk of water scarcity.

Biodiversity preservation emerges as another critical aspect of the review process, as it pertains to the maintenance of ecological diversity within agricultural and horticultural landscapes. The assessment encompasses strategies for promoting habitat diversity, supporting pollinator populations, and preserving natural habitats within production areas. Moreover, the review examines practices aimed at fostering ecosystem resilience, which involve enhancing the capacity of agroecosystems to withstand environmental stressors such as climate change, pests, and diseases.

In essence, the meticulous analysis of production plans requires individuals to navigate the intricate balance between agricultural productivity and environmental sustainability. By scrutinizing these plans through the lens of sustainable land use principles, stakeholders can identify opportunities for improvement and advocate for practices that promote ecological resilience and long-term viability. This critical review process lays the foundation for advancing sustainable agriculture and horticulture practices that safeguard the health of ecosystems and support the well-being of future generations.

A procedure for Reviewing Production and Management Plans for Sustainable Land Use Principles includes:

1. Establish Review Objectives:

   ◦ Define the purpose and scope of the review process.

   ◦ Determine specific sustainability goals and principles to be addressed, such as soil conservation, water efficiency, biodiversity preservation, and ecosystem resilience.

2. Gather Relevant Documentation:

   ◦ Collect production and management plans, including crop selection documents, cultivation techniques, irrigation schedules, pest management strategies, and harvesting practices.

   ◦ Obtain any supplementary materials related to environmental impact assessments, regulatory compliance reports, and previous sustainability initiatives.

3. Conduct Preliminary Assessment:

   ◦ Review the production plans to gain an understanding of the overall approach to land management.

   ◦ Identify key stakeholders involved in the development and implementation of the plans.

   ◦ Assess the availability of resources, including financial, human, and technical capacities, to support sustainable practices.

4. Evaluate Alignment with Sustainable Principles:

   ◦ Examine each component of the production and management plans to assess its adherence to sustainable land use principles.

   ◦ Evaluate soil conservation practices, such as tillage methods, cover cropping, and soil amendment strategies, to minimize erosion and enhance soil health.

   ◦ Analyse water management techniques, including irrigation systems and water recycling methods, to optimize water use efficiency and mitigate water-related risks.

   ◦ Assess biodiversity preservation efforts, such as habitat restoration, buffer zone establishment, and wildlife-friendly farming practices, to support ecosystem diversity.

   ◦ Review pest management strategies, focusing on integrated pest management (IPM) approaches that minimize reliance on chemical pesticides and promote biological control methods.

5. Identify Strengths and Weaknesses:

   ◦ Document areas of alignment with sustainable principles and commendable practices.

   ◦ Identify gaps, inconsistencies, or areas where improvements are needed to enhance sustainability outcomes.

   ◦ Consider the feasibility, effectiveness, and scalability of proposed practices and interventions.

6. Develop Recommendations:

   ◦ Generate actionable recommendations to address identified weaknesses and optimize sustainability outcomes.

   ◦ Prioritize recommendations based on their potential impact, feasibility of implementation, and alignment with organizational goals.

  ○ Consider the input and feedback of stakeholders involved in the review process to ensure buy-in and support for proposed recommendations.

7. Document Review Findings:

  ○ Compile a comprehensive report summarizing the review findings, including strengths, weaknesses, and recommendations.

  ○ Clearly articulate the rationale behind each recommendation and provide supporting evidence or references where applicable.

  ○ Ensure that the report is structured, well-organized, and accessible to relevant stakeholders.

8. Communicate Results and Implement Changes:

  ○ Present the review findings and recommendations to key stakeholders, including farm managers, decision-makers, and relevant staff members.

  ○ Facilitate discussions to solicit feedback, address concerns, and gain consensus on proposed changes.

  ○ Collaborate with stakeholders to develop an action plan for implementing recommended improvements and monitoring progress over time.

9. Monitor and Evaluate Progress:

  ○ Establish monitoring protocols and performance indicators to track the implementation of recommended changes.

  ○ Conduct regular evaluations to assess the effectiveness of implemented strategies and measure progress towards sustainability goals.

  ○ Adjust plans and strategies as needed based on monitoring results and evolving environmental conditions.

10. Continuously Improve:

  • Foster a culture of continuous improvement by actively seeking feedback, sharing lessons learned, and adapting practices in response to new challenges and opportunities.

  • Encourage ongoing dialogue and collaboration among stakeholders to drive innovation and promote best practices in sustainable land use management.

Following the above procedure, the outcomes can be represented in a template such as shown in Figure 26.

Outcome Record for Reviewing Production and Management Plans for Sustainable Land Use Principles:

1. **Review Objectives:**
   - Purpose and Scope Defined: [Insert details]
   - Sustainability Goals Identified: [Insert details]
2. **Relevant Documentation:**
   - Production and Management Plans Collected: [Insert details]
   - Supplementary Materials Obtained: [Insert details]
3. **Preliminary Assessment:**
   - Overall Approach to Land Management Reviewed: [Insert details]
   - Key Stakeholders Identified: [Insert details]
   - Resource Availability Assessed: [Insert details]
4. **Alignment with Sustainable Principles:**
   - Soil Conservation Practices Evaluated: [Insert details]
   - Water Management Techniques Analysed: [Insert details]
   - Biodiversity Preservation Efforts Assessed: [Insert details]
   - Pest Management Strategies Reviewed: [Insert details]
5. **Strengths and Weaknesses Identification:**
   - Areas of Alignment Documented: [Insert details]
   - Gaps and Areas for Improvement Identified: [Insert details]
   - Feasibility and Effectiveness Considered: [Insert details]
6. **Recommendations Development:**
   - Actionable Recommendations Generated: [Insert details]
   - Prioritization Criteria Applied: [Insert details]
   - Stakeholder Input Incorporated: [Insert details]
7. **Review Findings Documentation:**
   - Comprehensive Report Compiled: [Insert details]
   - Rationale for Recommendations Provided: [Insert details]
   - Report Structure Ensured: [Insert details]
8. **Results Communication and Changes Implementation:**
   - Findings Presented to Stakeholders: [Insert details]
   - Feedback Solicited and Addressed: [Insert details]
   - Action Plan Developed: [Insert details]
9. **Progress Monitoring and Evaluation:**
   - Monitoring Protocols Established: [Insert details]
   - Regular Evaluations Conducted: [Insert details]
   - Plans Adjusted as Needed: [Insert details]
10. **Continuous Improvement:**
    - Feedback Solicited and Utilized: [Insert details]
    - Ongoing Collaboration Encouraged: [Insert details]

[Additional comments or notes:] [Insert details]

Figure 26: Sample template for recording the review of production and management plans.

As an example of a review of production and management plans, and the template shown as Figure 26, the following shows the type of information collected and recorded:

Outcome Record for Reviewing Production and Management Plans for Sustainable Land Use Principles:

- *Review Objectives:*

  ◦ *Purpose and Scope Defined: The review aimed to assess the current production and management plans of the XYZ Farm to ensure alignment with sustainable land use principles and identify areas for improvement.*

  ◦ *Sustainability Goals Identified: The specific sustainability goals included soil conservation, water efficiency, biodiversity preservation, and ecosystem resilience.*

- *Relevant Documentation:*

  ◦ *Production and Management Plans Collected: The production and management plans for the XYZ Farm were obtained, including crop selection documents, irrigation schedules, and pest management strategies.*

  ◦ *Supplementary Materials Obtained: Environmental impact assessments, regulatory compliance reports, and previous sustainability initiatives related to the farm were gathered for review.*

- *Preliminary Assessment:*

  ◦ *Overall Approach to Land Management Reviewed: The review team examined the existing land management practices and strategies outlined in the production plans to gain insight into the farm's approach.*

  ◦ *Key Stakeholders Identified: Stakeholders involved in the development and implementation of the plans, including farm managers, agronomists, and environmental specialists, were identified.*

  ◦ *Resource Availability Assessed: The availability of resources, including financial, human, and technical capacities, was evaluated to determine the feasibility of implementing sustainable practices.*

- *Alignment with Sustainable Principles:*

  ◦ *Soil Conservation Practices Evaluated: Soil conservation practices such as cover cropping, reduced tillage, and erosion control measures were assessed for their effectiveness in minimizing soil erosion and maintaining soil health.*

  ◦ *Water Management Techniques Analysed: Irrigation systems, water recycling methods, and water conservation practices were analysed to determine their impact on water use efficiency and mitigating water-related risks.*

  ◦ *Biodiversity Preservation Efforts Assessed: Efforts to preserve biodiversity through habitat restoration, buffer zone establishment, and wildlife-friendly farming practices were evaluated to support ecosystem diversity.*

  ◦ *Pest Management Strategies Reviewed: Pest management strategies, including integrated pest management (IPM) approaches, were reviewed to minimize reliance on chemical pesticides and promote biological control methods.*

- *Strengths and Weaknesses Identification:*

  ◦ *Areas of Alignment Documented: Several commendable practices aligned with sustainable principles were identified, including the implementation of cover cropping and the adoption of IPM techniques.*

  ◦ *Gaps and Areas for Improvement Identified: Weaknesses such as inconsistent soil conservation measures and limited biodiversity preservation efforts were identified as areas requiring improvement.*

  ◦ *Feasibility and Effectiveness Considered: The feasibility and effectiveness of proposed interventions, such as enhancing soil conservation practices and implementing water-saving technologies, were assessed to ensure practicality and impact.*

- *Recommendations Development:*

  ◦ *Actionable Recommendations Generated: Actionable recommendations were generated to address identified weaknesses and optimize sustainability outcomes, including increasing cover crop diversity and implementing drip irrigation systems.*

    ◦ *Prioritization Criteria Applied: Recommendations were prioritized based on their potential impact, feasibility of implementation, and alignment with organizational goals, ensuring focus on high-impact initiatives.*

    ◦ *Stakeholder Input Incorporated: Feedback from stakeholders, including farm managers and environmental specialists, was incorporated into the development of recommendations to ensure buy-in and support.*

- *Review Findings Documentation:*

    ◦ *Comprehensive Report Compiled: A comprehensive report summarizing the review findings, strengths, weaknesses, and recommendations was compiled to provide a clear overview of the assessment outcomes.*

    ◦ *Rationale for Recommendations Provided: The rationale behind each recommendation was articulated, supported by evidence and references where applicable, to ensure transparency and credibility.*

    ◦ *Report Structure Ensured: The report was structured in a logical and organized manner, with clear sections addressing each aspect of the review process, making it accessible to relevant stakeholders.*

- *Results Communication and Changes Implementation:*

    ◦ *Findings Presented to Stakeholders: The review findings and recommendations were presented to key stakeholders, including farm managers, decision-makers, and relevant staff members, to facilitate discussions and decision-making.*

    ◦ *Feedback Solicited and Addressed: Feedback from stakeholders was solicited and addressed through facilitated discussions to ensure alignment with stakeholder needs and concerns.*

    ◦ *Action Plan Developed: An action plan for implementing recommended improvements and monitoring progress over time was collaboratively developed with stakeholders to ensure accountability and transparency.*

- *Progress Monitoring and Evaluation:*

    ◦ *Monitoring Protocols Established: Monitoring protocols and performance indicators were established to track the implementation of recommended changes, measure progress towards sustainability goals, and assess the effectiveness of implemented strategies.*

    ◦ *Regular Evaluations Conducted: Regular evaluations were conducted to assess the effectiveness of implemented strategies and measure progress towards sustainability goals, with adjustments made as needed based on monitoring results and changing environmental conditions.*

    ◦ *Plans Adjusted as Needed: Plans and strategies were adjusted as needed based on monitoring results and evolving environmental conditions to ensure continued progress towards sustainability objectives.*

- *Continuous Improvement:*

    ◦ *Feedback Solicited and Utilized: Feedback from stakeholders and ongoing evaluations were actively solicited and utilized to drive continuous improvement, with lessons learned shared and practices adapted in response to new challenges and opportunities.*

    ◦ *Ongoing Collaboration Encouraged: Ongoing collaboration among stakeholders was encouraged to drive innovation and promote best practices in sustainable land use management, fostering a culture of continuous improvement and learning.*

[Additional comments or notes:] Overall, the review process provided valuable insights into the current state of land management practices at the XYZ Farm and identified actionable recommendations to enhance sustainability outcomes. Moving forward, ongoing collaboration and monitoring will be essential to ensure the successful implementation of recommended changes and continued progress towards sustainability goals.

Assessing water resources, quality, and their impact on erosion requires a multifaceted approach that integrates various skills and knowledge domains. As an individual tasked with developing and implementing sustainable land use strategies, you must possess a deep understanding of hydrology, soil science, and erosion processes, coupled with strong analytical and problem-solving skills. Here's how you would approach this task:

1. Hydrological Understanding: Begin by conducting a comprehensive assessment of water resources available within the agricultural or horticultural enterprise. This involves analysing factors such as precipitation patterns, surface water availability from streams or rivers, groundwater recharge rates, and water storage capacity in reservoirs or tanks. Understanding the hydrological cycle and water dynamics specific to the region is crucial for making informed decisions about water management practices.

2. Water Quality Analysis: Evaluate the quality of available water sources to determine their suitability for agricultural or horticultural use. This assessment involves testing water samples for various parameters such as pH, nutrient levels (e.g., nitrogen, phosphorus), salinity, heavy metal contamination, and microbial pathogens. Assessing water quality ensures that irrigation practices do not inadvertently contribute to soil degradation or environmental pollution.

3. Erosion Risk Assessment: Assess the risk of soil erosion caused by water runoff from agricultural or horticultural activities. Evaluate factors such as slope steepness, soil texture and structure, vegetation cover, and land management practices that influence erosion susceptibility. Utilize tools such as the Universal Soil Loss Equation (USLE) or the Revised Universal Soil Loss Equation (RUSLE) to quantify erosion rates and identify erosion-prone areas within the landscape.

4. Impact Analysis: Analyse the impact of water resources and quality on erosion processes. Consider how factors such as excessive irrigation, poor drainage systems, land clearance, and intensive farming practices contribute to soil erosion and sedimentation in water bodies. Assessing these impacts helps prioritize areas for intervention and develop targeted strategies to mitigate erosion risks while optimizing water use efficiency.

5. Risk Mitigation Strategies: Based on the assessment findings, design and implement tailored strategies to mitigate erosion and protect water resources. This may involve implementing conservation practices such as contour ploughing, terracing, cover cropping, buffer strips, and grass waterways to reduce soil erosion and sediment transport. Additionally, adopting water-saving irrigation techniques such as drip irrigation or micro-sprinklers can minimize water runoff and soil disturbance.

6. Monitoring and Adaptive Management: Establish monitoring protocols to track changes in water resources, quality, and erosion rates over time. Regularly monitor key indicators such as soil moisture levels, water quality parameters, sediment deposition rates, and erosion severity. Use this data to evaluate the effectiveness of implemented strategies and make necessary adjustments to optimize outcomes. Adaptive management ensures that land use practices remain responsive to evolving environmental conditions and sustainability goals.

Assessing water resources, quality, and their impact on erosion requires a holistic understanding of hydrological processes, soil dynamics, and land management practices. By integrating scientific analysis with practical problem-solving skills, individuals can develop and implement sustainable land use strategies that enhance ecological sustainability while promoting efficient water management and erosion control.

Conducting a comprehensive assessment of water resources within an agricultural or horticultural enterprise is a fundamental step in understanding the availability and dynamics of water in the landscape. This assessment begins by analysing various factors that influence water availability and distribution. One crucial aspect is examining precipitation patterns, which provide insight into the amount and distribution of rainfall throughout the year. By studying historical precipitation data and trends, stakeholders can anticipate seasonal variations and plan irrigation schedules accordingly.

Surface water availability from streams or rivers is another critical consideration in water resource assessment. Analysing the flow rates, seasonal fluctuations, and reliability of surface water sources helps determine their suitability for irrigation or other agricultural uses. Additionally, assessing the quality of surface water sources is essential to ensure that irrigation practices do not inadvertently introduce contaminants or pollutants to the agricultural system.

Groundwater recharge rates play a significant role in sustaining water availability, particularly in areas where surface water sources may be limited or unreliable. Understanding the dynamics of groundwater recharge involves evaluating factors such as aquifer characteristics, infiltration rates, and land use practices that affect groundwater recharge processes. This information helps stakeholders assess the long-term sustainability of groundwater resources and develop strategies to minimize over-extraction or depletion.

Moreover, assessing the water storage capacity in reservoirs or tanks provides an indication of the enterprise's capacity to store and manage water resources effectively. Evaluating reservoir capacity, retention times, and infrastructure reliability ensures that adequate water supplies are available during periods of high demand or water scarcity. Implementing water storage solutions such as rainwater harvesting systems or on-farm reservoirs can help buffer against seasonal variability and ensure water security for agricultural operations.

Understanding the hydrological cycle and water dynamics specific to the region is crucial for making informed decisions about water management practices. By integrating data on precipitation, surface water availability, groundwater recharge rates, and water storage capacity, stakeholders can develop sustainable water management strategies tailored to the unique characteristics of the landscape. This comprehensive assessment lays the foundation for optimizing water use efficiency, minimizing environmental impacts, and ensuring the long-term viability of agricultural and horticultural enterprises. Figure 27 provides a checklist for conducting a comprehensive water resource assessment in agricultural or horticultural enterprises.

Checklist for Conducting a Comprehensive Water Resource Assessment in Agricultural or Horticultural Enterprises:

1. Precipitation Analysis:
   - Obtain historical precipitation data for the region.
   - Analyse precipitation patterns and trends over time.
   - Identify seasonal variations in rainfall amounts and distribution.
   - Assess the reliability of precipitation forecasts and climate projections.
2. Surface Water Availability:
   - Identify nearby streams, rivers, or other surface water sources.
   - Evaluate flow rates and seasonal fluctuations in surface water levels.
   - Determine the reliability and accessibility of surface water sources.
   - Assess water quality parameters, including pH, turbidity, and nutrient levels.
3. Groundwater Recharge Assessment:
   - Identify aquifers and groundwater sources within the enterprise's vicinity.
   - Evaluate aquifer characteristics, including permeability and storage capacity.
   - Assess groundwater recharge rates based on land use practices and infiltration rates.
   - Investigate potential impacts of groundwater extraction on aquifer sustainability.
4. Water Storage Capacity:
   - Evaluate existing water storage infrastructure, such as reservoirs or tanks.
   - Determine the capacity and retention times of water storage facilities.
   - Assess the reliability and condition of water storage infrastructure.
   - Identify opportunities for implementing additional water storage solutions, such as rainwater harvesting systems or on-farm reservoirs.
5. Hydrological Cycle Analysis:
   - Understand the hydrological cycle specific to the region, including evaporation, transpiration, and runoff processes.
   - Analyse surface runoff patterns and infiltration rates.
   - Consider the impacts of land use practices, such as deforestation or urbanization, on the hydrological cycle.
   - Integrate data on precipitation, surface water availability, groundwater recharge rates, and water storage capacity to develop a holistic understanding of water dynamics.
6. Stakeholder Engagement:
   - Involve relevant stakeholders, including farmers, landowners, water authorities, and local communities, in the assessment process.
   - Solicit input and feedback from stakeholders regarding water resource concerns, priorities, and objectives.
   - Foster collaboration and knowledge sharing among stakeholders to develop consensus-based water management strategies.
7. Data Integration and Analysis:
   - Compile and integrate data gathered from various sources, including historical records, monitoring stations, and on-site measurements.
   - Analyse the collected data to identify trends, patterns, and potential water resource challenges.
   - Use Geographic Information System (GIS) tools and hydrological models to visualize and interpret water resource data effectively.
8. Develop Water Management Strategies:
   - Based on the assessment findings, develop tailored water management strategies that address identified challenges and opportunities.
   - Prioritize strategies based on their potential impact, feasibility, and alignment with sustainability goals.
   - Consider implementing a combination of conservation practices, irrigation efficiency measures, and water reuse/recycling initiatives to optimize water use efficiency and minimize environmental impacts.
9. Monitoring and Adaptive Management:
   - Establish monitoring protocols and performance indicators to track the effectiveness of implemented water management strategies.
   - Regularly monitor key parameters such as water availability, quality, and usage patterns.
   - Implement adaptive management practices to adjust strategies as needed based on monitoring results and changing environmental conditions.
10. Documentation and Reporting:
    - Document the entire water resource assessment process, including data collection methods, analysis techniques, and assessment findings.
    - Prepare a comprehensive report summarizing the assessment results, key findings, and recommended water management strategies.
    - Ensure that the report is accessible to relevant stakeholders and serves as a valuable reference for future water resource planning and decision-making.

Figure 27: Checklist for Conducting a Comprehensive Water Resource Assessment in Agricultural or Horticultural Enterprises.

Assessing Water Resources, Quality and Impact on Erosion

In the process of evaluating water resources for agricultural or horticultural use, conducting a thorough analysis of water quality is paramount. This step delves into the examination of the composition and characteristics of available water sources to ascertain their suitability for various agricultural practices.

Water quality analysis encompasses a multifaceted approach, involving the collection and testing of water samples to assess several key parameters. One crucial aspect of this assessment is the examination of pH levels, which indicates the acidity or alkalinity of the water. Understanding pH levels is essential as it influences nutrient availability in the soil and can impact plant growth and development.

Furthermore, evaluating nutrient levels within the water samples, particularly nitrogen and phosphorus, is imperative. These nutrients play vital roles in plant growth and are essential for agricultural productivity. However, excessive levels of nitrogen and phosphorus in water bodies can lead to eutrophication, causing harmful algal blooms and degrading water quality. Thus, monitoring and managing nutrient levels in irrigation water are essential to prevent environmental harm and maintain ecosystem health.

Salinity is another critical parameter assessed during water quality analysis. High salinity levels in irrigation water can negatively affect soil structure and plant health by increasing soil salinity over time, leading to reduced crop yields and diminished agricultural productivity. By quantifying salinity levels in water sources, stakeholders can implement appropriate management practices to mitigate the adverse effects of salinity on agricultural lands.

Heavy metal contamination is also a concern when evaluating water quality for agricultural use. Heavy metals such as lead, arsenic, cadmium, and mercury can accumulate in soils and crops, posing risks to human health and environmental integrity. Testing water samples for heavy metal concentrations helps identify potential contamination sources and enables the implementation of remediation measures to safeguard agricultural produce and ecosystem health.

Moreover, assessing microbial pathogens in water sources is essential for preventing the spread of waterborne diseases and ensuring food safety in agricultural production. Contaminated water used for irrigation or crop spraying can introduce pathogens such as bacteria, viruses, and parasites to crops, posing risks to human health upon consumption. Therefore, monitoring and treating water sources for microbial pathogens are critical steps in maintaining food safety and minimizing health hazards associated with agricultural practices.

Water quality analysis plays a pivotal role in determining the suitability of available water sources for agricultural and horticultural use. By evaluating parameters such as pH, nutrient levels, salinity, heavy metal contamination, and microbial pathogens, stakeholders can make informed decisions regarding water management practices to safeguard soil, water, and environmental quality while promoting sustainable agriculture.

Ensuring a reliable and high-quality water supply is essential for the success of farming enterprises, as it directly impacts various aspects of agricultural operations. Adequate water availability is crucial for maintaining the health of livestock, maximizing plant production, and fulfilling household and garden needs (Agriculture Victoria, 2024a). However, the quality of farm water can vary significantly depending on its source, storage, and distribution methods.

One key consideration in assessing water resources is the examination of precipitation patterns and sources. Rainwater collected from clean surfaces, such as roofs, typically provides high-quality water suitable for agricultural use. Conversely, underground water sources in many areas of the state often contain high levels of salt, posing challenges for agricultural applications (Agriculture Victoria, 2024a).

Water quality not only affects its suitability for specific uses but also influences management strategies for collection, storage, and distribution. Storing underground water in dams, for instance, can lead to rapid deterioration of water quality due to evaporation and the concentration of dissolved salts. Additionally, unfenced catchment dams are susceptible to contamination from stock activities, while poorly vegetated catchment areas can introduce organic matter, manure, and fertilizers into water supplies (Agriculture Victoria, 2024a).

To maintain good water quality, various management strategies are recommended. These include storing saline water in tanks rather than dams, fencing off dams to prevent stock access, and protecting dam catchments with vegetation cover. Regular cleaning of troughs, establishment of windbreaks to reduce evaporation, and construction of sediment traps are also effective measures to preserve water quality (Agriculture Victoria, 2024a).

Several parameters are commonly assessed to determine water quality suitability for agricultural use, including salinity, turbidity, hardness, acidity (pH), and presence of algae and aquatic plants (Agriculture Victoria, 2024a). Salinity, measured in Electrical Conductivity

(EC) units, affects water's suitability for irrigation, livestock, and domestic use. Turbidity, caused by suspended solids, can lead to various physical impacts and elevated phosphorus levels, contributing to algal blooms.

Water salinity levels significantly impact plant growth and productivity, with different ranges of Electrical Conductivity (EC) indicating varying effects and suitability for irrigation (Agriculture Victoria, 2024a).

For water with an EC range of 0 to 300 µS/cm, categorized as low salinity, it is generally suitable for most crops and soils using various water application methods. There is minimal risk of developing salinity-related problems, making it an ideal choice for agricultural irrigation (Agriculture Victoria, 2024a). In the range of 300 to 800 µS/cm, classified as medium salinity water, it can still be used for irrigation if moderate leaching occurs. Plants with a medium salt tolerance can be cultivated without requiring special practices for salinity control, although monitoring is advisable to prevent potential issues.

High salinity water, falling within the range of 800 to 2500 µS/cm, poses challenges for irrigation, particularly on soils with restricted drainage. Even with adequate drainage, special management practices for salinity control may be necessary. The salt tolerance of the specific plants to be irrigated must be carefully considered to mitigate potential adverse effects.

Water with an EC range of 2500 to 5800 µS/cm is classified as very high salinity and is generally unsuitable for irrigation under ordinary conditions. However, if soils are permeable with adequate drainage, it may be used with excess water application to facilitate considerable leaching. Selecting salt-tolerant crops becomes essential in such cases to ensure minimal negative impacts on plant growth (Agriculture Victoria, 2024a).

In instances where water salinity exceeds 5800 µS/cm, categorized as extremely high salinity, it may only be used occasionally and as an emergency measure for salt-tolerant crops. Permeable, well-drained soils under effective management practices, including high rates of leaching, are required to minimize the adverse effects of extremely high salinity water on plant health and productivity.

Salinity tolerance levels vary among different types of livestock, with specific Electrical Conductivity (EC) ranges indicating when production decline begins and the maximum tolerable EC levels. For poultry, production decline typically begins at an EC level of 3100 µS/cm, with a maximum tolerable EC of 6250 µS/cm. Pigs exhibit similar tolerance levels, with production decline also commencing at 3100 µS/cm and a maximum tolerable EC of 6250 µS/cm (Agriculture Victoria, 2024a).

Horses have a higher tolerance level compared to poultry and pigs, with production decline starting at an EC of 6250 µS/cm and a maximum tolerable EC of 10,900 µS/cm. Similarly, dairy cattle have a comparable tolerance level to horses, with production decline beginning at 4700 µS/cm and a maximum tolerable EC of 9300 µS/cm. Beef cattle exhibit a higher tolerance level to salinity compared to dairy cattle, with production decline commencing at an EC of 6250 µS/cm and a maximum tolerable EC of 15,600 µS/cm (Agriculture Victoria, 2024a).

Lactating ewes and weaners have a lower tolerance to salinity compared to other livestock types, with production decline starting at an EC of 6000 µS/cm and a maximum tolerable EC of 10,000 µS/cm. Sheep being fed dry feed have the highest tolerance level among the listed livestock types, with production decline beginning at an EC of 9300 µS/cm and a maximum tolerable EC of 21,800 µS/cm (Agriculture Victoria, 2024a).

Hardness, indicating calcium and magnesium salt levels, affects water's lathering capacity and can lead to scaling in plumbing systems. pH levels determine water's acidity or alkalinity, with suitable ranges varying depending on intended use. Algae and aquatic plants, while natural components of water ecosystems, can become problematic when their excessive growth interferes with water quality and accessibility.

Hardness levels in water, measured in milligrams per litre (mg/L) of calcium carbonate (CaCO3), determine its suitability for various uses and its potential impact on equipment and plumbing. Water with a hardness level of less than 60 mg/L CaCO3 is considered soft, but it may still pose a risk of corrosion to certain materials due to its low mineral content. Hardness levels ranging from 60 to 200 mg/L CaCO3 indicate water that is considered hard but generally acceptable for most uses. However, it may still cause some scaling issues in equipment and plumbing over time (Agriculture Victoria, 2024a).

When water hardness exceeds 200 mg/L CaCO3 and falls within the range of 200 to 500 mg/L CaCO3, it is categorized as very hard. Water in this range is prone to causing increasing scale problems, leading to mineral buildup in pipes, appliances, and plumbing fixtures.

Water with a hardness level greater than 500 mg/L CaCO3 is classified as extremely hard. In such cases, severe scaling is expected, posing significant challenges for equipment and plumbing systems. The accumulation of mineral deposits can lead to reduced efficiency and increased maintenance costs (Agriculture Victoria, 2024a).

Overall, conducting a comprehensive assessment of water quality parameters and implementing appropriate management strategies are essential steps in ensuring a reliable and safe water supply for agricultural enterprises. By addressing water quality challenges effectively, farmers can optimize productivity, minimize environmental impacts, and promote sustainable agricultural practices (Agriculture Victoria, 2024a).

Assessing the risk of soil erosion is a critical component of sustainable land management practices in agricultural and horticultural enterprises. The assessment involves evaluating various factors that contribute to erosion susceptibility, particularly those related to water runoff. One key consideration is the steepness of the slope, as steeper slopes are more prone to erosion due to increased runoff velocity and reduced water infiltration. Assessing slope steepness provides insights into areas where erosion control measures may be most urgently needed to mitigate potential soil loss.

Soil texture and structure also play significant roles in erosion risk assessment. Soils with finer textures, such as clay, are more prone to erosion compared to soils with coarser textures like sand. Additionally, soil structure influences water infiltration rates and soil stability, with well-aggregated soils better able to resist erosion. Evaluating soil texture and structure helps identify areas vulnerable to erosion and informs decisions regarding soil management practices to improve soil stability and reduce erosion risk.

Vegetation cover is another crucial factor in erosion risk assessment. Plant roots help bind soil particles together, reducing the likelihood of erosion by water runoff. Assessing vegetation cover involves examining the type, density, and distribution of vegetation across the landscape. Areas with sparse or degraded vegetation cover are more susceptible to erosion and may require interventions such as reforestation, cover cropping, or grass buffer strips to enhance soil protection.

Furthermore, land management practices significantly influence erosion susceptibility. Practices such as tillage, bare soil exposure, and improper irrigation methods can exacerbate erosion risk by disrupting soil structure and increasing surface runoff. Evaluating current land management practices provides insights into potential erosion hotspots and opportunities for implementing erosion control measures.

To quantify erosion rates and identify erosion-prone areas more precisely, tools such as the Universal Soil Loss Equation (USLE) or the Revised Universal Soil Loss Equation (RUSLE) are commonly used. These equations consider various factors such as rainfall erosivity, soil erodibility, slope length, and land cover to estimate soil loss rates. By applying these tools, stakeholders can prioritize erosion control efforts and develop targeted strategies to minimize soil erosion and preserve soil health in agricultural and horticultural landscapes.

Analysing the impact of water resources and quality on erosion processes is essential for understanding the complex interplay between water management practices and soil erosion dynamics in agricultural and horticultural systems. Excessive irrigation, poor drainage systems, land clearance, and intensive farming practices are among the key factors that can significantly influence erosion processes and sedimentation in water bodies.

Excessive irrigation practices, characterized by over-application or inefficient water distribution, can lead to increased soil moisture content and surface runoff, exacerbating erosion rates. The continuous flow of water over the soil surface can dislodge soil particles, leading to sediment transport and deposition in downstream water bodies. Additionally, irrigation practices that result in waterlogging or ponding can degrade soil structure and reduce infiltration capacity, further exacerbating erosion risk.

Poorly designed or maintained drainage systems contribute to erosion by allowing water to accumulate and concentrate in low-lying areas, leading to increased runoff velocity and soil erosion. Inadequate drainage can result in waterlogging, soil compaction, and increased susceptibility to erosion during rainfall events. Addressing drainage issues through the installation of properly designed drainage infrastructure helps mitigate erosion risks by improving water flow and reducing surface runoff.

Land clearance activities, such as deforestation or clearing of natural vegetation for agricultural expansion, can have profound impacts on erosion processes. Removal of vegetative cover exposes soil to erosive forces, increases surface runoff, and reduces the soil's ability to absorb and retain water. Consequently, erosion rates escalate, leading to sedimentation in nearby water bodies and degradation of aquatic habitats. Implementing reforestation programs, establishing vegetative buffer strips, and adopting agroforestry practices can help restore vegetative cover and mitigate erosion impacts.

Intensive farming practices, including monoculture cropping, excessive tillage, and inadequate soil conservation measures, can accelerate erosion rates and sedimentation in water bodies. Continuous soil disturbance and removal of crop residues diminish soil structure and organic matter content, making the soil more susceptible to erosion. Furthermore, the use of agrochemicals such as fertilizers and pesticides can exacerbate erosion by altering soil properties and increasing runoff rates. Implementing conservation agriculture practices, such as conservation tillage, cover cropping, and contour farming, helps reduce erosion risks while promoting sustainable land management.

Assessing the impacts of water resources and quality on erosion processes enables stakeholders to prioritize intervention areas and develop targeted strategies to mitigate erosion risks while optimizing water use efficiency. By addressing the root causes of erosion and implementing appropriate management practices, agricultural and horticultural enterprises can minimize soil loss, protect water resources, and enhance overall landscape resilience.

Designing and implementing tailored strategies to mitigate erosion and protect water resources is crucial for maintaining the sustainability and resilience of agricultural and horticultural enterprises. These strategies are informed by comprehensive assessments that identify erosion risks and prioritize intervention areas. Various conservation practices can be implemented to mitigate soil erosion and sediment transport effectively, thereby safeguarding water quality and ecosystem health.

One of the primary strategies for erosion mitigation involves implementing conservation practices that promote soil conservation and water infiltration. Contour ploughing, for example, involves ploughing along the contours of the land rather than up and down slopes, which helps reduce water runoff and soil erosion by slowing down the flow of water across the landscape. Terracing is another effective technique that creates level platforms on sloping terrain, minimizing soil erosion by reducing the length and gradient of slopes. These practices help retain soil moisture, improve soil structure, and enhance vegetation cover, thereby reducing erosion risks.

Cover cropping is a widely adopted strategy for erosion control and soil protection. By planting cover crops during fallow periods or in between cash crops, soil erosion is minimized through the maintenance of vegetative cover and the enhancement of soil structure. Cover crops help prevent soil erosion by reducing surface runoff, improving water infiltration, and reducing soil compaction. Additionally, the root systems of cover crops help bind soil particles together, further stabilizing the soil and reducing erosion potential.

Buffer strips and grass waterways are essential components of erosion control strategies, especially in areas prone to runoff and sediment transport. Buffer strips consist of strips of vegetation, such as grass or native vegetation, planted along field edges or waterways. These strips act as barriers, trapping sediment and pollutants carried by runoff before they reach water bodies, thereby protecting water quality and aquatic habitats. Grass waterways are designed channels planted with grass or other vegetation, which help convey and filter runoff water, reducing erosion and sedimentation in downstream areas.

In addition to soil conservation practices, adopting water-saving irrigation techniques is essential for minimizing water runoff and soil disturbance while optimizing water use efficiency. Drip irrigation and micro-sprinklers are efficient irrigation methods that deliver water directly to the root zone of plants, minimizing surface runoff and evaporation losses. These methods help maintain soil moisture levels, promote healthy plant growth, and reduce erosion risks associated with conventional irrigation practices.

By integrating a combination of conservation practices and water-saving irrigation techniques, agricultural and horticultural enterprises can effectively mitigate erosion risks, protect water resources, and promote sustainable land management practices. These strategies not only enhance soil health and fertility but also contribute to the preservation of ecosystem integrity and long-term agricultural productivity.

Establishing robust monitoring protocols is essential for tracking changes in water resources, quality, and erosion rates over time within agricultural and horticultural enterprises. By regularly monitoring key indicators such as soil moisture levels, water quality parameters, sediment deposition rates, and erosion severity, stakeholders can gain valuable insights into the effectiveness of implemented strategies and make informed decisions to optimize outcomes.

Monitoring soil moisture levels provides critical information about the availability of water for plant growth and the risk of soil erosion. By using soil moisture sensors or probes, stakeholders can assess variations in soil moisture content across different field areas and depths, helping to inform irrigation scheduling and soil conservation practices. Understanding soil moisture dynamics enables stakeholders to adjust irrigation practices to minimize water runoff and optimize water use efficiency, thereby reducing erosion risks.

Assessing water quality parameters is another essential aspect of monitoring and adaptive management. Regular testing of water samples for parameters such as pH, nutrient levels, turbidity, and salinity helps ensure that water resources remain suitable for agricultural and horticultural use. Monitoring water quality allows stakeholders to identify any potential issues, such as contamination or nutrient imbalances, and take corrective actions to mitigate risks and protect ecosystem health.

Monitoring sediment deposition rates and erosion severity provides insights into the effectiveness of erosion control measures and the overall health of the landscape. By using sediment traps, erosion pins, or other monitoring tools, stakeholders can quantify sediment movement and erosion rates over time, identifying areas of erosion vulnerability and implementing targeted interventions as needed. Monitoring erosion severity helps track changes in land condition and prioritize conservation efforts to minimize soil loss and maintain soil fertility.

Adaptive management is a key principle that underpins effective monitoring and response strategies. By regularly reviewing monitoring data and evaluating the performance of implemented practices, stakeholders can identify emerging trends, challenges, and opportunities, and adjust land use practices accordingly. Adaptive management ensures that land management practices remain responsive to evolving environmental conditions and sustainability goals, enabling stakeholders to continuously improve their approaches and achieve long-term resilience and productivity.

Establishing monitoring protocols and adopting adaptive management practices are essential for effectively managing water resources, quality, and erosion within agricultural and horticultural enterprises. By monitoring key indicators and responding proactively to changing conditions, stakeholders can optimize outcomes, minimize risks, and promote sustainable land management practices that support both agricultural productivity and environmental stewardship.

Assessing Soil Health and Fertility Against Production Plan Requirements

Assessing soil health and fertility against production plan requirements is a another important aspect of developing and implementing sustainable land use strategies, especially for individuals managing agricultural and production horticulture enterprises. This process involves a comprehensive evaluation of the soil's physical, chemical, and biological properties to ensure they meet the specific needs outlined in the production plan.

Firstly, individuals must analyse the soil's physical characteristics, including texture, structure, and compaction. Soil texture refers to the relative proportions of sand, silt, and clay particles, which influence water retention, drainage, and aeration. Assessing soil structure involves examining the arrangement of soil particles into aggregates, which affects root penetration, nutrient availability, and microbial activity. Compacted soils may impede root growth and water infiltration, leading to reduced crop productivity. By conducting soil texture analysis and assessing soil structure and compaction, stakeholders can identify any physical constraints that may limit crop growth and productivity.

Secondly, evaluating the soil's chemical properties is essential for assessing its fertility and nutrient status. Soil pH, nutrient levels, and organic matter content are key indicators of soil fertility. Soil pH influences nutrient availability and microbial activity, with most crops preferring slightly acidic to neutral pH levels. Nutrient levels, including nitrogen, phosphorus, potassium, and micronutrients, must be balanced to support healthy plant growth and development. Organic matter content contributes to soil structure, water retention, and nutrient cycling, indicating the soil's overall health and productivity. By conducting soil tests and analysing nutrient levels, stakeholders can determine whether the soil meets the nutrient requirements outlined in the production plan and identify any deficiencies or imbalances that may require correction.

Furthermore, assessing the soil's biological properties, such as microbial diversity and activity, is critical for understanding its capacity to support plant growth and nutrient cycling. Soil microorganisms play essential roles in decomposing organic matter, releasing nutrients, and suppressing plant diseases. Analysing microbial biomass and activity provides insights into the soil's biological processes and its ability to sustainably support crop production. By assessing soil microbial communities, stakeholders can determine whether the soil ecosystem is functioning optimally and identify any imbalances or disturbances that may affect crop health and productivity.

Assessing soil health and fertility against production plan requirements involves a multidimensional analysis of the soil's physical, chemical, and biological properties. By evaluating soil texture, structure, compaction, pH, nutrient levels, organic matter content, and microbial activity, stakeholders can identify any constraints or deficiencies that may affect crop growth and productivity. This comprehensive assessment enables individuals to develop targeted strategies to improve soil health and fertility, ensuring the long-term sustainability and productivity of agricultural and horticultural enterprises.

Soil, the vital medium covering most of the Earth's land surface, is a complex blend of inorganic particles and organic matter (Agriculture Victoria, 2024b). It serves as the foundation for agriculture, providing structural support, water, and nutrients essential for plant growth. However, soils exhibit considerable variation in their chemical and physical properties due to processes such as leaching, weathering, and microbial activity, resulting in a diverse range of soil types, each with specific attributes and limitations for agricultural production.

The physical characteristics of soil encompass observable and tangible aspects such as texture, colour, depth, structure, porosity, and stone content. A well-structured soil is crucial for both soil and plant health, facilitating the movement of water and air throughout the soil profile, storing water for plant growth, and supporting machinery and livestock. While some soils possess naturally favourable structures, management practices can influence physical characteristics, making it imperative to monitor soil condition and ensure that activities do not degrade soil health. Excessive traffic, for instance, can compact soil, diminishing pore spaces and impeding essential air and water movement (Agriculture Victoria, 2024b).

Soil texture, structure, and drainage characteristics play pivotal roles in determining soil suitability for agricultural purposes. Texture, influenced by the relative proportions of sand, silt, clay, and organic matter, dictates soil behaviour and nutrient availability. Sandy soils, with their low water retention but good drainage, contrast with heavy clays, which retain water but drain poorly. Soil structure, defined by the arrangement of soil particles and aggregates, impacts water, air, and nutrient movement, thereby influencing plant growth. Furthermore, soil drainage, assessed by factors such as texture, stone content, and soil behaviour, is crucial for plant health, with poorly drained soils limiting oxygen availability to roots and potentially causing plant stunting or death.

Soil colour serves as a valuable indicator of various soil properties, including organic matter content, parent material, degree of weathering, and drainage characteristics. Darker soils typically signify higher organic matter content and better drainage, while reddish hues may indicate the presence of iron and good drainage. Conversely, yellowish tones may suggest moist conditions and restrictive drainage, whereas grey or blue-green hues may indicate poor drainage or waterlogging (Agriculture Victoria, 2024b).

The inorganic component of soils, primarily comprising mineral particles with specific physical and chemical properties, heavily influences soil texture, structure, density, and water retention. Soil texture, determined by the relative proportions of sand, silt, and clay particles, governs soil physical and chemical properties and plays a crucial role in plant growth and nutrient availability. Sand particles, dominated by quartz, exhibit low water retention and chemical activity, while clays, with their expansive surface areas, can retain nutrients and water, albeit with potential challenges such as soil slaking and compaction (Agriculture Victoria, 2024b).

Organic matter, constituting a smaller fraction of soil but playing a vital role in soil fertility and structure, stems from the decomposition of plant and animal residues. It contributes to stable soil aggregates, water retention, nutrient availability, and microbial activity, thereby fostering healthy soil ecosystems and supporting plant growth. The decomposition of organic matter by soil microorganisms releases nutrients essential for plant growth, with decomposition rates influenced by environmental factors such as temperature, moisture, aeration, and pH (Agriculture Victoria, 2024b).

Soil organisms, ranging from microscopic bacteria to larger soil animals like earthworms, significantly impact soil fertility, structure, and nutrient cycling. Beneficial soil organisms aid in organic matter decomposition, nitrogen fixation, nutrient transformation, and soil aggregation, enhancing soil health and plant productivity. However, certain soil organisms may compete with plants for nutrients or feed on plant roots, potentially hindering plant growth and productivity. Thus, understanding soil biology and managing soil ecosystems are essential for maintaining soil fertility and sustainability in agricultural systems.

A checklist for assessing soil health and fertility is shown as Figure 28.

Assessing Soil Health and Fertility Checklist:

1. **Analyse Physical Characteristics:**
   - ○ Conduct soil texture analysis to determine the relative proportions of sand, silt, and clay.
   - ○ Evaluate soil structure to understand the arrangement of soil particles and aggregates.
   - ○ Assess soil compaction to identify any constraints that may limit root growth and water infiltration.
2. **Evaluate Chemical Properties:**
   - ○ Test soil pH to gauge acidity or alkalinity, which influences nutrient availability.
   - ○ Analyse nutrient levels, including nitrogen, phosphorus, potassium, and micronutrients, to ensure they meet plant requirements.
   - ○ Measure organic matter content to assess soil fertility, structure, and microbial activity.
3. **Assess Biological Properties:**
   - ○ Analyse soil microbial diversity and activity to understand its role in nutrient cycling and plant health.
   - ○ Determine the abundance of beneficial microorganisms involved in organic matter decomposition, nitrogen fixation, and nutrient transformation.
4. **Interpret Soil Health and Fertility:**
   - ○ Compare assessment results with production plan requirements to identify any discrepancies or deficiencies.
   - ○ Consider the specific needs of crops or plants to ensure soil conditions align with optimal growth and productivity.
   - ○ Evaluate soil health indicators holistically to develop targeted strategies for improvement.
5. **Develop Targeted Strategies:**
   - ○ Implement soil management practices tailored to address identified constraints or deficiencies.
   - ○ Utilize techniques such as cover cropping, organic amendments, or microbial inoculants to enhance soil fertility and structure.
   - ○ Integrate sustainable practices like crop rotation or reduced tillage to promote soil health and resilience.
6. **Monitor and Adjust:**
   - ○ Establish monitoring protocols to track changes in soil health and fertility over time.
   - ○ Regularly assess soil parameters and adjust management practices as needed to optimize outcomes.
   - ○ Continuously evaluate the effectiveness of implemented strategies and make adjustments based on monitoring data.

Figure 28: Assessing Soil Health and Fertility Checklist.

Assessing Vegetation Types and Coverage, and Identifying Required Changes

Assessing vegetation types and coverage is a critical aspect of developing and implementing sustainable land use strategies, especially for individuals entrusted with the management of agricultural and production horticulture enterprises. This process involves a comprehensive evaluation of existing vegetation to determine its composition, distribution, and health status. By analysing vegetation types and coverage, stakeholders can gain insights into the ecological sustainability of the land under production and identify necessary changes to optimize land management practices.

To assess vegetation types and coverage and identify required changes for sustainable land use, it is essential to consider various factors such as land capability, sustainable scale, land evaluation methods, soil characteristics, carbon stock assessment, land certification, and sustainable land use modelling. Sarcinelli et al. (2022) emphasize the importance of land evaluations in determining sustainable scales for agriculture to avoid land degradation and ensure the provision of food, wood, energy, and ecosystem services over time. This highlights the significance of considering the long-term impact of land use on the environment and ecosystem services. Zhou et al. (2009) propose a specific method for evaluating sustainable land use, indicating the existence of established methods for assessing land use sustainability. Kabanda (2015) discusses the use of remote sensing, GIS, and geostatistics in land capability evaluation for crop production, emphasizing the importance of considering soil characteristics and their alignment to determine suitable land for sustainable usage. This highlights the relevance of advanced technologies in assessing land suitability for sustainable land use.

Firstly, individuals must conduct thorough surveys to identify the various vegetation types present on the land. This may involve mapping out different plant communities, ranging from native vegetation to cultivated crops or managed horticultural plantings. Understanding the diversity of vegetation types allows stakeholders to assess the overall ecological integrity of the landscape and identify any areas of concern, such as the presence of invasive species or monoculture cropping systems.

Once vegetation types have been identified, the next step is to assess vegetation coverage across the landscape. This involves quantifying the extent of vegetation cover, including both natural and cultivated vegetation. High vegetation coverage is often indicative of healthy ecosystems, providing numerous benefits such as soil stabilization, water retention, and habitat for wildlife. Conversely, low vegetation coverage may signal degradation or land use practices that are not environmentally sustainable.

After assessing existing vegetation types and coverage, individuals must analyse the findings to identify any required changes to improve ecological sustainability. This may involve implementing measures to enhance biodiversity, restore degraded ecosystems, or promote more sustainable land management practices. For example, if monoculture cropping is dominant in certain areas, transitioning to diversified cropping systems or incorporating agroforestry practices could improve ecological resilience and productivity.

Furthermore, stakeholders must consider the ecological functions provided by different vegetation types and coverage levels. Native vegetation, for instance, often plays a crucial role in supporting biodiversity, regulating water cycles, and sequestering carbon. Therefore, efforts to conserve and restore native vegetation should be prioritized to enhance overall ecological sustainability.

In addition to assessing existing vegetation, individuals must also anticipate future changes in vegetation patterns due to factors such as climate change, land use intensification, or invasive species encroachment. By proactively identifying potential shifts in vegetation dynamics, stakeholders can develop adaptive strategies to mitigate risks and capitalize on emerging opportunities.

Overall, assessing vegetation types and coverage is integral to the development and implementation of sustainable land use strategies. By understanding the composition and distribution of vegetation across the landscape, stakeholders can identify opportunities to enhance ecological sustainability, improve land productivity, and mitigate environmental risks. Through careful analysis, planning, and implementation, individuals can contribute to the long-term health and resilience of agricultural and horticultural enterprises while preserving the integrity of natural ecosystems.

Gnanavelrajah et al. (2007) stress the importance of examining changes in soil organic carbon (SOC) over time for different land uses to ensure agricultural land-use sustainability. This underscores the significance of monitoring soil health and carbon management in sustainable land use practices. Mengesha et al. (2019) highlight the role of agroforestry trees as indicators of sustainable land use practices, emphasizing the economic and ecological sustainability they provide through the provisioning of timber and non-timber forest products.

Furthermore, Widjonarko and Maryono (2022) discuss the use of spatial regression methods in sustainable land use modelling, indicating the application of advanced statistical techniques in assessing and modelling sustainable land use. Jiang et al. (2018) emphasize the need for integrating biophysical, socioeconomic, quantitative, and qualitative analysis for sustainable marginal land identification and use, highlighting the multidimensional approach required for sustainable land use assessment. Luo et al. (2010) emphasize the importance of land types and their combination patterns in designing sustainable land use patterns, indicating the significance of considering spatial patterns and GIS-based mapping for sustainable land use planning. Niu et al. (2022) stress the importance of sustainable land resource use assessment to ensure land use is on a sustainable development trajectory, highlighting the need for continuous assessment and monitoring of land use practices.

Ahmadi et al. (2017) emphasize the ecological approach in identifying suitable future land uses for sustainable land use planning, indicating the importance of considering ecological factors in land suitability evaluation. Liebig et al. (2017) highlight the need for innovative management systems to ensure sustainable use of agricultural land, emphasizing the importance of developing and applying sustainable land management practices.

To practically implement the assessment of vegetation types and coverage and identify required changes for sustainable land use, farm managers can follow a structured approach that integrates various factors and methodologies as outlined by researchers in the field.

1. Land Capability Assessment: Begin by conducting a comprehensive evaluation of the land's capability, considering factors such as soil characteristics, topography, and climate. This assessment, as emphasized by Sarcinelli et al. (2022), helps determine the sustainable scale for agriculture and ensures the prevention of land degradation while optimizing the provision of food, wood, energy, and ecosystem services over time.

2. Utilization of Advanced Technologies: Incorporate advanced technologies such as remote sensing, GIS, and geostatistics, as suggested by Kabanda (2015), to assess land capability for crop production accurately. These tools enable farm managers to analyse soil characteristics and their alignment, facilitating the identification of suitable land for sustainable usage.

3. Monitoring Soil Health and Carbon Management: Implement regular monitoring of soil organic carbon (SOC) changes over time for different land uses, as emphasized by Gnanavelrajah et al. (2007). This ensures the sustainability of agricultural land-use practices by maintaining soil health and effective carbon management strategies.

4. Integration of Agroforestry Practices: Recognize the role of agroforestry trees as indicators of sustainable land use practices, as highlighted by Mengesha et al. (2019). Integrate agroforestry systems into land management strategies to promote economic and ecological sustainability through the provision of timber and non-timber forest products.

5. Application of Spatial Regression Methods: Apply spatial regression methods in sustainable land use modelling, as discussed by Widjonarko and Maryono (2022). Utilize advanced statistical techniques to assess and model sustainable land use, enabling farm managers to make informed decisions based on spatial data analysis.

6. Multidimensional Approach to Land Use Assessment: Integrate biophysical, socioeconomic, quantitative, and qualitative analysis for sustainable marginal land identification and use, as advocated Jiang et al. (2018). Adopt a multidimensional approach to land use assessment to consider various factors influencing sustainable land management practices comprehensively.

7. Spatial Patterns and GIS-based Mapping: Consider land types and their combination patterns in designing sustainable land use patterns, as emphasized by Luo et al. (2010). Utilize GIS-based mapping to analyse spatial patterns and develop optimized land use plans that prioritize ecological sustainability.

8. Continuous Assessment and Monitoring: Implement continuous assessment and monitoring of land use practices to ensure they remain on a sustainable development trajectory, as highlighted by Niu et al. (2022). Regularly review and adjust land management strategies based on ongoing assessments to maintain ecological integrity and productivity.

9. Ecological Approach to Land Use Planning: Adopt an ecological approach in identifying suitable future land uses for sustainable land use planning, as suggested by Ahmadi et al. (2017). Consider ecological factors alongside other criteria in evaluating land suitability to promote environmental sustainability.

10. Innovative Management Systems: Develop and apply innovative management systems to ensure the sustainable use of agricultural land, as recommended by Liebig et al. (2017). Implement sustainable land management practices that optimize resource use efficiency and minimize environmental impacts while maximizing agricultural productivity.

By integrating these methodologies and approaches into their land management practices, farm managers can effectively assess vegetation types and coverage and identify necessary changes to promote sustainable land use and ensure long-term ecological sustainability.

Remote sensing, Geographic Information Systems (GIS), and geostatistics are indispensable tools when combined to accurately assess land capability for crop production. Each of these technologies plays a crucial role in the assessment process.

Remote sensing involves collecting information about the Earth's surface from a distance, typically using satellites or aircraft equipped with sensors. These sensors capture data in various wavelengths of electromagnetic radiation, including visible, infrared, and microwave. Remote sensing data can be utilized to assess vegetation health, identify crop types, and detect anomalies such as stress, disease, or pest infestation. By analysing vegetation indices derived from satellite imagery, such as NDVI (Normalized Difference Vegetation Index), farm managers can evaluate the overall health and vigour of crops across large areas. Additionally, remote sensing imagery can be processed to classify land cover types, distinguishing between different vegetation types, land uses, and soil types, providing valuable information for assessing land capability and suitability for specific crops.

Figure 29: Aerial view of agriculture land with a highway running through the middle. Land uses include circle irrigation, contoured ploughing, non-irrigated and other irrigated fields. Pacific Southwest Region 5, Public domain, via Wikimedia Commons.

The Normalized Difference Vegetation Index (NDVI) is a widely-utilized measure for assessing the health and density of vegetation based on sensor data. It is derived from spectrometric data captured in two specific bands: red and near-infrared, typically obtained from remote sensors like satellites.

Renowned for its accuracy, the NDVI exhibits a strong correlation with the actual condition of vegetation on the ground. Its interpretation is straightforward: NDVI values range from -1 to 1. A value of zero indicates areas devoid of vegetation, while values closer to one signify dense and healthy vegetation. Conversely, negative NDVI values suggest non-vegetated regions, with bodies of water yielding an NDVI of -1.

In precision agriculture, NDVI data serves as a crucial indicator of crop health. Modern applications of NDVI often involve agricultural drones equipped with sensors to monitor and analyse crop health efficiently. Companies like PrecisionHawk and Sentera offer agricultural drones capable of swiftly capturing and processing NDVI data, significantly reducing the turnaround time compared to traditional methods.

Recent advancements have demonstrated the feasibility of obtaining NDVI images using standard digital RGB cameras with minor modifications, yielding results comparable to those obtained from multispectral cameras. This innovation holds promise for integrating NDVI-based monitoring systems into crop health assessment practices more effectively.

Mobile applications leveraging NDVI data for crop health monitoring have seen a surge in popularity. Platforms like Doktar's Orbit provide farmers with NDVI data integrated into health maps, facilitating the detection of anomalies in their fields. These applications aim to streamline farming practices by digitalizing field scouting and enabling more efficient irrigation management, consequently helping farmers reduce fuel costs by minimizing the need for physical field visits.

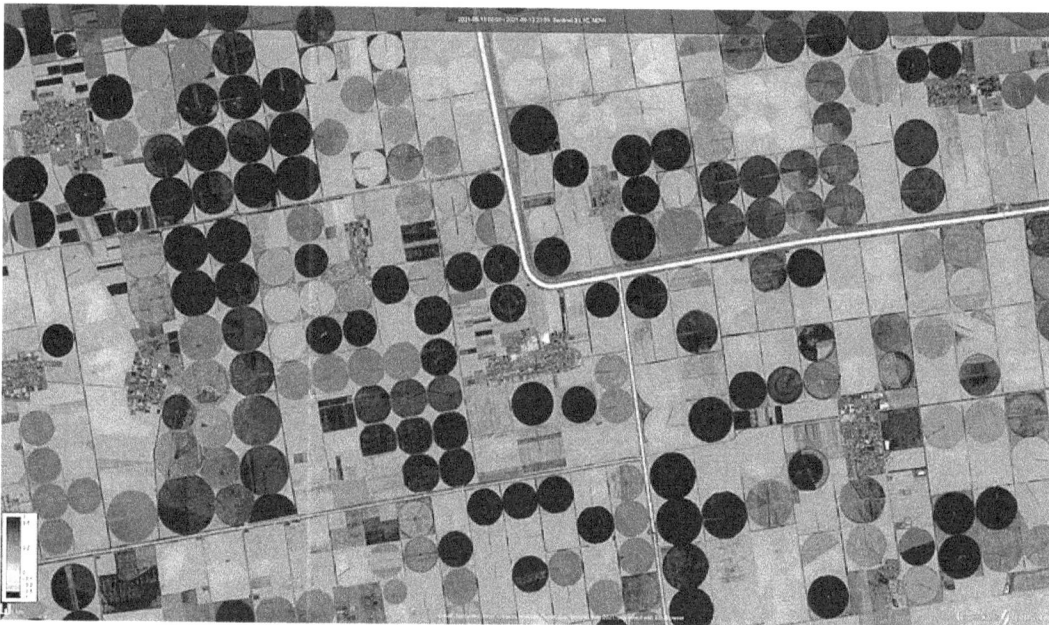

Figure 30: The map of the distribution of the NDVI of irrigated fields in the Kherson region in the south of Ukraine was obtained on September 13, 2021 by the Sentinel-2 satellite. Tomchenko Olha, CC BY 4.0, via Wikimedia Commons.

GIS allows for the integration, visualization, analysis, and interpretation of spatial data related to land resources. By overlaying layers of information such as soil properties, terrain characteristics, climate data, and land use patterns, GIS facilitates the identification of suitable areas for crop production based on various criteria. It enables the creation of detailed maps and visualizations that display key factors affecting land capability, such as soil types, slope gradients, elevation, drainage patterns, and proximity to water sources. These maps provide farm managers with valuable insights into the spatial distribution of resources and constraints across their agricultural landscapes. Moreover, GIS-based decision support systems help farm managers make informed decisions about crop selection, planting strategies, irrigation management, and land use planning by integrating diverse datasets and spatial analysis tools.

Geostatistical techniques, such as kriging and inverse distance weighting, are employed to interpolate and predict values at unsampled locations based on observations from nearby sampling points. These methods are valuable for estimating soil properties, crop yields, and other spatially distributed variables across agricultural landscapes. Geostatistics also provides tools for quantifying and analysing the uncertainty associated with spatial predictions and interpolation results. By assessing the reliability and accuracy of spatial estimates, farm managers can make more informed decisions and account for uncertainty in their land capability assessments. Furthermore, geostatistical analysis helps farm managers understand the spatial variability of soil properties and other environmental factors within their fields, informing precision agriculture practices such as variable rate fertilization, seeding, and irrigation.

Kriging, for example, is a sophisticated method that considers the spatial autocorrelation of the variable being studied. It takes into account not just the distances between sampled points, but also the variability or similarity of the variable being measured between those points. By analysing the spatial structure of the data, kriging can provide more accurate predictions and also quantifies the uncertainty associated with those predictions.

Inverse distance weighting, on the other hand, is a simpler technique that assigns weights to nearby sample points based on their distance from the location being estimated. Closer points are given more weight, implying that they have a stronger influence on the estimated value. However, unlike kriging, inverse distance weighting does not consider the spatial structure or variability of the data beyond distance.

Both methods aim to provide estimates for unsampled locations by leveraging information from nearby sampled points. This interpolation process allows researchers and practitioners to create continuous surfaces or maps of the variable being studied, providing valuable insights into spatial patterns and distributions across a landscape.

The integration of remote sensing, GIS, and geostatistics allows farm managers to accurately assess land capability for crop production by leveraging spatially explicit information, analysing landscape dynamics, and making data-driven decisions to optimize agricultural practices and enhance productivity.

Monitoring soil organic carbon (SOC) involves a systematic approach to measure and assess the quantity of organic carbon present in the soil. The process encompasses several key steps:

Sampling Design: The first step is to develop a sampling plan that ensures representative coverage of the area of interest. Factors such as soil type, land use, topography, and historical management practices are considered in designing the sampling strategy. Depending on the scale and objectives of the monitoring program, random or systematic sampling methods may be employed.

Sample Collection: Soil samples are collected using appropriate tools such as soil augers or corers from various depths. Consistency in sampling depth intervals is crucial to allow for comparisons over time. Each sample is accurately labelled and recorded with relevant metadata, including location, depth, and date of collection.

Sample Preparation: Upon collection, soil samples undergo preparation to remove debris, roots, and stones. Air-drying the samples to a consistent moisture content ensures uniformity before further analysis. Additionally, samples may be sieved to a specific particle size if necessary for laboratory analysis.

Laboratory Analysis: Soil samples are analysed for organic carbon content using suitable laboratory methods such as dry combustion, wet oxidation, or infrared spectroscopy. These methods determine the concentration of organic carbon in the soil, typically reported as a percentage of soil organic carbon by weight.

Quality Control: Quality control measures are implemented to ensure the accuracy and reliability of the analysis. This may involve analysing replicate samples, using certified reference materials, and regularly calibrating instruments. Standardized protocols and procedures are followed to minimize variability and maintain consistency across samples.

Data Interpretation: Soil organic carbon data are interpreted in the context of monitoring objectives and relevant environmental factors. Spatial and temporal trends are analysed to identify patterns, changes, and potential drivers of SOC dynamics. Factors such as land management practices, climate variability, and soil properties are considered in interpreting the results.

Data Management: A comprehensive database is maintained to store and manage soil organic carbon data along with associated metadata. The data are organized in a structured format that facilitates retrieval, analysis, and reporting. Data integrity, security, and accessibility are ensured to support decision-making and future research efforts.

Reporting and Communication: The findings of the soil organic carbon monitoring program are effectively communicated to relevant stakeholders, including land managers, policymakers, and researchers. Reports, summaries, and visualizations are prepared to convey key findings, trends, and implications for land management and environmental stewardship.

Soil organic carbon monitoring programs can provide valuable insights into soil health, carbon sequestration, and ecosystem dynamics. These insights support sustainable land management practices and contribute to efforts in climate change mitigation.

Applying spatial regression methods in sustainable land use modelling involves a systematic approach to analyse spatially distributed data and predict land use changes or patterns based on various environmental, socioeconomic, and land management factors.

Data Collection is the initial step, requiring the gathering of spatially referenced data on relevant variables influencing land use patterns, such as land cover, soil properties, climate, topography, land tenure, infrastructure, population density, economic activities, and policy regulations. These data may originate from remote sensing, GIS databases, field surveys, socioeconomic surveys, government agencies, or research institutions.

Data Preprocessing follows, involving the cleaning and preprocessing of collected data to ensure consistency, accuracy, and compatibility. This may include standardizing units, resolving inconsistencies, handling missing values, and converting data into a common spatial resolution and projection system.

Spatial Regression Model Selection is crucial, where an appropriate spatial regression model is chosen based on data characteristics and research objectives. Common techniques include spatial autoregressive models (SAR), spatial error models (SEM), spatial lag models (SLM), geographically weighted regression (GWR), and spatial panel data models.

Model Specification entails defining the spatial regression model by specifying the dependent variable (e.g., land use/land cover classes or changes), independent variables (e.g., environmental, socioeconomic, and land management factors), and spatial weights matrix (to account for spatial autocorrelation). Consideration is given to spatial scale, heterogeneity, and relationships among variables.

Model Estimation involves estimating the parameters of the spatial regression model using appropriate techniques like maximum likelihood estimation (MLE), generalized method of moments (GMM), or Bayesian inference. Adjustments are made for spatial autocorrelation, multicollinearity, and other statistical issues during model estimation.

Model Validation is essential to assess the goodness-of-fit, predictive accuracy, and robustness of the spatial regression model. Cross-validation techniques like k-fold cross-validation or leave-one-out cross-validation are employed to evaluate model performance and identify potential sources of bias or error.

Spatial Prediction utilizes the validated spatial regression model to predict land use patterns or changes across the study area. This step involves generating spatially explicit maps or scenarios of future land use distributions based on estimated model parameters and input data.

Sensitivity Analysis is conducted to examine the sensitivity of model outputs to changes in parameters, input data, and assumptions. This helps in assessing uncertainties and limitations of the spatial regression model and its predictions, thereby informing decision-making and scenario planning.

Interpretation and Communication involve interpreting the results of the spatial regression analysis in the context of sustainable land use planning, policy development, and natural resource management. Findings, implications, and uncertainties of the land use model are communicated to stakeholders, policymakers, and the broader community using visualizations, reports, and presentations.

Integrating biophysical, socioeconomic, quantitative, and qualitative analysis for sustainable marginal land identification and use involves a comprehensive approach that considers various factors affecting land suitability and management decisions.

In the initial phase, a biophysical analysis is conducted to comprehend the physical characteristics of the land. This includes assessing soil quality, topography, hydrology, vegetation cover, and climate conditions through methods such as soil surveys, remote sensing data, GIS, and field assessments. The aim is to evaluate soil fertility, drainage patterns, erosion risks, and the suitability for different types of vegetation.

Subsequently, socioeconomic factors influencing land use decisions and practices are taken into account. This involves analysing demographic trends, land tenure systems, market dynamics, policy frameworks, and socio-cultural preferences. Data from surveys, interviews, census records, and government reports are gathered to understand the social and economic context of the area.

Quantitative methods are then employed to analyse numerical data related to land characteristics, environmental variables, economic indicators, and land use patterns. Statistical analysis, spatial modelling, economic modelling, and geospatial analysis are utilized to quantify relationships and trends, predict land suitability, productivity, and potential impacts of various land use scenarios.

Qualitative methods complement quantitative analysis by providing insights into subjective aspects of land use decision-making and stakeholder perspectives. Interviews, focus groups, participatory workshops, and observations are conducted to capture qualitative data on local knowledge, cultural values, community aspirations, and perceptions of marginal lands, enhancing the understanding of human dimensions in land management.

Findings from these analyses are integrated to develop a holistic understanding of sustainable marginal land identification and use. Synergies, trade-offs, and potential conflicts between biophysical constraints, socioeconomic opportunities, and sustainability goals are identified. Data and insights from different disciplines inform decision-making processes and prioritize interventions.

Multidisciplinary collaboration is crucial, fostering communication among experts from diverse fields such as agronomy, ecology, economics, sociology, anthropology, and geography. Interdisciplinary dialogue bridges gaps between biophysical and socioeconomic perspectives, engaging stakeholders including farmers, landowners, policymakers, researchers, and community members to ensure inclusivity and transparency.

Implementing adaptive management strategies allows for continuous learning, feedback, and adjustment based on monitoring and evaluation results. The effectiveness of land use interventions is monitored, social and environmental impacts assessed, and management practices adapted accordingly. A participatory approach engages stakeholders in the process, empowering local communities to take ownership of sustainable land use initiatives.

By integrating these analyses, stakeholders develop informed and context-specific strategies for identifying and utilizing marginal lands sustainably. This approach promotes resilience, equity, and environmental stewardship in land management practices.

Adopting an ecological approach in identifying suitable future land uses for sustainable land use planning encompasses a holistic consideration of the intricate interplay between ecological processes, biodiversity, ecosystem services, and human activities. To embark on this approach, several steps can be followed:

Firstly, commence with an in-depth ecological assessment of the targeted area. This entails a meticulous examination of the existing ecosystems, habitats, species diversity, and ecological functions. Identify pivotal ecological features such as wetlands, forests, grasslands, rivers, and biodiversity hotspots, while gauging their ecological significance and susceptibility to disturbance.

Subsequently, conduct an evaluation of ecosystem services provided by various land uses and ecosystems within the area. This evaluation encompasses provisioning services like food, water, and timber, as well as regulating services such as climate regulation and water purification. Furthermore, cultural services like recreation and spiritual values, along with supporting services like soil formation and nutrient cycling, should be assessed. Gauge the capacity of ecosystems to sustainably provide these services and their contribution to human well-being.

Analyse landscape connectivity and ecological corridors to pinpoint areas of elevated ecological value and potential habitat fragmentation. Consider the spatial arrangement and connectivity of ecosystems, wildlife habitats, migration routes, and green infrastructure networks. Strive to enhance connectivity between natural areas to facilitate species movement, foster genetic diversity, and fortify ecosystem resilience.

Assess the vulnerability of ecosystems and land uses to climate change impacts, including temperature variations, precipitation fluctuations, extreme weather events, and sea-level rise. Identify climate-resilient land uses and adaptation strategies such as habitat restoration, green infrastructure development, and land-use zoning tailored to projected climate scenarios.

Apply adaptive management principles to sustainable land use planning, prioritizing flexibility, learning, and responsiveness to evolving ecological conditions. Formulate adaptive management strategies enabling iterative decision-making, continuous monitoring, and adjustment based on new insights and feedback from stakeholders and ecological monitoring programs.

Engage stakeholders, encompassing local communities, indigenous peoples, landowners, government agencies, NGOs, scientists, and businesses, throughout the land use planning process. Incorporate diverse perspectives, traditional ecological knowledge, and community priorities into decision-making to uphold social equity, cultural sensitivity, and participatory governance.

Prioritize conservation endeavours and land management interventions guided by ecological significance, biodiversity conservation targets, and ecosystem services priorities. Identify areas suitable for habitat restoration, protected area expansion, sustainable agriculture, agroforestry, and urban green spaces to bolster ecological resilience and foster sustainable land use practices.

Implement monitoring and evaluation programs to gauge the effectiveness of land use planning strategies, conservation measures, and ecosystem management interventions. Monitor shifts in ecological indicators, biodiversity metrics, ecosystem functions, and socio-economic outcomes over time to chart progress towards sustainability goals and adapt management approaches as required.

Through adopting an ecological approach in identifying suitable future land uses for sustainable land use planning, stakeholders can champion biodiversity conservation, bolster ecosystem resilience, and advance human well-being, thus tackling the multifaceted challenges of environmental degradation, climate change, and land degradation.

Planning and Prioritising Land and Water Protection Measures and Structures

To effectively plan and prioritize land and water protection measures for sustainable land use, individuals must possess a comprehensive understanding of ecological sustainability principles, land management practices, and the interplay between land and water dynamics. This involves a series of key steps that require analytical thinking, strategic decision-making, and effective communication. The process begins with a comprehensive assessment of the land and water resources, encompassing factors such as soil quality, vegetation cover, hydrological patterns, water availability, erosion risks, and pollution sources. This detailed analysis helps identify areas vulnerable to degradation, habitat loss, water contamination, and other environmental threats (Liuzzo & Freni, 2018). Based on the assessment findings, prioritization of the most pressing land and water protection needs is essential, considering factors such as environmental risks, ecosystem services, regulatory requirements, stakeholder interests, and long-term sustainability goals (Chen et al., 2020).

The selection of appropriate protection measures and structures tailored to the specific needs and challenges identified during the assessment phase is crucial (Anaba et al., 2017). These measures may include erosion control practices, vegetative measures, and best management practices (BMPs) such as precision agriculture techniques and efficient irrigation systems (Chen et al., 2020). Infrastructure development to support land and water protection efforts, such as the construction of physical structures to intercept runoff and filter pollutants, should be evaluated for feasibility, cost-effectiveness, and environmental compatibility (Anaba et al., 2017).

Detailed implementation plans outlining specific actions, timelines, responsibilities, and resources required to execute land and water protection measures effectively are necessary, along with monitoring protocols to track the performance of implemented protection measures and structures over time (Anaba et al., 2017). Proactive communication and outreach efforts are also vital to raise awareness, build consensus, and mobilize support for land and water protection initiatives (Anaba et al., 2017).

Undertaking infrastructure development to bolster land and water protection endeavours necessitates a systematic approach that encompasses several key steps. Firstly, individuals must conduct a comprehensive assessment to understand the specific challenges and vulnerabilities pertaining to land and water protection within the targeted area. This involves identifying sources of erosion, sedimentation, pollution, as well as areas susceptible to flooding or waterlogging, while evaluating the potential risks posed to soil and water resources, biodiversity, and human activities.

Subsequently, based on the findings of the assessment, individuals need to identify potential infrastructure options capable of effectively addressing the identified challenges. This entails considering various physical structures such as sediment traps, retention ponds, check dams, and vegetative filter strips, with each option serving a distinct purpose depending on the specific context and objectives.

Following the identification of infrastructure options, a thorough feasibility analysis must be conducted for each option to determine its practicality and viability. Factors such as site conditions, hydrological characteristics, available land, construction materials, and technical requirements should be taken into account to assess whether the proposed infrastructure can be implemented successfully within the given constraints and timeframe.

Furthermore, an evaluation of the cost-effectiveness of the proposed infrastructure projects is essential. This involves estimating the total costs involved in planning, design, construction, operation, and maintenance, and comparing them against the expected benefits and outcomes. Projects offering the greatest return on investment and significant contributions to land and water protection goals should be prioritized.

Additionally, an environmental compatibility assessment must be conducted to ensure that the proposed infrastructure projects do not result in adverse impacts on natural ecosystems, biodiversity, or cultural heritage. Environmental impact assessments (EIAs) and ecological risk assessments (ERAs) should be carried out to identify and mitigate potential negative effects, with input from relevant stakeholders, regulatory agencies, and environmental experts being sought to address concerns and minimize environmental risks.

Community engagement and consultation are integral aspects of the infrastructure development process, requiring active involvement with local communities, landowners, and other stakeholders to solicit feedback, address concerns, and build support. This involves consulting with indigenous peoples, traditional land users, and marginalized groups to ensure their perspectives and rights are respected, thereby fostering participatory decision-making and collaborative governance to enhance project transparency, accountability, and social acceptance.

Lastly, ensuring regulatory compliance is paramount, necessitating adherence to applicable laws, regulations, and permitting requirements governing infrastructure development, land use, and environmental protection. Obtaining necessary permits, licenses, and approvals from regulatory authorities and adhering to relevant standards and guidelines are imperative to uphold environmental standards and safeguard public health and safety.

By meticulously following these steps, individuals can effectively evaluate the need for infrastructure development to support land and water protection efforts, select appropriate projects, and ensure their implementation in a manner that maximizes benefits while minimizing adverse environmental impacts. This approach not only promotes sustainable land management practices but also enhances ecosystem resilience and fosters long-term environmental stewardship.

Chapter Three

Carrying Out Structural Improvements to Address Threats to Sustainability

U ndertaking structural improvements to mitigate threats to sustainability is a multifaceted endeavour that necessitates adherence to specific protocols and meticulous attention to detail. To begin, it is imperative to adhere to workplace health and safety policies and procedures rigorously, particularly during fieldwork activities. This ensures the protection of personnel involved in the process and mitigates potential risks or hazards that may arise during structural improvement initiatives.

Furthermore, conducting a comprehensive inspection of the site is paramount to identify key features and pinpoint areas where improvements are urgently needed. This meticulous examination helps in assessing the extent of the threats to sustainability and provides crucial insights into the areas requiring immediate attention. By meticulously recording observations and highlighting areas for improvement directly on-site, stakeholders can create a clear roadmap for subsequent actions.

Moreover, determining the materials required for construction from meticulously prepared plans and specifications is crucial to the success of the structural improvement efforts. Careful consideration of materials ensures that they are suitable for the intended purpose, environmentally sustainable, and aligned with project specifications. Additionally, obtaining and organizing the necessary materials, personnel, and equipment in advance streamlines the construction process and minimizes delays or disruptions.

Prior to commencing construction activities, it is essential to select and thoroughly inspect equipment to ensure its safe and environmentally sustainable operation. This involves verifying that equipment is in optimal working condition, properly calibrated, and equipped with necessary safety features. Conducting routine checks and maintenance procedures mitigates the risk of accidents or environmental damage during the construction process.

Subsequently, the construction of structural works can proceed in accordance with established plans and specifications. This may involve building erosion control structures, installing retention ponds, constructing check dams, or implementing other measures to mitigate threats to sustainability effectively. Additionally, revegetating the area post-construction helps restore ecological balance and enhance the site's resilience against erosion and other environmental stressors. Securing the area from livestock intrusion further safeguards the structural improvements and ensures their long-term effectiveness in protecting the site's sustainability. Through meticulous planning, diligent execution, and strict adherence to safety and environmental standards, structural improvements can effectively address threats to sustainability and contribute to the long-term health and resilience of the ecosystem.

Following Workplace Health and Safety Policies and Procedures in Field Work

Farmers are tasked with an array of responsibilities beyond simply understanding the inherent risks of farming. They must also possess a multifaceted skill set that encompasses roles as diverse as business management, animal and crop expertise, mechanics, construction, transportation, and environmental stewardship. The farm is not merely a workplace but often serves as both home and livelihood for farm families, presenting additional challenges in maintaining safety for children and ensuring the well-being of working parents and siblings (Cruickshank, 2022).

Unlike many other industries characterized by repetitive tasks, farming presents a dynamic environment where daily activities can vary significantly. From managing livestock to working in packing sheds, operating machinery, or servicing windmills at heights, each day brings a unique set of challenges and risks that farmers must contend with to ensure both productivity and safety (Cruickshank, 2022).

Figure 31: Livestock are one of the many hazards in agricultural settings.

One significant concern for farmers is the safety of children on farms, as rural properties pose numerous hazards, including machinery, dams, chemicals, and livestock. To mitigate these risks, it is essential to keep children supervised, educate them about potential dangers, and ensure that farm buildings and vehicles are properly secured.

Another prevalent issue is asbestos exposure, as many farms contain asbestos-containing materials used in buildings and structures. Farmers must be vigilant in managing these materials to minimize exposure risks and protect their health and that of their families and workers.

The management of hazardous chemicals, such as herbicides and pesticides, also presents challenges, as improper use or spills can lead to health problems and environmental harm. Farmers must carefully follow safety protocols, including reading labels, using appropriate personal protective equipment, and avoiding spraying on windy days (Cruickshank, 2022).

Extreme weather conditions, such as sun exposure, pose risks to farmers, necessitating protective measures such as wearing appropriate clothing and sunscreen and taking regular breaks in the shade. Falls are another common hazard on farms, particularly when working at heights or in cluttered environments, highlighting the importance of proper footwear, awareness of surroundings, and good housekeeping practices.

Manual tasks involving heavy lifting or handling stock also carry risks of injury, emphasizing the need for proper lifting techniques, mechanical aids, and adequate rest breaks. Firearm use, electrical hazards, vehicle operation, livestock handling, respiratory hazards, confined space work, and remote and isolated work further underscore the diverse range of risks farmers must manage daily (Cruickshank, 2022).

To address these risks effectively, farmers can implement robust health and safety management systems tailored to their specific operations. These systems help identify and mitigate risks, improve regulatory compliance, reduce costs associated with accidents, and enhance the overall safety culture of the farm. By prioritizing safety and implementing proactive measures, farmers can ensure a safer and more sustainable working environment for themselves, their families, and their workers (Cruickshank, 2022).

In adhering to workplace health and safety policies and procedures during fieldwork activities related to the development and implementation of sustainable land use strategies, several key steps should be followed. Firstly, it is imperative to conduct a comprehensive assessment of the work site to identify potential hazards and risks that may pose a threat to the health and safety of individuals involved. This assessment should encompass factors such as terrain conditions, weather patterns, presence of hazardous materials, and any other relevant environmental considerations.

Following the assessment, it is essential to develop and implement appropriate safety protocols and procedures tailored to the specific needs and challenges of the work site. This may involve establishing clear guidelines for personal protective equipment (PPE) usage, safe equipment operation, emergency response procedures, and communication protocols. Additionally, ensuring that all personnel are adequately trained in these safety procedures and protocols is paramount to mitigating risks and preventing accidents or injuries.

Furthermore, regular monitoring and supervision of fieldwork activities are essential to maintain compliance with health and safety regulations and to address any emerging hazards or concerns promptly. This may entail assigning designated safety officers or supervisors responsible for overseeing field operations and ensuring that safety protocols are strictly adhered to by all personnel.

In addition to proactive measures, it is essential to maintain open lines of communication and encourage reporting of any safety incidents, near misses, or concerns that may arise during fieldwork. Establishing a culture of safety awareness and accountability among team members fosters a collaborative approach to identifying and addressing potential risks, ultimately enhancing overall safety performance and reducing the likelihood of accidents or injuries.

Finally, periodic review and evaluation of safety protocols and procedures should be conducted to identify areas for improvement and ensure ongoing compliance with relevant regulations and standards. By continuously striving to enhance safety practices and promote a culture of safety consciousness, individuals can effectively mitigate risks and create a safer working environment for themselves and their colleagues while carrying out fieldwork activities in support of sustainable land use strategies.

Inspecting Site, Recording Key Features and Indicating on the Site where Improvements are Required

As an individual responsible for developing and implementing sustainable land use strategies, inspecting the site, recording key features, and indicating where improvements are required is a critical step in the process. This task involves a thorough assessment of the land to identify existing conditions, potential challenges, and opportunities for improvement.

Firstly, conduct a comprehensive site inspection, carefully examining the various elements that contribute to land sustainability. This includes assessing soil quality, water availability, vegetation cover, biodiversity, and existing infrastructure. Take note of any visible signs of erosion, degradation, or pollution, as well as areas with high ecological value or sensitivity.

During the inspection, record key features of the site using appropriate documentation tools and techniques. This may involve taking photographs, sketching maps, or using digital mapping technologies such as GIS (Geographic Information Systems). Document the location of natural resources, land contours, water bodies, vegetation types, and any existing land use patterns.

As you identify areas in need of improvement, clearly indicate them on the site documentation. This can be done by marking specific locations on maps or photographs and providing detailed descriptions of the issues observed. Use standardized symbols or annotations to distinguish between different types of improvements required, such as erosion control measures, vegetation restoration areas, or infrastructure upgrades.

Prioritize the identified improvement areas based on their significance, urgency, and feasibility. Consider factors such as the severity of the problem, potential environmental impact, available resources, and alignment with sustainability objectives. This prioritization process will help focus efforts and resources on addressing the most critical issues first.

Collaborate with relevant stakeholders, including landowners, government agencies, environmental experts, and community members, to validate the findings of the site inspection and gather additional insights. Engage in dialogue to ensure a shared understanding of the challenges and opportunities identified, as well as the proposed solutions.

Finally, use the information gathered during the site inspection to inform the development of sustainable land use strategies. Integrate the identified improvement areas into comprehensive action plans that outline specific interventions, timelines, responsibilities, and performance indicators. Communicate these strategies effectively to stakeholders, emphasizing the importance of collective action and ongoing monitoring and evaluation to achieve ecological sustainability goals.

By conducting thorough site inspections, recording key features, and indicating where improvements are required, you can lay the groundwork for the successful development and implementation of sustainable land use strategies. This proactive approach not only helps address current land management challenges but also fosters long-term resilience and stewardship of natural resources. Figure 32 provides a sample sustainable land use site inspection recording template that can be used or adapted to record the outcome of the inspection.

Sustainable Land Use Site Inspection
Date of Inspection: [Insert Date]
Location: [Insert Location]
Inspector: [Insert Name]

1. Assessment of Existing Conditions:
- Soil Quality:
  - Description:
  - Observations:
- Water Availability:
  - Description:
  - Observations:
- Vegetation Cover:
  - Description:
  - Observations:
- Biodiversity:
  - Description:
  - Observations:
- Existing Infrastructure:
  - Description:
  - Observations:

2. Recording Key Features:
- Documentation Tools Used:
  - Photographs:
  - Sketch Maps:
  - Digital Mapping Technologies (e.g., GIS):
- Recorded Features:
  - Natural Resources:
  - Land Contours:
  - Water Bodies:
  - Vegetation Types:
  - Existing Land Use Patterns:

3. Identification of Areas for Improvement:
- Description of Areas:
  - Erosion Control Measures Needed:
  - Vegetation Restoration Areas:
  - Infrastructure Upgrades Required:
- Indication on Documentation:
  - Marking Locations on Maps/Photographs:
  - Detailed Descriptions of Issues Observed:
  - Standardized Symbols/Annotations Used:

4. Prioritization of Improvement Areas:
- Significance:
  - Severity of the Problem:
  - Potential Environmental Impact:
  - Available Resources:
- Urgency:
  - Immediate Needs:

- o Short-Term Goals:
- o Long-Term Objectives:
- Feasibility:
  - o Technical Considerations:
  - o Financial Constraints:
  - o Stakeholder Input:

5. Collaboration with Stakeholders:
- Relevant Stakeholders:
  - o Landowners:
  - o Government Agencies:
  - o Environmental Experts:
  - o Community Members:
- Validation of Findings:
  - o Shared Understanding of Challenges:
  - o Additional Insights Gathered:
  - o Consensus on Proposed Solutions:

6. Development of Sustainable Land Use Strategies:
- Integration of Improvement Areas:
  - o Action Plans Developed:
  - o Specific Interventions Outlined:
  - o Timelines Established:
- Responsibilities and Performance Indicators:
  - o Assigned Responsibilities:
  - o Performance Metrics Defined:
  - o Monitoring and Evaluation Plan:
- Communication of Strategies:
  - o Stakeholder Engagement Strategy:
  - o Emphasis on Collective Action:
  - o Importance of Ongoing Monitoring and Evaluation:

7. Summary and Next Steps:
- Summary of Inspection Findings:
  - o Key Challenges Identified:
  - o Opportunities for Improvement:
- Next Steps:
  - o Immediate Actions Required:
  - o Long-Term Strategy Implementation:
  - o Follow-Up Inspections and Monitoring:

8. Conclusion:
By conducting this thorough site inspection and recording key features while indicating areas for improvement, we are taking proactive steps towards the development and implementation of sustainable land use strategies. This approach not only addresses current land management challenges but also contributes to the long-term resilience and stewardship of natural resources.

Inspector's Signature: _____

Date: _____

Figure 32: Sustainable Land Use Site Inspection Recording Template.

Using digital mapping technologies such as GIS (Geographic Information Systems) involves a series of steps aimed at efficiently gathering, analysing, and visualizing spatial data. The following provides a guide on how to utilize GIS effectively:

- Data Collection: Begin by gathering relevant spatial data from a variety of sources, including satellite imagery, aerial photographs, GPS devices, and existing GIS databases. It's essential to ensure that the collected data is accurate, up-to-date, and aligned with the objectives of your project.

- Data Preparation: Once the data is collected, the next step involves cleaning and preprocessing it to ensure consistency and compatibility for GIS analysis. Tasks in this stage may include georeferencing images, digitizing features, and standardizing data formats. Additionally, organize the data into layers based on thematic categories such as land use, vegetation, and water bodies.

- Data Analysis: Utilize GIS software to conduct spatial analysis and extract meaningful insights from the collected data. Common analytical techniques include overlay analysis, proximity analysis, interpolation, and spatial statistics. By analysing spatial patterns, relationships, and trends, you can gain a deeper understanding of the factors influencing land use and environmental conditions.

Figure 33: Geographic Information System. Emilio Gómez Fernández, CC BY-SA 3.0, via Wikimedia Commons.

- Map Design: Design visually engaging and informative maps to effectively communicate the results of your GIS analysis. Customize map layouts, symbology, labels, and legends to convey key information clearly. Consider the preferences and requirements of your target audience when designing maps to ensure their effectiveness.

- Visualization: Visualize spatial data and analysis results using GIS software to create maps, charts, graphs, and 3D visualizations. Employ various visualization techniques, such as choropleth maps, pie charts, and scatter plots, to present information in a compelling and understandable manner.

- Interpretation: Interpret the outputs of GIS analysis and map visualizations to draw meaningful conclusions and make informed decisions. Identify patterns, trends, and spatial relationships revealed by the data analysis, and evaluate their implications for land use planning, resource management, and environmental conservation.

- Integration: Integrate GIS data and analysis results with other relevant information sources, such as socio-economic data, environmental reports, and stakeholder feedback. Combining GIS outputs with non-spatial data facilitates a comprehensive understanding of land use dynamics and sustainability challenges.

- Communication: Effectively communicate the findings and insights derived from GIS analysis to stakeholders, decision-makers, and the broader community. Prepare reports, presentations, and interactive web maps to share the results in a clear and accessible format. Use storytelling techniques to engage the audience and convey the significance of the findings.

- Validation: Validate the accuracy and reliability of GIS analysis by comparing the results with ground truth data, field observations, and expert knowledge. Ensure the consistency of spatial relationships and patterns identified through GIS with real-world conditions. Incorporate feedback from stakeholders to refine the analysis and ensure its relevance.

- Continuous Improvement: Continuously update and refine GIS analysis as new data becomes available and project requirements evolve. Stay abreast of advancements in GIS technology and methodologies to enhance the effectiveness and efficiency of spatial analysis workflows. Iterate on the analysis process based on lessons learned and feedback received from stakeholders.

Integrating the information gathered during the site inspection into the development of sustainable land use strategies involves several key steps.

Firstly, it requires synthesizing the collected data to identify areas for improvement and prioritize them based on their significance and urgency. This involves compiling and analysing the data obtained during the site inspection to gain insights into the existing conditions of the land.

Next, the insights gained from the site inspection are utilized to formulate comprehensive action plans for sustainable land use. These plans outline specific interventions, such as erosion control measures, vegetation restoration, or infrastructure upgrades, aimed at addressing the identified improvement areas.

Additionally, clear timelines need to be established for implementing the proposed interventions, considering factors such as seasonality, resource availability, and project dependencies. Realistic timelines help ensure the timely execution of the action plans.

Responsibilities for implementing each intervention are then assigned to relevant stakeholders or team members. Clearly defining roles and duties ensures accountability and effective coordination throughout the implementation process.

Furthermore, measurable performance indicators are defined to track the progress and effectiveness of the implemented interventions. These indicators may include metrics such as soil erosion rates, water quality parameters, vegetation cover, or biodiversity indices.

A communication strategy is developed to effectively convey the sustainable land use strategies to stakeholders, emphasizing the importance of collective action and the role of each stakeholder in achieving ecological sustainability goals.

Stakeholders are engaged throughout the strategy development process to gather feedback, address concerns, and foster buy-in for the proposed interventions. Collaboration and participation are encouraged to ensure that the strategies reflect diverse perspectives and interests.

Lastly, a robust monitoring and evaluation framework is implemented to assess the progress and outcomes of the sustainable land use strategies. Regular monitoring of key performance indicators, collecting feedback from stakeholders, and adjusting strategies as needed based on the evaluation results are essential for achieving ecological sustainability goals.

Synthesizing the collected data to identify areas for improvement and prioritize them involves several steps and considerations:

1. Data Compilation: Gather all the data collected during the site inspection, including observations, measurements, photographs, and any other relevant information. Ensure that the data is organized and documented systematically for easy analysis.

2. Data Analysis: Analyse the collected data to gain insights into the existing conditions of the land. This analysis may involve various techniques depending on the nature of the data, such as statistical analysis, spatial analysis, trend analysis, or comparative analysis.

3. Identifying Patterns and Trends: Look for patterns, trends, and anomalies in the data that may indicate areas of concern or potential opportunities for improvement. This may include identifying recurring issues such as soil erosion hotspots, water quality degradation zones, or areas of low vegetation cover.

4. Assessing Significance and Urgency: Evaluate the significance and urgency of the identified issues based on their potential impact on land sustainability, ecosystem health, and human activities. Consider factors such as the severity of the problem, the extent

of its influence, and the immediacy of action required.

5. Prioritization Criteria: Establish criteria for prioritizing areas for improvement, taking into account factors such as ecological sensitivity, regulatory requirements, stakeholder priorities, and available resources. Determine which issues are most critical and require immediate attention.

6. Stakeholder Input: Seek input from relevant stakeholders, including landowners, government agencies, environmental experts, and local communities, to validate the findings of the data analysis and gather additional insights. Consider stakeholders' perspectives and priorities in the prioritization process.

7. Documentation and Reporting: Document the results of the data analysis, including the identified areas for improvement and the rationale for prioritization decisions. Prepare clear and concise reports or presentations to communicate the findings to stakeholders and decision-makers.

In analysing the data obtained during the site inspection, several key aspects should be considered:

1. Soil Quality: Assess soil properties such as texture, structure, pH, nutrient levels, and compaction to determine soil health and fertility.

2. Water Availability and Quality: Evaluate water resources, including availability, quality, and usage patterns. Identify sources of contamination, erosion, or depletion.

3. Vegetation Cover and Biodiversity: Examine the type, distribution, and health of vegetation to assess ecosystem diversity and habitat quality.

4. Infrastructure and Land Use: Analyse existing infrastructure, land use patterns, and management practices to identify areas of inefficiency or environmental impact.

5. Erosion and Degradation: Identify signs of erosion, degradation, or loss of natural features such as wetlands, forests, or riparian zones.

6. Ecological Sensitivity: Consider the ecological sensitivity of the landscape, including areas with high biodiversity, unique habitats, or susceptibility to environmental disturbances.

7. Human Activities: Evaluate the impact of human activities such as agriculture, urbanization, mining, or industrial development on the landscape and natural resources.

By carefully analysing these aspects of the data collected during the site inspection, one can gain a comprehensive understanding of the existing conditions of the land and prioritize areas for improvement effectively.

Determining Materials for Construction from Plans and Specifications

Determining materials for construction from plans and specifications involves several key steps to ensure that the chosen materials align with sustainability goals and project requirements.

Firstly, it requires a comprehensive understanding of the plans and specifications provided for the construction project. This may include reviewing architectural drawings, engineering plans, and environmental assessments to identify the scope of work and specific requirements for materials.

Construction plans, referred to interchangeably as blueprints or drawings, constitute indispensable documents for projects of all sizes in the construction industry. They serve as a comprehensive guide for stakeholders, ranging from project owners to on-site workers, providing a visual representation of the intended structure (GoCodes, 2024). Therefore, proficiency in deciphering construction plans is essential for all involved parties.

A construction plan typically comprises two-dimensional drawings, commonly known as blueprints, illustrating the envisioned structure upon completion. These drawings incorporate vital information in the form of dimensions, symbols, abbreviations, and line types, facilitating a clear understanding of the construction project's scope and specifications. Regardless of the specific type of construction drawings being reviewed, each sheet typically features a title block, positioned prominently to convey crucial project details (GoCodes, 2024).

The title block, positioned either in the bottom right corner or along the right vertical or lower horizontal edge of the sheet, serves as a pivotal reference point for pertinent project information. It typically includes essential elements such as the project name, client's name, architect, designer, or engineer's name, drawing title, issue date, scale, and sheet number. Moreover, the title block may incorporate additional details, such as the drawing or print number and a revision block, allowing for comprehensive documentation of any modifications made over time (GoCodes, 2024).

After familiarizing oneself with the title block, the next step in reading construction plans involves reviewing the plan legend. Given the complexity inherent in construction projects, plan legends play a crucial role in guiding stakeholders through the diverse array of drawings included in a set of construction plans. These plans encompass various types of drawings, including floor plans, section and elevation drawings, structural drawings, electrical drawings, HVAC drawings, plumbing drawings, and finishing drawings (GoCodes, 2024).

Each type of drawing relies on a multitude of standardized or customized symbols and abbreviations to convey information effectively within the limited space available. Plan legends serve as indispensable guides, providing clarity on the meaning and interpretation of these symbols and abbreviations. By deciphering the plan legend, stakeholders can gain insights into critical details, such as building materials, dimensions, and construction methods, enabling informed decision-making throughout the project lifecycle (GoCodes, 2024).

Furthermore, thorough inspection of the drawings is imperative, with particular attention to gridlines and scale considerations. Gridlines serve as essential reference points, aiding in pinpointing specific areas on the drawing and facilitating coordinated communication among project teams. Understanding the scale to which a blueprint is drawn is equally crucial, as it ensures accurate interpretation and visualization of project dimensions and proportions (GoCodes, 2024).

Moreover, determining the orientation of buildings or structures on the construction site and within the floor plan is essential for effective project execution. Building orientation plays a significant role in optimizing energy efficiency, utilizing solar and wind power to minimize heating, cooling, and lighting costs. By aligning building elements with favourable solar exposure and natural ventilation, stakeholders can enhance both environmental sustainability and occupant comfort (GoCodes, 2024).

Lastly, attention should be given to any notes accompanying the construction plans, as they provide essential explanations and tie each drawing sheet to relevant specifications. These notes, categorized into general notes, general discipline notes, and general sheet notes, offer valuable insights into project-specific requirements and design considerations. Overall, a comprehensive understanding of the various elements present in construction plans equips stakeholders with the knowledge needed to navigate complex construction projects effectively (GoCodes, 2024).

Once the project requirements are understood, the next step is to assess the environmental impact of various construction materials. This involves considering factors such as resource depletion, energy consumption, greenhouse gas emissions, and waste generation associated with the production, transportation, and disposal of materials.

In line with sustainability principles, preference should be given to materials that are locally sourced, renewable, recyclable, or low in embodied energy. This helps minimize the carbon footprint of the construction project and reduce its overall environmental impact.

Furthermore, it's essential to prioritize materials that have minimal negative effects on ecosystem health and biodiversity. Avoiding materials that contain hazardous chemicals or contribute to habitat destruction helps preserve the ecological integrity of the land under production.

Additionally, considering the durability, longevity, and maintenance requirements of materials is crucial for ensuring the long-term sustainability of the constructed infrastructure. Choosing high-quality, resilient materials can reduce the need for frequent replacements or repairs, thereby minimizing resource consumption and waste generation over time.

Collaboration with suppliers, contractors, and other stakeholders is essential in the material selection process. Engaging with professionals who specialize in sustainable construction practices can provide valuable insights and expertise in choosing the most appropriate materials for the project.

Finally, effective communication of the chosen materials to all project stakeholders is vital for ensuring alignment with sustainability objectives. Clearly documenting the rationale behind material selection decisions and highlighting their environmental benefits can help foster support and buy-in from stakeholders.

By carefully considering the environmental impact, resource efficiency, and ecological compatibility of construction materials, individuals can contribute to the development and implementation of sustainable land use strategies that promote ecological sustainability and resilience.

To thoroughly comprehend the plans and specifications for a construction project, it's essential to follow a structured approach. Start by delving into the architectural drawings furnished for the project. These drawings encompass floor plans, elevations, sections, and intricate details of the proposed structures. Pay meticulous attention to the layout, dimensions, and spatial organization, keeping an eye out for any distinctive features, design elements, or specified material requirements embedded within the drawings.

Proceed to scrutinize the engineering plans linked to the project. These plans encompass structural drawings, site layouts, utility designs, and drainage schematics. Evaluate the structural elements like foundations, beams, columns, and roofing systems to grasp their material prerequisites and specifications. Assess the site layout, grading, and drainage arrangements to pinpoint any supplementary material necessities for site preparation and infrastructure enhancement.

Take into account any environmental assessments or impact studies conducted concerning the project. These assessments may shed light on ecological considerations, vulnerable habitats, or regulatory demands that could influence material selection. Pay heed to recommendations advocating sustainable practices, resource preservation measures, or environmentally sound alternatives proposed in the assessments.

Define the scope of work outlined in the plans and specifications. This entails understanding the overarching objectives of the construction project alongside the specific tasks and deliverables anticipated. Clarify the intended utilization of the constructed facilities, performance criteria, and functional requisites to steer material selection determinations.

Identify the specific material requirements for the project based on insights gleaned from the architectural drawings, engineering plans, and environmental assessments. Factor in considerations like structural integrity, thermal efficiency, acoustic properties, fire resistance, and aesthetic appeal when assessing materials. Collaborate with design professionals, engineers, and environmental experts as needed to ensure adherence to project specifications and sustainability objectives.

Thoroughly document your findings and observations from the review process. Generate a comprehensive summary or report encapsulating pivotal aspects of the plans and specifications, including material requisites, design nuances, and environmental considerations. This documentation serves as a pivotal reference for material selection deliberations and decision-making throughout the construction endeavour.

If any aspects of the plans or specifications appear unclear or ambiguous, don't hesitate to seek clarification from project stakeholders or the design team. Engage in effective communication to address discrepancies, resolve conflicts, or procure additional information requisite for making informed material selection decisions.

Once the project requirements have been fully grasped, delving into the assessment of environmental impacts associated with different construction materials emerges as a critical endeavour. Commence by gathering relevant data on the various construction materials under consideration. This entails collecting information on their composition, manufacturing processes, transportation methods, and end-of-life disposal options. Utilize resources such as material suppliers, manufacturers' specifications, environmental product declarations (EPDs), and life cycle assessments (LCAs) to obtain comprehensive insights.

Next, consider a spectrum of environmental factors that contribute to the overall impact of construction materials. These factors encompass resource depletion, energy consumption, greenhouse gas emissions, and waste generation. Assess the depletion of finite resources like timber or minerals, energy consumption during production and transportation, carbon dioxide emissions during manufacturing, and the generation of packaging waste and construction debris.

Subsequently, conduct a life cycle assessment (LCA) for each construction material to evaluate its environmental impact across the entire life cycle. This comprehensive analysis encompasses the extraction of raw materials, production processes, transportation, use, and disposal. Utilize specialized LCA software or consult existing LCAs and EPDs to quantify environmental impacts in terms of carbon footprint, energy consumption, water usage, and other relevant metrics.

Compare the environmental impacts of different construction materials to identify the most sustainable options. Consider factors such as embodied carbon, embodied energy, water usage, toxicity, recyclability, and durability when evaluating alternatives. Prioritize materials with lower environmental footprints and higher levels of resource efficiency.

Assess the environmental benefits of sourcing construction materials locally to minimize transportation-related emissions. Evaluate the feasibility of sourcing materials from nearby suppliers or utilizing recycled and reclaimed materials to reduce the environmental impact of transportation and support local economies.

Explore certifications from reputable third-party certification schemes for construction materials that attest to their environmental sustainability. Examples include Forest Stewardship Council (FSC) certification for responsibly sourced wood products, Cradle to Cradle certification for products designed for circularity, and ENERGY STAR certification for energy-efficient building materials.

Engage stakeholders, including architects, engineers, contractors, and clients, in the decision-making process regarding construction materials. Seek input from diverse perspectives to ensure that environmental considerations are adequately addressed and balanced with other project requirements.

Document the findings of the environmental assessment process, including the environmental impacts of different construction materials, key considerations, and rationale behind material selection decisions. Maintain clear and transparent records to facilitate communication and decision-making throughout the project lifecycle.

Continuously monitor and reassess the environmental performance of construction materials throughout the project lifecycle. Explore opportunities for optimization, innovation, and improvement to minimize environmental impacts further and enhance the overall sustainability of the project.

By exercising careful consideration in our selection of materials and processes, we can effectively reduce the ecological impact during both the construction phase of a building and its long-term use. Energy plays a significant role in this process, being consumed during the extraction or harvesting of raw materials used in construction products. Subsequently, additional energy is expended to transport these materials to factories, where further energy is utilized to transform them into finished construction products. At each stage of this journey, energy becomes "trapped" or embodied in the final product, a concept often referred to as "cradle to gate," signifying the journey from raw material source to factory gate (Paul McAlister Architects, 2024).

Embodied energy emerges as a pivotal factor in assessing the sustainability of construction materials or products. Sustainable options typically exhibit lower levels of embodied energy, with locally sourced and minimally processed materials presenting lower levels (Paul McAlister Architects, 2024). Conversely, materials with high embodied energy levels are generally considered unsustainable and should be avoided where feasible. Despite the significance of embodied energy, it's essential to acknowledge that the energy-saving potential of a material throughout a building's lifetime often outweighs its embodied energy considerations. Thus, our aim should always be to utilize materials that fulfill all aspects of environmental sustainability.

The selection of materials for various elements of a building is a complex task that demands meticulous consideration. Key principles in assessing the embodied energy of materials underscore the importance of understanding the energy consumed in producing, transporting, installing, maintaining, and disposing of construction materials and products. This becomes increasingly critical as more energy-efficient buildings are designed and constructed, with the proportion of energy attributed to building materials sometimes reaching as high as 50% (Paul McAlister Architects, 2024).

To facilitate a clear distinction between natural and processed building materials, it's beneficial to categorize materials into five main groupings: those of short-term renewable origin, extracted or mined materials, extracted and further processed materials, extracted and highly processed materials, and recycled or reclaimed materials (Paul McAlister Architects, 2024). Key principles for selecting sustainable construction materials emphasize factors such as low energy use, minimal reliance on new resources, utilization of whole unprocessed materials, low embodied energy, potential for reuse, and contribution to a healthy indoor environment.

While no product or material is perfect in every aspect, architects must consider a wide range of information and possibilities when selecting the best solution. This includes sourcing heavyweight materials locally to minimize transportation-related fuel consumption, sourcing lightweight materials globally to mitigate the energy expended during their manufacture, and prioritizing materials with high potential for reuse and recycling to minimize waste and environmental impact. Life cycle assessment (LCA) emerges as a vital tool in calculating the environmental impact of materials or products, considering factors such as energy consumption, greenhouse gas emissions, resource consumption, waste, and pollution. Both embodied energy and LCA must be carefully evaluated when specifying materials and products for a project to ensure the overall sustainability of the scheme (Paul McAlister Architects, 2024).

The Constructor (2024) presents selected data extracted from the Inventory of Carbon and Energy (ICE) compiled by the University of Bath (UK). They provide information on various construction materials, including their energy consumption per kilogram (MJ/kg), carbon emissions per kilogram (kg CO_2/kg), and material density in kilograms per cubic meter (kg/m3).

Aggregate, a commonly used material, has an energy consumption of 0.083 MJ/kg and emits 0.0048 kg CO_2/kg, with a material density of 2240 kg/m3. Concrete, with a mix ratio of 1:1.5:3, requires 1.11 MJ/kg of energy and emits 0.159 kg CO_2/kg, with a material density of 2400 kg/m3. Common bricks have an energy consumption of 3 MJ/kg and emit 0.24 kg CO_2/kg, with a density of 1700 kg/m3.

Other materials listed include concrete blocks, aerated blocks, limestone blocks, marble, cement mortar, and various types of insulation such as cellulose, cork, glass fibre, flax, rockwool, expanded polystyrene, polyurethane, and recycled wool. Each material is characterized by its energy consumption, carbon emissions, and density, providing valuable insights into their environmental impact.

Additionally, the table includes data on roofing materials like mineral fibre roofing tiles, slate, and clay tiles, as well as metals such as steel, stainless steel, and aluminium. Wood-based materials like medium-density fibreboard, plywood, and plasterboard are also listed, along with glass, PVC, vinyl flooring, and ceramic and terrazzo tiles. Cement, aluminium, and steel production stand out among building materials for their substantial consumption of non-renewable energy, emphasizing the need for cautious utilization during construction endeavours (The Constructor, 2024).

Efforts to diminish embodied energy within a building involve employing locally accessible materials, designing for minimal maintenance, enhancing adaptability in usage, and tailoring the building's design to suit prevailing climatic conditions.

Understanding the environmental impact of construction materials is essential for making informed decisions in building design and construction. By examining factors such as energy consumption and carbon emissions, stakeholders can assess the sustainability of materials and strive to minimize their ecological footprint.

Obtaining and Organising Materials, Personnel and Equipment

In the pursuit of developing and implementing sustainable land use strategies within the realm of agricultural and production horticulture enterprises, obtaining and organizing materials, personnel, and equipment is paramount. This process is multifaceted and requires a comprehensive approach that aligns with the overarching goal of enhancing ecological sustainability while optimizing productivity and efficiency.

To begin, it's imperative to assess the specific material requirements necessary for implementing sustainable land use practices. This involves sourcing materials that are environmentally friendly, ethically sourced, and conducive to long-term ecological health. For instance, when selecting fertilizers or pesticides, opting for organic or natural alternatives can mitigate harm to soil biodiversity and aquatic ecosystems. Additionally, utilizing recycled or locally sourced materials wherever possible reduces carbon emissions associated with transportation and supports the local economy.

Simultaneously, attention must be given to the procurement and organization of personnel with the requisite skills and expertise to execute sustainable land management practices effectively. This entails assembling a team of individuals knowledgeable in various aspects of agriculture, horticulture, ecology, and sustainable land management. Moreover, fostering a culture of continuous learning and professional development ensures that personnel remain abreast of the latest advancements and best practices in sustainable agriculture.

Furthermore, acquiring and organizing appropriate equipment is essential for the successful implementation of sustainable land use strategies. This involves investing in tools and machinery that minimize environmental impact, such as precision agriculture technologies that optimize resource utilization and reduce waste. Additionally, prioritizing equipment maintenance and upkeep ensures operational efficiency while extending the lifespan of assets, thereby minimizing resource consumption and waste generation.

In essence, obtaining and organizing materials, personnel, and equipment for sustainable land use practices necessitates a holistic and proactive approach. By prioritizing the use of environmentally friendly materials, assembling a skilled and knowledgeable workforce, and investing in sustainable technologies and equipment, agricultural and horticultural enterprises can effectively develop and implement strategies that promote ecological sustainability while driving productivity and profitability.

To obtain and organize materials, personnel, and equipment to carry out structural improvements aimed at addressing threats to sustainability, a systematic approach is essential.

Firstly, begin by conducting a thorough assessment of the threats to sustainability facing your structure or project. This assessment may encompass factors such as environmental degradation, resource depletion, climate change impacts, or inefficient energy usage.

Next, identify sustainable solutions that can mitigate the identified threats. This may involve exploring renewable materials, energy-efficient technologies, waste reduction strategies, or implementing eco-friendly construction practices.

In terms of material sourcing, prioritize suppliers offering eco-friendly and sustainable building materials, such as recycled materials, certified wood products, or low-impact construction materials. Consider the lifecycle impact of materials, including extraction, production, transportation, installation, and disposal.

Recruit personnel with expertise in sustainable construction practices or provide training to existing staff. Ensure that workers are knowledgeable about sustainable building techniques, green building certifications, and environmental regulations.

Select equipment that minimizes environmental impact and maximizes efficiency. Opt for energy-efficient machinery, low-emission vehicles, and equipment with minimal resource consumption. Consider renting or leasing equipment to reduce the environmental footprint associated with ownership.

Develop a comprehensive logistics plan to coordinate the delivery of materials, deployment of personnel, and scheduling of equipment. Prioritize efficient transportation routes, consolidate deliveries to reduce emissions, and optimize resource utilization to minimize waste.

Create a phased implementation strategy for carrying out structural improvements, prioritizing projects based on urgency, feasibility, and potential environmental impact. Integrate sustainability goals into the project timeline, budget, and performance metrics.

Implement monitoring and evaluation mechanisms to track progress and measure the effectiveness of sustainability initiatives. Collect data on energy consumption, resource usage, waste generation, and environmental performance to identify areas for improvement and demonstrate the benefits of sustainable practices.

Foster a culture of continuous improvement by soliciting feedback from stakeholders, analysing performance data, and adapting strategies based on lessons learned. Stay informed about advancements in sustainable construction technologies and best practices to continually enhance sustainability efforts. Through these steps, you can effectively promote environmental stewardship and long-term resilience in your construction projects.

Selecting and Checking Equipment for Safe and Environmentally Sustainable Operation

Selecting and checking equipment for safe and environmentally sustainable operation is a crucial aspect of developing and implementing sustainable land use strategies, particularly for individuals managing agricultural and production horticulture enterprises. These professionals play a significant role in ensuring that the equipment used on their land not only facilitates efficient operations but also minimizes environmental impact and promotes long-term ecological sustainability.

Firstly, individuals in this role need to possess a deep understanding of the equipment required for various agricultural and horticultural activities. This includes machinery such as tractors, harvesters, irrigation systems, and pesticide applicators. They must be knowledgeable about the specific needs of their operations and the optimal equipment for carrying out tasks effectively while minimizing resource consumption and environmental harm.

In selecting equipment, consideration should be given to factors such as energy efficiency, emissions, resource usage, and maintenance requirements. Opting for equipment with high energy efficiency ratings and low emissions can help reduce the carbon footprint of agricultural and horticultural activities. Additionally, choosing equipment that minimizes water usage and pesticide application can contribute to sustainable land management practices.

Moreover, individuals must ensure that the equipment selected meets safety standards to protect both workers and the environment. This involves conducting thorough inspections and maintenance checks to identify any potential hazards or malfunctions. Regular maintenance not only prolongs the lifespan of the equipment but also prevents leaks, spills, or emissions that could harm soil, water, or air quality.

Furthermore, promoting environmentally sustainable operation involves implementing best practices for equipment usage. This includes proper calibration of machinery to avoid over-application of inputs such as fertilizers or pesticides, which can lead to soil and water contamination. Additionally, employing precision agriculture techniques, such as GPS-guided systems, can help minimize unnecessary passes and optimize resource use.

Individuals in this role should also prioritize training and educating staff on safe and sustainable equipment operation practices. This involves providing instruction on equipment usage, maintenance protocols, and environmental stewardship principles. By fostering a culture of environmental responsibility among workers, these professionals can ensure that sustainable practices are consistently followed.

Finally, ongoing monitoring and evaluation of equipment performance and environmental impact are essential. This enables individuals to identify areas for improvement and implement corrective actions as needed. By continuously refining equipment selection and operation practices, agricultural and horticultural enterprises can contribute to the overall ecological sustainability of the land under production while achieving their production goals.

When it comes to selecting sustainable and environmentally friendly equipment and technology, there are several steps you can take to ensure that your choices align with your goals and contribute to improved ecological sustainability.

Firstly, assess your needs thoroughly. Before making any investments, take the time to understand the current situation of your plant operations and your overarching objectives. Identify the main environmental impacts, legal requirements, and standards that need to be met. Additionally, consider the opportunities and challenges you face, and establish methods for measuring and monitoring performance. By addressing these questions, you can pinpoint your needs and priorities, thus avoiding wasteful spending on unnecessary or ineffective solutions.

Once you have a clear understanding of your needs, you can compare alternatives to find the best options within your budget and specifications. This involves evaluating different types of equipment and technology based on various criteria such as energy efficiency, water consumption, waste generation, emissions, maintenance, lifespan, and reliability. You can also consider the social and economic impacts of your choices, including worker safety, customer satisfaction, and community benefits. Utilizing tools like life cycle assessment (LCA) or cost-benefit analysis (CBA) can assist in evaluating the environmental and financial implications of each alternative.

Seeking expert advice is another valuable step in selecting sustainable equipment and technology. Consult with professionals, including suppliers, manufacturers, consultants, researchers, or industry associations, who possess the knowledge and experience relevant to your field. These experts can provide reliable information, recommendations, and best practices, drawing from their expertise and experience.

Learning from others who have successfully implemented similar solutions can offer valuable insights and feedback, helping you to avoid common pitfalls and find innovative solutions.

Before finalizing your decision and purchasing equipment or technology, it's essential to test and verify its performance and suitability for your plant operations. Conducting trials, demonstrations, or pilot projects allows you to observe how the equipment or technology functions in real conditions, its impact on processes and outputs, and how well it aligns with your expectations and requirements. Additionally, ensure that the equipment or technology meets relevant certifications, warranties, and guarantees, and complies with applicable standards and regulations. Testing and verifying in this manner help to ensure quality, reliability, and safety.

Once you've selected and purchased your equipment or technology, proper implementation and monitoring are necessary to ensure effectiveness and efficiency. Plan and manage the installation, operation, maintenance, and disposal of the equipment or technology, providing adequate training to staff on correct and safe usage. Regularly monitor and measure its performance and impact on environmental and operational objectives, making adjustments or improvements as needed. Utilize tools like environmental management systems (EMS) or key performance indicators (KPIs) to track and report progress and results.

Selecting sustainable and environmentally friendly equipment and technology is an ongoing process that requires careful planning, evaluation, consultation, verification, and improvement. By following these steps, you can make informed and responsible choices that benefit both your plant operations and the environment, contributing to a greener future.

Here are examples of sustainable and environmentally friendly equipment that can be used in construction of structures in agricultural settings:

- Hybrid Excavators combine diesel engines with electric motors or battery packs, resulting in reduced fuel consumption and emissions compared to conventional diesel-powered excavators. These excavators are ideal for construction projects where sustainability and environmental impact are significant considerations.

- Bamboo Scaffolding serves as a sustainable alternative to traditional steel scaffolding. Bamboo is a rapidly renewable resource that offers strength, durability, and versatility. Its use in scaffolding reduces reliance on steel, thereby minimizing the environmental impact associated with steel production and contributing to sustainable construction practices.

- Recycled Aggregates, such as crushed concrete or reclaimed asphalt pavement (RAP), when incorporated into construction projects, help reduce the demand for virgin materials. By utilizing recycled aggregates, construction activities can conserve natural resources, divert waste from landfills, and lower the overall environmental footprint of the project.

- Low-Emission Construction Vehicles, powered by alternative fuels like compressed natural gas (CNG), liquefied natural gas (LNG), or hydrogen fuel cells, offer a sustainable solution to reduce emissions and pollutants. These vehicles emit fewer greenhouse gases and pollutants compared to traditional diesel-powered counterparts, contributing to improved air quality and reduced environmental impact at construction sites.

- Prefabricated Construction Materials, manufactured off-site and assembled on-site, offer several environmental benefits. By reducing construction waste and minimizing transportation-related emissions, prefabricated materials contribute to sustainable construction practices. Additionally, prefabricated construction can expedite project timelines, resulting in reduced energy consumption and lower overall environmental impact.

These examples highlight how the integration of sustainable and environmentally friendly equipment in construction practices can contribute to resource conservation, waste reduction, and environmental stewardship. By prioritizing such equipment, construction projects can align with sustainability goals and promote a greener, more sustainable built environment.

Constructing Structural Works, Revegetating Area and Securing Area from Livestock

Constructing structural works, revegetating areas, and securing areas from livestock are integral components of developing and implementing sustainable land use strategies, particularly for individuals tasked with managing agricultural and production horticulture enterprises. These tasks require a comprehensive understanding of land management practices, environmental conservation principles, and the ability to analyse complex problems to devise effective solutions that enhance ecological sustainability.

To begin, constructing structural works involves the careful planning and implementation of physical infrastructure to support sustainable land use practices. This may include the construction of erosion control structures such as terraces, retaining walls, or sediment ponds to prevent soil erosion and mitigate the loss of valuable topsoil. Additionally, the construction of irrigation systems, water retention structures, or drainage channels can help optimize water use efficiency, improve soil moisture retention, and support healthy plant growth. Individuals overseeing these tasks must possess expertise in engineering principles, construction techniques, and environmental regulations to ensure that structural works are designed and implemented in a manner that minimizes environmental impact and maximizes long-term sustainability.

Revegetating areas involves restoring and enhancing vegetation cover in degraded or disturbed landscapes to improve ecological resilience and biodiversity. This may involve planting native vegetation, establishing cover crops, or implementing agroforestry practices to enhance soil fertility, provide habitat for wildlife, and mitigate the impacts of climate change. Individuals responsible for revegetation efforts must possess knowledge of plant ecology, soil science, and restoration ecology to select appropriate plant species, design planting schemes, and implement management practices that promote successful establishment and growth of vegetation. Furthermore, incorporating community engagement and participatory approaches can foster local stewardship of revegetated areas and ensure the long-term success of restoration efforts.

Securing areas from livestock is essential for protecting sensitive ecosystems, minimizing soil degradation, and maintaining the integrity of agricultural and horticultural production systems. This may involve the construction of fencing, barriers, or exclusion zones to prevent livestock access to ecologically sensitive areas such as riparian zones, wetlands, or native vegetation reserves. Individuals tasked with securing areas from livestock must consider factors such as animal behaviour, grazing patterns, and landscape connectivity to design and implement effective management strategies. Additionally, integrating alternative grazing practices such as rotational grazing, managed intensive grazing, or silvopasture systems can promote sustainable livestock management while minimizing environmental impacts and enhancing ecosystem services.

Constructing structural works, revegetating areas, and securing areas from livestock are essential components of developing and implementing sustainable land use strategies. Individuals managing agricultural and production horticulture enterprises must possess advanced skills and knowledge in land management practices to analyse complex problems, design innovative solutions, and communicate effectively with stakeholders to achieve ecological sustainability and long-term resilience. By prioritizing these tasks, land managers can contribute to the conservation of natural resources, enhancement of ecosystem services, and promotion of sustainable agricultural production systems.

Figure 34: Farm Fencing - Saxlingham Green: New livestock fencing, Hall Farm by Michael Garlick, CC BY-SA 2.0, via Wikimedia Commons.

Securing areas from livestock is crucial for protecting sensitive ecosystems, minimizing soil degradation, and maintaining the integrity of agricultural and horticultural production systems. Several methods can be employed to achieve this goal:

1. Fencing: Erecting physical barriers such as fences or enclosures is one of the most common methods for securing areas from livestock. Fences can be constructed using various materials, including wood, wire mesh, electric wire, or synthetic materials. The design of the fence should consider the size and behaviour of the livestock species being excluded, as well as the terrain and vegetation of the area.

2. Exclusion Zones: Designating specific areas as exclusion zones where livestock are not permitted to graze can help protect sensitive habitats, riparian zones, wetlands, or native vegetation reserves. Exclusion zones can be delineated using fencing, signage, or natural barriers such as water bodies or topographic features.

3. Grazing Management: Implementing rotational grazing, managed intensive grazing, or cell grazing systems can help control livestock access to specific areas while promoting sustainable grazing practices. By strategically rotating livestock between different paddocks or pastures, land managers can prevent overgrazing, allow vegetation to recover, and maintain soil health.

4. Herding and Stockmanship: Utilizing trained livestock guardians or employing skilled stockmen to manage and control livestock movement can help prevent unauthorized access to restricted areas. Effective herding techniques and stockmanship practices can ensure that livestock are guided away from sensitive habitats or designated exclusion zones.

5. Livestock Deterrents: Installing deterrents such as visual barriers, sound devices, or odor repellents can discourage livestock from entering restricted areas. For example, scarecrows, reflective tape, or predator decoys can be used to visually deter livestock, while sound-emitting devices or motion-activated alarms can startle animals and discourage them from approaching certain areas.

6. Water Provision: Providing alternative water sources such as troughs, tanks, or water points in areas where livestock are permitted to graze can help prevent them from congregating around sensitive habitats or water bodies. By strategically locating water

sources, land managers can influence livestock movement patterns and reduce the risk of overgrazing or trampling in ecologically sensitive areas.

7. Community Engagement: Engaging with local communities, stakeholders, and land users to raise awareness about the importance of securing areas from livestock and promoting responsible grazing practices can foster support and cooperation. Collaborative approaches involving participatory decision-making, stakeholder consultation, and shared management responsibilities can help achieve consensus and ensure the long-term success of livestock exclusion efforts.

By employing a combination of these methods, land managers can effectively secure areas from livestock while promoting sustainable land use practices, protecting natural resources, and maintaining ecosystem health. It's essential to consider site-specific factors such as terrain, vegetation, climate, and livestock behaviour when selecting and implementing appropriate livestock exclusion measures. Additionally, regular monitoring and adaptive management are crucial to assess the effectiveness of measures and make necessary adjustments to achieve desired conservation outcomes.

Chapter Four

Treating Areas of Land Degradation

Addressing areas of land degradation requires a systematic approach that encompasses various measures aimed at restoring ecological balance and preventing further deterioration. To initiate this process, it is essential to assess the classification of land classes meticulously and ensure that the alignment of fences and soil conservation works is consistent with these classifications. By aligning these structures appropriately, stakeholders can effectively manage soil erosion, water runoff, and other forms of degradation, thereby promoting the sustainable use of land resources.

Moreover, protecting and maintaining contour banks is crucial to mitigating degradation and preserving soil integrity. Contour banks help to control soil erosion by reducing water velocity and directing runoff along the land's contours, thereby minimizing the risk of soil loss. Regular inspection and maintenance of these structures are essential to ensure their effectiveness in preventing erosion and maintaining soil stability over time.

Furthermore, the construction and maintenance of water-carrying structures play a vital role in managing water resources and preventing land degradation. Properly designed and maintained structures, such as drainage channels, culverts, and bunds, help to regulate water flow, prevent soil erosion, and mitigate the risk of flooding. Regular monitoring and upkeep of these structures are necessary to ensure their functionality and effectiveness in safeguarding against degradation.

Establishing shelter belts for crop and stock protection is another critical measure in combating land degradation. Shelter belts, composed of trees or shrubs planted along field boundaries or exposed areas, serve as windbreaks and provide protection against soil erosion, wind damage, and livestock trampling. By strategically planting shelter belts, stakeholders can enhance soil stability, conserve moisture, and create microclimates conducive to plant growth.

Additionally, reviewing and amending soil cultivation and planting practices is essential to prevent erosion and promote soil health. Implementing conservation tillage techniques, such as no-till or reduced tillage, helps to minimize soil disturbance and maintain soil structure, thereby reducing erosion risk. Furthermore, adopting practices such as cover cropping and crop rotation enhances soil fertility, reduces erosion, and improves overall land productivity.

Planning grazing strategies to avoid localized soil degradation is also crucial in sustainable land management. Rotational grazing systems, proper stocking rates, and strategic pasture management help to prevent overgrazing, soil compaction, and vegetation degradation. By implementing grazing strategies that promote sustainable land use practices, stakeholders can maintain soil health, enhance biodiversity, and preserve ecosystem integrity.

Finally, it is imperative to ensure that all works and plans comply with environmental protection legislation, regulations, and codes of practice. This involves conducting thorough assessments to identify potential environmental impacts, obtaining necessary permits and approvals, and implementing measures to mitigate adverse effects on the environment. Compliance with regulatory requirements helps to uphold environmental standards, protect natural resources, and ensure the long-term sustainability of land management practices. Through the diligent implementation of these measures, stakeholders can effectively address areas of land degradation, promote sustainable land use practices, and safeguard the health and resilience of ecosystems for future generations.

Assessing Land Classes and Aligning Fences and Soil Conservation Works Consistent with Class

In the context of developing and implementing sustainable land use strategies, assessing land classes and aligning fences and soil conservation works consistent with these classes is a critical task. It requires a comprehensive understanding of land management principles, ecological processes, and soil conservation techniques to ensure that land is utilized in a manner that maximizes productivity while minimizing environmental impact.

To begin, assessing land classes involves categorizing land parcels based on their inherent characteristics, such as soil type, topography, vegetation cover, and ecological sensitivity. This classification helps identify areas with distinct ecological functions, such as prime agricultural land, sensitive habitats, water catchment areas, or erosion-prone slopes. Individuals responsible for land assessment must possess advanced skills in land surveying, remote sensing, GIS technology, and ecological modelling to accurately delineate land classes and assess their ecological significance.

Once land classes have been identified, the next step is aligning fences and soil conservation works consistent with these classes. This involves designing and implementing physical barriers, erosion control structures, or land management practices tailored to the specific needs and characteristics of each land class. For example:

- Prime Agricultural Land: Fences may be aligned to delineate field boundaries and manage livestock access, while soil conservation works such as contour ploughing, terracing, or cover cropping may be implemented to minimize soil erosion and maintain soil fertility.

- Sensitive Habitats: Fences can be strategically positioned to exclude livestock from sensitive habitats such as riparian zones, wetlands, or native vegetation reserves, while soil conservation measures such as buffer strips or vegetative buffers may be established to protect water quality and wildlife habitat.

- Erosion-Prone Slopes: Soil conservation works such as erosion control blankets, check dams, or vegetative stabilization techniques may be implemented to mitigate soil erosion and slope instability, while fences may be aligned to prevent livestock access and reduce trampling damage.

Individuals tasked with aligning fences and soil conservation works must possess expertise in erosion control techniques, hydrology, soil science, and ecological restoration to design effective solutions that address site-specific conditions and achieve desired conservation outcomes. Additionally, collaboration with landowners, stakeholders, and environmental agencies is essential to ensure that land use practices align with regulatory requirements and conservation objectives.

Furthermore, communication and education are vital aspects of implementing sustainable land use strategies. Land managers must effectively communicate the rationale behind fence alignment and soil conservation works to stakeholders, land users, and the broader community. This may involve conducting outreach activities, providing technical assistance, and fostering collaboration to build consensus and support for sustainable land management practices.

Assessing land classes and aligning fences and soil conservation works consistent with these classes are fundamental steps in developing and implementing sustainable land use strategies. By integrating ecological principles, soil conservation techniques, and stakeholder engagement, land managers can optimize land productivity, enhance ecosystem resilience, and promote long-term sustainability in agricultural and horticultural enterprises.

Protecting and Maintaining Contour Banks to Avoid Degradation

Protecting and maintaining contour banks to avoid degradation is a crucial aspect of developing and implementing sustainable land use strategies, especially for individuals managing agricultural and production horticulture enterprises. Contour banks are earthen embankments constructed along the contour lines of sloping land to reduce soil erosion and retain water runoff. Proper management of contour banks is essential for maintaining soil fertility, preventing land degradation, and promoting ecological sustainability.

Firstly, individuals tasked with protecting and maintaining contour banks must possess advanced skills in land management practices, erosion control techniques, and hydrological principles. They need to understand the factors contributing to soil erosion, such as slope steepness, soil type, rainfall intensity, and land use practices. By analysing these factors, land managers can assess the vulnerability of contour banks to degradation and implement appropriate management measures.

To protect contour banks from degradation, several strategies can be employed:

1. Vegetative Cover: Maintaining vegetative cover on contour banks helps stabilize the soil, reduce erosion, and enhance water infiltration. Planting grasses, legumes, or erosion-resistant vegetation along the banks helps bind soil particles together, preventing them from being washed away by runoff. Additionally, establishing vegetation with deep root systems can improve soil structure and stability.

2. Maintenance of Bank Structure: Regular inspection and maintenance of contour banks are essential to ensure their structural integrity and effectiveness in controlling erosion. This includes repairing erosion damage, reinforcing bank slopes, and clearing vegetation or debris that may obstruct water flow. Proper maintenance helps prevent bank failure and ensures continuous erosion control.

3. Water Management: Managing water runoff and drainage around contour banks is crucial for minimizing erosion and maintaining soil moisture levels. Techniques such as installing diversion channels, water bars, or drop structures can help redirect runoff away from vulnerable areas and prevent concentrated flow that can lead to bank erosion. Additionally, implementing water conservation practices such as mulching or rainwater harvesting reduces runoff volume and protects contour banks from erosion.

4. Soil Conservation Practices: Implementing soil conservation practices such as conservation tillage, cover cropping, and agroforestry helps improve soil health and reduce erosion risk on sloping land. These practices promote soil organic matter accumulation, enhance soil structure, and increase water infiltration, thus reducing the erosive forces acting on contour banks.

Furthermore, effective communication and education are essential for promoting the importance of protecting and maintaining contour banks among stakeholders, land users, and the broader community. Land managers must convey the benefits of erosion control measures, provide technical guidance on proper management practices, and engage stakeholders in collaborative efforts to conserve soil and water resources.

Protecting and maintaining contour banks to avoid degradation requires a holistic approach that integrates ecological principles, erosion control techniques, and stakeholder engagement. By implementing effective management strategies and promoting sustainable land use practices, individuals can safeguard soil health, preserve water quality, and promote ecological sustainability in agricultural and horticultural enterprises.

Figure 35: Contour banks near Adelaide at O'Halloran Hill, South Australia. CSIRO, CC BY 3.0, via Wiki-media Commons.

Maintaining the integrity of banks involves employing earth-moving equipment to relocate sediment and soil from the channel onto the contour bank, thereby enhancing its elevation. This process is most effective when the soil is dry and has undergone previous cultivation or ripping. In expansive contour-banked paddocks, specialized tractor-drawn grader blades are commonly utilized to craft a suitable channel shape. Conversely, on smaller paddocks, machinery such as dozers or motorized graders is often employed. Dozers prove particularly beneficial for addressing significant contour bank breaches or gullies that require repair, while scrapers play a role in removing sediment from the channel, subsequently utilizing it to level rills and gullies within the contour bay above the bank.

In grazing lands, numerous paddocks in densely populated regions were previously cultivated and contour banked but have since transitioned to permanent pastures due to diminishing returns from cropping endeavors. While contour banks remain a necessity in sloping and cropped paddocks, a well-managed pasture typically does not necessitate them. In such cases, while upgrading is a feasible option, allowing existing banks to gradually diminish may be acceptable, provided the pasture is adequately maintained. However, in instances where contour banks in grazing lands fall below specifications and induce issues through breakages or overflow, upgrading or complete removal is warranted.

Key considerations for maintaining contour banks include ensuring substantial surface coverage within contour bays to mitigate silt accumulation in bank channels, regularly inspecting bank outlets and areas where banks intersect old gully lines, rectifying rills, gully lines, and old fence lines, promptly addressing any broken contour banks, and eliminating silt blockages from channels following a runoff event. Optimal maintenance should occur during dry spells when the land lies fallow.

In terms of erosion control in cropped lands, implementing conservation cropping practices such as minimum and zero tillage is essential to maintain soil cover. During the fallow period, modern tillage implements are employed to eradicate weeds without burying stubble, and herbicides are utilized to minimize tillage frequency. Additionally, managing runoff concentration through structural measures like contour banks in upland regions or strip cropping on floodplains is crucial. Meticulous planning is imperative for runoff systems, including coordination of flow between properties and across roads and railways. Collaboration with neighbours and seeking professional guidance is encouraged, especially in situations where runoff water can impact neighbouring properties or infrastructure. Contour banks play a vital role in trapping approximately 80% of soil lost due to insufficient cover, channelling runoff at low speeds into grassed waterways. Maintaining good surface cover between contour banks and in waterways ensures their stability and significantly reduces soil deposition in watercourses, while strip cropping on floodplains serves to disperse flood flows, preventing concentration.

Constructing and Maintaining Water Carrying Structures

Constructing and maintaining water carrying structures is a critical aspect of developing and implementing sustainable land use strategies, particularly for individuals managing agricultural and production horticulture enterprises. These structures play a vital role in managing water resources efficiently, preventing soil erosion, and promoting ecological sustainability.

Water carrying structures are man-made features designed to manage the flow of water across landscapes, particularly in agricultural and horticultural settings. These structures are essential for controlling erosion, preventing waterlogging, and optimizing water distribution for crops. Some common types of water carrying structures include:

1. Terraces: Terraces are horizontal or gently sloping platforms constructed along the contour lines of hillsides to reduce soil erosion and retain water. They help slow down the flow of water, allowing it to infiltrate into the soil rather than causing runoff.

2. Contour Banks: Contour banks are ridges of soil constructed along the contour lines of slopes to intercept and redirect surface runoff. They help prevent soil erosion by slowing down the flow of water and promoting infiltration. Contour banks are typically vegetated to enhance their stability and erosion control capabilities.

3. Diversion Channels: Diversion channels are constructed to redirect the flow of water away from sensitive areas or to distribute water evenly across fields. They are often used to channel excess runoff away from fields during heavy rainfall events or to divert water from higher elevation areas to lower-lying fields.

4. Drainage Systems: Drainage systems, including subsurface tile drains and open ditches, are designed to remove excess water from the soil and prevent waterlogging. They help improve soil structure, reduce compaction, and enhance crop productivity by maintaining optimal soil moisture levels.

5. Irrigation Canals: Irrigation canals are artificial waterways designed to transport water from a water source, such as rivers or reservoirs, to fields for crop irrigation. They can be constructed using various materials, including concrete, earth, or lined channels, and may be equipped with gates, weirs, or control structures to regulate water flow.

6. Retention Ponds: Retention ponds, also known as detention basins or stormwater ponds, are designed to capture and temporarily store excess runoff from rainfall events. They help reduce the risk of flooding downstream by slowing down the flow of water and allowing sediment to settle out before the water is discharged.

These water carrying structures are essential components of sustainable land management practices, helping to conserve soil, prevent erosion, and optimize water resources for agricultural and horticultural production. Proper design, construction, and maintenance of these structures are critical for their effectiveness in managing water flow and promoting ecological sustainability.

Figure 36: Irrigation ditch or canal. USDA NRCS Montana, Public domain, via Wikimedia Commons.

Individuals responsible for constructing and maintaining water carrying structures must possess advanced skills and knowledge in land management practices, hydrology, and engineering principles. They need to understand the dynamics of water flow, soil erosion processes, and the impact of water management practices on ecosystem health. By analysing these factors, they can design and implement effective water carrying structures tailored to the specific needs of their land and crops.

Constructing water carrying structures involves several steps, including site assessment, design, construction, and ongoing maintenance. Site assessment involves evaluating the topography, soil types, and existing drainage patterns to identify areas prone to waterlogging, erosion, or runoff. Based on this assessment, individuals can design appropriate structures such as terraces, contour banks, diversion channels, or underground drainage systems to manage water flow effectively.

During the construction phase, attention to detail and adherence to engineering standards are essential to ensure the structural integrity and longevity of water carrying structures. Proper installation techniques, quality materials, and appropriate sizing of structures are crucial to withstand the forces of water flow and weather conditions.

Once constructed, ongoing maintenance is necessary to preserve the functionality and effectiveness of water carrying structures. Regular inspections, cleaning of channels, repairing erosion damage, and vegetation management are essential tasks to prevent blockages, sediment buildup, or structural failures. Additionally, monitoring water quality and flow rates can help identify potential issues and implement corrective measures promptly.

Incorporating sustainable practices into the construction and maintenance of water carrying structures is essential for minimizing environmental impact and promoting long-term resilience. This may include using locally sourced and environmentally friendly materials, implementing erosion control measures, and integrating habitat enhancement features to support biodiversity.

Furthermore, effective communication and collaboration with stakeholders, such as neighbouring landowners, environmental agencies, and community groups, are crucial for ensuring the success and sustainability of water management initiatives. By engaging stakeholders in the planning, implementation, and monitoring processes, individuals can foster greater support and cooperation towards achieving shared water management goals.

Constructing and maintaining water carrying structures requires a comprehensive understanding of land management principles, hydrological processes, and engineering techniques. By incorporating sustainable practices and engaging stakeholders, individuals can develop effective water management strategies that enhance ecological sustainability, mitigate risks, and optimize resource utilization on agricultural and horticultural enterprises.

Establishing Shelter Belts for Crop and Stock Protection

Establishing shelter belts for crop and stock protection is another important component of developing sustainable land use strategies, particularly for individuals managing agricultural and production horticulture enterprises. Shelter belts, also known as windbreaks or hedgerows, are linear plantings of trees, shrubs, or other vegetation strategically positioned to mitigate the impacts of wind, reduce soil erosion, provide habitat for wildlife, and enhance microclimate conditions within agricultural landscapes.

To successfully establish shelter belts, individuals must possess a comprehensive understanding of land management practices, ecological principles, and the specific needs of their crops and livestock. Analysing information related to prevailing wind patterns, soil types, topography, and existing vegetation cover is essential for selecting suitable species, determining optimal placement, and designing effective shelter belt configurations.

The selection of plant species for shelter belts should consider their adaptability to local environmental conditions, growth characteristics, resilience to pests and diseases, and potential benefits for biodiversity and ecosystem services. Native species that are well-suited to the region and provide additional ecological benefits, such as food and habitat for pollinators and wildlife, are often preferred for inclusion in shelter belt plantings.

When establishing shelter belts, individuals must exercise judgment in site preparation, planting techniques, and ongoing management practices to ensure the long-term success and effectiveness of the windbreaks. Proper site preparation may involve soil testing, weed control, and appropriate land grading to optimize growing conditions for the selected plant species.

Planting techniques such as spacing, row orientation, and tree density are critical considerations for maximizing the windbreak's efficacy in providing protection for crops and livestock while minimizing competition for resources and ensuring sufficient light penetration. Utilizing diverse species compositions and mixed planting arrangements can enhance the structural integrity and ecological functionality of shelter belts.

Figure 37: Shelter belt, Broughton. Shelter belt, Broughton by Richard Webb, CC BY-SA 2.0, via Wikimedia Commons.

Ongoing maintenance of shelter belts is essential for promoting healthy growth, structural stability, and ecosystem resilience. This may include irrigation during establishment, weed control, pruning to remove dead or diseased branches, and periodic assessments of windbreak performance and condition. Additionally, monitoring the impact of shelter belts on microclimate conditions, soil erosion rates, and biodiversity can inform adaptive management strategies and optimize the benefits derived from these landscape features.

Effective communication and collaboration with stakeholders, including neighbouring landowners, conservation organizations, and agricultural extension services, can facilitate the development and implementation of shelter belt projects. By engaging stakeholders in the planning, implementation, and monitoring processes, individuals can leverage collective expertise, resources, and support to enhance the ecological sustainability and resilience of agricultural landscapes through the establishment of shelter belts.

Shelterbelts function as vegetative barriers strategically designed to mitigate wind speed and create sheltered areas on both the leeward and windward sides (Agriculture Victoria, 2017). As wind encounters the shelterbelt, it diverts around the ends, passes through, or flows over the top. This interaction generates variations in air pressure, with higher pressure on the windward side and lower pressure on the leeward side, driving the shelter effect and influencing wind speed reduction and turbulence levels (Agriculture Victoria, 2017).

The effectiveness of a shelterbelt in reducing wind speed depends on its structural characteristics, including height, density, number of rows, species composition, and spacing between trees or shrubs. Altering these features can modify the shelterbelt's structure and its capacity to attenuate wind speed and turbulence. For instance, denser shelterbelts create greater air pressure differentials, resulting in more significant wind speed reductions (Agriculture Victoria, 2017).

When designing a shelterbelt, it is essential to define its purpose, whether it is to protect crops, livestock, or mitigate erosion. Factors such as height, length, density, location, number of rows, and species selection play crucial roles in achieving the desired outcomes (Agriculture Victoria, 2017). Maximizing shelterbelt height and length enhances its effectiveness in creating protected areas, while ensuring uninterrupted length and minimizing gaps optimize windbreak performance.

Density, or the proportion of solid material within the shelterbelt, influences the level of shelter provided and wind turbulence. Higher density shelterbelts offer more shelter over a shorter distance, while lower density ones allow more air to pass through, extending the downwind protected area. Achieving a medium density of around 40-60% cover is often recommended for balanced effectiveness (Agriculture Victoria, 2017).

The arrangement of rows within a shelterbelt affects its resilience to gaps and non-uniform growth. Multiple rows are generally more effective than single rows, with taller species in the centre and lower-growing species on the sides creating a uniform density profile (Agriculture Victoria, 2017). Strategic row design, incorporating fast-growing species for quicker benefits, can optimize shelterbelt performance.

Careful consideration of plant location and spacing is essential to ensure optimal growth and density. Proper spacing between rows and individual plants allows for unrestricted growth and prevents livestock interference. Additionally, selecting species adapted to the local climate and soil conditions, including native species, promotes establishment success and enhances ecological value.

The orientation of shelterbelts relative to prevailing wind directions influences their effectiveness in providing protection. Shelterbelts should be placed perpendicular to prevailing winds to maximize wind speed reduction (Agriculture Victoria, 2017). Orientation also impacts sunlight exposure, livestock movement, and erosion control, with north-south and east-west orientations offering comprehensive benefits.

Site preparation is crucial for shelterbelt establishment, involving weed control, soil preparation, pest animal management, and fencing. Thorough weed removal and deep ripping promote favourable growing conditions for tubestock plantings, while pest animal control measures protect young plants from damage. Livestock-proof fencing is essential to prevent grazing damage during establishment (Agriculture Victoria, 2017).

Managing existing shelterbelts involves a series of proactive measures aimed at optimizing their ecological function and structural integrity. One crucial aspect is maintaining fences to regulate grazing pressure, as reduced grazing levels facilitate increased vegetation cover, accumulation of leaf litter, reduction of bare ground, and enhancement of biodiversity, including various bird and reptile species.

Regular addition of new plants is essential to replace those that may have perished during the early stages of establishment. This practice ensures the maintenance of structural diversity within the shelterbelt, promoting a balanced mix of trees, shrubs, and groundcovers to sustain habitat complexity and ecological resilience.

Utilizing fallen timber within shelterbelts contributes significantly to habitat enhancement. Rather than disposing of fallen logs or branches elsewhere on the farm, incorporating them into shelterbelts provides valuable habitat for wildlife, such as the yellow-footed antechinus, enriching the biodiversity of the area.

Effective management of pests and weeds is imperative within shelterbelts, presenting an opportunity to concentrate pest control efforts and mitigate potential ecological disruptions caused by invasive species. By addressing pest and weed infestations, shelterbelts can maintain their ecological integrity and continue to provide valuable ecosystem services.

Figure 38: View along a shelter belt. View along a shelter belt by Fernweh, CC BY-SA 2.0, via Wikimedia Commons.

Additionally, rejuvenating older shelterbelts is crucial to restoring their effectiveness and ecological functionality. Over time, older shelterbelts may deteriorate due to factors such as fence damage, stock access, or inadequate structural design. Rejuvenation efforts may involve repairing fences, enhancing understory vegetation, or diversifying tree species composition to improve windbreak efficiency and maximize ecological benefits. Despite their age, rejuvenated shelterbelts have the potential to become vital wildlife habitats and contribute substantially to farm productivity and sustainability with proper management interventions.

Reviewing and Amending Soil Cultivation and Planting Practices to Prevent Erosion

To develop and implement sustainable land use strategies that lead to improved ecological sustainability of land under production, it is crucial to review and amend soil cultivation and planting practices to prevent erosion. This is essential for individuals managing agricultural and production horticulture enterprises, as it requires a deep knowledge of land management practices and the ability to analyse information and exercise judgment to complete a range of advanced skilled activities (Edwards et al., 2019).

Soil erosion can have detrimental effects on plant performance and increase greenhouse gas emitting microbes, impacting the ecological sustainability of land under production (Edwards et al., 2019). Agricultural practices have been shown to affect microbial biodiversity in arable soils, with a reduction of dominant bacterial operational taxonomic units in cultivated soils (Wolińska et al., 2017). Additionally, extensive cultivation has been linked to a significant increase in the mean annual soil erosion rate, emphasizing the need for effective erosion prevention strategies (Gashaw et al., 2020).

Furthermore, the impact of land management practices on soil erosion has been evaluated, indicating that the bacterial community structure and microbial activity in traditional organic farming systems are influenced by soil moisture conditions (Moreno-Espíndola et al., 2018). Additionally, the fertility status of cultivated land has been found to be lower in terms of organic carbon and total nitrogen compared to forest and grazing lands, highlighting the need for soil fertility management in cultivated areas (Seyoum, 2016).

Long-term grazing and cultivation have differential effects on soil microbial communities, emphasizing the importance of understanding the long-term impacts of land use practices on soil sustainability (He et al., 2017). Shifting cultivation has been a topic of debate in scientific and institutional communities, underscoring the need for consensus on sustainable land use practices (Filho et al., 2013). Mapping cultivated organic soils for targeting greenhouse gas mitigation has been proposed as a strategy to reduce emissions and increase the sustainability of agriculture (Kekkonen et al., 2019).

Application of organic matter to cultivated soil has been identified as an important practice for maintaining adequate soil organic matter, which is crucial for sustainable land use (Chang et al., 2014). Changes in land-use management practices have been shown to impact soil physicochemical properties, necessitating the implementation of appropriate land conservation strategies based on land-use structure and slope variation (Assefa et al., 2020). Severe soil erosion caused by agricultural expansion and poor management has worsened soil nutrient depletion in cultivated outfields, highlighting the urgency of effective soil management practices (Bufebo & Elias, 2020).

As such, as a practitioner responsible for developing and implementing sustainable land use strategies, reviewing and amending soil cultivation and planting practices to prevent erosion is a critical aspect of your role. Erosion poses significant threats to soil health, productivity, and environmental sustainability, making it imperative to adopt practices that minimize erosion risk while promoting sustainable land management.

The first step in reviewing soil cultivation and planting practices involves conducting a comprehensive assessment of current methods and their impact on erosion. This assessment should encompass various factors such as soil type, slope gradient, climate conditions, vegetation cover, and land use patterns. By analysing these factors, you can identify areas of vulnerability to erosion and prioritize them for intervention.

Once the assessment is complete, the next step is to identify and implement erosion control measures tailored to specific soil cultivation and planting practices. This may involve adopting conservation tillage techniques such as minimum tillage or no-till farming, which help to preserve soil structure, minimize soil disturbance, and reduce erosion risk. Conservation tillage practices also promote soil organic matter retention, moisture conservation, and biodiversity enhancement, contributing to long-term soil health and resilience.

In addition to conservation tillage, implementing soil conservation practices such as contour farming, terracing, and strip cropping can help mitigate erosion by reducing the speed and volume of surface water runoff. Contour farming, for example, involves planting crops along the contour lines of the land, effectively creating natural barriers that intercept and slow down runoff, minimizing soil erosion. Terracing, on the other hand, involves constructing level or gently sloping platforms on steep terrain to reduce the length and gradient of slopes, thereby preventing erosion and promoting water infiltration.

Furthermore, integrating cover crops into cropping systems can provide valuable erosion control benefits by protecting soil surfaces from erosive forces, improving soil structure, and enhancing organic matter content. Cover crops help to maintain soil cover throughout the year, reducing the exposure of bare soil to wind and water erosion during fallow periods and between cash crop rotations.

As part of your role in developing and implementing sustainable land use strategies, ongoing monitoring and evaluation of soil cultivation and planting practices are essential. Regular assessment of erosion risk, soil health indicators, and vegetation cover will allow for adjustments and refinements to management practices as needed, ensuring continued progress towards erosion prevention and sustainable land management objectives.

Ultimately, by reviewing and amending soil cultivation and planting practices to prevent erosion, you play a pivotal role in safeguarding soil resources, preserving ecosystem integrity, and promoting the long-term viability of agricultural and horticultural enterprises in a sustainable manner.

Planning Grazing Strategies to Avoid Localised Soil Degradation

Effective grazing management plays a crucial role in preserving soil health, preventing erosion, and promoting the long-term productivity of grazing lands. The first step in planning grazing strategies is to conduct a thorough assessment of the landscape, considering factors such as soil type, topography, vegetation composition, and climatic conditions. By understanding the unique characteristics of the land, you can identify areas prone to soil degradation and tailor grazing practices accordingly.

One key strategy to avoid localized soil degradation is rotational grazing, which involves dividing grazing areas into smaller paddocks or pastures and rotating livestock through these areas at regular intervals. Rotational grazing allows for rest periods between grazing sessions, enabling vegetation to recover, soil to regenerate, and soil compaction to be minimized. By preventing overgrazing in specific areas and promoting uniform utilization of pasture resources, rotational grazing helps to mitigate soil degradation and maintain ecosystem balance.

Furthermore, implementing stocking rate management is essential for preventing soil degradation. Stocking rates should be adjusted based on factors such as forage availability, carrying capacity, and seasonal variations in pasture growth. By matching stocking rates to carrying capacity, you can prevent overgrazing, reduce soil compaction, and minimize erosion risk, thereby safeguarding soil health and productivity.

Integrating riparian management practices into grazing strategies is also critical for preventing soil degradation in riparian areas. Riparian zones are ecologically sensitive areas that provide essential habitat, water quality regulation, and erosion control functions. By implementing riparian fencing, establishing buffer zones, and managing livestock access to waterways, you can protect riparian vegetation, stabilize stream banks, and prevent soil erosion, thereby preserving water quality and ecosystem integrity.

Additionally, adopting holistic grazing management approaches, such as adaptive multi-paddock grazing or planned grazing systems, can further enhance soil health and mitigate degradation risks. These approaches prioritize regenerative practices, emphasize soil carbon sequestration, and promote biodiversity, contributing to ecosystem resilience and sustainability.

Continuous monitoring and evaluation of grazing impacts on soil health are essential components of effective grazing management. Regular soil testing, vegetation assessments, and monitoring of erosion indicators allow for timely adjustments to grazing strategies based on evolving environmental conditions and management objectives.

Communication and collaboration with stakeholders, including landowners, livestock producers, conservation agencies, and community members, are also essential for successful implementation of grazing strategies to avoid localized soil degradation. By engaging stakeholders in decision-making processes, sharing knowledge and best practices, and fostering partnerships, you can create a supportive environment for sustainable land management practices and promote stewardship of natural resources.

Planning grazing strategies to avoid localized soil degradation requires a holistic approach that integrates ecological principles, adaptive management practices, and stakeholder engagement. By implementing rotational grazing, stocking rate management, riparian management, and holistic grazing approaches, while emphasizing continuous monitoring and collaboration, you can effectively preserve soil health, promote ecosystem resilience, and ensure the long-term sustainability of agricultural and horticultural enterprises.

A comprehensive grazing management plan involves several key components aimed at optimizing pasture utilization and maintaining sustainable land use practices. Initially, it requires conducting a thorough analysis of the current environmental conditions and resources, encompassing factors such as climate, pasture quality, soil health, water availability, human resources, financial considerations, and the size of the livestock population. This analysis also involves calculating the long-term carrying capacity of the land.

The following provides an example of a Grazing Management Plan.

Sample Grazing Management Plan

1. Current Situation Analysis:

- *Climate: Assess the local climate patterns, including temperature fluctuations, precipitation levels, and seasonal variations. Consider how these factors impact pasture growth and livestock management practices.*

- *Pasture Quality: Conduct a comprehensive evaluation of pasture composition, species diversity, and nutritional value. Identify any areas of overgrazing or underutilization.*

- *Soil Condition: Evaluate soil health and fertility, including factors such as pH levels, nutrient content, and compaction. Determine soil erosion risks and erosion control measures.*

- *Water Availability: Assess water sources, availability, and quality for livestock consumption. Consider strategies for water conservation and efficient distribution.*

- *Human Resources: Evaluate the skill levels and availability of labour for implementing grazing management practices. Determine training needs and responsibilities for personnel.*

- *Finances: Review the budget allocated for grazing management activities, including expenses related to pasture maintenance, infrastructure development, and supplementary feeding.*

- *Livestock Population: Inventory the current livestock population, including species, breeds, ages, and reproductive status. Estimate carrying capacity based on current stocking rates and forage availability.*

2. Strategies for Optimization:

- *Rainfall Utilization: Implement water management strategies such as contour ploughing, water harvesting, and irrigation to maximize rainfall utilization for pasture growth.*

- *Pasture Utilization: Develop rotational grazing plans to ensure even utilization of pasture resources while allowing for adequate rest periods to promote regrowth and soil health.*

3. Evaluation of Grazing Strategies:

- *Continuous Grazing: Assess the impact of continuous grazing on livestock productivity, pasture condition, and soil erosion rates. Consider alternatives to mitigate negative effects.*

- *Rotational Grazing: Evaluate the effectiveness of rotational grazing in improving pasture productivity, enhancing biodiversity, and reducing soil erosion. Monitor changes in livestock performance and land health.*

- *Resting Paddocks: Determine the benefits of resting paddocks for pasture regeneration, seed dispersal, and weed control. Measure changes in pasture quality and species composition over time.*

- *Time-Controlled Grazing: Analyse the outcomes of time-controlled grazing on forage utilization, livestock performance, and pasture recovery. Compare results with other grazing strategies.*

4. Monitoring and Recording Methods:

- *Pasture Assessments: Conduct regular pasture assessments to monitor changes in vegetation composition, ground cover, and soil health indicators. Use tools such as transect surveys, biomass sampling, and soil tests.*

- *Livestock Performance: Track livestock performance metrics such as weight gain, body condition scores, and reproductive rates. Use individual animal records and herd management software for data collection.*

- *Environmental Monitoring: Monitor environmental parameters such as water quality, erosion rates, and biodiversity indicators. Utilize remote sensing technology, weather stations, and ecological surveys.*

- *Financial Analysis: Maintain detailed records of grazing management expenses and income generated from livestock production. Calculate key financial indicators such as cost per head, return on investment, and net profit.*

By implementing these elements of the grazing management plan, we aim to optimize pasture utilization, enhance livestock productivity, improve land health, and achieve sustainable financial outcomes. Regular monitoring and evaluation will enable us to adapt our strategies to changing conditions and continuously improve our grazing management practices.

Subsequently, the plan should identify strategies to maximize rainfall and pasture utilization, thereby enhancing overall productivity. This includes evaluating various grazing strategies such as continuous grazing, rotational grazing, resting paddocks, and time-controlled (cell) grazing. Each strategy has unique implications for livestock productivity, pasture condition, land health, and financial outcomes, necessitating careful consideration and evaluation.

Total grazing pressure, which represents the ratio of pasture demand to available pasture supply, is a critical metric that should be monitored and managed effectively. Factors influencing total grazing pressure include rainfall patterns, livestock breeding rates, stock sales, and the presence of native or feral animals. Strategies such as rotational grazing, stock sales, and agisting livestock can help maintain total grazing pressure at reasonable levels.

Implementing a forage budget is another essential aspect of grazing management, enabling the matching of stock numbers with available pasture resources to maintain or improve land condition. This involves assessing pasture availability at the end of the wet season and adjusting stock numbers accordingly to meet residual pasture yield and ground cover targets tailored to specific land types and average rainfall.

Successful grazing management often necessitates employing a flexible approach known as tactical grazing, which combines various grazing methods such as set stocking and rotational grazing to meet diverse animal and pasture objectives over different time frames. This adaptive approach balances feed supply with livestock demands, thereby enhancing overall productivity and sustainability.

Furthermore, grazing management plays a crucial role in influencing the productivity and persistence of pasture species. Effective grazing practices can target undesirable species to prevent seed set or promote desirable species to encourage seed production, thereby optimizing pasture composition and productivity.

Supplementary feeding can complement grazing strategies by addressing seasonal fluctuations in pasture quality or nutritional deficiencies in livestock diets. However, the cost-effectiveness of supplementary feeding should be carefully evaluated, and strategies to reduce costs may include adjusting target markets, management calendars, species or enterprise mix, or utilizing high-genetic merit animals.

Effective grazing management requires careful planning, monitoring, and adaptation to achieve sustainable land use and optimal livestock production. By implementing appropriate grazing strategies, monitoring total grazing pressure, conducting forage budgeting, adopting tactical grazing approaches, managing pasture species, and utilizing supplementary feeding when necessary, land managers can optimize pasture productivity while preserving ecosystem health and resilience.

Confirming Works and Plans Comply With Environmental Protection Legislation, Regulations and Codes of Practice

As a professional tasked with developing and implementing sustainable land use strategies, ensuring compliance with environmental protection legislation, regulations, and codes of practice is paramount. This involves a multifaceted approach that integrates legal knowledge, environmental expertise, and practical implementation skills.

Firstly, it is important to possess a comprehensive understanding of relevant environmental protection laws and regulations at the local, regional, and national levels. This includes familiarity with statutes governing land use, water management, biodiversity conservation, pollution control, and habitat preservation. By staying informed about current legislation and regulatory updates, you can ensure that your land use strategies align with legal requirements and avoid potential violations.

Moreover, conducting thorough assessments of proposed works and plans against applicable environmental standards is essential. This involves meticulously reviewing project details, site conditions, and environmental impact assessments to identify potential risks and compliance issues. By conducting detailed evaluations, you can proactively address any discrepancies or concerns and modify plans as needed to meet regulatory requirements.

In addition to legal compliance, it is essential to consider industry best practices and codes of practice related to environmental protection. These guidelines provide valuable insights into sustainable land management techniques, conservation strategies, and mitigation measures to minimize environmental harm. By incorporating these principles into your land use strategies, you can enhance ecological sustainability while complying with industry standards.

Furthermore, effective communication and collaboration with relevant stakeholders are essential for ensuring compliance with environmental protection regulations. This includes engaging with government agencies, environmental organizations, community groups, and indigenous communities to solicit feedback, address concerns, and incorporate diverse perspectives into decision-making processes. By fostering open dialogue and transparency, you can build consensus around sustainable land use practices and facilitate compliance with regulatory requirements.

Ultimately, developing and implementing sustainable land use strategies requires a proactive approach to environmental compliance that integrates legal compliance, environmental stewardship, stakeholder engagement, and continuous improvement. By prioritizing environmental protection and adhering to legislative requirements, you can create land management plans that promote ecological sustainability, preserve natural resources, and safeguard the environment for future generations.

Ensuring compliance with environmental protection legislation, regulations, and codes of practice is inherently linked to the principles of sustainable agriculture, particularly in terms of biodiversity conservation, land and water use management, soil health, and environmental stewardship.

Biodiversity plays a critical role in farming systems as it supports ecosystem resilience, nutrient cycling, pest control, and pollination. Sustainable agriculture recognizes the importance of preserving biodiversity within farming landscapes to maintain ecosystem services and enhance agricultural productivity. By implementing methods for preserving biodiversity, such as incorporating diverse crop rotations, maintaining hedgerows and buffer zones, and conserving natural habitats, farmers can create more resilient and productive farming systems.

Sustainable land and water use principles emphasize the importance of balancing agricultural production with environmental conservation to ensure long-term sustainability. This includes recognizing and managing different land classes based on their ecological characteristics and suitability for specific uses. By adopting sustainable soil management practices, such as minimizing soil disturbance, improving soil organic matter content, and implementing erosion control strategies, farmers can mitigate land degradation and preserve soil health for future generations.

Grazing strategies also play a crucial role in sustainable land use by reducing surface cover on pastures and embankments, thereby minimizing soil erosion and preserving water quality. Implementing rotational grazing, rest periods for pasture recovery, and maintaining appropriate stocking rates help to optimize pasture health and reduce soil erosion risks.

Environmental controls and codes of practice applicable to agricultural businesses are essential for minimizing environmental impacts, managing water resources responsibly, and protecting biodiversity. Compliance with legislation and regulations for soil and water degradation, chemical use, and structural works helps to ensure that farming activities are conducted in an environmentally responsible manner.

Furthermore, workplace health and safety and environmental protection legislation underscore the importance of integrating biosecurity measures into agricultural operations to prevent the spread of pests and diseases that can harm both agricultural productivity and natural ecosystems.

In summary, sustainable agriculture encompasses a holistic approach to farming that prioritizes environmental conservation, biodiversity preservation, and responsible land and water use practices. By adhering to environmental protection legislation, regulations, and codes of practice, farmers can contribute to the long-term sustainability of agricultural systems while safeguarding natural resources and ecosystems.

Part 3 - Developing Sustainable Agricultural Practices that Utilise Renewable Energy and Recycling Systems

This section outlines a structured process for identifying, developing, and implementing strategies to use renewable energy and recycling products within an enterprise. It focuses on reducing greenhouse gas emissions, improving sustainability, and ensuring compliance with relevant regulations and safety standards.

Overall, this part of the book provides a systematic approach to integrating renewable energy and recycling practices into business operations, aiming to achieve environmental sustainability, cost savings, and regulatory compliance. It emphasizes the importance of thorough planning, effective implementation, and continuous monitoring and evaluation to ensure the success of renewable energy initiatives.

Chapter Five

Identifying Opportunities to Use Renewable Energy

Identifying Areas of Enterprise where Renewable Energy, Recycling Products or Improving Work Practices Could be Utilised to Reduce Greenhouse Gas Emissions

Many farmers are already engaged in renewable energy production, utilizing resources such as corn to produce ethanol. An increasing number of farmers and ranchers are expanding their income streams by harnessing wind energy blowing across their land to generate electricity. Moreover, new opportunities in the realm of renewable energy are emerging.

The synergy between renewable energy and agriculture proves to be highly beneficial. Wind, solar, and biomass energy offer sustainable, perpetual sources of income for farmers, ensuring long-term financial stability. Renewable energy can be utilized directly on farms to replace conventional fuels or sold as a lucrative commodity.

In developing sustainable agricultural practices that utilize renewable energy and recycling systems, confirming that works and plans comply with environmental protection legislation, regulations, and codes of practice is paramount. This process involves a multifaceted approach that requires a comprehensive understanding of legal requirements, environmental considerations, and best practices in sustainable resource management. Sustainable agriculture necessitates the efficient utilization of nutrients and recycling of energy to preserve natural ecosystem resources under climate change (Bhattacharyya et al., 2016). The recycling of phosphorus in agricultural residues is critical for sustainability, dominated by the route of direct land application (Dai et al., 2016).

Agricultural sustainability can only be accomplished using a whole-systems approach that thoroughly considers nutrient stocks, removals, exports, and recycling (Jones et al., 2013). The design of the ideal integrated farming system can increase system efficiency by making use of indigenous natural materials and waste reuse/recycling (Thảo et al., 2020). Contemporary agricultural systems are transitioning from linear to circular, adopting concepts of recycling, repurposing, and regeneration (Messina et al., 2022). Agriculture presents a major opportunity for the utilization of wastes, by-products, and co-products in the development of a circular economy via the design of circular agricultural production systems and the creation of new sustainable value chains (Rauw et al., 2022).

The capture and recycling of phosphorus from human excreta is particularly important as an alternative organic fertilizer source for agriculture (Metson et al., 2017). The recycling of organic wastes in agriculture contributes to closing nutrient cycles and improving soil chemical fertility, leading to mineral fertilizer savings for the same crop yield (Levavasseur & Houot, 2022). The main sources of biofertilizers are residues from agriculture, forestry, fishery/aquaculture, and food processing, sewage, and municipal waste (Foereid, 2019). Recycling and application of organic waste products in agricultural soils provide an innovative practice that benefits circular agriculture, leading to the preservation of agricultural soils, restoration of biodiversity, and sustainability of agroecosystems (Janati et al., 2021). Urban agriculture has been proposed as a means to improve phosphorus management by recycling cities' phosphorus-rich waste back into local food production (Metson & Bennett, 2015). The agricultural sector generates approximately 1300 million tonnes of waste annually, where up to 50% comprising of raw material are discarded without treatment (Amran et al., 2021). Based on this, the agricultural production changed from the traditional mode of "resources–products–wastes" to the circulation pattern of "resources–products–renewable resources–products" (Liu et al., 2023).

The development of sustainable agricultural practices that utilize renewable energy and recycling systems requires a comprehensive understanding of legal requirements, environmental considerations, and best practices in sustainable resource management. The recycling of nutrients and energy, particularly phosphorus, plays a critical role in sustainable agriculture, contributing to the preservation of agricultural soils, restoration of biodiversity, and the creation of new sustainable value chains.

Firstly, it is essential to conduct a thorough review of relevant environmental protection legislation, regulations, and codes of practice applicable to agricultural operations. This includes familiarizing oneself with national, regional, and local laws governing environmental conservation, waste management, water usage, air quality, and biodiversity protection. Consulting with legal experts or environmental professionals may be necessary to ensure a comprehensive understanding of the legal framework.

Once the regulatory landscape is understood, the next step is to assess how proposed works and plans align with these legal requirements. This involves conducting detailed environmental impact assessments to identify potential risks and mitigation measures associated with renewable energy projects, recycling systems, waste management practices, and other sustainability initiatives. Environmental impact assessments may involve evaluating factors such as land use changes, water usage, air emissions, noise levels, and habitat disturbance.

In addition to legal compliance, it is essential to consider broader environmental stewardship principles and sustainability goals when developing works and plans. This includes minimizing environmental impacts, conserving natural resources, protecting sensitive ecosystems, and promoting biodiversity conservation. Integrating these principles into planning strategies helps ensure that agricultural practices are not only legally compliant but also environmentally responsible and socially beneficial.

Consultation with external experts, such as alternative energy consultants, waste management specialists, and environmental engineers, can provide valuable insights and expertise in developing sustainable agricultural practices. These experts can offer technical guidance, innovative solutions, and recommendations for optimizing renewable energy systems, recycling infrastructure, and waste reduction initiatives to suit local conditions and maximize environmental benefits.

Furthermore, collaboration with planning authorities and government departments responsible for administering subsidies and providing regulatory guidance is essential for navigating the regulatory landscape and accessing financial incentives for sustainable agricultural projects. Engaging with these stakeholders early in the planning process can help streamline approvals, access funding opportunities, and ensure that works and plans comply with regulatory requirements.

Confirming that works and plans comply with environmental protection legislation, regulations, and codes of practice requires a proactive and interdisciplinary approach that integrates legal compliance, environmental stewardship, stakeholder engagement, and technical expertise. By adhering to regulatory requirements, adopting best practices, and leveraging renewable energy and recycling systems, land-based businesses can develop sustainable agricultural practices that contribute to environmental conservation, resource efficiency, and long-term viability.

The utilization of renewable energies in agricultural practices presents a multifaceted approach to sustainability, with applications spanning from distributed electricity generation to space cooling and heating systems, desalination of saltwater, water pumping and irrigation systems, drying of agricultural products, and solar-powered agricultural machinery and farm robots (Gorjian et al., 2022).

Beginning with distributed electricity generation, solar energy emerges as a versatile renewable energy source not only for food production but also for electricity generation within agricultural settings. Agrivoltaic systems integrate photovoltaic modules into agricultural land, optimizing sunlight distribution to crops while simultaneously generating electricity. Studies have demonstrated that such systems, when applied to rice fields in Japan, can preserve significant rice yields while fulfilling a substantial portion of the energy demand for rice production (Gorjian et al., 2022).

Figure 39: Aerial shot of the agrivoltaics pilot plant by Fraunhofer Institute for Solar Energy Systems (ISE) located at Heggelbach farming community in the south of Germany. Tobi Kellner, CC BY-SA 4.0, via Wikimedia Commons.

Additionally, agrivoltaic systems offer benefits beyond electricity generation, including water demand reduction and increased water productivity in certain crops (Gorjian et al., 2022). Furthermore, the cultivation of various crops under agrivoltaic systems has shown substantial increases in land productivity, highlighting the potential of this technology to address contemporary challenges such as climate change and land scarcity.

Moreover, wind energy systems and biomass-based hybrid configurations have been explored as cost-effective and environmentally friendly alternatives for rural and agricultural electrification, providing further avenues for sustainable energy adoption.

In the realm of agricultural cultivation within greenhouses, the integration of solar technologies presents promising solutions for enhancing crop yields and product quality. Techniques such as photovoltaic-thermal modules offer dual functionality by simultaneously generating electrical and thermal energy, thereby improving greenhouse performance and product outcomes.

Furthermore, innovative approaches like biomass-based dual-mode systems and wind turbine installations in greenhouse areas demonstrate the feasibility of sustainable energy integration for agricultural heating, CO_2 enrichment, and electricity supply, contributing to enhanced energy efficiency and environmental sustainability.

In addressing water pumping and irrigation needs, studies have explored off-grid hybrid renewable energy sources, such as photovoltaic and wind turbine systems, for irrigation applications. These systems offer techno-economic optimization potential, with considerations for factors like system size, solar radiation, and wind speed variations.

Desalination of saltwater presents another critical area where renewable energy technologies offer sustainable solutions. Approaches like solar-powered desalination systems and tidal energy-based reverse osmosis systems demonstrate significant water cost savings and environmental benefits compared to conventional methods, paving the way for future sustainable desalination practices.

Additionally, the drying of agricultural products using solar technologies has been investigated, showcasing the potential for hybrid solar dryers to outperform conventional drying methods in terms of efficiency and product quality preservation.

Finally, the integration of solar-powered agricultural machinery and farm robots signifies a paradigm shift towards electrification and renewable energy adoption in the agricultural sector. Despite challenges such as initial costs and environmental sensitivity, advancements in solar-powered electric tractors and field robots offer promising avenues for reducing emissions and enhancing operational efficiency in agricultural practices.

The applications of renewable energies in agriculture hold immense potential for addressing sustainability challenges, improving productivity, and mitigating environmental impacts across various aspects of agricultural operations. Through continued research, innovation, and adoption, renewable energy technologies can play a pivotal role in fostering a more sustainable and resilient agricultural sector.

Agrivoltaic systems, also referred to as agrivoltaics or solar farming, represent a pioneering approach to agriculture by merging solar panel electricity generation with traditional farming practices on the same parcel of land. These systems leverage photovoltaic (PV) modules installed on structures either above or adjacent to crops, facilitating the dual-purpose utilization of land for both agricultural production and renewable energy generation. Typically, solar panels are mounted on structures elevated above crops, such as poles or racks, or integrated into canopies or greenhouse structures. This setup allows the panels to capture sunlight and convert it into electricity through the photovoltaic effect, while concurrently providing shaded space beneath for agricultural activities such as crop cultivation or livestock grazing. This dual-use strategy optimizes land efficiency and resource utilization, fostering sustainability in land management practices.

In terms of functionality, agrivoltaic systems are designed to ensure an optimal distribution of sunlight to both solar panels and the crops beneath. Various mechanisms, including adjustable tilt angles or panel spacing, may be employed to regulate sunlight exposure, thereby balancing energy generation with crop growth requirements. Careful planning and design are paramount to maximizing energy yields from solar panels while mitigating potential shading effects on crops.

Crop selection plays a crucial role in the success of agrivoltaic systems. A diverse range of crops, including vegetables, fruits, grains, and specialty crops, can be cultivated within these systems. Factors such as local climate, soil conditions, water availability, market demand, and compatibility with shading from solar panels influence crop choices. Shade-tolerant crops or those with lower light requirements are often favored to ensure optimal growth and yield within agrivoltaic environments.

The environmental benefits of agrivoltaic systems are manifold, encompassing reduced land competition between agriculture and solar energy production, enhanced land use efficiency, improved soil moisture retention, and mitigation of heat stress on crops. By harnessing solar energy and promoting agricultural sustainability, these systems contribute to climate change mitigation and ecosystem conservation.

Agrivoltaic systems exhibit versatility in geographical suitability, with applicability ranging from sunny regions with high solar irradiance to temperate climates with moderate sunlight exposure. Regions possessing ample sunlight and available agricultural land are particularly well-suited for agrivoltaic projects. Moreover, areas grappling with land-use conflicts or scarcity issues stand to benefit significantly from the dual-use approach offered by agrivoltaic systems.

Ongoing research and development endeavours are focused on refining agrivoltaic system designs, evaluating their economic viability, and assessing their long-term impacts on crop productivity, soil health, and ecosystem services. Collaborative initiatives involving researchers, farmers, energy companies, and policymakers are essential for advancing agrivoltaic technology and fostering its widespread adoption.

Agrivoltaic systems represent a sustainable solution for integrating solar energy production with agricultural practices, thereby promoting land-use efficiency, resource conservation, and climate resilience. While adaptable to various crops and geographic regions, diligent planning, ongoing research, and stakeholder engagement are imperative to unlock the full potential of agrivoltaic systems and address potential challenges.

Cultivation greenhouses are enclosed structures designed to create controlled environments for plant growth, enabling year-round cultivation regardless of external weather conditions. These structures typically feature transparent or translucent walls and roofs, allowing sunlight to penetrate and provide the necessary light energy for photosynthesis. Cultivation greenhouses operate on the principle of trapping solar radiation within the structure, creating a warm and stable climate conducive to plant growth.

Within cultivation greenhouses, various environmental factors such as temperature, humidity, light intensity, and ventilation are carefully regulated to optimize plant growth and yield. Heating systems, such as boilers or heaters, are often employed to maintain optimal temperatures during colder periods, while ventilation systems, including fans and vents, help regulate air circulation and prevent overheating. Additionally, irrigation systems deliver water and nutrients directly to plants, ensuring consistent hydration and nutrient uptake.

The design and functionality of cultivation greenhouses make them well-suited for a wide range of agricultural activities, including the production of fruits, vegetables, flowers, and ornamental plants. They are particularly advantageous in regions with adverse climatic conditions, such as extreme temperatures, high winds, or heavy rainfall, where traditional outdoor farming may be challenging or impractical. By providing a controlled environment, cultivation greenhouses offer growers greater flexibility and reliability in crop production, enabling them to extend growing seasons, mitigate weather-related risks, and optimize resource use.

Cultivation greenhouses are especially beneficial in urban and peri-urban areas where land availability may be limited or land prices are high. Their compact footprint allows for efficient land use, making them ideal for urban agriculture initiatives aimed at promoting local food production and reducing food miles. Additionally, cultivation greenhouses can be customized to accommodate specific crop requirements, allowing growers to cultivate a diverse range of crops tailored to local market demands.

Figure 40: Tomato cultivation greenhouse in France. Mouh2jijel, CC BY SA 3.0, via Wikimedia Commons.

Furthermore, cultivation greenhouses are increasingly being integrated with advanced technologies such as automated climate control systems, hydroponic or aquaponic growing systems, and LED lighting to further enhance productivity and resource efficiency. These innovations enable growers to optimize growing conditions, minimize resource inputs, and maximize crop yields, thereby contributing to sustainable agriculture practices.

Cultivation greenhouses provide a controlled and conducive environment for year-round plant cultivation, offering numerous advantages such as weather protection, resource efficiency, and crop diversification. They are well-suited for a variety of agricultural applications, from traditional farming to urban agriculture, and play a crucial role in ensuring food security, promoting local food production, and supporting sustainable agriculture initiatives.

A semi-transparent photovoltaic-thermal (PVT) system, often referred to as GiSPVT (Glass-integrated Solar Photovoltaic-Thermal), is a multifunctional solar energy harvesting technology that combines photovoltaic (PV) and solar thermal components into a single unit. These systems typically consist of a transparent cover (glass or polymer) over PV cells, allowing sunlight to pass through while generating electricity. Meanwhile, the absorbed sunlight not converted to electricity is harvested for thermal energy through a heat exchanger integrated into the system.

The GiSPVT system, in conjunction with a ground-air heat exchanger (GAHE), operates by harnessing renewable energy sources to provide heating, cooling, and ventilation for buildings. The GiSPVT system integrates photovoltaic (PV) panels with solar thermal collectors, allowing for the simultaneous generation of electricity and heat from sunlight. This hybrid system can be coupled with a GAHE, which utilizes the stable temperature of the ground to pre-heat or pre-cool ventilation air, thereby reducing the energy consumption of traditional HVAC systems (Congedo et al., 2014). The GAHE works by exchanging heat with the ground to pre-heat or pre-cool ventilation air, thus reducing the energy required for conditioning the air (Skotnicka-Siepsiak, 2020). The combination of these systems offers the potential to significantly reduce the size of the ground heat exchanger, thereby lowering the overall cost of the system (Bottarelli & Gallero, 2020).

The semi-transparent PVT system, with its ability to generate both electricity and heat, is well-suited for buildings where space for separate PV panels and solar thermal collectors is limited. By integrating these technologies, the system can maximize energy production within a constrained area, making it particularly suitable for urban environments or buildings with limited roof space. Additionally, the GAHE is well-suited for regions with moderate to extreme climates, as it can efficiently pre-heat or pre-cool ventilation air using the stable temperature of the ground, contributing to energy savings in both heating and cooling seasons (Mostafaeipour et al., 2021; Skotnicka-Siepsiak, 2020). Furthermore, the GAHE can be integrated into conventional HVAC systems, making it a versatile solution for various building types, including residential, commercial, and industrial structures (Brata et al., 2019).

Furthermore, the integration of GAHE enhances the performance of GiSPVT systems in climates with fluctuating temperature extremes, as it provides a means of thermal storage and regulation, contributing to improved system reliability and performance throughout the year. Overall, GiSPVT systems with GAHE represent an innovative and versatile approach to solar energy harvesting, offering both economic and environmental benefits across various applications and geographical locations.

Solar desalination is a sustainable method of producing freshwater from seawater or brackish water using solar energy as the primary power source. The process typically involves utilizing solar thermal energy or photovoltaic (PV) electricity to drive the desalination process, which removes salts and other impurities from the water, making it suitable for drinking, irrigation, or industrial purposes.

One common approach to solar desalination is solar stills. Solar stills use the sun's energy to heat water, causing it to evaporate and leave behind salts and contaminants. The vapor then condenses on a cooler surface, such as a glass or plastic cover, and collects as freshwater, while the remaining brine is discarded. This method is relatively simple and does not require complex infrastructure, making it suitable for remote or off-grid areas where access to freshwater is limited.

Another method of solar desalination involves solar-powered membrane distillation systems. In membrane distillation, solar thermal energy or electricity is used to heat the saline water, creating vapor that passes through a semi-permeable membrane, leaving behind concentrated brine. The vapor is then condensed on the other side of the membrane to produce freshwater. This approach offers higher efficiency and scalability compared to solar stills, making it suitable for larger-scale desalination plants and applications where higher water output is required.

Solar desalination systems are particularly well-suited to regions with abundant sunlight and limited freshwater resources, such as arid and coastal areas. These regions often face water scarcity and rely heavily on energy-intensive methods like conventional desalination or long-distance water transport. Solar desalination offers a sustainable alternative by harnessing freely available solar energy to produce freshwater locally, reducing reliance on fossil fuels and mitigating environmental impacts associated with conventional desalination methods.

Additionally, solar desalination systems can be deployed in remote or off-grid areas where access to centralized water infrastructure is limited or non-existent. They offer a decentralized and cost-effective solution for providing clean water to communities, agricultural operations, and industrial facilities without the need for extensive infrastructure investments.

Furthermore, solar desalination systems can be integrated with existing water treatment facilities or combined with renewable energy systems like wind or hydroelectric power to enhance their reliability and efficiency. This integrated approach can help offset energy costs and reduce greenhouse gas emissions associated with water desalination, contributing to sustainable development and climate change mitigation efforts.

Overall, solar desalination represents a promising technology for addressing water scarcity challenges in a sustainable and environmentally friendly manner, offering a viable solution for communities and industries facing water stress around the world.

Wind power has been traditionally utilized on farms for water pumping and electricity generation. However, recent advancements have seen large wind turbines installed on farms and ranches across several states to supply power to electric utilities and consumers. Developers may offer substantial annual payments, ranging from $2,000 to $5,000 per turbine installed, particularly in areas with robust wind resources. With each turbine occupying less than half an acre, farmers can continue agricultural activities up to the turbine's base. Some farmers opt to purchase wind turbines individually, while others are forming wind power cooperatives.

Figure 41: Wind turbine, trig point and sheep. Gareth James, CC BY-SA 2.0, via Wikimedia Commons.

Biomass energy derives from plants and organic waste, including crops, trees, and residues. Energy crops can be cultivated on a large scale akin to food crops. While corn is presently the predominant energy crop, native prairie grasses like switchgrass and fast-growing trees such as poplar and willow are poised to gain popularity due to their lower maintenance requirements and sustainability.

Crops and biomass residues can be converted into energy on farms or sold to energy companies producing fuel for vehicles, heating, and electricity. According to the U.S. Department of Energy, a threefold increase in biomass energy utilization in the U.S. could yield up to $20 billion in additional income for farmers and rural communities while significantly reducing global warming emissions, equivalent to removing 70 million cars from the roads (Union of Concerned Scientists, 2008). Various federal and state incentives may be available to facilitate the realization of these benefits.

As such, renewable energy offers various specific applications in agriculture, helping farmers reduce costs, increase efficiency, and minimize environmental impact. Some specific applications include:

1. Solar Water Pumping: Solar-powered water pumps can provide a reliable and cost-effective solution for irrigation in remote or off-grid agricultural areas. These pumps utilize photovoltaic panels to power the pump, enabling farmers to draw water from wells, rivers, or ponds without relying on grid electricity or fossil fuels.

2. Solar Heating and Drying: Solar thermal collectors can be used to heat water for agricultural processes such as cleaning equipment, sanitizing produce, and providing hot water for livestock. Solar dryers can also be employed to dry crops, grains, and herbs, reducing moisture content and preserving quality without the need for fossil fuels.

3. Wind Power: Wind turbines can generate electricity for on-farm use or to sell back to the grid, providing a reliable and renewable energy source. Wind energy can also be used for water pumping, aeration in ponds, and powering electric fencing systems.

4. Biomass Energy: Agricultural residues such as crop residues, animal manure, and forestry waste can be converted into biomass energy through processes like anaerobic digestion, combustion, or gasification. This energy can be used for heating, electricity generation, or as biofuel for vehicles and machinery.

5. Biofuels Production: Biofuels such as ethanol and biodiesel can be produced from crops like corn, sugarcane, soybeans, and oilseeds. These biofuels can be used to power farm equipment, vehicles, and machinery, reducing dependence on fossil fuels and lowering greenhouse gas emissions.

6. Geothermal Heating and Cooling: Geothermal energy systems can provide heating and cooling for greenhouses, livestock buildings, and other agricultural facilities. These systems utilize the natural heat stored in the Earth's crust to maintain consistent temperatures year-round, improving crop yields and animal comfort.

7. Micro-hydro Power: In areas with flowing water resources such as streams or rivers, micro-hydro systems can be installed to generate electricity for on-farm use. These systems harness the kinetic energy of flowing water to power small turbines, providing a reliable and renewable energy source.

8. Energy-Efficient Technologies: Beyond direct energy production, renewable energy can also be integrated into agricultural practices through energy-efficient technologies such as LED lighting, energy-efficient appliances, and improved insulation in buildings. These measures help reduce energy consumption and lower operating costs on farms.

Overall, the adoption of renewable energy technologies in agriculture offers numerous benefits, including cost savings, energy independence, reduced environmental impact, and enhanced sustainability. By harnessing the power of renewable resources, farmers can improve productivity, resilience, and long-term viability in an increasingly uncertain climate and energy landscape.

Identifying areas within a land-based business where renewable energy, recycling products, or improved work practices can be effectively utilized to reduce greenhouse gas emissions requires a comprehensive understanding of the operation's current processes and potential areas for improvement. Firstly, conducting a thorough assessment of energy consumption and waste production across various aspects of the enterprise is essential. This could involve analysing energy bills, waste disposal records, and production processes to identify areas of inefficiency and opportunities for optimization.

In agricultural operations, for instance, energy-intensive activities such as irrigation, heating, and transportation are prime targets for renewable energy integration. Assessing the feasibility of installing solar panels for on-farm electricity generation or transitioning to electric or biofuel-powered machinery can significantly reduce reliance on fossil fuels and mitigate emissions. Similarly, implementing energy-efficient technologies and practices, such as precision agriculture techniques or improved building insulation, can further enhance energy conservation and reduce greenhouse gas emissions.

Moreover, evaluating waste streams generated within the enterprise presents opportunities for recycling and waste reduction. This could involve implementing composting systems for organic waste, establishing recycling programs for packaging materials, or exploring opportunities for repurposing waste materials as inputs for other processes. For example, agricultural residues such as crop stubble or animal manure can be converted into bioenergy through anaerobic digestion or biomass combustion, providing both energy and waste management benefits.

Furthermore, improving work practices to minimize resource consumption and emissions is crucial for overall sustainability. This could include optimizing production schedules to reduce idle machinery time, implementing conservation tillage practices to improve soil health and carbon sequestration, or adopting integrated pest management strategies to minimize pesticide use and environmental impact. Additionally, training and capacity-building programs for employees can promote awareness and engagement in sustainable practices throughout the organization.

In identifying these areas for improvement, collaboration with external experts such as alternative energy consultants, waste management specialists, and government agencies is invaluable. These experts can provide technical expertise, feasibility assessments, and guidance on available subsidies or incentives for implementing renewable energy and waste management initiatives. By leveraging their knowledge and support, land-based businesses can develop comprehensive planning strategies tailored to their specific circumstances and effectively address sustainability challenges while maximizing economic and environmental benefits.

Identifying Available Government Subsidies for The Implementation of Strategies Using Renewable Energy or Recycling and Potential Cost Savings to the Business in the Longer Term

Identifying available government subsidies for implementing strategies utilizing renewable energy or recycling systems is a crucial step in developing sustainable agricultural practices. It requires a thorough understanding of the various subsidies, incentives, and financial assistance programs offered by government agencies at the local, regional, and national levels. Firstly, engaging with experts such as alternative energy consultants and government departments responsible for administering subsidies is essential to stay informed about the latest programs and eligibility criteria.

Government subsidies for renewable energy initiatives often include financial incentives, tax credits, grants, or low-interest loans to offset the upfront costs of implementing renewable energy systems such as solar panels, wind turbines, or biomass energy facilities. These subsidies aim to encourage the adoption of clean energy technologies and accelerate the transition to a low-carbon economy. Additionally, subsidies may also be available for energy efficiency improvements, such as upgrading equipment or retrofitting buildings with energy-saving technologies.

Similarly, government subsidies for recycling initiatives can include grants or financial assistance to support the development of recycling infrastructure, purchase of recycling equipment, or implementation of waste reduction and recycling programs. These subsidies are designed to incentivize businesses to adopt sustainable waste management practices and reduce their environmental footprint. For example, subsidies may cover a portion of the costs associated with installing composting facilities, purchasing recycling bins, or implementing waste sorting systems.

In assessing the potential cost savings to the business in the longer term, it is important to consider both the upfront investment required and the expected returns or savings over the lifespan of the renewable energy or recycling system. While the initial capital costs of implementing renewable energy or recycling systems may be significant, government subsidies can help offset these costs and improve the return on investment. Moreover, renewable energy systems typically offer long-term cost savings through reduced energy bills, increased energy independence, and potential revenue generation through excess energy production or participation in incentive programs such as net metering.

Similarly, recycling systems can result in cost savings by reducing waste disposal fees, lowering purchasing costs for raw materials through recycling and reuse, and enhancing resource efficiency. Additionally, adopting sustainable practices can also lead to intangible benefits such as improved brand reputation, increased customer loyalty, and compliance with environmental regulations, which can further contribute to the overall economic viability of the business.

By leveraging available government subsidies and understanding the potential cost savings associated with renewable energy and recycling initiatives, land-based businesses can develop effective planning strategies to transition towards sustainable practices while maximizing economic benefits in the longer term. Collaboration with external experts and government agencies can provide valuable support and guidance throughout the process, ensuring the successful implementation of sustainable agricultural practices.

Finding available government subsidies for the implementation of strategies using renewable energy or recycling typically involves several steps. Firstly, it's crucial to research government websites, starting with relevant agencies at the local, regional, and national levels. These agencies, responsible for energy, environment, agriculture, and economic development, often provide detailed information on available subsidies, grants, tax credits, and financial assistance programs for renewable energy and recycling initiatives. This entails exploring dedicated sections or pages related to energy incentives, environmental programs, or business assistance.

Next, it's important to explore specific renewable energy programs offered by government agencies such as the Department of Energy, Environmental Protection Agency, Department of Agriculture, or state energy offices. These programs may encompass incentives for various renewable energy sources like solar, wind, biomass, geothermal, or hydropower projects, as well as energy efficiency improvements. Attention should be paid to eligibility criteria, application deadlines, and required documentation.

Similarly, checking recycling and waste management programs provided by government agencies responsible for environmental protection and waste management is essential. These programs often offer grants, rebates, or technical assistance to support recycling infrastructure, waste reduction initiatives, composting projects, or the purchase of recycling equipment. Valuable resources can be found from agencies like the EPA, state environmental departments, or local waste management authorities.

Utilizing online tools and databases that aggregate information on available government subsidies and incentives for renewable energy and recycling is also recommended. Websites like the US Database of State Incentives for Renewables & Efficiency (DSIRE) provide comprehensive listings of state, local, and federal incentives for renewable energy and energy efficiency projects. Additionally, resources like the EPA's Grants and Funding page offer information on funding opportunities for environmental projects, including recycling initiatives.

Seeking guidance from experts and organizations specializing in renewable energy, recycling, and sustainability is another important step. Alternative energy consultants, waste management specialists, industry associations, and non-profit organizations may offer insights into available subsidies and assistance programs, as well as strategies for navigating the application process. Networking with peers and attending relevant workshops, conferences, or webinars can also provide valuable information and resources.

Finally, reaching out directly to relevant government agencies, energy offices, environmental departments, or economic development agencies to inquire about available subsidies and incentives is key. Contact information is often provided on agency websites, and government representatives can offer personalized assistance, answer questions, and provide guidance on accessing subsidies and navigating the application process. By conducting thorough research, utilizing online resources, consulting with experts, and reaching out to government agencies, land-based businesses can identify available subsidies and incentives for implementing renewable energy and recycling strategies. It's important to carefully review eligibility requirements, application procedures, and deadlines to maximize the chances of securing financial assistance for sustainable initiatives.

Identifying the Specified Standards of Quality, Licensing, Regulatory Requirements, Government Legislation and Safety Issues for the Introduction of any Renewable Energy Resources and Recycled Products

Identifying the specified standards of quality, licensing, regulatory requirements, government legislation, and safety issues for the introduction of renewable energy resources and recycled products is a important aspect of developing sustainable agricultural practices. As individuals responsible for managing land-based businesses, it's essential to navigate this complex landscape to ensure compliance and mitigate potential risks effectively.

First and foremost, conducting comprehensive research and consulting with external experts such as alternative energy consultants, suppliers of alternative energy equipment, and government departments administering subsidies is paramount. These experts can provide valuable insights into the standards and regulations governing renewable energy technologies and recycled products within the agricultural sector.

Quality standards play a vital role in ensuring the reliability and performance of renewable energy systems and recycled products. It is essential to familiarize oneself with relevant certifications, such as ISO standards for energy management systems, product quality certifications, and licensing requirements for the installation and operation of renewable energy infrastructure. Adhering to these standards not only ensures the effectiveness of the systems but also enhances credibility and trust among stakeholders.

Moreover, staying abreast of regulatory requirements and government legislation is crucial to remain compliant with environmental, energy, and safety regulations. Government agencies at various levels may impose specific requirements for the installation, operation, and maintenance of renewable energy systems, as well as the production and utilization of recycled products. This may include zoning ordinances, building codes, environmental permits, and safety standards aimed at protecting public health and the environment.

Addressing safety issues associated with renewable energy and recycling systems is paramount to safeguarding both personnel and the environment. This involves implementing appropriate safety measures, conducting risk assessments, and adhering to industry best practices for installation, operation, and maintenance. Safety considerations may include electrical hazards, structural integrity, fire safety, and proper handling and disposal of recycled materials to prevent accidents and minimize environmental impact.

Integrating sustainability principles into business operations requires a proactive approach to compliance and risk management. It entails conducting regular audits, monitoring performance metrics, and implementing corrective actions to address any deviations from established standards or regulatory requirements. By proactively identifying and addressing quality, licensing, regulatory, and safety issues, land-based businesses can ensure the successful implementation of renewable energy and recycling systems while minimizing legal and operational risks.

The worldwide examples of standards, regulations, and safety measures pertaining to the introduction of renewable energy resources and recycled products in agriculture demonstrate the global efforts to promote sustainability and compliance with legal requirements (Jebli & Youssef, 2017). In the European Union (EU), the Renewable Energy Directive (RED) sets binding targets for renewable energy and establishes sustainability criteria for biofuels, biomass, and biogas production, while the Circular Economy Package aims to increase resource efficiency and promote the use of renewable energy through recycling and waste diversion (Jebli & Youssef, 2017). Additionally, European Standards (EN) ensure the quality, safety, and environmental performance of renewable energy systems and recycling processes (Jebli & Youssef, 2017).

In the United States, the Renewable Fuel Standard (RFS) mandates the blending of renewable fuels into transportation fuels to reduce greenhouse gas emissions, and the Environmental Protection Agency (EPA) enforces air and water quality standards, emission limits, and waste management requirements for renewable energy projects and recycling facilities (Jebli & Youssef, 2017)). The Occupational Safety and Health Administration (OSHA) regulations ensure workplace safety in renewable energy installations and recycling operations, covering aspects such as equipment safety, hazardous materials handling, and worker training (Jebli & Youssef, 2017).

China's Renewable Energy Law promotes the development and utilization of renewable energy sources through incentives, subsidies, and regulatory support, while National Standards (GB) set technical standards and specifications for renewable energy equipment and recycling processes to ensure quality, reliability, and safety (Jebli & Youssef, 2017). The Environmental Protection Law establishes environmental quality standards, pollution control measures, and waste management requirements for renewable energy and recycling activities (Jebli & Youssef, 2017).

In India, the National Action Plan on Climate Change aims to promote renewable energy deployment and energy efficiency measures to reduce greenhouse gas emissions and enhance sustainability in agriculture and other sectors (Jebli & Youssef, 2017). The Bureau of Indian Standards (BIS) develops national standards for renewable energy technologies, recycled products, and waste management practices to ensure quality, safety, and performance, while the Electricity Act and Renewable Energy Policies enforce licensing requirements, regulatory approvals, and safety standards for renewable energy projects (Jebli & Youssef, 2017).

Australia's Renewable Energy Target (RET) sets a goal to achieve a specified percentage of electricity generation from renewable sources by a certain deadline, and the Australian Renewable Energy Agency (ARENA) administers funding programs, grants, and subsidies for renewable energy projects, including research, development, and commercialization initiatives (Jebli & Youssef, 2017). The Work Health and Safety (WHS) Regulations ensure workplace safety in renewable energy installations and recycling facilities, covering aspects such as risk assessments, safety training, and emergency procedures (Jebli & Youssef, 2017).

These examples highlight the diverse range of standards, regulations, and safety measures implemented worldwide to promote the adoption of renewable energy resources and recycled products in agriculture while ensuring quality, sustainability, and compliance with legal requirements (Jebli & Youssef, 2017).

Specifically related to sustainable agriculture, in the European Union (EU), specific standards and regulations bolster sustainable agriculture practices. The Renewable Energy Directive (RED) mandates that biofuels, biomass, and biogas meet stringent sustainability criteria. For instance, biofuels must demonstrate greenhouse gas emission savings compared to fossil fuels, and biomass sourcing must not contribute to deforestation or environmental degradation. Complementing this directive, the Circular Economy Package encourages sustainable farming by setting targets for recycling and waste diversion. Farmers are urged to adopt practices like composting agricultural residues and using recycled materials for packaging and infrastructure, thus promoting resource efficiency. Furthermore, European Standards (EN) ensure the quality, safety, and environmental performance of renewable energy systems and recycling processes, such as establishing best practices for biogas production through anaerobic digestion processes.

In the United States, agricultural sustainability is bolstered by regulations like the Renewable Fuel Standard (RFS), which promotes the production and use of renewable fuels derived from agricultural feedstocks like corn and soybeans, aiding rural economies and curbing greenhouse gas emissions from transportation. Additionally, Environmental Protection Agency (EPA) regulations enforce air and water quality standards, emission limits, and waste management requirements for renewable energy projects and recycling facilities in agriculture, safeguarding water resources and mitigating pollution risks. Occupational Safety and Health Administration (OSHA) regulations ensure workplace safety in renewable energy installations and recycling operations, safeguarding farm workers and the environment.

In China, sustainability in agriculture is championed through the Renewable Energy Law, which incentivizes the development of renewable energy sources like solar, wind, and biomass. National Standards (GB) establish technical specifications for renewable energy equipment and recycling processes, ensuring safety and performance. Moreover, the Environmental Protection Law sets standards for pollution control, waste management, and environmental quality, safeguarding air, water, and soil quality in rural areas.

In India, the National Action Plan on Climate Change promotes renewable energy deployment and energy efficiency measures in agriculture to mitigate greenhouse gas emissions and enhance sustainability. The Bureau of Indian Standards (BIS) develops national standards for renewable energy technologies and recycling processes, ensuring quality and safety. Additionally, regulations like the Electricity Act and Renewable Energy Policies enforce licensing requirements, regulatory approvals, and safety standards for renewable energy projects in agriculture, ensuring safe and reliable operation.

In Australia, sustainability efforts in agriculture are facilitated by the Renewable Energy Target (RET), which sets goals for electricity generation from renewable sources. The Australian Renewable Energy Agency (ARENA) administers funding programs and subsidies for renewable energy projects in agriculture, fostering innovation and improving energy efficiency. Work Health and Safety (WHS) regulations ensure workplace safety in renewable energy installations and recycling facilities, protecting the health and safety of farm workers and communities.

In relation to a range of specific sustainable agricultural methods:

Organic Farming:

1. European Union (EU): The Renewable Energy Directive (RED) ensures that biofuels, biomass, and biogas used for energy production in sustainable agriculture meet strict sustainability criteria. For instance, biofuels produced from agricultural crops must demonstrate greenhouse gas emission savings compared to fossil fuels, aligning with the principles of organic farming that emphasize reducing chemical inputs and promoting environmental sustainability. Additionally, the Circular Economy Package encourages sustainable agriculture by setting targets for recycling and waste diversion. Farmers practicing organic farming are incentivized to adopt practices such as composting agricultural residues and using recycled materials for packaging and

infrastructure, thus reducing waste and promoting resource efficiency, in line with EU regulations.

2. United States: In the U.S., the Renewable Fuel Standard (RFS) promotes the production and use of renewable fuels such as ethanol and biodiesel, derived from agricultural feedstocks like corn and soybeans, supporting rural economies and reducing greenhouse gas emissions from transportation. This aligns with the principles of organic farming that focus on reducing reliance on synthetic inputs and promoting renewable resources. Additionally, Environmental Protection Agency (EPA) regulations enforce air and water quality standards, emission limits, and waste management requirements for renewable energy projects and recycling facilities in agriculture, ensuring that organic farming practices are in harmony with environmental regulations.

3. China: China's Renewable Energy Law promotes the development of renewable energy sources in agriculture, such as solar, wind, and biomass, through incentives, subsidies, and regulatory support. This encourages farmers to invest in renewable energy systems, which is complementary to organic farming practices that prioritize reducing reliance on fossil fuels and promoting environmental sustainability. Moreover, China's National Standards (GB) establish technical specifications for renewable energy equipment and recycling processes used in agriculture, ensuring the safety and performance of renewable energy systems, which is essential for organic farming operations.

Permaculture:

1. European Union (EU): The European Standards (EN) ensure the quality, safety, and environmental performance of renewable energy systems and recycling processes used in agriculture. For instance, EN standards for biogas production outline best practices for anaerobic digestion processes, aligning with permaculture principles that emphasize diversity, integration, and self-sufficiency. Permaculture designs often incorporate agroforestry, which promotes biodiversity enhancement and ecosystem resilience, in line with EU standards for sustainable agriculture.

2. United States: In the U.S., the Environmental Protection Agency (EPA) regulations enforce air and water quality standards, emission limits, and waste management requirements for renewable energy projects and recycling facilities in agriculture. These regulations ensure that permaculture practices, which prioritize maximizing productivity while minimizing environmental impact, comply with environmental standards and promote ecosystem health. Additionally, Occupational Safety and Health Administration (OSHA) regulations ensure workplace safety in renewable energy installations and recycling operations in agriculture, which is essential for implementing permaculture designs that may involve diverse and integrated farming systems.

3. China: China's Environmental Protection Law sets environmental quality standards, pollution control measures, and waste management requirements for renewable energy and recycling activities in agriculture, ensuring that permaculture practices align with environmental regulations. Farmers must adhere to these regulations to minimize environmental impacts and protect air, water, and soil quality in rural areas, which is essential for implementing permaculture designs that aim to enhance ecosystem services and promote sustainability.

Agroecology:

1. European Union (EU): The Renewable Energy Directive (RED) ensures that biofuels, biomass, and biogas used for energy production in sustainable agriculture meet strict sustainability criteria, aligning with agroecological principles that focus on enhancing biodiversity, soil health, and ecosystem services. For instance, biofuels produced from agricultural crops must demonstrate greenhouse gas emission savings compared to fossil fuels, promoting environmental sustainability in line with agroecological practices. Additionally, the Circular Economy Package encourages sustainable agriculture by setting targets for recycling and waste diversion, which is complementary to agroecology's emphasis on minimizing waste and promoting resource efficiency.

2. United States: In the U.S., the Renewable Fuel Standard (RFS) promotes the production and use of renewable fuels such as

ethanol and biodiesel, derived from agricultural feedstocks like corn and soybeans, supporting rural economies and reducing greenhouse gas emissions from transportation. This aligns with agroecological principles that prioritize enhancing ecosystem services and promoting environmental sustainability. Additionally, Environmental Protection Agency (EPA) regulations enforce air and water quality standards, emission limits, and waste management requirements for renewable energy projects and recycling facilities in agriculture, ensuring that agroecological practices comply with environmental standards and promote ecosystem health.

3. China: China's Renewable Energy Law promotes the development of renewable energy sources in agriculture, such as solar, wind, and biomass, through incentives, subsidies, and regulatory support, which is complementary to agroecological practices that prioritize enhancing biodiversity and promoting environmental sustainability. Moreover, China's Environmental Protection Law sets environmental quality standards, pollution control measures, and waste management requirements for renewable energy and recycling activities in agriculture, ensuring that agroecological practices align with environmental regulations and promote ecosystem resilience.

Identifying Potential Income Generated by On Selling Energy Excesses or Recycled Products

Identifying potential income generated by on-selling energy excesses or recycled products involves a comprehensive assessment of available resources, market demand, and regulatory frameworks to maximize profitability while minimizing environmental impact.

Firstly, it is essential to conduct a thorough analysis of the energy generation potential on the agricultural property. This includes evaluating the feasibility of implementing renewable energy systems such as solar panels, wind turbines, or biomass generators. Factors such as available land area, local climate conditions, and energy consumption patterns should be considered to determine the most suitable technologies. Consulting with alternative energy consultants can provide valuable insights into the most cost-effective and efficient solutions for harnessing renewable energy resources.

Once renewable energy systems are in place, monitoring and management tools should be employed to track energy production and consumption patterns accurately. This data can be used to identify periods of excess energy generation when the agricultural operation produces more energy than it consumes. By leveraging net metering programs or participating in energy markets, surplus energy can be sold back to the grid or directly to consumers, generating additional revenue streams for the business.

In addition to energy excesses, attention should be given to opportunities for recycling products generated during agricultural operations. This may include organic waste from crop residues, livestock manure, or packaging materials from agricultural products. Implementing recycling systems such as composting, anaerobic digestion, or material recovery facilities can help convert these waste streams into valuable resources.

To identify potential income from recycled products, it is essential to assess market demand and pricing for recycled materials. Collaboration with suppliers of alternative energy equipment and recycling facilities can provide valuable insights into market dynamics and potential revenue streams. Furthermore, engaging with planning authorities and government departments administering subsidies or incentives for sustainable practices can help maximize profitability and offset initial investment costs.

Overall, by proactively identifying opportunities to on-sell energy excesses or recycled products, land-based businesses can enhance their financial resilience while contributing to environmental sustainability. Through strategic planning and collaboration with external experts, agricultural operations can develop sustainable practices that optimize resource utilization and minimize waste generation, ultimately fostering long-term viability and profitability.

To conduct a comprehensive analysis of the energy generation potential on your agricultural property, several sequential steps should be followed.

Firstly, begin by assessing the available land area suitable for installing renewable energy systems. This involves surveying your agricultural property to identify locations with unobstructed access to sunlight or wind, depending on the renewable energy technology being considered.

Secondly, evaluate local climate conditions by researching relevant climate data. Factors such as sunlight availability, wind patterns, and biomass resources should be considered. This assessment helps gauge the suitability of different renewable energy options based on average annual sunshine hours, prevailing wind direction and speed, and biomass availability.

Thirdly, analyse historical energy consumption data for your agricultural operation. Understanding your current energy needs and consumption patterns is essential. Identify peak energy demand periods and areas where energy efficiency improvements can be made to optimize energy use.

Next, determine suitable technologies based on the available land area, local climate conditions, and energy consumption patterns. Evaluate the feasibility of implementing various renewable energy systems such as solar photovoltaic (PV) panels, wind turbines, or biomass generators. Consider factors such as energy density, upfront costs, maintenance requirements, and regulatory constraints.

Subsequently, engage with alternative energy consultants specializing in renewable energy systems for agricultural applications. These experts can provide valuable advice and conduct site assessments to determine the most suitable technologies for your specific needs. Discussing energy goals, budget constraints, and regulatory considerations with consultants can help develop a tailored renewable energy strategy.

Following this, evaluate the cost-effectiveness and efficiency of different renewable energy options. Compare installation costs, ongoing maintenance expenses, energy production potential, and potential revenue streams from selling excess energy. Conduct financial analyses, including payback period calculations and return on investment (ROI) assessments, to determine the most cost-effective technology option.

Finally, based on the analysis and consultation with alternative energy consultants, develop a comprehensive renewable energy plan. This plan should outline the proposed system design, equipment specifications, installation timeline, and budget estimates. Ensure that the plan incorporates considerations for regulatory compliance, permitting requirements, and available incentives or subsidies for renewable energy projects.

By following these steps and collaborating with experts in the field, you can conduct a thorough analysis of the energy generation potential on your agricultural property and develop a sustainable renewable energy strategy tailored to your specific needs and conditions.

Once renewable energy systems are installed, monitoring and managing their performance becomes paramount in optimizing energy production and consumption. A range of specific tools and technologies are available to facilitate this process.

Energy Monitoring Systems serve as foundational tools, collecting real-time data on energy production and consumption. These systems incorporate sensors, meters, and data loggers to track various parameters, including electricity generation from solar panels or wind turbines, energy usage within buildings or equipment, and levels of battery storage.

Smart Meters offer detailed insights into energy consumption patterns and enable remote monitoring and control of energy systems. With the ability to track energy usage at a granular level, users can pinpoint areas for energy efficiency improvements and fine-tune energy management strategies accordingly.

Remote Monitoring Software platforms provide convenient access to energy data and system performance metrics from any location with an internet connection. These platforms often feature customizable dashboards, alerts for abnormal energy usage or system faults, and tools for historical data analysis.

Energy Management Software (EMS) facilitates the analysis of energy data, enabling users to identify trends and implement strategies for optimizing energy use. EMS platforms may include features such as load forecasting, demand response optimization, and integration with renewable energy systems to maximize self-consumption and minimize reliance on the grid.

Performance Monitoring Tools tailored for renewable energy systems offer specialized monitoring solutions for different types of installations, such as solar PV monitoring software or wind turbine monitoring systems. These tools track parameters like energy production, system efficiency, and weather conditions to assess the performance of renewable energy assets accurately.

Building Energy Management Systems (BEMS) integrate with renewable energy systems to optimize energy usage within buildings or facilities. These systems regulate HVAC systems, lighting, and other energy-consuming equipment based on real-time energy data, occupancy schedules, and weather conditions, thereby minimizing energy waste and maximizing cost savings.

Data Analytics and Predictive Maintenance tools utilize advanced algorithms to analyse large volumes of energy data, identifying patterns, trends, and anomalies. Predictive maintenance algorithms anticipate equipment failures or performance degradation in renewable energy systems, enabling proactive maintenance to minimize downtime and optimize system reliability.

Energy Storage Management Systems (ESMS) optimize the operation of energy storage components within renewable energy systems, such as batteries. ESMS platforms manage charging and discharging cycles, energy flows, and maximize the economic value of stored energy. They can also integrate with demand-side management strategies to balance supply and demand in real-time.

By leveraging these monitoring and management tools, users can effectively track energy production and consumption patterns, optimize system performance, and maximize the benefits of their renewable energy investments.

As a practical example of potential income generated by on-selling energy excesses or recycled products, Tanigawa (2017) discusses to use of biogas in US farms. Biogas, generated through the breakdown of organic materials by bacteria in an oxygen-free environment known as anaerobic digestion, is a valuable resource for energy production and soil enrichment. This process, which occurs naturally in environments like landfills and certain livestock management systems, can be optimized and controlled through the use of anaerobic digesters. These systems convert organic materials into biogas, composed mainly of methane and carbon dioxide, along with trace gases. The byproduct, known as digestate, is a nutrient-rich material beneficial for soil amendment. Different organic wastes vary in their ease of breakdown; food waste and fats are more readily digested compared to livestock waste. Co-digestion, blending multiple wastes, and maintaining warmer temperatures within digesters aid in enhancing biogas yields.

Once captured, biogas can be utilized for heat and electricity production, either through engines, microturbines, or fuel cells. It can also be upgraded into biomethane and injected into natural gas pipelines or used as vehicle fuel. In the United States, there are thousands of operational biogas systems across all states, with significant potential for further expansion.

Biogas offers numerous benefits, including serving as a clean and renewable baseload power source, replacing fossil fuels, and reducing methane emissions. Its utilization can contribute to significant greenhouse gas emission reductions and cost savings. For instance, diverting organic waste from landfills to anaerobic digestion not only generates energy revenue but also mitigates methane emissions. Furthermore, biogas production has the potential to create jobs, lower waste remediation costs, and benefit local economies.

Figure 42: Anaerobic Digestion Biogas Plant with Composting for energetic and material recycling of Municipal Biowaste in Sundern, Germany. Thzorro77, CC BY-SA 4.0, via Wikimedia Commons.

Various feedstocks, including food waste, landfill gas, livestock waste, wastewater treatment byproducts, and crop residues, contribute to biogas production. Each feedstock presents opportunities for energy recovery and waste reduction. Additionally, biogas can be utilized in different forms, such as raw biogas, digestate, or upgraded renewable natural gas (RNG), for heating, electricity generation, or vehicle fuel.

Figure 43: Biogas collection feedlot. Robert Basic from Germany, CC BY-SA 2.0 , via Wikimedia Commons.

Federal policies, such as the Renewable Fuel Standard and programs under the Farm Bill, play a crucial role in supporting the biogas industry's growth. These initiatives provide funding and incentives for biogas projects, encouraging innovation and investment in renewable energy infrastructure.

Tanigawa (2017) concludes that biogas systems offer a sustainable solution to waste management, energy production, and environmental stewardship. With proper policy support and investment, biogas has the potential to significantly contribute to the United States' energy diversification and climate change mitigation efforts.

Measuring Improvement Outcomes from the Introduction of Renewable Energy Sources and Recycling Products in Relation to Achieving Greater Sustainability

To effectively measure improvement outcomes from the introduction of renewable energy sources and recycling products in relation to achieving greater sustainability in agricultural practices, a comprehensive approach is necessary. As someone tasked with developing planning strategies to address sustainability issues in land-based businesses, you must collaborate closely with various stakeholders, including alternative energy consultants, suppliers, planning authorities, and government departments. By leveraging their expertise and resources, you can design and implement robust measurement frameworks tailored to rural, regional, and local conditions.

Firstly, establish clear sustainability goals and performance indicators that align with the overarching objectives of utilizing renewable energy and recycling systems in agricultural operations. These goals should encompass environmental, economic, and social dimensions of sustainability, considering factors such as energy efficiency improvements, waste reduction, greenhouse gas emissions mitigation, cost savings, and community engagement.

Next, conduct baseline assessments to quantify the current state of energy consumption, waste generation, and resource utilization within the agricultural operation. This baseline data will serve as a reference point for evaluating the effectiveness of interventions and tracking progress over time. Utilize tools such as energy audits, waste audits, and life cycle assessments to gather comprehensive data on resource inputs and outputs.

Implement renewable energy systems, such as solar panels, wind turbines, or biomass generators, and integrate recycling processes for organic and non-organic waste streams. Monitor the installation and performance of these systems closely, ensuring compliance with regulatory requirements and industry standards. Collaborate with alternative energy consultants and suppliers to optimize system design, operation, and maintenance practices for maximum efficiency and reliability.

Utilize advanced monitoring and data analytics tools to track key performance metrics related to energy generation, consumption, and savings. Implement real-time monitoring systems and remote sensing technologies to capture granular data on renewable energy production, energy usage patterns, waste generation rates, and material flows. Analyse this data regularly to identify trends, identify areas for improvement, and make informed decisions.

Engage stakeholders, including employees, local communities, and regulatory agencies, in the monitoring and evaluation process. Foster transparency and accountability by communicating progress updates, sharing insights, and soliciting feedback from stakeholders. Encourage participation in sustainability initiatives and seek input on potential innovations or improvements.

Periodically assess the economic viability and return on investment of renewable energy and recycling initiatives. Conduct financial analyses to evaluate the cost-effectiveness of sustainability interventions, considering factors such as upfront capital costs, operational expenses, revenue streams from energy sales or waste recycling, and long-term savings. Adjust strategies and priorities based on the results of these assessments to ensure the allocation of resources aligns with sustainability goals.

Lastly, document and report on the outcomes and impacts of renewable energy and recycling initiatives to demonstrate progress towards greater sustainability. Prepare comprehensive sustainability reports detailing key performance indicators, achievements, challenges, and lessons learned. Share these reports with stakeholders, policymakers, and the broader community to showcase the benefits of sustainable agricultural practices and inspire further action.

By implementing a systematic approach to measuring improvement outcomes from the introduction of renewable energy sources and recycling products, you can effectively assess the sustainability performance of land-based businesses and drive continuous improvement towards a more sustainable future.

Establishing clear sustainability goals and performance indicators is crucial for measuring the effectiveness of renewable energy and recycling initiatives in agricultural practices. These metrics should encompass environmental, economic, and social dimensions of sustainability. Here are some examples of metrics that can be used to establish such goals and indicators:

1. Energy Efficiency:

   ◦ Energy consumption per unit of agricultural output (e.g., per crop yield, per livestock product).

   ◦ Percentage reduction in energy consumption compared to baseline levels.

   ◦ Efficiency of renewable energy systems (e.g., solar panels, wind turbines) in converting natural resources into usable energy (e.g., kilowatt-hours produced per unit of sunlight or wind).

   ◦ Energy intensity, measured as energy used per unit area of land or per unit of agricultural product.

2. Renewable Energy Generation:

   ◦ Total renewable energy produced (e.g., kilowatt-hours of electricity generated from solar, wind, or biomass sources).

   ◦ Percentage of total energy demand met by renewable sources.

- ○ Capacity factor of renewable energy systems (i.e., actual energy output compared to maximum potential output).

- ○ Renewable energy penetration rate, indicating the proportion of energy derived from renewable sources relative to total energy consumption.

3. Waste Reduction and Recycling:

- ○ Quantity of organic waste recycled through anaerobic digestion or composting.

- ○ Percentage of total waste diverted from landfills through recycling or reuse.

- ○ Reduction in greenhouse gas emissions associated with waste management practices.

- ○ Recovery rates for recyclable materials (e.g., plastics, paper, glass) separated from agricultural waste streams.

- ○ Cost savings achieved through waste reduction and recycling initiatives.

4. Environmental Impact:

- ○ Greenhouse gas emissions intensity, measured as emissions per unit of agricultural output.

- ○ Carbon footprint of agricultural activities, including emissions from energy use, livestock, and land use changes.

- ○ Water usage efficiency, expressed as water consumption per unit of agricultural output.

- ○ Biodiversity conservation measures implemented (e.g., habitat restoration, preservation of native species).

5. Economic Viability:

- ○ Return on investment (ROI) for renewable energy systems and recycling infrastructure.

- ○ Cost savings achieved through energy efficiency improvements and waste reduction measures.

- ○ Revenue generated from the sale of renewable energy or recycled products (e.g., biogas, compost).

- ○ Net present value (NPV) of sustainability investments over their lifecycle.

6. Social and Community Benefits:

- ○ Stakeholder engagement levels and participation in sustainability initiatives.

- ○ Employment opportunities created through renewable energy and recycling projects.

- ○ Health and well-being improvements for workers and local communities (e.g., reduced air and water pollution).

- ○ Contributions to local economic development and resilience.

These metrics can be tailored to specific agricultural operations and regional contexts, reflecting the unique challenges and opportunities faced by land-based businesses. By defining clear sustainability goals and performance indicators across these dimensions, stakeholders can effectively monitor progress, evaluate outcomes, and drive continuous improvement towards greater sustainability.

Various organizations and institutions have developed and implemented numerous frameworks for evaluating the outcomes of renewable energy initiatives. For instance, the International Renewable Energy Agency (IRENA) has introduced the Renewable Energy Benefits

(REB) framework, offering a comprehensive and standardized method to gauge the environmental, social, and economic advantages of renewable energy deployment. The Greenhouse Gas Protocol (GHG Protocol), devised by the World Resources Institute (WRI) and the World Business Council for Sustainable Development (WBCSD), stands as a widely recognized framework for assessing and reporting greenhouse gas emissions from renewable energy projects. Moreover, the Multi-Tier Framework (MTF) by the World Bank presents a holistic and nuanced approach to measuring and monitoring aspects such as energy access, affordability, quality, reliability, and safety across various energy sources and technologies. Additionally, the Social Return on Investment (SROI) framework developed by the SROI Network serves to measure and appraise the social, environmental, and economic outcomes of renewable energy initiatives from the perspective of stakeholders.

The Renewable Energy Benefits (REB) Framework, developed by the International Renewable Energy Agency (IRENA), provides a structured and standardized method for evaluating the positive impacts of renewable energy deployment (Gilbert & Sovacool, 2016). The framework involves several key steps, beginning with the identification and categorization of the various benefits associated with renewable energy deployment, typically falling into environmental, social, and economic domains (Gilbert & Sovacool, 2016). Following this, data collection ensues to quantify and qualify these benefits, encompassing aspects such as energy generation, emissions reductions, job creation, community development, and other pertinent factors (Gilbert & Sovacool, 2016). The REB Framework employs specific methodologies tailored to assess each category of benefits, such as evaluating environmental benefits through metrics like greenhouse gas emissions reductions, air quality enhancements, and water conservation (Gilbert & Sovacool, 2016). Social benefits could involve examining factors like job creation, community empowerment, and health improvements, while economic benefits may include cost savings, economic growth, and increased market competitiveness (Gilbert & Sovacool, 2016).

To ensure quantification and standardization, the framework endeavours to express benefits in a consistent and comparable format across different renewable energy projects or technologies, involving the use of common metrics and units of measurement to facilitate clear comparison (Gilbert & Sovacool, 2016). Subsequently, the quantified benefits undergo evaluation to gauge the overall impact of renewable energy deployment, which may entail comparing benefits against associated costs or trade-offs associated with the project (Gilbert & Sovacool, 2016). Finally, the results of the assessment are reported to stakeholders, policymakers, and other interested parties to inform decision-making and encourage investment in renewable energy (Gilbert & Sovacool, 2016).

Overall, the REB Framework offers a comprehensive and systematic approach to assessing the multitude of benefits associated with renewable energy deployment, thereby illustrating its value proposition and supporting its continued growth and adoption (Gilbert & Sovacool, 2016).

The Greenhouse Gas Protocol (GHG Protocol) stands as a framework devised collaboratively by the World Resources Institute (WRI) and the World Business Council for Sustainable Development (WBCSD). It serves as a widely acknowledged methodology for the assessment and reporting of greenhouse gas (GHG) emissions, encompassing those emanating from renewable energy projects. The process of the GHG Protocol generally unfolds in several steps.

Firstly, the GHG Protocol initiates by defining the scope of emissions to be measured. This entails identifying the sources of GHG emissions linked with the renewable energy project, encompassing both direct emissions (Scope 1) and indirect emissions from purchased electricity (Scope 2).

Subsequently, data collection ensues following the determination of the scope. This phase involves the gathering of information on energy consumption, fuel usage, and other pertinent activities associated with the renewable energy endeavour.

Using the collected data, emissions are computed utilizing specific emission factors and calculation methodologies outlined within the GHG Protocol. For renewable energy projects, emissions could encompass those correlated with various phases such as construction, operation, decommissioning, as well as upstream emissions from the supply chain.

Upon the calculation of emissions, they are reported employing standardized formats and units of measurement as prescribed by the GHG Protocol. This standardization facilitates consistency and comparability of emissions data across diverse projects or organizations.

Depending on stakeholder or regulatory requisites, emissions data may undergo verification by independent third parties to ensure its accuracy and reliability.

Lastly, based on the emissions data and insights derived from the GHG Protocol assessment, organizations can formulate action plans aimed at reducing emissions, enhancing energy efficiency, and mitigating climate impacts associated with their renewable energy projects.

Overall, the GHG Protocol offers a systematic and rigorous approach to the measurement and reporting of greenhouse gas emissions from renewable energy projects, thereby assisting organizations in comprehending and managing their carbon footprint effectively. Additionally, it fosters transparency and accountability in climate-related disclosures, thereby supporting broader endeavours to combat climate change and transition towards a low-carbon economy.

The Multi-Tier Framework (MTF) developed by the World Bank provides a comprehensive approach to measuring and monitoring various aspects related to energy access, affordability, quality, reliability, and safety across different energy sources and technologies (Manogaran et al., 2019). The framework begins with a comprehensive assessment of energy access and related factors, considering not only the availability of energy sources but also their quality, reliability, affordability, and safety (Manogaran et al., 2019). It employs a tiered classification system to categorize households or communities based on their level of energy access, ranging from Tier 0 (no access) to Tier 5 (high-quality, reliable, and affordable access) (Manogaran et al., 2019). The MTF utilizes a set of indicators and metrics to assess different dimensions of energy access and service quality, such as electricity availability, duration of power outages, energy expenditure as a percentage of income, energy source reliability, and safety standards (Manogaran et al., 2019).

Data is collected through surveys, interviews, and other research methods to gather information on energy access and related indicators, involving collaboration with national statistical agencies, utilities, and other stakeholders (Manogaran et al., 2019). Once collected, the data is analysed to determine the energy access status and tier classification of households or communities, and the findings are reported in a clear and understandable manner, often using visualizations and dashboards to communicate results to policymakers, practitioners, and other stakeholders (Manogaran et al., 2019). The insights generated from the MTF analysis can inform the development of policies, programs, and interventions aimed at improving energy access and service quality, including targeted investments in infrastructure, technology, and capacity building efforts to address specific gaps identified through the framework (Manogaran et al., 2019). Furthermore, the MTF provides a framework for ongoing monitoring and evaluation of energy access initiatives over time, allowing stakeholders to track progress, identify emerging challenges, and adjust strategies as needed to ensure sustainable improvements in energy access and service delivery (Manogaran et al., 2019).

The Social Return on Investment (SROI) Framework, developed by the SROI Network, provides a structured and participatory approach to assessing the holistic impact of renewable energy initiatives (Guerry et al., 2015). The framework begins with stakeholder engagement, involving the identification and involvement of relevant stakeholders impacted by the renewable energy project. This participatory process allows for the mapping out of intended outcomes and impacts of the project across social, environmental, and economic dimensions. Subsequently, the framework employs various valuation techniques to assign monetary values to the identified outcomes, which may involve market prices, cost savings, or willingness-to-pay assessments. After valuing the outcomes, the framework assesses the extent to which the renewable energy project has achieved its intended outcomes and impacts by collecting data and evidence to measure the changes brought about by the project. The SROI Framework then calculates the Social Return on Investment (SROI) ratio, representing the social and environmental value generated per unit of investment in the renewable energy project. Additionally, the framework conducts sensitivity analysis to assess the robustness of the SROI calculation, accounting for uncertainties and variability in the data. Finally, the findings and insights derived from the analysis are communicated to stakeholders, policymakers, investors, and other interested parties through SROI reports, case studies, or presentations.

The SROI Framework aligns with the principles of natural capital and ecosystem services, providing a methodological approach to inform decisions related to the conservation of natural resources and ecology. It emphasizes the valuation of outcomes, which is essential in understanding the economic benefits generated by renewable energy projects. Moreover, the framework's focus on stakeholder engagement and collaborative outcome mapping resonates with the influence of clients on valuations, highlighting the importance of incorporating the perspectives of various stakeholders in the valuation process. Additionally, the SROI Framework's emphasis on assessing the holistic impact of renewable energy initiatives aligns with the concept of economic valuation and the commodification of ecosystem services, as it seeks to quantify the social, environmental, and economic benefits of such projects.

Conducting Risk Assessment for Work Health and Safety Hazards Associated with Renewable Energy and Recycling Initiatives

Ensuring the safety of your team in the unique fields of renewable energy and recycling initiatives begins with a thorough understanding of both common and specific hazards inherent in these occupations. Employers must be diligent in recognizing potential risks and implementing solutions in compliance with pertinent health and safety regulations.

One significant hazard to consider is confined space, particularly prevalent in biofuels production. Ethanol and biodiesel, two primary forms of biofuels, entail processes that can generate hazardous materials such as acids, bases, and flammable substances. Ethanol, for instance, is highly flammable and poses risks of ignition under ordinary conditions. Similarly, biodiesel production involves the use of potentially caustic and combustible materials, necessitating careful management to protect workers.

Another area of concern is geothermal energy, which relies on harnessing heat from the earth's interior. Hazards faced by geothermal workers mirror those in construction, including exposure to dangerous chemicals like glycol or hydrogen sulfide used in energy extraction processes. Moreover, risks associated with drilling and excavation equipment, coupled with inadequate safety measures for workers entering trenches or wellheads, must be addressed.

In the realm of green roofs, workers face hazards associated with industrial equipment like cranes, derricks, and hoists used during installation. Exposure to electrical equipment and power lines further compounds the risk. Additionally, the hardscaping process may expose workers to silica dust, known to cause debilitating respiratory conditions like silicosis.

Hydrogen fuel cells, employed to generate electricity, bring their own set of hazards, including fire, explosion, freeze burns, arc flashes, and electrocution. Meanwhile, recycling initiatives, integral to sustainability efforts, present risks akin to those in construction, such as exposure to hazardous materials, traffic safety concerns, and slips, trips, and falls.

In the solar energy sector, workers encounter hazards throughout manufacturing, installation, and maintenance processes. These include arc flashes, electrical shock, falls, thermal burns, and extreme weather conditions. Industrial equipment like cranes and hoists also pose risks if not properly inspected and maintained.

Weatherization activities involve inspecting buildings, improving air quality, and reducing utility usage through insulation and sealing. Crews working with materials like fiberglass and spray foam face risks of bodily irritation, respiratory issues, and exposure to electrical hazards and fire.

Finally, wind energy, while offering numerous benefits, presents hazards ranging from confined space work to electrical arcs, falls, respiratory issues, and machinery-related injuries. Workers must be equipped with proper training, personal protective equipment (PPE), and safety measures to mitigate these risks effectively.

Renewable energy has been identified as having a significant impact on agriculture, particularly in reducing CO_2 emissions and improving agricultural production (Jebli & Youssef, 2017). The integration of renewable energy systems in agriculture presents a great opportunity to mitigate climatic risks and enhance agricultural sustainability (Acosta-Silva et al., 2019). However, the use of renewable energy in agriculture also poses unique challenges to health and safety, especially in the context of climate change and automation (Comi et al., 2023). Despite these challenges, the use of renewable energy in agriculture offers the potential to solve various challenges related to the use of fossil fuels and reduce environmental pollutants and greenhouse gas emissions (Sobczak & Sobczak, 2022).

There is a growing body of literature that addresses health and safety risks in agriculture, emphasizing the need for comprehensive studies and integrated approaches to address these challenges (Aksüt & Eren, 2023). Furthermore, renewable energy can be utilized in agriculture for heating, cooling, and biofuel production, highlighting its potential to transform energy use in the agricultural sector (Rokicki et al., 2021). The significant untapped potential for renewable energy in agriculture, particularly in the production of biofuels, presents an economically and environmentally efficient opportunity for the industry (Yakubiv et al., 2019).

Conducting risk assessment for work health and safety hazards associated with renewable energy and recycling initiatives involves a systematic process to identify, evaluate, and mitigate potential risks to ensure the safety and well-being of workers and stakeholders involved in land-based businesses. As individuals responsible for developing planning strategies to address sustainability issues, including renewable energy and recycling systems, it's imperative to approach risk assessment with diligence and expertise.

Firstly, it's essential to establish a comprehensive understanding of the specific renewable energy and recycling initiatives being implemented within the agricultural context. This includes identifying the types of renewable energy technologies being utilized, such as solar panels, wind turbines, or biomass systems, as well as the recycling processes involved in managing agricultural waste streams.

Once the initiatives are identified, the next step is to conduct a thorough assessment of the potential work health and safety hazards associated with each aspect of the renewable energy and recycling systems. This may include hazards related to equipment operation, installation, maintenance, and exposure to hazardous materials or substances.

To effectively assess risks, collaboration with external experts such as alternative energy consultants and suppliers of alternative energy equipment is crucial. These experts can provide specialized knowledge and insights into the specific hazards and risks associated with renewable energy technologies, as well as best practices for mitigating them.

Additionally, consulting with planning authorities and government departments administering subsidies and providing advice on sustainability initiatives can offer valuable regulatory guidance and compliance requirements related to work health and safety standards.

The risk assessment process should involve thorough site inspections and observations to identify potential hazards and assess their severity and likelihood of occurrence. This may include evaluating factors such as equipment integrity, installation practices, environmental conditions, and human factors.

Furthermore, it's essential to engage with workers and stakeholders directly involved in the implementation and operation of renewable energy and recycling systems. Their firsthand knowledge and experiences can provide valuable insights into potential hazards and risks that may not be immediately apparent during initial assessments.

Once hazards are identified and evaluated, appropriate control measures and risk mitigation strategies should be developed and implemented. This may include implementing engineering controls, administrative controls, personal protective equipment (PPE), training programs, and emergency response procedures to minimize risks and ensure the safety of workers and stakeholders.

Regular monitoring and review of the implemented control measures are essential to ensure their effectiveness and identify any emerging hazards or changes in risk levels. Continuous improvement processes should be integrated to adapt to evolving circumstances and maintain a safe working environment.

In summary, conducting risk assessment for work health and safety hazards associated with renewable energy and recycling initiatives requires a collaborative and systematic approach, involving expertise from external consultants, regulatory authorities, and stakeholders. By prioritizing safety and implementing robust risk management practices, land-based businesses can develop sustainable agricultural practices that utilize renewable energy and recycling systems effectively while safeguarding the well-being of workers and communities.

Chapter Six

Developing Strategies to Use Renewable Energy

Developing Strategies to Reduce Greenhouse Gas Emissions and Use Renewable Energy Resources, Recycled Products and Improved Work Practices

Developing strategies to reduce greenhouse gas emissions and utilize renewable energy resources, recycled products, and improved work practices in the agricultural sector requires a multifaceted approach and a deep understanding of sustainability principles. As someone tasked with managing land-based businesses and consulting with various stakeholders, including alternative energy experts, planning authorities, and government departments, your role is crucial in spearheading these initiatives.

One key strategy is to prioritize the adoption of renewable energy technologies such as solar panels, wind turbines, and biomass systems. By harnessing the power of renewable energy sources, agricultural operations can significantly reduce their reliance on fossil fuels and decrease greenhouse gas emissions associated with traditional energy production methods. Collaborating with alternative energy consultants and suppliers will facilitate the selection and implementation of the most suitable renewable energy solutions tailored to the specific needs and conditions of rural, regional, and local agricultural settings.

Additionally, integrating energy-efficient practices and technologies into daily operations can further enhance sustainability efforts and mitigate greenhouse gas emissions. This may include optimizing irrigation systems, improving building insulation, and upgrading equipment to minimize energy consumption. By investing in energy-efficient infrastructure and adopting best practices, land-based businesses can achieve substantial reductions in greenhouse gas emissions while simultaneously lowering operating costs.

Furthermore, incorporating recycled products and materials into agricultural processes can significantly reduce the carbon footprint associated with resource extraction, manufacturing, and disposal. Partnering with suppliers of recycled products and implementing waste management strategies that prioritize recycling and reuse will not only contribute to greenhouse gas mitigation but also promote circular economy principles within the agricultural sector. This may involve composting organic waste, using recycled water for irrigation, and sourcing recycled materials for infrastructure development and packaging.

Improved work practices play a pivotal role in achieving sustainability goals and minimizing greenhouse gas emissions throughout the agricultural supply chain. Providing training and education to employees on energy conservation, waste reduction, and sustainable farming techniques empowers them to implement environmentally responsible practices on the ground. Additionally, fostering a culture of innovation and continuous improvement encourages the adoption of new technologies and methodologies that further enhance sustainability performance.

Collaboration with planning authorities and government departments responsible for administering subsidies and providing advice on sustainable energy initiatives is essential for securing financial support and navigating regulatory requirements. By leveraging available incentives and resources, land-based businesses can overcome financial barriers and accelerate the transition to renewable energy and sustainable practices.

Developing strategies to reduce greenhouse gas emissions and embrace renewable energy resources, recycled products, and improved work practices in the agricultural sector requires proactive leadership, collaboration, and innovation. By leveraging your expertise and consulting with external experts and stakeholders, you can drive meaningful change and pave the way for a more sustainable future in rural, regional, and local agricultural communities.

Reducing greenhouse gas (GHG) emissions stemming from agricultural activities necessitates more efficient management of carbon and nitrogen flows within agricultural systems. Recognized as beneficial management practices, these approaches aim to lower GHG emissions while enhancing agricultural sustainability (Government of British Columbia, 2022).

Livestock and manure management represent significant contributors to agricultural GHG emissions. Implementing certain practices can effectively sequester carbon and mitigate GHG emissions associated with livestock and manure. These practices include the use of livestock feed additives, rotational grazing to sequester carbon in the soil, selecting high-quality feed to reduce methane released from enteric fermentation, and managing manure to minimize methane and nitrous oxide emissions (Government of British Columbia, 2022). Strategies such as covering manure storage facilities, optimizing manure use with nutrient management plans, and capturing and combusting methane from manure storage are instrumental in this regard.

Soil conservation and carbon sequestration play a pivotal role in mitigating GHG emissions from agricultural ecosystems. Various farm practices promote carbon sequestration by either increasing carbon storage or reducing the loss of stored carbon. These practices encompass enhancing nitrogen management through nutrient management planning, reducing tillage, decreasing bare fallow, returning crop residues to the soil, establishing agroforestry systems, increasing cover cropping, and implementing rotational grazing (Government of British Columbia, 2022).

Energy conservation and fuel switching offer opportunities for reducing GHG emissions within farm operations. Conducting on-farm, all-fuel energy assessments can identify energy-saving opportunities, while ensuring the maintenance of heating and cooling systems can enhance efficiency. Employing timers, sensors, or variable speed drives on ventilation, heating, cooling, and lighting systems further contributes to energy conservation. Additionally, replacing fossil-fuel-powered equipment with electrical pumps and motors can facilitate a transition towards cleaner energy sources.

On-farm energy production through renewable sources presents a viable solution to displace fossil fuel use and decrease GHG emissions both on and off the farm. Moreover, these renewable energy technologies offer economic diversification opportunities for agricultural producers while reducing reliance on energy sources with volatile prices. Anaerobic digestion, geothermal systems, solar thermal infrastructure, wind turbines, solar panels (photovoltaic), and rechargeable batteries are among the renewable energy technologies suitable for on-farm use, contributing to a more sustainable agricultural sector (Government of British Columbia, 2022).

The following provides a sample strategy based on the principles outlined above.

Sample Strategy: Implementing Sustainable Practices in Agriculture

Our strategy revolves around reducing greenhouse gas emissions and embracing renewable energy resources, recycled products, and improved work practices. This strategy encompasses a multifaceted approach aimed at maximizing environmental stewardship, operational efficiency, and economic viability.

1. ***Prioritizing Renewable Energy Adoption:***

   ○ *Identify suitable renewable energy technologies such as solar panels, wind turbines, and biomass systems for integration into agricultural operations.*

   ○ *Collaborate with alternative energy experts and suppliers to assess feasibility, select appropriate solutions, and oversee implementation.*

   ○ *Leverage available incentives and subsidies provided by government departments to offset initial investment costs and accelerate adoption.*

2. *Enhancing Energy Efficiency:*

  ○ *Implement energy-efficient practices and technologies across all facets of operations, including irrigation systems, buildings, and equipment.*

  ○ *Conduct energy audits to identify areas for improvement and prioritize investments in energy-saving measures.*

  ○ *Provide training and support to employees to promote energy conservation and optimize resource utilization.*

3. *Integrating Recycled Products:*

  ○ *Partner with suppliers offering recycled products and materials to substitute conventional inputs in agricultural processes.*

  ○ *Develop waste management strategies focused on recycling and reuse to minimize the carbon footprint associated with resource extraction and disposal.*

  ○ *Implement composting programs for organic waste and explore opportunities to utilize recycled water for irrigation purposes.*

4. *Promoting Sustainable Work Practices:*

  ○ *Offer comprehensive training programs to employees on sustainable farming techniques, waste reduction, and environmental stewardship.*

  ○ *Foster a culture of innovation and continuous improvement by encouraging employee engagement and participation in sustainability initiatives.*

  ○ *Implement performance monitoring mechanisms to track progress, identify areas for optimization, and celebrate achievements in sustainability.*

5. *Collaborating with Stakeholders:*

  ○ *Engage with planning authorities, government departments, and regulatory bodies to stay informed about sustainable energy initiatives, subsidies, and regulatory requirements.*

  ○ *Leverage partnerships and collaborations with external stakeholders to access additional resources, expertise, and funding opportunities.*

  ○ *Advocate for supportive policies and incentives that incentivize the adoption of renewable energy and sustainable practices in agriculture.*

By implementing these strategies, our aim is to lead the transition towards a more sustainable agricultural sector that minimizes its environmental impact, maximizes resource efficiency, and ensures long-term resilience and prosperity for rural, regional, and local communities. Through proactive leadership, collaboration, and innovation, we can drive meaningful change and pave the way for a greener and more sustainable future in agriculture.

Estimating Plant, Material, Labour and Other Associated Costs in Consultation with Appropriate Person or Organisation

Estimating plant, material, labour, and other associated costs stands as a pivotal aspect in the development of sustainable agricultural practices that leverage renewable energy and recycling systems. As a manager of land-based businesses tasked with liaising with diverse stakeholders, including alternative energy experts, suppliers, planning authorities, and governmental departments, your role holds significant weight in ensuring precise cost estimation and efficient resource distribution.

To begin, it's imperative to identify the relevant stakeholders and experts in renewable energy systems, recycling technologies, and sustainable agricultural practices. This may encompass engaging with alternative energy consultants, renewable energy equipment suppliers, waste management specialists, and agricultural experts. Collaborating with these stakeholders allows for valuable insights into the costs involved in implementing tailored sustainable solutions suitable for rural, regional, and local agricultural contexts.

Moving forward, a thorough assessment of renewable energy options is crucial. Working closely with alternative energy consultants and suppliers facilitates the evaluation of various renewable energy technologies such as solar panels, wind turbines, biomass systems, and anaerobic digesters. Factors such as installation costs, equipment maintenance, energy output, and potential utility bill savings should be carefully considered. Conducting comprehensive cost-benefit analyses aids in identifying the most economically viable renewable energy solutions for agricultural operations.

Simultaneously, analysing recycling systems and waste management technologies is essential. Collaboration with waste management experts enables the assessment of recycling infrastructure options such as composting facilities, anaerobic digesters, and material recovery facilities. Evaluating associated costs, including equipment procurement, facility construction, operational maintenance, and regulatory compliance, provides clarity on the financial implications of implementing recycling initiatives.

Estimating plant and material costs requires close collaboration with suppliers and contractors. Obtaining quotations for renewable energy equipment, recycling machinery, and other necessary materials enables an accurate assessment of upfront expenses. Considerations such as installation costs, transportation expenses, and additional fees should be factored in to ensure a comprehensive estimation of plant and material costs.

Likewise, calculating labour costs demands collaboration with agricultural specialists and labour experts. Estimating the labour hours required for the installation, operation, and maintenance of renewable energy systems, recycling facilities, and agricultural infrastructure is essential. Determining wages, salaries, and benefits for employees involved in sustainable agricultural activities, including training and supervision, ensures an accurate estimation of labour costs.

Consulting with planning authorities and government departments plays a crucial role in accessing financial support and navigating regulatory requirements. Seeking guidance on available subsidies, grants, tax incentives, and funding opportunities helps offset the costs associated with implementing sustainable practices. Collaboration with these entities ensures compliance with regulations and optimizes resource allocation.

Conducting comprehensive cost-benefit analyses is paramount in evaluating the economic viability of sustainable agricultural practices. Comparing the estimated costs of renewable energy systems, recycling infrastructure, and waste management technologies against projected benefits, including energy savings and environmental impact mitigation, informs decision-making. Leveraging financial modelling tools and feasibility studies facilitates informed prioritization of investments based on their potential for positive outcomes.

In essence, by estimating plant, material, labour, and associated costs in collaboration with relevant stakeholders, you can develop robust planning strategies to address sustainability challenges. Your collaboration with experts, stakeholders, and regulatory authorities ensures accurate cost estimation, efficient resource allocation, and successful implementation of sustainable initiatives tailored to rural, regional, and local agricultural contexts.

Collaborating with suppliers and contractors to estimate the costs of purchasing renewable energy equipment, recycling machinery, and other necessary materials involves several key steps. First, you need to identify reputable suppliers and contractors who specialize in the relevant products and services for sustainable agricultural practices. This entails conducting thorough research to ensure you choose reliable partners with the expertise and resources needed for your project.

Once you've identified potential suppliers and contractors, the next step is to reach out to them and request quotations for the required equipment and materials. Providing detailed specifications and quantities is essential to ensure accurate quotations that align with your project requirements. Clear communication regarding your specific needs, including any preferences for brands, models, or performance criteria, helps suppliers and contractors tailor their quotations to meet your expectations.

After receiving quotations from various suppliers and contractors, it's crucial to evaluate them carefully. This involves reviewing factors such as pricing, quality, warranty, delivery timelines, and any other terms and conditions outlined in the quotations. Additionally, you should consider potential additional costs beyond the purchase price, such as installation expenses, transportation fees, taxes, and permit fees.

Comparing the options presented by different suppliers and contractors allows you to assess the overall value proposition offered by each. Factors to consider include cost-effectiveness, reliability, technical support, and after-sales service. Engaging in negotiations with suppliers and contractors can help you secure competitive pricing and favourable terms, including any available discounts, incentives, or package deals.

Once you've selected preferred suppliers and contractors, it's time to finalize contracts or purchase orders detailing the agreed-upon terms, pricing, delivery schedules, payment terms, and any other relevant conditions. Coordination with suppliers and contractors is essential to arrange for the timely delivery of the equipment and materials to your agricultural site. Proper scheduling and logistics planning minimize disruptions and delays during the implementation process.

Throughout the implementation phase, it's essential to monitor progress closely to ensure that the purchased equipment and materials are installed and integrated into your sustainable agricultural practices according to plan. Addressing any issues or concerns promptly helps maintain project timelines and quality standards, ensuring the successful realization of your sustainability goals. Following these steps enables effective collaboration with suppliers and contractors to estimate costs and procure the necessary resources for implementing sustainable agricultural practices.

Estimating labour costs for implementing sustainable practices on a farm involves a methodical process that entails collaboration with agricultural specialists and labour experts. Here's a breakdown of the typical steps involved:

To begin, it's crucial to engage closely with agricultural specialists well-versed in sustainable farming practices. These experts offer valuable insights into the labour needs for installing renewable energy systems, managing recycling facilities, and executing other sustainable initiatives on your farm.

Together with agricultural specialists and labour experts, conduct a thorough assessment of the labour hours required for various tasks, including installation, operation, and maintenance of renewable energy systems, recycling facilities, and agricultural infrastructure. This assessment considers factors such as project scope, task complexity, and requisite skill sets.

Next, break down project tasks into specific activities such as site preparation, equipment installation, monitoring, maintenance, and troubleshooting. Estimate the labour resources needed for each task, drawing upon industry standards and best practices to ensure accuracy.

Determine appropriate wages, salaries, and benefits for employees engaged in sustainable agricultural activities. Consult labour experts or industry benchmarks to establish competitive compensation rates that attract skilled workers and incentivize performance.

Factor in the costs associated with training and supervising employees involved in sustainable practices. Training programs may be necessary to equip workers with the requisite skills and knowledge for operating renewable energy systems, managing recycling facilities, and implementing sustainable farming techniques safely and effectively.

Consider additional benefits provided to employees, such as health insurance and retirement plans, which contribute to overall labour costs associated with sustainable practices on the farm.

Integrate estimated labour costs into the overall budget for the sustainable agriculture project, ensuring strategic allocation of resources to cover labour expenses while meeting other project requirements.

Continuously monitor labour costs throughout the project implementation phase, reviewing labour schedules, productivity levels, and budget allocations regularly. This ongoing review allows for identification of any deviations from initial estimates, enabling necessary adjustments to optimize resource utilization.

By following this comprehensive approach and collaborating with agricultural specialists and labour experts, you can effectively estimate labour costs for implementing sustainable practices on your farm. This ensures efficient allocation of labour resources and facilitates smooth progress toward achieving sustainability goals. As an example, let's apply the process of estimating labour costs for implementing sustainable practices to a hydroponic farm:

1. **Collaboration with Specialists:**

   ○ Engage with agricultural specialists experienced in hydroponic farming methods. These specialists can provide insights into the labour requirements specific to hydroponic systems, such as setup, maintenance, and crop management.

2. **Assessment of Labor Requirements:**

   ○ Work with hydroponic experts to assess the labour hours needed for tasks like constructing hydroponic infrastructure (e.g., setting up grow beds, installing irrigation systems), planting and transplanting crops, monitoring nutrient levels and pH, managing pests and diseases, and harvesting produce.

3. **Task Breakdown:**

   ○ Break down the project tasks into specific activities. For instance, installation tasks may include setting up grow beds, connecting irrigation lines, and installing nutrient delivery systems. Ongoing maintenance tasks may involve monitoring water quality, adjusting nutrient levels, pruning plants, and replacing or repairing equipment as needed.

4. **Wage Determination:**

   ○ Determine appropriate wages for employees based on the skill level required for each task. Skilled labour, such as experienced hydroponic technicians or crop specialists, may command higher wages than general labourers responsible for routine maintenance tasks.

5. **Training and Supervision:**

   ○ Allocate resources for training programs to ensure that workers are proficient in hydroponic farming techniques and equipment operation. Supervision may be required, especially during the initial setup phase, to ensure that tasks are performed correctly and efficiently.

6. **Consideration of Benefits:**

   ○ Factor in additional benefits such as health insurance, retirement plans, and bonuses to attract and retain skilled workers. Offering competitive benefits can enhance employee satisfaction and productivity, contributing to the overall success of the hydroponic farm.

7. **Budgeting and Financial Planning:**

   ○ Integrate estimated labour costs into the farm's overall budget, taking into account other expenses such as equipment purchases, utilities, and operational costs. Allocate sufficient funds to cover labour expenses while maintaining financial sustainability.

8. **Regular Review and Adjustment:**

   ○ Continuously monitor labour costs throughout the operation of the hydroponic farm. Adjust labour schedules and resource allocations as needed to optimize productivity and ensure efficient use of labour resources.

By following these steps and collaborating with specialists in hydroponic farming, you can accurately estimate labour costs and effectively manage resources to implement sustainable practices on your hydroponic farm. This systematic approach helps ensure the success and profitability of the operation while promoting sustainability and environmental stewardship.

Developing a Budget for Estimated Plant, Material, Labour and Other Associated Costs

Developing a budget for estimated plant, material, labour, and other associated costs is a methodical process aimed at ensuring precise financial planning and resource allocation. Here's a step-by-step guide on how to develop such a budget:

Firstly, start by gathering detailed cost estimates for plant, material, labour, and other associated costs from relevant stakeholders, suppliers, contractors, and specialists. Obtain quotations and pricing information for all required resources and services to establish a baseline.

Next, categorize the estimated costs into different expense categories. These may include plant and equipment costs (such as renewable energy equipment and recycling machinery), material costs (including construction materials and supplies), labour costs (encompassing wages, salaries, and benefits), and other associated costs like permits, transportation, and overhead expenses.

Break down the budget estimates by project phases or activities, if applicable. This helps in allocating resources efficiently and tracking expenses throughout the project lifecycle. For example, separate budgets may be created for installation, operation, and maintenance phases.

Include contingency funds in the budget to account for unexpected expenses or changes in project scope. Allocating a certain percentage of the total budget (e.g., 5-10%) as contingency helps mitigate risks and uncertainties that may arise during the project.

Factor in overhead costs associated with project management, administration, and indirect expenses. This may include salaries for project managers, office rent, utilities, insurance, and other administrative expenses that support the project's execution.

Account for taxes, permit fees, and other regulatory expenses that may apply to the project. Ensure that all applicable taxes and fees are included in the budget estimates to avoid any financial surprises during project execution.

Review and validate the cost estimates with relevant stakeholders and experts to ensure accuracy and completeness. Verify that the estimates align with project requirements, timelines, and quality standards to minimize discrepancies.

Document the assumptions, methodologies, and sources used to develop the budget estimates. This provides transparency and accountability, allowing stakeholders to understand the basis of the budget and any underlying assumptions made.

Consolidate all cost estimates into a comprehensive budget document. Ensure that the budget is well-organized, clearly formatted, and easy to understand. Seek approval from appropriate authorities or stakeholders before finalizing the budget for implementation.

Once the budget is finalized, establish systems and processes to monitor and track expenses throughout the project lifecycle. Regularly review actual expenditures against budgeted amounts and make adjustments as necessary to stay within budget constraints.

By following these steps meticulously, you can develop a comprehensive budget for estimated plant, material, labour, and other associated costs, ensuring effective financial management and successful project execution.

The following is a hypothetical sample budget for developing estimated plant, material, labour, and other associated costs for a sustainability project:

1. *Gather Cost Estimates*:

    ○ *Renewable Energy Equipment: $50,000*

    ○ *Recycling Machinery: $30,000*

    ○ *Construction Materials: $20,000*

    ○ *Supplies: $5,000*

  ○ *Labor Costs (including wages, salaries, benefits): $80,000*

  ○ *Permits and Fees: $3,000*

  ○ *Contingency (10% of total budget): $18,000*

2. **Categorize Expenses**:

  ○ *Plant and Equipment Costs: $80,000 (Renewable Energy Equipment + Recycling Machinery)*

  ○ *Material Costs: $25,000 (Construction Materials + Supplies)*

  ○ *Labor Costs: $80,000*

  ○ *Other Associated Costs: $21,000 (Permits and Fees + Contingency)*

3. **Breakdown by Project Phases**:

  ○ *Installation Phase Budget: $60,000*

  ○ *Operation Phase Budget: $50,000*

  ○ *Maintenance Phase Budget: $46,000*

4. **Consider Contingencies**:

  ○ *Contingency Funds: $18,000 (10% of total budget)*

5. **Factor in Overhead Costs**:

  ○ *Project Management and Administration: $10,000*

  ○ *Indirect Expenses: $8,000*

6. **Account for Taxes and Fees**:

  ○ *Taxes: $2,000*

  ○ *Permit Fees: $1,000*

7. **Review and Validate Estimates**:

  ○ *Review with Relevant Stakeholders and Experts: Done*

8. **Document Assumptions and Methodologies**:

  ○ *Assumptions and Methodologies Documented: Yes*

9. **Finalize the Budget**:

  ○ *Total Budget: $206,000*

  ○ *Budget Document: Reviewed and Approved*

10. *Monitor and Track Expenses:*

  ○ *Establish Systems and Processes for Expense Monitoring: Implemented*

This sample budget provides a comprehensive breakdown of estimated costs, categorized by expense type and project phase. It includes provisions for contingencies, overhead costs, taxes, and fees, ensuring effective financial management throughout the project lifecycle. Regular monitoring and tracking of expenses will help stay within budget constraints and ensure successful project execution.

Developing a Work Plan for the Introduction of Renewable Energy Resources and Recycled Products

Developing a work plan for the introduction of renewable energy resources and recycled products in the context of sustainable agricultural practices requires a comprehensive and strategic approach. As someone tasked with managing land-based businesses and collaborating with various stakeholders, including alternative energy consultants, suppliers, planning authorities, and government departments, your role is pivotal in orchestrating this transition towards sustainability.

Firstly, conducting an initial assessment of the current energy consumption patterns and waste management practices on your agricultural operation is crucial. Set clear goals and objectives for integrating renewable energy resources and recycled products into your agricultural practices, while defining key performance indicators (KPIs) to measure progress and success.

Engage with relevant stakeholders, including alternative energy consultants, suppliers of renewable energy equipment, waste management experts, and regulatory authorities. Collaborate with these stakeholders to gather insights, expertise, and support for the introduction of renewable energy and recycling initiatives.

Identify suitable renewable energy resources and recycled products that align with the specific needs and conditions of your agricultural operation. Procure necessary resources and equipment from reputable suppliers, ensuring quality, reliability, and compatibility with existing infrastructure.

Conduct a thorough site assessment to determine the feasibility of implementing renewable energy systems and recycling infrastructure on your agricultural property. Develop site-specific plans for the installation and integration of renewable energy and recycling facilities, considering factors such as available sunlight, wind patterns, land topography, water sources, and waste generation rates.

Familiarize yourself with relevant regulations, permits, and codes governing the installation and operation of renewable energy systems and recycling facilities in your jurisdiction. Obtain necessary permits and approvals from regulatory authorities to ensure compliance with environmental and safety standards.

Execute the installation and implementation of renewable energy systems and recycling infrastructure according to the established plans and timelines. Coordinate with contractors, technicians, and labourers to ensure smooth execution of the work plan, adhering to industry best practices and safety protocols throughout the installation process.

Provide training and capacity building opportunities for employees involved in operating and maintaining renewable energy systems and recycling facilities. Equip them with the necessary skills and knowledge to effectively manage and optimize these technologies for long-term sustainability.

Implement monitoring and evaluation mechanisms to track the performance and effectiveness of renewable energy and recycling initiatives over time. Collect data on energy production, waste diversion rates, cost savings, and environmental impact indicators. Utilize this information to identify areas for improvement and refine future strategies.

Foster a culture of continuous improvement and adaptation within your agricultural operation. Stay informed about advancements in renewable energy technologies and recycling innovations, seeking feedback from stakeholders and employees to identify opportunities for optimization and innovation.

The following is an example of a hypothetical sustainable agriculture project.

Work Plan for Introduction of Renewable Energy Resources and Recycled Products

Objective: *Implement sustainable agricultural practices by integrating renewable energy resources and recycled products into operations, reducing environmental impact and promoting long-term viability.*

Duration: *12 months*

Phase 1: Initial Assessment and Planning (Month 1-2)

- *Conduct comprehensive assessment of current energy consumption and waste management practices.*

- *Define specific goals and objectives for integrating renewable energy and recycled products.*

- *Identify key stakeholders and establish communication channels for collaboration.*

Phase 2: Stakeholder Engagement and Resource Identification (Month 3-4)

- *Engage with alternative energy consultants, suppliers, and waste management experts.*

- *Gather insights and expertise on available renewable energy options and recycled products.*

- *Procure necessary resources and equipment from reputable suppliers.*

Phase 3: Site Assessment and Regulatory Compliance (Month 5-6)

- *Conduct site assessments to evaluate feasibility of renewable energy system installation.*

- *Obtain necessary permits and approvals from regulatory authorities.*

- *Develop site-specific plans for installation and integration of renewable energy and recycling facilities.*

Phase 4: Installation and Implementation (Month 7-9)

- *Execute installation of renewable energy systems and recycling infrastructure according to established plans.*

- *Coordinate with contractors, technicians, and labourers to ensure smooth implementation.*

- *Adhere to industry standards and safety protocols during installation process.*

Phase 5: Training and Capacity Building (Month 10-11)

- *Provide training programs for employees on operation and maintenance of renewable energy systems.*

- *Equip employees with necessary skills and knowledge to optimize use of recycled products.*

- *Foster culture of sustainability and environmental stewardship among staff.*

Phase 6: Monitoring and Evaluation (Month 12)

- *Implement monitoring and evaluation mechanisms to track performance of renewable energy and recycling initiatives.*

- *Collect data on energy production, waste diversion rates, and cost savings.*

- *Analyse findings to identify areas for improvement and refine strategies as needed.*

Conclusion *By following this work plan, we aim to successfully integrate renewable energy resources and recycled products into our agricultural operations, thereby reducing environmental impact and promoting sustainability for our land-based business.*

Investigating Sources and Availability of Materials and Human Resources Needed to Complete the Work Plan

Investigating the sources and availability of materials and human resources necessary for the completion of the work plan aimed at developing sustainable agricultural practices demands a meticulous and methodical approach. As an individual tasked with overseeing land-based businesses and collaborating with a diverse array of stakeholders, including alternative energy consultants, suppliers, planning authorities, and government departments, your role is indispensable in identifying and securing the requisite resources for the project's success.

To begin, a comprehensive assessment of the materials required to implement renewable energy systems and recycling infrastructure is imperative. This entails identifying the types and quantities of materials necessary for various aspects of the project, encompassing construction, installation, and ongoing operation. Consulting with experts in renewable energy and waste management aids in discerning the most suitable materials for the endeavour.

Subsequently, identifying potential suppliers emerges as a crucial step. Through diligent research and scrutiny, viable suppliers capable of providing the required materials are pinpointed. Factors such as reputation, reliability, product quality, and proximity to the project site are carefully considered. Engaging with multiple suppliers facilitates a comparison of prices, availability, and delivery timelines, ultimately enabling the establishment of partnerships with suppliers capable of furnishing materials in a timely and cost-effective manner.

In line with the project's sustainability goals, exploring alternative material sources is advocated. This entails investigating options for sourcing recycled or sustainable materials, with an emphasis on local procurement to mitigate transportation emissions and support the regional economy. Collaboration with recycling facilities and manufacturers is sought to procure recycled products and materials conducive to the project's objectives.

Simultaneously, an assessment of human resource needs is imperative. Evaluating the requisite skills and expertise for tasks such as construction, installation, maintenance, and operation of renewable energy systems and recycling facilities guides the identification of necessary personnel. Whether within the organization or through outsourcing to specialized contractors, securing qualified personnel is paramount.

To ensure personnel are equipped with the requisite skills and knowledge, a comprehensive plan for training and capacity building is developed. Training programs encompassing renewable energy technologies, waste management practices, safety protocols, and sustainable agricultural techniques are devised to enhance workforce capabilities continually.

Moreover, collaboration with relevant stakeholders is emphasized. Engagement with government agencies, educational institutions, industry associations, and community organizations facilitates access to additional resources and expertise. Leveraging partnerships and networks enables tapping into funding opportunities, grants, and support programs essential for the project's viability.

Conducting a rigorous risk assessment and implementing appropriate mitigation strategies are integral components of the process. Identifying potential challenges related to material sourcing and human resource management allows for the development of contingency plans to address disruptions and unforeseen issues effectively.

Developing Risk Management Strategies Including Contingencies to Maintain Supply and Deliver Contract On Time, and Incorporate Into The Plan

Developing risk management strategies, including contingencies, to ensure uninterrupted supply chains and timely contract delivery is crucial for the advancement of sustainable agricultural practices employing renewable energy and recycling systems. As an individual equipped with expertise in navigating sustainability challenges and collaborating with external stakeholders, your role is pivotal in orchestrating effective risk mitigation measures within the project plan.

Initiating the risk management process involves identifying potential threats and vulnerabilities that could disrupt the supply chain or hinder contract delivery. These risks may encompass delays in material procurement, equipment malfunctions, adverse weather conditions, regulatory changes, or unforeseen economic fluctuations. Collaboration with relevant stakeholders, such as suppliers, contractors, and industry experts, is essential to conduct a comprehensive risk assessment and prioritize risks based on their likelihood and potential impact.

Once risks are identified and assessed, the next step involves developing contingency plans to mitigate their adverse effects and uphold project momentum. Contingency planning necessitates identifying alternative courses of action to promptly address potential disruptions. For instance, establishing backup suppliers or vendors for critical materials can mitigate supply chain disruptions. Implementing redundant systems or backup equipment can minimize the impact of equipment failures, while developing alternative work schedules or resource allocation strategies enables adaptation to changing circumstances.

Seamless integration of risk management strategies and contingencies into the project plan is vital for effective implementation and monitoring throughout the project lifecycle. Specific risk mitigation measures, responsibilities, and timelines should be incorporated into the project schedule and budget. Assigning dedicated resources or personnel to oversee contingency plans ensures prompt responses to emerging risks. Regular review and updating of the risk management plan in response to new risks or evolving project conditions are imperative.

Fostering open communication and collaboration among project stakeholders facilitates proactive risk management. Maintaining regular dialogue with suppliers, contractors, and other key stakeholders aids in anticipating potential challenges and coordinating mitigation efforts effectively. Establishing clear channels for reporting and escalating risks ensures that all stakeholders are informed and involved in decision-making processes. Encouraging a culture of transparency and accountability fosters collective ownership of risk management initiatives.

Implementing robust monitoring and evaluation mechanisms is essential to track the effectiveness of risk management strategies and contingencies. Regular assessment of identified risks, monitoring of key performance indicators, and evaluation of contingency plan implementation are crucial. Periodic reviews and lessons learned sessions help identify areas for improvement and refine risk management processes. Adjusting contingency plans as necessary to address evolving risks maintains project resilience and ensures successful development of sustainable agricultural practices utilizing renewable energy and recycling systems.

Developing Work Health and Safety Risk Control Measures and Establishing Procedures

Developing work health and safety risk control measures and establishing procedures is a critical aspect of implementing sustainable agricultural practices that utilize renewable energy and recycling systems. As someone equipped with the skills and knowledge to address sustainability issues and collaborate with external experts and stakeholders, your role in ensuring the safety of workers and the environment is paramount.

Engaging in farming is undoubtedly fulfilling, yet it also entails inherent risks. Prioritizing farm safety isn't just crucial for averting accidents; it's also vital for upholding productivity and safeguarding your livelihood. By adopting appropriate safety measures, you can significantly diminish the probability of incidents and foster a safety-oriented ethos on your farm (Digi Clip, 2024).

Creating comprehensive farm safety policies and procedures is imperative for ensuring the welfare of farmers, workers, and visitors, while also protecting the overall success of farm operations. These protocols provide a framework for instilling a safety-oriented culture, identifying potential hazards, and implementing effective controls. The benefits of formulating farm safety policies and procedures encompass (Digi Clip, 2024):

Farm safety policies and procedures prioritize safeguarding human life and well-being. By establishing explicit guidelines and protocols, farms can mitigate the risk of accidents, injuries, and occupational ailments. This not only ensures physical safety but also enhances mental well-being, productivity, and job satisfaction (Digi Clip, 2024). Developing farm safety policies and procedures cultivates a culture of safety within the farm. It underscores that safety is paramount and underscores the collective responsibility of all stakeholders. When safety becomes ingrained in the farm's ethos and operations, it evolves into an integral aspect of daily functions, leading to a proactive approach to identifying hazards, mitigating risks, and continual improvement.

Farm safety policies and procedures facilitate a methodical approach to identifying hazards and managing risks. Through comprehensive risk assessments, farms can pinpoint potential hazards unique to their operations, such as machinery-related risks, chemical exposures, or environmental hazards. This allows for the implementation of suitable control measures, encompassing engineering controls, administrative measures, and the utilization of personal protective equipment (PPE).

Developing farm safety policies and procedures aids farms in adhering to pertinent regulations and standards. Agriculture, akin to any other industry, is subject to specific legal requirements concerning worker safety, chemical handling, equipment maintenance, and more (Digi Clip, 2024). By formulating policies aligned with these regulations, farms can ensure compliance and mitigate the risk of penalties or legal entanglements.

Farm safety policies and procedures furnish a framework for training workers on safe practices and protocols. Regular safety training and education equip workers with the requisite knowledge and skills to recognize hazards, respond to emergencies, and make informed decisions regarding safety. This instils a sense of ownership and accountability among the workforce (Digi Clip, 2024).

Developing farm safety policies and procedures fosters a culture of continuous improvement and adaptability. Farms evolve over time, with new technologies, equipment, and practices being introduced. Regular reviews and updates to safety policies ensure their relevance and efficacy in addressing emerging hazards and evolving industry standards.

Developing procedures includes:

1. Risk Identification and Assessment: Begin by identifying potential health and safety risks associated with implementing renewable energy and recycling systems in agricultural operations. These risks may include electrical hazards from solar panels or wind turbines, exposure to hazardous materials during recycling processes, ergonomic risks from manual handling tasks, and risks related to working at heights during installation and maintenance activities. Conduct a thorough risk assessment to evaluate the likelihood and severity of these risks.

2. Control Measures Selection: Once risks are identified, select appropriate control measures to mitigate or eliminate them. Control measures may include engineering controls (e.g., installing safety guards on machinery, implementing fall protection systems), administrative controls (e.g., developing safe work procedures, providing training and supervision), and personal protective equipment (e.g., safety helmets, gloves, goggles). Consider the hierarchy of controls, prioritizing measures that eliminate or minimize risks at the source.

3. Implementation of Control Measures: Implement the selected control measures systematically and comprehensively across all stages of the project. Ensure that workers are adequately trained in the use of control measures and understand their roles and responsibilities in maintaining a safe work environment. Regularly inspect and maintain equipment and machinery to ensure they are in safe working condition. Monitor compliance with safety procedures and address any non-compliance promptly.

4. Emergency Preparedness and Response: Develop procedures for responding to emergencies and incidents that may arise during the implementation of renewable energy and recycling systems. Establish protocols for evacuating workers in the event of fire, electrical faults, or other emergencies. Provide training on emergency response procedures and ensure that emergency contact information is readily accessible to all workers.

5. Communication and Consultation: Foster open communication and consultation with workers, contractors, and relevant stakeholders regarding health and safety matters. Encourage workers to report hazards, near misses, and safety concerns promptly. Consult with external experts, such as occupational health and safety consultants, to ensure that control measures are effective and comply with regulatory requirements.

6. Review and Continuous Improvement: Regularly review and evaluate the effectiveness of health and safety control measures and procedures. Conduct inspections, audits, and incident investigations to identify areas for improvement. Use feedback from workers and stakeholders to refine control measures and enhance safety procedures continuously. Stay informed about emerging risks and technological advancements to adapt safety practices accordingly.

By developing robust work health and safety risk control measures and establishing procedures, you can create a safe and healthy working environment for all personnel involved in implementing sustainable agricultural practices utilizing renewable energy and recycling systems. Your proactive approach to managing risks ensures the success and sustainability of the project while safeguarding the well-being of workers and the environment.

In farming environments, various hazards exist that endanger personal well-being, livestock, and overall farm operations. To effectively mitigate these risks, it's essential to systematically identify and address potential hazards. A comprehensive farm safety checklist serves as a valuable tool in achieving this goal. By creating such a checklist, encompassing hazard identification and the implementation of suitable controls, a proactive and safety-conscious culture can be established on the farm.

Here are key areas to consider when developing your farm safety checklist:

Mobile Plant and Vehicle Movement: Establish designated traffic routes, well-marked and avoiding areas of common worker activity. Enforce speed limits and install clear signage to alert workers and visitors of moving vehicle presence. Conduct regular vehicle inspections for maintenance issues, ensuring operators are adequately trained and licensed. Ensure machinery is equipped with appropriate guards and protective barriers to prevent accidental contact with moving parts.

Quad Bike Safety: Offer comprehensive training on quad bike operation and safety, including proper riding techniques and personal protective equipment usage. Regularly maintain quad bikes, including tire pressure, brakes, and lights checks. Implement safe usage guidelines, prohibiting passengers, riding on steep slopes, and exceeding recommended weight limits.

Working in Isolation: Develop clear protocols and procedures for working alone or in isolated areas, including regular communication check-ins with designated contacts. Equip workers with personal safety devices, like personal alarms or satellite communication tools, enabling them to request help if necessary. Conduct risk assessments for isolated tasks, establishing emergency response plans for various scenarios.

Livestock Safety: Install proper handling facilities and equipment to minimize injuries when working with livestock. Train workers on safe livestock handling techniques, including proper restraints usage and maintaining a safe distance. Ensure awareness among workers about specific livestock risks, including behaviour and potential aggression.

Fuel Storage: Store fuel in designated, well-ventilated areas away from ignition sources, such as open flames or electrical equipment. Utilize appropriate storage containers, properly labelled and secured. Establish protocols for fuel handling, emphasizing personal protective equipment use and spill response procedures.

Manual Handling: Evaluate tasks involving manual handling, implementing strategies like mechanical aids or team lifting to reduce risks. Train workers on proper lifting techniques, emphasizing the importance of using leg muscles, maintaining back straightness, and avoiding twisting motions. Encourage regular breaks and stretching during physically demanding tasks.

Slips, Trips, and Falls: Maintain clear walkways, adequately lit work areas, and promptly address uneven or slippery surfaces. Provide slip-resistant footwear and emphasize its importance to workers.

Grain Movement and Storage: Develop safe grain handling procedures, employing appropriate equipment like augers or conveyors. Regularly inspect grain storage facilities for structural integrity and ensure proper ventilation and monitoring systems. Train workers on grain bin entry procedures and associated risks.

Confined Spaces: Identify all confined spaces, assessing hazards like toxic gases or lack of oxygen. Use signage to mark confined spaces and communicate risks. Implement a confined space entry permit system and train personnel accordingly.

Machinery Safety: Regularly inspect machinery for wear, damage, or malfunction and address issues promptly. Ensure machinery has proper guards and protective barriers. Provide comprehensive training to machinery operators on safe operating procedures and lockout/tag-out procedures.

Chemicals and Pesticides: Store chemicals in designated areas away from food, water sources, and livestock. Provide workers with appropriate personal protective equipment and train them on safe handling, application, and disposal practices.

Falls from Height: Identify potential fall areas and implement fall protection measures. Provide proper fall protection equipment and ensure workers are trained in their usage. Regularly inspect and maintain elevated work areas.

Biological Hazards: Implement measures to prevent the spread of diseases, such as proper waste management and hygiene protocols. Train workers on personal hygiene practices and establish protocols for handling biological waste.

Environmental Hazards: Identify potential environmental hazards and develop emergency response procedures. Provide appropriate protective equipment and train workers on mitigating risks.

Fatigue Management: Promote rest and adequate sleep among workers and establish schedules allowing regular breaks. Encourage workers to recognize signs of fatigue and report them promptly.

Electrical Safety: Regularly inspect electrical systems and ensure workers are trained on safe practices.

Ergonomics: Identify tasks involving repetitive movements and provide ergonomic training. Implement job rotation and use assistive devices.

Vibration Exposure: Maintain machinery and provide vibration-damping gloves or tools.

Building/Structural Integrity: Regularly inspect buildings and address structural issues promptly.

Fire Safety: Install and test fire safety equipment and train workers on fire safety procedures.

Ventilation and Air Quality: Ensure proper ventilation and provide respiratory protection when necessary.

By implementing and regularly reviewing this comprehensive farm safety checklist, you can effectively identify and mitigate potential hazards, fostering a safer working environment for all involved in farm operations. An example Farm Safety Procedure follows.

Farm Safety Procedure

Objective: *The objective of this farm safety procedure is to ensure the safety of all personnel and operations on the farm by implementing preventive measures and protocols to mitigate potential hazards.*

1. Mobile Plant and Vehicle Movement:

- *Establish designated traffic routes, well-marked and avoiding areas of common worker activity.*

- *Enforce speed limits and install clear signage to alert workers and visitors of moving vehicle presence.*

- *Conduct regular vehicle inspections for maintenance issues, ensuring operators are adequately trained and licensed.*

- *Ensure machinery is equipped with appropriate guards and protective barriers to prevent accidental contact with moving parts.*

2. Quad Bike Safety:

- *Offer comprehensive training on quad bike operation and safety, including proper riding techniques and personal protective equipment usage.*

- *Regularly maintain quad bikes, including tire pressure, brakes, and lights checks.*

- *Implement safe usage guidelines, prohibiting passengers, riding on steep slopes, and exceeding recommended weight limits.*

3. Working in Isolation:

- *Develop clear protocols and procedures for working alone or in isolated areas, including regular communication check-ins with designated contacts.*

- *Equip workers with personal safety devices, like personal alarms or satellite communication tools, enabling them to request help if necessary.*

- *Conduct risk assessments for isolated tasks, establishing emergency response plans for various scenarios.*

4. Livestock Safety:

- Install proper handling facilities and equipment to minimize injuries when working with livestock.

- Train workers on safe livestock handling techniques, including proper restraints usage and maintaining a safe distance.

- Ensure awareness among workers about specific livestock risks, including behavior and potential aggression.

5. Fuel Storage:

- Store fuel in designated, well-ventilated areas away from ignition sources, such as open flames or electrical equipment.

- Utilize appropriate storage containers, properly labeled and secured.

- Establish protocols for fuel handling, emphasizing personal protective equipment use and spill response procedures.

6. Manual Handling:

- Evaluate tasks involving manual handling, implementing strategies like mechanical aids or team lifting to reduce risks.

- Train workers on proper lifting techniques, emphasizing the importance of using leg muscles, maintaining back straightness, and avoiding twisting motions.

- Encourage regular breaks and stretching during physically demanding tasks.

7. Slips, Trips, and Falls:

- Maintain clear walkways, adequately lit work areas, and promptly address uneven or slippery surfaces.

- Provide slip-resistant footwear and emphasize its importance to workers.

8. Grain Movement and Storage:

- Develop safe grain handling procedures, employing appropriate equipment like augers or conveyors.

- Regularly inspect grain storage facilities for structural integrity and ensure proper ventilation and monitoring systems.

- Train workers on grain bin entry procedures and associated risks.

9. Confined Spaces:

- Identify all confined spaces, assessing hazards like toxic gases or lack of oxygen.

- Use signage to mark confined spaces and communicate risks.

- Implement a confined space entry permit system and train personnel accordingly.

10. Machinery Safety:

- Regularly inspect machinery for wear, damage, or malfunction and address issues promptly.

- Ensure machinery has proper guards and protective barriers.

- Provide comprehensive training to machinery operators on safe operating procedures and lockout/tag-out procedures.

11. Chemicals and Pesticides:

- Store chemicals in designated areas away from food, water sources, and livestock.

- *Provide workers with appropriate personal protective equipment and train them on safe handling, application, and disposal practices.*

12. Falls from Height:

- *Identify potential fall areas and implement fall protection measures.*

- *Provide proper fall protection equipment and ensure workers are trained in their usage.*

- *Regularly inspect and maintain elevated work areas.*

13. Biological Hazards:

- *Implement measures to prevent the spread of diseases, such as proper waste management and hygiene protocols.*

- *Train workers on personal hygiene practices and establish protocols for handling biological waste.*

14. Environmental Hazards:

- *Identify potential environmental hazards and develop emergency response procedures.*

- *Provide appropriate protective equipment and train workers on mitigating risks.*

15. Fatigue Management:

- *Promote rest and adequate sleep among workers and establish schedules allowing regular breaks.*

- *Encourage workers to recognize signs of fatigue and report them promptly.*

16. Electrical Safety:

- *Regularly inspect electrical systems and ensure workers are trained on safe practices.*

17. Ergonomics:

- *Identify tasks involving repetitive movements and provide ergonomic training.*

- *Implement job rotation and use assistive devices.*

18. Vibration Exposure:

- *Maintain machinery and provide vibration-damping gloves or tools.*

19. Building/Structural Integrity:

- *Regularly inspect buildings and address structural issues promptly.*

20. Fire Safety:

- *Install and test fire safety equipment and train workers on fire safety procedures.*

21. Ventilation and Air Quality:

- *Ensure proper ventilation and provide respiratory protection when necessary.*

In the realm of farm safety, being equipped to handle emergencies is paramount. Given the inherent risks and potential accidents on farms, it's imperative to have adequate emergency equipment and a meticulously devised emergency plan. These measures serve to mitigate the consequences of emergencies and ensure the safety of farm workers and visitors alike. Following are the key facets to consider regarding emergency equipment and planning in farm settings (Digi Clip, 2024).

To begin with, maintaining fully stocked first aid kits strategically positioned across the farm is imperative. Regular checks and replenishment of supplies guarantee their availability during critical moments.

Fire extinguishers should be installed in areas prone to fire hazards, such as machinery sheds, fuel storage areas, or barns. Ensuring regular inspection, servicing, and clear identification of these extinguishers is essential for effective firefighting.

Equipping workers with reliable communication devices, be it two-way radios or mobile phones, facilitates prompt reporting of emergencies and seeking assistance as needed.

Installing backup lighting systems, such as battery-powered lights or emergency lighting fixtures, is crucial to ensure visibility during power outages or other emergencies.

Lastly, it's imperative to provide readily available and properly maintained personal protective equipment (PPE), including gloves, goggles, helmets, or high-visibility vests.

Developing a comprehensive emergency response plan is vital. This plan should delineate roles, responsibilities, and protocols for various emergency scenarios, encompassing fires, accidents, natural disasters, or hazardous substance spills.

Designating specific assembly points for workers during emergencies ensures accountability and streamlines headcounts, facilitating a swift and organized response.

Clear marking of evacuation routes, coupled with regular reviews and communication of evacuation procedures, is essential for ensuring swift and safe evacuation of all farm personnel.

Maintaining an updated list of emergency contact numbers for local emergency services, healthcare providers, veterinarians, and utility companies is crucial. Access to this information should be readily available to all workers.

Conducting regular training sessions and emergency drills is instrumental in familiarizing workers with emergency procedures, including evacuation protocols, first aid, and the operation of emergency equipment.

Regular inspections, testing, and maintenance of emergency equipment, such as fire extinguishers and emergency lighting, are essential to ensure their functionality when needed most.

In conclusion, by ensuring the availability of necessary emergency equipment and implementing a meticulously planned emergency response strategy, farms can significantly enhance their preparedness and response capabilities. This, in turn, minimizes the potential impacts of emergencies and safeguards the well-being of all individuals present on the farm (Digi Clip, 2024).

Chapter Seven

Implementing Strategies to use Renewable Energy

Implementing and Monitor the Plan for Renewable Energy and Recycling Products to Ensure On Time Supply of Plant and Materials

To implement and monitor a plan for renewable energy and recycling products it is important to adopt a strategic approach that encompasses various aspects of sustainability. This involves integrating renewable energy sources and recycling systems into the operational framework while ensuring timely supply of necessary plants and materials.

Firstly, it's essential to conduct a comprehensive assessment of the current energy consumption and waste generation patterns within the agricultural operations. This assessment should identify opportunities for integrating renewable energy sources such as solar, wind, or biomass, as well as potential areas for implementing recycling systems to minimize waste.

Once the assessment is complete, a detailed plan outlining the specific strategies for incorporating renewable energy and recycling systems should be developed. This plan should include specific targets, timelines, and resource allocations for each component of the strategy. It's important to consult with external experts such as alternative energy consultants and suppliers of recycling equipment to ensure that the chosen systems are suitable for the rural, regional, and local conditions.

In parallel with the development of the plan, arrangements should be made to secure a reliable and timely supply of necessary plants and materials for implementing renewable energy and recycling systems. This may involve establishing partnerships with suppliers who specialize in renewable energy equipment and recycling products, as well as exploring options for accessing subsidies and grants offered by government departments to support sustainable initiatives.

Once the plan is implemented, it's essential to monitor its progress regularly to ensure that targets are being met and adjustments can be made as needed. This may involve tracking energy consumption and waste generation levels, as well as evaluating the performance of renewable energy systems and recycling processes. Regular reviews should be conducted to identify any challenges or areas for improvement, and external experts should be consulted as needed to provide guidance and support.

By adopting a proactive approach to implementing and monitoring the plan for renewable energy and recycling products, agricultural businesses can effectively reduce their environmental footprint while also improving their long-term sustainability and resilience.

As a contextual example of implementing and monitor the plan for renewable energy and recycling products to ensure on time supply of plant and materials, from the perspective of biodynamic farming, implementing and monitoring a plan for renewable energy and recycling products aligns closely with the holistic principles of sustainability inherent in this agricultural approach. Biodynamic farming emphasizes the interconnectedness of all elements within the farming system, including energy usage and waste management. Therefore, adopting a strategic approach that integrates renewable energy sources and recycling systems is fundamental to maintaining the health and vitality of the farm ecosystem.

To begin, conducting a comprehensive assessment of current energy consumption and waste generation patterns is in line with the biodynamic principle of observation. This assessment allows farmers to understand the existing dynamics of energy usage and waste production within their operations. By identifying opportunities for integrating renewable energy sources such as solar, wind, or biomass, and implementing recycling systems, farmers can work towards minimizing their environmental impact while maximizing resource efficiency.

Once the assessment is complete, developing a detailed plan becomes essential. This plan should not only outline specific strategies for incorporating renewable energy and recycling systems but also set clear targets, timelines, and resource allocations. In the context of Biodynamic farming, this planning process emphasizes the importance of working in harmony with natural cycles and rhythms. Consulting with external experts such as alternative energy consultants and suppliers of recycling equipment ensures that chosen systems align with the principles of sustainability and are suitable for the unique rural, regional, and local conditions of the farm.

In tandem with the development of the plan, arrangements must be made to secure a reliable and timely supply of necessary plants and materials. Biodynamic farming emphasizes the use of natural inputs and materials, so sourcing renewable energy equipment and recycling products that are in harmony with this approach is crucial. Establishing partnerships with suppliers who share a commitment to sustainability and exploring avenues for accessing government subsidies and grants further supports the implementation of environmentally friendly initiatives.

Once the plan is put into action, ongoing monitoring and evaluation are paramount. This process involves tracking energy consumption and waste generation levels, as well as assessing the performance of renewable energy systems and recycling processes. Regular reviews allow farmers to identify any challenges or areas for improvement and make necessary adjustments to ensure the plan's effectiveness. Engaging external experts for guidance and support ensures that the farm remains aligned with biodynamic principles and continues to evolve towards greater sustainability and resilience over time.

In summary, adopting a proactive approach to implementing and monitoring renewable energy and recycling initiatives not only reduces the environmental footprint of agricultural operations but also aligns with the holistic principles of Biodynamic farming. By working in harmony with natural cycles and ecosystems, farmers can cultivate a more sustainable and resilient farming system that supports the health and vitality of the land, plants, animals, and humans alike.

Monitoring the Progress of Strategies to Reduce Greenhouse Gas Emissions and Recycling Against Schedule, Quality Requirements and Budget

Monitoring the progress of strategies to reduce greenhouse gas emissions and implement recycling initiatives in agricultural practices requires a systematic approach that integrates planning, coordination, and evaluation. As land-based businesses strive to develop sustainable practices utilizing renewable energy and recycling systems, it's imperative to ensure that efforts align with schedule, quality requirements, and budget constraints. This involves continuous monitoring and adjustment of strategies to achieve desired outcomes effectively.

Firstly, it's essential to establish clear benchmarks and performance indicators to measure the success of greenhouse gas emission reduction and recycling efforts. These indicators should be specific, measurable, achievable, relevant, and time-bound (SMART), allowing for meaningful assessment of progress against predetermined goals. Additionally, stakeholders should be engaged in the development of these indicators to ensure buy-in and accountability throughout the process.

Regular monitoring of greenhouse gas emissions and recycling activities should be conducted to track progress against established benchmarks. This may involve collecting data on energy consumption, waste generation, recycling rates, and other relevant metrics. Utilizing technology such as sensors, meters, and data analytics can facilitate real-time monitoring and provide valuable insights into the effectiveness of implemented strategies.

As part of the monitoring process, it's important to conduct periodic reviews to evaluate the performance of renewable energy systems and recycling initiatives. This includes assessing adherence to schedule timelines, compliance with quality standards, and adherence to allocated budgets. Any deviations from planned targets should be promptly identified and addressed through corrective actions to ensure alignment with overall sustainability objectives.

Effective communication and collaboration among stakeholders are essential for successful monitoring of greenhouse gas emission reduction and recycling strategies. This includes regular engagement with alternative energy consultants, suppliers of renewable energy equipment, planning authorities, government departments, and other relevant parties. By fostering open dialogue and sharing progress updates, potential challenges can be identified early, and collaborative solutions can be developed to overcome them.

Furthermore, ongoing training and capacity-building initiatives should be provided to stakeholders involved in implementing renewable energy and recycling systems. This ensures that they have the necessary skills and knowledge to effectively monitor and manage these initiatives. Additionally, periodic reviews of budget allocations and expenditure should be conducted to ensure that financial resources are being utilized efficiently and effectively.

As a contextual example, in this case let us consider the practice of hydroponics. From the perspective of Hydroponics, monitoring the progress of strategies aimed at reducing greenhouse gas emissions and implementing recycling initiatives in agricultural practices is vital for ensuring sustainability and efficiency. As hydroponic farming focuses on soil-less cultivation methods, which often rely heavily on controlled environments and energy inputs, adopting a systematic approach to monitoring and adjusting strategies becomes even more critical.

To begin with, it's crucial to establish clear benchmarks and performance indicators to gauge the effectiveness of greenhouse gas emission reduction and recycling efforts. These indicators should follow the SMART criteria—specific, measurable, achievable, relevant, and time-bound—to provide meaningful insights into progress. Engaging stakeholders in the development of these indicators fosters ownership and commitment to achieving the set goals.

Regular monitoring of greenhouse gas emissions and recycling activities is necessary to track progress against established benchmarks. This entails collecting data on energy consumption, waste generation, recycling rates, and other pertinent metrics. Utilizing technology such as sensors, meters, and data analytics allows for real-time monitoring, providing valuable insights into the efficiency of implemented strategies.

Periodic reviews are essential components of the monitoring process, enabling evaluation of the performance of renewable energy systems and recycling initiatives. These reviews assess adherence to schedule timelines, compliance with quality standards, and alignment with allocated budgets. Any deviations from planned targets should be promptly identified and addressed through corrective actions to ensure that sustainability objectives are met.

Effective communication and collaboration among stakeholders are paramount for successful monitoring of greenhouse gas emission reduction and recycling strategies in hydroponic farming. Regular engagement with alternative energy consultants, suppliers of renewable energy equipment, and relevant authorities fosters an open dialogue where challenges can be identified early, and collaborative solutions can be developed to address them.

Furthermore, ongoing training and capacity-building initiatives are essential for stakeholders involved in implementing renewable energy and recycling systems. This ensures that they possess the necessary skills and knowledge to effectively monitor and manage these initiatives. Additionally, regular reviews of budget allocations and expenditure help ensure the efficient and effective utilization of financial resources, contributing to the overall sustainability of hydroponic farming practices.

Adopting a systematic approach to monitoring and adjusting strategies for reducing greenhouse gas emissions and implementing recycling initiatives is essential for promoting sustainability and efficiency in hydroponic farming. By establishing clear benchmarks, leveraging technology for real-time monitoring, and fostering collaboration among stakeholders, hydroponic farms can effectively achieve their sustainability goals while minimizing environmental impact.

Reviewing Plan Against Required Inputs and Rectifying Any Anomalies to Meet and Supply Contractual Arrangements On Time

Reviewing plans against required inputs and rectifying any anomalies to meet and supply contractual arrangements on time is a critical aspect of developing sustainable agricultural practices that utilize renewable energy and recycling systems. This process involves a comprehensive evaluation of the planning strategies, ensuring alignment with sustainability objectives, and addressing any discrepancies to uphold contractual obligations.

To begin, land-based businesses must assess the sustainability goals outlined in the planning strategies. This includes reviewing the objectives related to renewable energy utilization, waste reduction, and recycling initiatives. By understanding these goals, stakeholders can identify the necessary inputs and resources required to execute the plan effectively.

Next, a thorough examination of the available inputs and resources is essential to determine their adequacy and suitability for meeting contractual arrangements. This may involve assessing the availability of renewable energy sources such as solar, wind, or biomass, as well as evaluating the infrastructure and equipment needed for recycling systems. Any discrepancies between the planned requirements and available resources must be identified and addressed promptly.

Consultation with external experts, including alternative energy consultants, suppliers of renewable energy equipment, and government departments administering subsidies, can provide valuable insights into optimizing resource allocation and addressing any gaps in the plan. These experts can offer recommendations on alternative technologies, funding opportunities, and regulatory considerations to ensure compliance with contractual arrangements.

In cases where anomalies are identified, proactive measures must be taken to rectify them and ensure timely fulfillment of contractual obligations. This may involve adjusting the planning strategies, revising timelines, or reallocating resources to mitigate potential delays. Effective communication and collaboration among stakeholders are essential to facilitate timely decision-making and implementation of corrective actions.

Continuous monitoring and review of the plan against required inputs are essential throughout the implementation process. Regular assessments enable stakeholders to track progress, identify emerging challenges, and make necessary adjustments to stay on course. By maintaining flexibility and adaptability, land-based businesses can overcome obstacles and successfully meet supply contractual arrangements on time while advancing sustainability goals.

Furthermore, documenting the review process and any corrective actions taken is crucial for accountability and transparency. This documentation provides a record of compliance with contractual obligations and serves as a basis for future improvements and optimizations.

Monitoring Work Health and Safety Risk Control Measures and Procedures and Implementing Changes

Monitoring work health and safety (WHS) risk control measures and procedures is essential for those managing land-based businesses aiming to develop sustainable agricultural practices utilizing renewable energy and recycling systems. This responsibility involves a thorough understanding of sustainability principles, risk management strategies, and regulatory requirements to ensure the safety and well-being of workers and stakeholders involved in agricultural operations.

First and foremost, it is imperative to establish comprehensive WHS risk control measures and procedures tailored to the specific needs and conditions of the agricultural environment. This includes identifying potential hazards associated with renewable energy installations, recycling processes, and other agricultural activities. Hazards may include exposure to machinery, handling of hazardous materials, working at heights, and environmental risks.

Once identified, risk control measures must be implemented to mitigate the identified hazards effectively. This may involve implementing engineering controls such as installing safety guards on machinery, administrative controls such as developing safe work procedures, and providing personal protective equipment (PPE) to workers. Additionally, training programs should be developed to ensure that workers understand the risks associated with their tasks and are equipped with the necessary skills to perform their duties safely.

Regular monitoring of WHS risk control measures and procedures is essential to assess their effectiveness and identify any areas for improvement. This can be achieved through routine inspections, hazard assessments, and ongoing consultation with workers and WHS representatives. Any deficiencies or shortcomings in existing controls should be promptly addressed to prevent accidents or injuries from occurring.

Furthermore, staying abreast of changes in regulations, industry standards, and best practices is crucial for ensuring WHS compliance and continuous improvement. This may involve consulting with external experts such as WHS consultants, regulatory authorities, and industry associations to access up-to-date information and guidance on emerging risks and control measures.

In the event that changes to WHS risk control measures and procedures are necessary, proactive steps should be taken to implement these changes promptly. This may involve revising policies and procedures, providing additional training to workers, or investing in new equipment or technology to enhance safety performance. Effective communication and consultation with all relevant stakeholders are essential to ensure that changes are understood and implemented effectively.

Ultimately, the goal of monitoring WHS risk control measures and procedures is to create a safe and healthy working environment for all individuals involved in agricultural operations. By proactively identifying and addressing risks, implementing effective control measures, and continuously monitoring and adapting to changing conditions, land-based businesses can minimize the likelihood of workplace incidents and injuries while promoting sustainability and productivity.

From the perspective of a sustainable farm implementing polycultures and crop rotation, as an example, monitoring work health and safety (WHS) risk control measures and procedures is crucial for fostering a safe and sustainable agricultural environment. As farms strive to develop practices that utilize renewable energy and recycling systems, ensuring the safety and well-being of workers and stakeholders become paramount considerations.

Polycultures and crop rotation, despite their advantages in enhancing soil health and biodiversity, introduce certain hazards and risks to agricultural operations. One concern is pest and disease management, as growing a variety of crops in proximity can attract a diverse range of pests and diseases, potentially leading to crop damage and reduced yields without proper management practices. Additionally, weed competition poses a risk, as different crops may have varying growth rates and competitive abilities against weeds, potentially resulting in reduced crop yields if not effectively managed.

Moreover, soil nutrient imbalance is a significant risk associated with crop rotation practices, as different crops have varying nutrient requirements. Without adequate monitoring and management, nutrient imbalances can occur, leading to deficiencies or excesses that affect plant growth and productivity. Furthermore, the introduction of different crops with varying root structures and canopy cover in a polyculture system can impact soil stability, increasing the risk of soil erosion, especially on sloping terrain, if improper management practices are employed.

Figure 44: Polyculture between the vines in the Australian wine region of the Hunter Valley in New South Wales. F Delventhal, CC BY 2.0, via Wikimedia Commons.

In addition to agronomic challenges, polycultures and crop rotation also present economic and logistical risks. Diversifying crops can provide resilience against market fluctuations and climate variability, but farmers must navigate market demand, price volatility, and input costs, which can affect the economic viability of the farming operation. Moreover, implementing polycultures and crop rotation effectively requires a deep understanding of crop interactions, agronomic practices, and ecological principles. Lack of knowledge or experience in managing diverse cropping systems can increase the risk of suboptimal outcomes and production challenges. Therefore, careful assessment and management of hazards and risks are crucial for farmers to maximize the success and resilience of their farming systems, involving appropriate management practices, crop performance monitoring, and continuous learning and adaptation.

To begin with, it's essential to establish comprehensive WHS risk control measures and procedures tailored to the unique needs and conditions of the agricultural setting. This involves identifying potential hazards associated with various activities, including renewable energy installations, recycling processes, and agricultural tasks such as machinery operation and handling of hazardous materials. Hazards may also include environmental risks and working at heights, among others.

Once hazards are identified, effective risk control measures must be implemented to mitigate them. This may include a combination of engineering controls, administrative measures, and the provision of personal protective equipment (PPE) to workers. Additionally, comprehensive training programs should be developed to ensure that workers are aware of the risks involved in their tasks and equipped with the necessary skills to perform them safely.

Regular monitoring of WHS risk control measures and procedures is essential to evaluate their effectiveness and identify areas for improvement. Routine inspections, hazard assessments, and ongoing consultation with workers and WHS representatives facilitate the identification of deficiencies or shortcomings in existing controls, allowing for timely corrective actions to be taken.

Staying informed about changes in regulations, industry standards, and best practices is crucial for maintaining WHS compliance and driving continuous improvement. This may involve seeking guidance from external experts such as WHS consultants, regulatory authorities, and industry associations to access up-to-date information on emerging risks and control measures.

In cases where changes to WHS risk control measures and procedures are warranted, proactive steps should be taken to implement these changes promptly. This may include updating policies and procedures, providing additional training to workers, or investing in new equipment or technology to enhance safety performance. Effective communication and consultation with all relevant stakeholders are essential to ensure that changes are understood and implemented effectively.

Ultimately, the overarching goal of monitoring WHS risk control measures and procedures is to cultivate a safe and healthy working environment for all individuals involved in agricultural operations. By proactively identifying and addressing risks, implementing effective control measures, and continuously monitoring and adapting to changing conditions, farms can minimize the likelihood of workplace incidents and injuries while promoting sustainability and productivity.

Evaluating the Improvement Outcomes and Documenting Appropriate Corrective Actions

Evaluating improvement outcomes and documenting appropriate corrective actions is a crucial aspect of developing sustainable agricultural practices that utilize renewable energy and recycling systems. For those managing land-based businesses, this process requires a comprehensive understanding of sustainability principles, renewable energy technologies, waste management practices, and regulatory requirements.

To begin, it is essential to establish clear metrics and performance indicators to assess the effectiveness of sustainability initiatives and identify areas for improvement. These metrics may include energy consumption, waste generation, resource efficiency, greenhouse gas emissions, and overall environmental impact. By quantifying these factors, land-based businesses can track their progress towards sustainability goals and measure the success of implemented measures.

Once performance data has been collected and analysed, the next step is to evaluate the improvement outcomes against predefined targets and objectives. This involves comparing actual results with expected outcomes to determine whether sustainability initiatives are achieving the desired impact. Additionally, it may involve conducting cost-benefit analyses to assess the economic viability of implemented measures and their contribution to overall business performance.

During the evaluation process, it is essential to involve key stakeholders, including employees, suppliers, customers, regulatory agencies, and community members. Their input and feedback can provide valuable insights into the effectiveness of sustainability initiatives and help identify areas for further improvement. Open communication and collaboration are essential for fostering a culture of continuous improvement and innovation within the organization.

Based on the evaluation findings, appropriate corrective actions should be identified and documented to address any shortcomings or areas of non-compliance. This may involve revising policies and procedures, implementing additional training programs, upgrading equipment or infrastructure, or exploring new technologies or approaches to sustainability. Corrective actions should be tailored to the specific needs and circumstances of the business and aligned with overall sustainability objectives.

It is crucial to document all corrective actions taken, including the rationale behind each decision, the responsible parties, timelines for implementation, and expected outcomes. This documentation serves as a record of the organization's commitment to continuous improvement and accountability to stakeholders. It also provides a basis for tracking progress over time and demonstrating compliance with regulatory requirements and industry standards.

Evaluating improvement outcomes and documenting appropriate corrective actions are essential steps in the process of developing sustainable agricultural practices. By systematically assessing the effectiveness of sustainability initiatives, identifying areas for improvement, and taking proactive measures to address deficiencies, land-based businesses can enhance their environmental performance, reduce resource consumption, and contribute to a more sustainable future.

Overall, in exploring the integration of renewable energy sources, reduction of greenhouse emissions, and recycling opportunities within various agricultural practices such as permaculture, biodynamic farming, hydroponics, aquaponics, urban agriculture, agroforestry, food forests, polycultures, crop rotation, and integrated pest management, several key considerations and principles of sustainability come into play (Chel & Kaushik, 2011; Mostefaoui & Amara, 2019).

Firstly, identifying key parts of the enterprise operation that can be converted to renewable energy sources involves assessing the energy needs of the operation and identifying areas where renewable energy technologies such as solar panels, wind turbines, or biomass generators can be implemented. This process should consider factors such as energy consumption patterns, available resources, and site-specific conditions.

Moreover, calculating the potential savings associated with implementing renewable energy improvements is crucial for assessing the economic viability and benefits of such initiatives. This may involve conducting cost-benefit analyses, considering factors such as initial investment costs, ongoing operational expenses, and potential long-term savings from reduced energy bills and environmental benefits.

Implementing renewable energy improvements requires careful planning, coordination, and investment in appropriate technologies and infrastructure. Strategies may include installing solar panels on rooftops or unused land, integrating wind turbines into agricultural landscapes, or implementing energy-efficient systems for irrigation, heating, and cooling.

Evaluating the outcomes of renewable energy improvements involves assessing their impact on energy efficiency, greenhouse gas emissions reduction, cost savings, and overall sustainability performance. This may include monitoring energy consumption levels, tracking greenhouse gas emissions, and soliciting feedback from stakeholders involved in the implementation process.

Furthermore, incorporating principles of sustainability into renewable energy and recycling systems involves minimizing resource consumption, reducing waste generation, and promoting environmental stewardship. This may entail implementing practices such as composting organic waste, utilizing recycled materials in construction and infrastructure projects, and adopting eco-friendly packaging and production methods.

Legislation, regulatory requirements, and licensing associated with renewable energy and recycling vary depending on the jurisdiction and type of operation. Compliance with relevant laws and regulations, obtaining necessary permits, and adhering to industry standards are essential for ensuring legal and ethical operation of renewable energy and recycling systems.

Subsidies and incentives offered by governments and organizations can help offset the costs of adopting renewable energy systems and implementing recycling initiatives. These may include tax credits, grants, rebates, or feed-in tariffs designed to incentivize investment in renewable energy technologies and sustainable practices.

Effective environmental and work health and safety risk management strategies are essential for mitigating potential hazards and ensuring the safety of workers, stakeholders, and the surrounding community. This may involve conducting risk assessments, implementing safety protocols and procedures, providing training and education on safe practices, and maintaining compliance with relevant regulations and standards.

In conclusion, integrating renewable energy sources, reducing greenhouse emissions, and promoting recycling opportunities within various agricultural practices require a holistic approach that considers economic, environmental, and social factors. By identifying opportunities, developing strategies, implementing improvements, and evaluating outcomes, agricultural enterprises can enhance their sustainability performance while contributing to the transition towards a more resilient and environmentally friendly food system (Chel & Kaushik, 2011; Mostefaoui & Amara, 2019).

Part 4 - Developing Climate Risk Management Strategies

This section of the book delineates the essential competencies and expertise necessary for formulating climate risk management strategies tailored to agricultural, horticultural, or land management enterprises. It encompasses three primary components: reviewing climate and enterprise data, identifying and analysing climate-related risks and opportunities, and preparing comprehensive climate risk management strategies.

Initially, proficiency in reviewing climate and enterprise data is imperative. This involves a thorough examination of historical climate patterns, including temperature fluctuations, precipitation levels, and extreme weather events, alongside pertinent enterprise data such as crop yields, soil conditions, and water availability. By assimilating and comprehending this information, practitioners can gain insights into the interplay between climatic factors and enterprise operations.

Subsequently, the skill set extends to identifying and analysing climate risks and opportunities. This entails a meticulous assessment of potential hazards posed by climate variability and change, such as droughts, floods, heatwaves, or pest outbreaks, and their implications for agricultural productivity and sustainability. Simultaneously, practitioners must remain vigilant in recognizing opportunities arising from climate trends, such as shifts in growing seasons or the introduction of new crop varieties suited to changing conditions.

Lastly, proficiency in preparing climate risk management strategies is paramount. This involves synthesizing the findings from climate and enterprise data reviews and risk analyses to devise proactive and adaptive strategies aimed at mitigating identified risks and capitalizing on opportunities. These strategies may encompass diverse interventions, ranging from implementing resilient agricultural practices and investing in climate-resilient infrastructure to diversifying crops or income streams.

By honing these skills and knowledge areas, stakeholders in agricultural, horticultural, or land management enterprises can enhance their capacity to navigate the complexities of climate variability and change, fortifying their operations against risks while capitalizing on emerging opportunities for sustainable growth and resilience.

Chapter Eight

Reviewing Climate and Enterprise Data

Climate Change and Agriculture

Climate change has numerous effects on agriculture, posing challenges to global food security. Rising temperatures and shifting weather patterns contribute to decreased crop yields due to factors like water scarcity resulting from droughts, heatwaves, and floods. Such impacts also heighten the risk of simultaneous crop failures in various regions, potentially disrupting the global food supply. Additionally, the proliferation of pests and plant diseases is anticipated, threatening agricultural productivity and food security worldwide. Livestock face similar challenges, including heat stress, feed shortages, and increased susceptibility to diseases and parasites.

The rise in atmospheric CO_2 levels due to human activities triggers a fertilization effect, somewhat mitigating the adverse effects of climate change on agriculture. However, this effect is limited for certain crops like maize and may lead to decreased levels of essential micronutrients. Coastal areas may witness the loss of agricultural land due to sea-level rise, while melting glaciers could reduce the availability of irrigation water. Conversely, thawing frozen land may expand arable areas. Other consequences include erosion, changes in soil fertility, and alterations in growing seasons, with potential negative impacts on food safety.

Extensive research has examined climate change's impact on staple crops like maize, rice, wheat, and soybeans, which play a crucial role in global food consumption. However, uncertainties persist, including future population growth and separate challenges like soil erosion and groundwater depletion. Despite these challenges, advancements in agricultural practices, known as the Green Revolution, have significantly boosted yields per unit of land area. Nevertheless, projections suggest that climate change may exacerbate food insecurity, affecting millions of people by 2050. To address these challenges, adaptation measures such as changes in management practices and agricultural innovation are crucial for creating a sustainable food system alongside efforts to mitigate global warming.

Climate change has become a significant concern for the agricultural sector, as it poses various challenges and impacts on agricultural productivity and sustainability. The impact of climate change on agriculture is multifaceted, affecting crop yields, livestock production, and overall food security. Farmers are increasingly recognizing the need to adapt their agricultural practices to mitigate the adverse effects of climate change (Jena, 2021; Joshi & Bhandari, 2023; Khajuria & Ravindranath, 2012; Maho et al., 2019; Xuan et al., 2021).

Climate change has been observed to lead to changes in temperature, precipitation patterns, and extreme weather events, which directly impact agricultural systems. These changes can result in reduced crop yields, increased vulnerability to pests and diseases, and shifts in suitable agricultural areas (Jena, 2021; Joshi & Bhandari, 2023; Khajuria & Ravindranath, 2012; Maho et al., 2019; Xuan et al., 2021). For instance, studies have shown that climate change may lead to a reduction in the production of staple crops such as wheat, rice, and corn, affecting international food trade (Xuan et al., 2021). Additionally, the negative implications of climate change on the agricultural sector are unequivocal, particularly for smallholder farmers within the tropics (Dawid & Workalemahu, 2021).

Furthermore, climate change affects not only crop production but also livestock and fish production, thereby influencing human livelihoods and food security (Mwalukasa, 2013; Ogunbameru et al., 2013). The adverse effects of climate change on farming outputs have been observed to lead to household food insecurity, low profits, income, and welfare of farm families (Ifeanyi-obi & Henri-Ukoha, 2022; Ogunbameru et al., 2013). Additionally, climate change can exacerbate the existing challenges faced by farmers, such as drought and water scarcity, further impacting agricultural productivity (Tadesa, 2020).

In response to these challenges, farmers have been adopting various adaptation strategies to cope with the impacts of climate change on agriculture. These strategies include changes in cultivation practices, adoption of climate-resilient agricultural technologies, and the implementation of conservation agriculture and climate-smart agricultural practices (Jena, 2021; Joshi & Bhandari, 2023; Khajuria & Ravindranath, 2012; Maho et al., 2019; Xuan et al., 2021). Moreover, it has been emphasized that the capacity of farmers to adapt to climate change can be enhanced through education, skill training, strengthening social networks, and increasing investment in climate-resilient agricultural technologies and practices (Joshi & Bhandari, 2023).

Observed changes in adverse weather conditions, such as increasing temperatures and shifting precipitation patterns, have significant implications for crop production and livestock management. Extreme weather events, including heatwaves, droughts, and floods, have become more frequent and severe, leading to substantial agricultural losses globally. For instance, heatwaves in Europe and floods in the Midwestern United States have caused significant crop damage, impacting food supply and prices.

Changes in temperature and weather patterns are projected to further alter agricultural landscapes, affecting areas suitable for farming and crop yields. Rising temperatures and more frequent heatwaves pose challenges to crop growth, with certain crops failing to reproduce at high temperatures. Moreover, changes in precipitation patterns, including increased heavy rainfall and evapotranspiration, affect water availability and reliability for irrigation, exacerbating droughts and floods in various regions. These changes in agricultural water availability have far-reaching consequences, including crop failures, loss of pasture grazing land, and increased water scarcity, particularly in developing countries.

Livestock are also vulnerable to climate change impacts, with rising temperatures and heatwaves increasing the risk of heat stress, affecting animal welfare and productivity. Moreover, changes in water resources, such as glacier retreat in mountainous regions, threaten irrigation-dependent agriculture and hydropower generation, posing challenges to food and water security for millions of people. While adaptation measures, such as changes in management practices and climate-smart agriculture, can help mitigate some of these impacts, addressing the root causes of climate change remains critical for ensuring the long-term sustainability of agricultural systems and global food security.

The effects of increasing atmospheric CO_2 and methane levels have significant impacts on plants, influencing various aspects of their physiology and growth. Elevated atmospheric carbon dioxide has been observed to enhance crop yields and growth by stimulating photosynthesis and reducing water loss through stomatal closure. This phenomenon, known as the CO_2 fertilization effect, varies depending on factors such as plant species, temperature, and water availability. While increased CO_2 levels generally boost net primary productivity, the extent of growth enhancement differs among plant species and ecosystems. However, the exact ecosystem processes associated with this effect remain uncertain and challenging to model accurately.

Studies have demonstrated that elevated CO_2 concentrations may reduce the nutritional quality of some crops, particularly C3 plants like wheat and rice. This decrease in nutritional quality includes lower levels of protein and essential minerals such as zinc and iron, which are crucial for human health. Furthermore, empirical evidence suggests that increasing CO_2 levels result in lower concentrations of various minerals in plant tissues, potentially worsening the nutritional content of food crops and leading to deficiencies in essential nutrients for humans.

Additionally, anthropogenic methane emissions contribute to surface ozone formation, which acts as a significant air pollutant detrimental to plant physiology and crop yield. Increased tropospheric ozone levels have been associated with reduced physiological functions in plants, leading to lower crop yields and quality. Studies estimate that ozone-induced yield losses in major crops account for a substantial portion of the negative impacts of climate change on agriculture, offsetting much of the CO_2 fertilization effect. Therefore, addressing methane emissions and mitigating ozone pollution is crucial for minimizing the adverse effects of climate change on plant health and agricultural productivity.

Changes in the extent and quality of agricultural land are influenced by various factors, including coastal vulnerability to saltwater intrusion, erosion, soil fertility, agricultural land loss from sea level rise, and the thawing of potentially arable land due to climate change.

Coastal regions in the United States, for instance, face the risk of saltwater infiltration into freshwater wells, with some areas having more than half of their wells below sea level. This vulnerability exacerbates the challenges of maintaining freshwater resources for agricultural use.

The warming trend in atmospheric temperatures contributes to a more intense hydrological cycle, leading to increased erosion and soil degradation. Soil fertility is also impacted by global warming, with the potential for significant losses in soil carbon over time due to anthropogenic factors.

Sea level rise poses a significant threat to agricultural lands, particularly in regions like South East Asia, where low-lying areas are susceptible to inundation and salinization of water tables. Countries such as Bangladesh, India, and Vietnam face substantial risks to their rice crops, especially in areas like the Mekong Delta, which are vital for rice cultivation.

Additionally, the thawing of frozen land in regions like Siberia may present new opportunities for agriculture, potentially extending farmable lands northward. However, conflicting projections suggest possible productivity losses and increased drought risks, highlighting the complex and uncertain impacts of climate change on agricultural land use. Despite these challenges, the Arctic region could see increased opportunities for agriculture and forestry as temperatures rise and frozen lands become more accessible.

Response of insects, plant diseases, and weeds to climate change will result in significant alterations in their distributions and behaviours, potentially leading to reduced crop yields of staple crops such as wheat, soybeans, and corn.

Warmer temperatures are likely to accelerate the metabolic rate and breeding cycles of insect populations, as historically cold temperatures have served to limit their numbers. Consequently, the proliferation of fungal plant diseases like wheat rusts and soybean rust is expected to increase due to warmer and wetter winters.

Insect populations are anticipated to experience significant shifts, with some species facing extinction while others, particularly agricultural pests and disease vectors, may benefit from the changing conditions. Areas with temperate climates and higher latitudes could witness substantial transformations in insect populations, potentially leading to outbreaks of pests like the Mountain Pine Beetle in Canada or the spread of pests like the fall armyworm across Sub-Saharan Africa.

Similarly, invasive plant pests such as the fall armyworm are projected to expand their range due to climate change, posing significant threats to global agriculture. Weeds, with their genetic diversity and fast growth rates, are expected to thrive in changing climates, outcompeting cultivated crops and potentially developing increased tolerance to herbicides.

Plant pathogens, responsible for significant global harvest losses, are also likely to thrive in altered climatic conditions, leading to increased risks of crop diseases. Climate change may accelerate the development and spread of plant pathogens such as those causing potato blackleg disease or late blight, further exacerbating agricultural challenges.

Overall, climate change is anticipated to disrupt the delicate balance of insect populations, plant diseases, and weeds, posing significant challenges to global food security and agricultural sustainability.

Climate change has had a significant impact on crop yields globally, with both positive and negative effects observed across different regions and crops. The IPCC Sixth Assessment Report from 2022 highlights the predominantly adverse effects of climate change on crop yields and quality, particularly in low-latitude areas such as tropical regions, where crops like maize and wheat have experienced reductions in yield. Conversely, some high-latitude regions have seen positive impacts on crops like maize, wheat, and sugar beets.

Studies have consistently shown negative temperature impacts on crop yield at the global scale, generally underpinned by similar impacts at country and site scales (Zhao et al., 2017). For instance, between 1981 and 2008, global warming led to a 5.5% decrease in wheat yield in tropical regions, emphasizing the negative influence of climate change on crop production (Farooq et al., 2023). A comprehensive study in 2019 assessed the impact of climate change on 10 crops globally and found decreases in crop yields across Europe, Sub-Saharan

Africa, and Australia, with impacts varying by crop, such as oil palm experiencing a reduction of up to 13.4% and soybean yields decreasing by 3.5% (Ray et al., 2019).

Furthermore, projections for the future indicate further challenges for crop yields under different climate scenarios. A study in 2021 estimated declines in global yields of major crops like maize, rice, wheat, and soybeans, with losses ranging from 3% to 25% by the year 2100 under the most intense climate change scenario. Among the four major crops, maize is considered the most vulnerable to warming, with each 1°C increase resulting in a 7.4% reduction in yield. Similarly, rice, wheat, and soybeans are projected to experience declines in yield with warming, although the extent varies depending on factors such as precipitation and CO2 fertilization effect (Zhao et al., 2017).

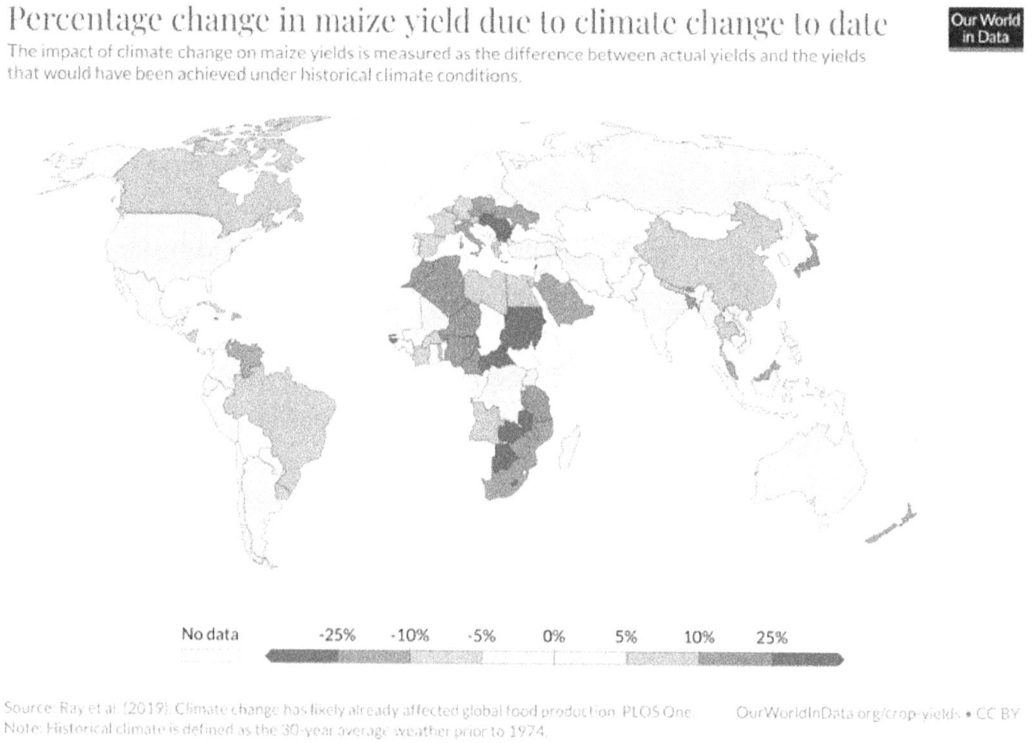

Figure 45: Percentage change in maize yield due to climate change to date. The impact of climate change on maize yields is measured as the difference between actual yields and the yields that would have been achieved under historical climate conditions. Our World in Data, CC BY 4.0, via Wikimedia Commons.

The effects of climate change on livestock rearing are manifold and interconnected. Livestock farming, while being significantly impacted by climate change, also contributes substantially to anthropogenic climate change through its greenhouse gas emissions. As of 2011, approximately 400 million people relied on livestock for their livelihood, highlighting the socio-economic importance of this sector, estimated to be valued at nearly $1 trillion.

Climate change is already leading to a variety of adverse effects on livestock production. These include reductions in the quantity and quality of animal feed, often due to factors such as drought or as a secondary consequence of the CO2 fertilization effect. Additionally, animal parasites and diseases transmitted by vectors are expanding their range, with evidence indicating superior data quality compared to estimates of human pathogen spread.

With the rise in global surface temperatures, heat stress on livestock is becoming more widespread, posing lethal and sublethal risks. Heatwaves have resulted in mass livestock mortality, while sublethal impacts include decreased productivity such as lower milk yield and increased susceptibility to lameness or reproductive impairments. Continued global warming is projected to exacerbate feed-growing challenges, potentially reducing global livestock populations by 7–10% by mid-century. While some regions may avoid extreme heat stress, others may become unsuitable for livestock farming as early as mid-century.

Sub-Saharan Africa is particularly vulnerable to food security shocks due to climate change impacts on livestock, with over 180 million people in the region expected to experience significant declines in the suitability of rangelands by mid-century. Conversely, countries like Japan, the United States, and those in Europe are considered less vulnerable, reflecting differences in human development index, national resilience, and the relative importance of pastoralism in national diets.

Our understanding of how climate change influences global food security has evolved over time. The most recent IPCC Sixth Assessment Report (2022) indicates that by 2050, the number of people at risk of hunger is expected to rise across all scenarios, with estimates ranging from 8 to 80 million. The majority of these individuals are projected to be in Sub-Saharan Africa, South Asia, and Central America. However, it's crucial to note that these projections are relative to a world without climate change, and thus, they do not preclude the possibility of an overall reduction in hunger risk compared to present-day conditions.

An earlier Special Report on Climate Change and Land suggested that under a high emission scenario (RCP6.0), the cost of cereals could increase by 1–29% by 2050, depending on the socioeconomic pathway. This escalation in prices could put between 1–181 million low-income individuals at risk of hunger compared to a scenario without climate change.

Predicting the impact of climate change on food utilization, including food preservation and nutritional absorption, poses challenges. Nevertheless, modelling studies from 2016 suggest that under the most severe climate change scenario, global per capita food availability could decrease by 3.2% by mid-century, with notable reductions in red meat, fruit, and vegetable consumption. These changes could result in approximately 529,000 deaths by 2050, primarily due to micronutrient deficiencies. However, efforts to mitigate climate change could substantially reduce these projections.

As of 2017, approximately 821 million people worldwide suffered from hunger, representing about 11% of the global population. Vulnerability to food insecurity varied regionally, with Sub-Saharan Africa, the Caribbean, and South Asia facing significant challenges.

In 2020, research suggested that under the baseline scenario (SSP2), the number of undernourished individuals globally could decline to 122 million by 2050, despite a growing population. However, the negative impact of climate change could potentially increase this figure by up to 80 million. Implementing measures such as facilitating food trade could mitigate this impact.

A meta-analysis of 57 studies in 2021 painted a more pessimistic picture, projecting that by 2050, around 500 million people could be at risk of hunger under SSP2. In scenarios with high climate change and inequitable global development, this number could rise even further.

Looking ahead, climate change is anticipated to exacerbate heat stress in agricultural regions like the North China Plain, potentially rendering outdoor work impossible on hot summer days by the end of the century. Extreme weather events, such as floods and droughts, can devastate crops and disrupt agricultural activities, leading to economic hardships and job losses. Moreover, increased food prices and decreased production could exacerbate undernutrition and micronutrient deficiencies, posing significant challenges to global food security.

Adaptation in agriculture often emerges from farmers' individual decisions in response to their circumstances. Among these, changes in management practices stand out as perhaps the most crucial adaptation strategy. Additionally, alterations in agricultural locations and international trade of food commodities may also contribute significantly to adaptation efforts.

In confronting the challenges posed by climate change, agricultural innovation proves indispensable. This entails enhanced management of soil, the adoption of water-saving technologies, the cultivation of crops suitable for specific environments, the introduction of diverse crop varieties, the implementation of crop rotations, judicious use of fertilization, and the promotion of community-based adaptation strategies. At governmental and global levels, research and investments are imperative to gain a comprehensive understanding of the issues at hand and to identify the most effective approaches to address them. Government policies and programs should encompass environmentally sensitive subsidies, educational initiatives, economic incentives, as well as provisions for funds, insurance, and safety nets for vulnerable populations. Furthermore, the establishment of early warning systems and the dissemination of accurate weather forecasts to underserved or remote areas can enhance preparedness.

Several proposed solutions aim to address the proliferation of pest populations. One approach involves increasing the application of pesticides on crops, which, despite being cost-effective and straightforward, may prove ineffective due to the development of pesticide resistance among many pest insects. Alternatively, utilizing biological control agents, such as planting rows of native vegetation among crops, offers environmental benefits by preventing the buildup of pesticide resistance in pest insects. However, this method requires additional space for planting native plants.

A sole emphasis on agricultural technology falls short of addressing the multifaceted challenges of climate change. Efforts are underway to facilitate and finance institutional changes, as well as to formulate dynamic policies for long-term climate change adaptation in agriculture. For instance, a 2013 study by the International Crops Research Institute for the Semi-Arid Tropics aimed to identify science-based, pro-poor approaches and techniques that would enable Asia's agricultural systems to cope with climate change, benefitting poor and vulnerable farmers. Recommendations from the study encompassed improving the utilization of climate information in local planning, enhancing weather-based agro-advisory services, promoting diversification of rural household incomes, and incentivizing farmers to adopt natural resource conservation measures aimed at enhancing forest cover, replenishing groundwater, and utilizing renewable energy sources.

Climate-smart agriculture (CSA), also known as climate-resilient agriculture, presents an integrated approach to land management aimed at adapting agricultural practices, livestock, and crops to the impacts of climate change. Additionally, CSA endeavours to mitigate climate change by reducing greenhouse gas emissions from agriculture, all while addressing the imperative of global food security in the face of a growing world population. The focus of CSA extends beyond carbon farming and sustainable agriculture to encompass strategies for increasing agricultural productivity.

The three pillars of CSA include enhancing agricultural productivity and incomes, building resilience to climate change, and reducing or eliminating greenhouse gas emissions from agriculture. Various actions are proposed to tackle future challenges for crops and plants, such as developing heat-tolerant crop varieties, implementing mulching and water management techniques, establishing shade structures and boundary trees, sequestering carbon, and optimizing housing and spacing for livestock. By stabilizing crop production and mitigating the adverse impacts of climate change, CSA aims to maximize food security.

Climate-resilient agriculture is an approach aimed at addressing the challenges posed by climate change in agricultural practices (Fallah-Alipour et al., 2018). The primary goal of climate-resilient agriculture is twofold: first, to enable agricultural systems to adapt to changing climate conditions, and second, to contribute to efforts aimed at mitigating climate change (Talukder et al., 2017). This approach encompasses several critical components, including adaptation, mitigation, maintaining or enhancing agricultural productivity, and sustainability (Agbaje et al., 2001)

Adaptation in climate-resilient agriculture focuses on implementing practices that assist agricultural systems in adjusting to the impacts of climate change. This involves strategies such as the utilization of drought-resistant crop varieties, improved water management techniques to cope with alterations in precipitation patterns, and adjusting planting schedules in response to shifting growing seasons (Janker et al., 2018). Mitigation, on the other hand, involves reducing greenhouse gas emissions from agricultural activities. This includes practices such as decreasing fossil fuel use, implementing conservation tillage methods to sequester carbon in the soil, and optimizing fertilizer application to minimize emissions of nitrous oxide, a potent greenhouse gas (Al-Subaiee et al., 2005).

Furthermore, climate-resilient agriculture aims to maintain or enhance agricultural productivity despite the challenges posed by climate change. This involves adopting new technologies and practices that improve crop yields, enhance soil health, and increase the resilience of livestock to extreme weather events (Williams & Wise, 1997). Sustainability is also a fundamental aspect of climate-resilient agriculture, emphasizing the importance of implementing sustainable agricultural practices that minimize negative environmental impacts and conserve natural resources. This includes promoting soil conservation, enhancing water efficiency, and conserving biodiversity (Lv et al., 2019).

Climate change impacts various agricultural practices and techniques differently. Permaculture, for instance, faces both challenges and benefits from climate change. Increased temperatures, altered precipitation patterns, and extreme weather events may disrupt traditional permaculture designs. However, permaculture principles such as diversity, resilience, and regenerative practices can help mitigate some of these impacts by creating ecosystems better able to adapt to changing conditions.

Biodynamic farming is also affected by climate change. It may experience shifts in growing seasons, precipitation patterns, and pest and disease pressures. Nonetheless, its holistic management approach and focus on soil health, biodiversity, and ecosystem balance may offer some resilience to these changes.

Hydroponics and aquaponics systems are impacted by climate change through alterations in water availability, temperature, and nutrient availability. Increased temperatures may necessitate more energy for cooling in hydroponic systems, while changes in precipitation patterns may affect water availability for aquaponic operations. Yet, these systems can adapt through improved water management and climate-controlled environments.

Urban agriculture faces challenges from climate change such as urban heat islands, altered water availability, and extreme weather events. Despite these challenges, urban agriculture also presents opportunities for climate adaptation and mitigation, such as utilizing green infrastructure for managing stormwater and carbon sequestration.

Agroforestry and food forests experience both benefits and challenges due to climate change. Changes in temperature and precipitation patterns may affect the growth and distribution of tree crops and other perennial plants. However, they can offer resilience to climate change by enhancing soil health, increasing biodiversity, and sequestering carbon.

Polycultures and crop rotation methods are impacted by climate change through altered pest and disease pressures, growing seasons, and precipitation patterns. Diversified cropping systems may offer resilience by reducing the risk of crop failure and improving soil health. Additionally, crop rotation can help manage pests and diseases while enhancing soil fertility.

Integrated pest management (IPM) strategies may need adaptation due to climate change. Changes in pest distribution and abundance, as well as timing and intensity of outbreaks, require adjustments in IPM practices. Utilizing climate-based pest forecasting models and incorporating additional biological control methods may be necessary.

Overall, climate change presents both challenges and opportunities for these agricultural practices and techniques. Adaptation and innovation are essential to ensuring their continued success in a changing climate.

Climate and Enterprise Data

From the perspective of the skills and knowledge required to develop climate risk management strategies for an agricultural, horticultural, or land management enterprise, an agribusiness manager would need to develop proficiency in a number of tasks.

Firstly, the manager would obtain and interpret historical climate data, including any natural disasters, from a range of sources. They would utilize various sources such as meteorological agencies, government databases, historical records, and climate modelling platforms to gather comprehensive historical climate data. This data would be analysed to understand past climate patterns, trends, and occurrences of natural disasters like droughts, floods, or extreme weather events.

Next, the manager would identify weather and climate risk factors using the historical climate data gathered. This involves assessing various weather and climate risk factors that could impact the enterprise, such as the frequency, intensity, and duration of climatic events like heatwaves, frosts, storms, or prolonged dry spells. Understanding these factors helps in evaluating their potential implications for agricultural production and land management practices.

Additionally, the manager would collect information on normal and significant climate events and their impact on natural and rural systems. This entails compiling data on both typical climate patterns and significant climate events that have occurred in the region over time. By assessing the impact of these events on natural ecosystems, rural communities, and agricultural production systems, the manager gains insights into the vulnerabilities and resilience of the enterprise to different climate scenarios.

Moreover, the manager would detail current and historical property and enterprise production. This involves gathering detailed information on the current and historical production activities of the enterprise, including crop yields, livestock performance, land use practices, and resource management techniques. Analysing this data provides insights into the enterprise's exposure to climate risks and its capacity to adapt to changing environmental conditions.

Furthermore, the manager would review short and long-term enterprise goals to ensure they fit within climatic constraints. This entails evaluating the enterprise's goals in the context of climatic constraints and uncertainties. By assessing the feasibility and sustainability of existing business plans and practices in light of potential climate impacts, the manager identifies opportunities for adaptation and innovation to enhance the enterprise's resilience.

Lastly, the manager would source, present, and update climate and enterprise data according to enterprise requirements. This involves sourcing relevant climate and enterprise data, presenting it in a format suitable for decision-making and communication within the enterprise, and ensuring that it is regularly updated and reviewed to reflect changing climatic conditions and business priorities. Effective data management, analysis, and communication skills are crucial to supporting informed decision-making and proactive climate risk management strategies.

Obtaining and interpreting historical climate data, including any natural disasters, involves a systematic approach to ensure thorough understanding and analysis. The process typically unfolds through several key steps.

Firstly, the manager initiates the process by identifying relevant sources from which historical climate data can be sourced. These sources encompass a range of entities, including meteorological agencies, government databases, academic institutions, historical archives, and climate modelling platforms. Each source offers varying types of data, such as temperature records, precipitation levels, wind patterns, and accounts of past natural disasters.

Once the sources are identified, the manager proceeds to gather pertinent historical climate data spanning the desired time period. This data encompasses a broad spectrum of meteorological variables, including temperature, rainfall, humidity, atmospheric pressure, and more. Additionally, data on previous natural disasters, such as droughts, floods, hurricanes, tornadoes, and heatwaves, are collected to comprehend their frequency, intensity, and impact on the region.

Ensuring the quality and reliability of the collected data is paramount. Thus, the manager meticulously verifies the data for inconsistencies, errors, or missing information. Employing data validation techniques helps ensure the accuracy and completeness of the datasets.

With the collected data at hand, the manager embarks on analysing past climate patterns and trends. This involves scrutinizing long-term trends in temperature, precipitation, and other meteorological variables to detect any significant changes or anomalies over time. Statistical methods like trend analysis, time series analysis, and regression analysis may be employed to quantify and visualize these patterns.

In addition to climate patterns, the manager also examines occurrences of natural disasters recorded in the historical data. By analysing the frequency, severity, and spatial distribution of events such as droughts, floods, storms, and wildfires, insights into the region's vulnerability to different natural hazards are gained.

Finally, the manager interprets the findings of the data analysis to grasp the historical climate dynamics and natural disaster occurrences in the region comprehensively. This involves identifying recurring weather patterns, climate cycles, and extreme events that have influenced the local climate over time. These insights serve as the foundation for decision-making processes related to climate risk management and adaptation strategies for agricultural, horticultural, or land management enterprises.

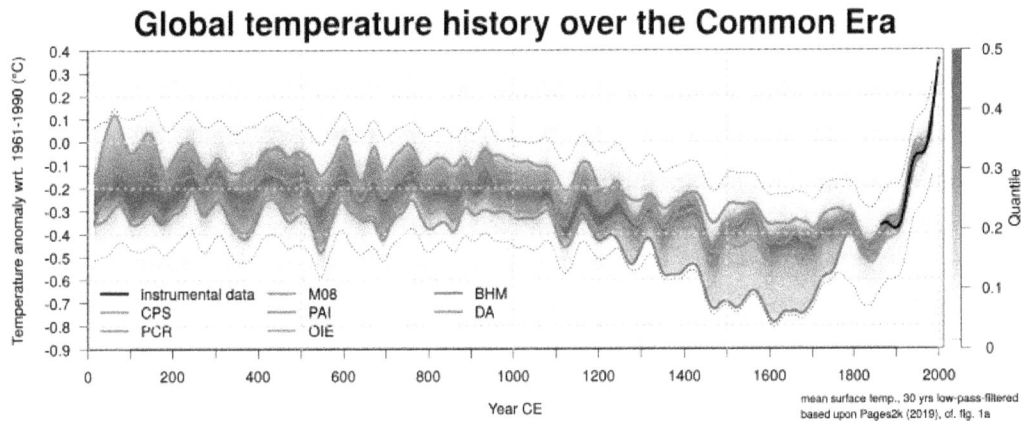

Figure 46: Example data, History of global mean surface temperatures over the Common Era. Data and computation based upon Pages2k (2019), doi:10.1038/s41561-019-0400-0, fig. 1a. DeWikiMan, CC BY-SA 4.0, via Wikimedia Commons.

Historical climate data can be sourced from various sources, including:

1. Meteorological Agencies: National meteorological agencies, such as the National Oceanic and Atmospheric Administration (NOAA) in the United States, the Met Office in the United Kingdom, and the Australian Bureau of Meteorology, often maintain extensive databases of historical weather observations. These agencies collect data from weather stations, satellites, and other monitoring instruments.

2. Government Databases: Many governments maintain databases of historical climate data, often in collaboration with meteorological agencies. These databases may include records of temperature, precipitation, wind speed, humidity, and other meteorological variables. They are typically accessible through government websites or data portals.

3. Academic Institutions: Universities and research institutions may compile and analyse historical climate data for research purposes. Researchers often publish their findings in academic journals or make datasets available for public access through institutional repositories.

4. Historical Archives: Historical archives, including libraries, museums, and archives, may contain records of past weather observations, climate reports, and other relevant documents. These archives can provide valuable insights into historical climate patterns and events.

5. Climate Modelling Platforms: Climate modelling platforms, such as the Coupled Model Intercomparison Project (CMIP) and the Climate Data Store (CDS), provide access to climate model simulations and projections. While these platforms primarily focus on future climate scenarios, they may also include historical climate data for model validation and calibration purposes.

6. International Organizations: International organizations, such as the World Meteorological Organization (WMO) and the Intergovernmental Panel on Climate Change (IPCC), compile and disseminate global climate datasets and assessments. These organizations often collaborate with national meteorological agencies and research institutions to collect and analyse climate data from around the world.

7. Open Data Portals: Many countries and organizations provide access to historical climate data through open data portals. These portals offer a wide range of datasets and tools for exploring and analysing climate information. Examples include the NOAA National Centers for Environmental Information (NCEI) Climate Data Online portal and the European Climate Data Explorer.

Overall, historical climate data can be obtained from a variety of sources, each offering unique datasets and insights into past weather and climate conditions.

Identifying weather and climate risk factors entails a systematic process of analysing historical climate data to evaluate potential threats to an enterprise. Initially, the manager gathers comprehensive historical climate data from reputable sources such as meteorological agencies, government databases, or academic institutions. This dataset encompasses various meteorological variables, including temperature, precipitation, wind speed, humidity, and extreme weather events, spanning a significant time period.

Subsequently, the manager reviews the collected historical climate data to discern recurring weather patterns and trends. This analytical phase entails examining long-term averages, seasonal variations, and extreme weather conditions. Through understanding past climate patterns, the manager can discern potential risk factors that may impact the enterprise's operations.

Drawing upon the historical climate data, the manager proceeds to identify and evaluate various weather and climate risk factors that could affect the enterprise. This entails assessing factors such as the frequency, intensity, and duration of climatic events like heatwaves, frosts, storms, or prolonged dry spells. Moreover, the manager gauges the likelihood of these events occurring in the future based on historical trends and climate projections.

Once the risk factors are identified, the manager proceeds to assess their potential implications for agricultural production and land management practices. This involves considering how each risk factor could influence crop yields, soil moisture levels, irrigation requirements, livestock health, and other operational aspects. Additionally, the manager evaluates the vulnerability of different crops, livestock, and land management practices to specific weather and climate hazards.

Finally, based on the identified risk factors and their implications, the manager formulates appropriate risk management strategies to mitigate potential impacts on the enterprise. This may encompass implementing adaptive measures such as crop diversification, irrigation upgrades, soil conservation practices, or infrastructure improvements to bolster resilience to weather and climate risks. Furthermore, the manager integrates risk management into the enterprise's overall business planning and decision-making processes to ensure proactive and effective responses to changing climate conditions.

Some specific examples of recurring weather patterns and trends that can be identified through the analysis of historical climate data:

1. Seasonal Variations: Historical climate data often reveals recurring seasonal variations in temperature and precipitation. For example, in temperate regions, there is typically a pattern of warmer temperatures and higher precipitation levels in the spring and summer months, followed by cooler temperatures and lower precipitation in the fall and winter.

2. Droughts and Dry Spells: Historical climate data can show periods of prolonged droughts or dry spells, characterized by below-average precipitation and reduced soil moisture levels. These events often occur in certain regions during specific times of the year and can have significant impacts on agriculture, water resources, and ecosystems.

3. Heatwaves: Analysis of historical climate data may reveal patterns of extreme heatwaves, where temperatures rise significantly above normal for an extended period. Heatwaves often occur during the summer months and can lead to heat stress in crops, livestock, and humans, as well as increased energy demand for cooling.

4. Cold Snaps and Frosts: Similarly, historical climate data may show patterns of cold snaps or frosts, where temperatures drop sharply below freezing, particularly during the winter months. Cold snaps can damage crops and sensitive vegetation, disrupt transportation and infrastructure, and pose health risks to humans and animals.

5. Storm Events: Historical climate data can identify patterns of severe weather events such as storms, hurricanes, tornadoes, and thunderstorms. These events may occur seasonally or sporadically throughout the year and can cause significant damage to crops, infrastructure, and communities.

6. Floods: Analysis of historical climate data may reveal trends in flooding events, characterized by heavy rainfall or snowmelt leading to overflowing rivers, flash floods, and inundation of low-lying areas. Floods can have devastating impacts on agriculture,

property, and public safety, particularly in flood-prone regions.

These are just a few examples of recurring weather patterns and trends that can be identified through the analysis of historical climate data. Understanding these patterns is important for assessing climate risks and developing effective strategies for climate adaptation and resilience in agriculture and other sectors.

Collecting information on normal and significant climate events and their impact on natural and rural systems involves a systematic approach to gather comprehensive data and assess their implications. The manager initiates this task by defining the data requirements, specifying the types of climate events to be documented, such as temperature extremes, precipitation anomalies, droughts, floods, storms, and other relevant phenomena. Additionally, the manager identifies the spatial and temporal scales at which the data will be collected, considering both short-term weather events and long-term climate trends.

Next, the manager identifies sources of information on climate events and their impacts, including meteorological agencies, government reports, academic studies, historical records, local knowledge from community members, and firsthand observations from stakeholders such as farmers, land managers, and environmental organizations. By leveraging a diverse range of sources, the manager can obtain comprehensive and reliable data on past climate events.

The manager then compiles historical data on normal and significant climate events that have occurred in the region over time. This may involve accessing meteorological records, climate databases, historical archives, satellite imagery, and other relevant sources of information. The data collection process aims to document the frequency, duration, intensity, and spatial distribution of climate events, as well as their impacts on natural ecosystems, rural communities, and agricultural production systems.

Once the data on climate events is compiled, the manager assesses their impact on natural ecosystems, rural communities, and agricultural production systems. This involves analysing how different types of climate events, such as heatwaves, droughts, floods, and storms, have affected vegetation, soil moisture, water resources, infrastructure, livelihoods, and socioeconomic activities in the region. By understanding the vulnerabilities and resilience of the enterprise to different climate scenarios, the manager can identify adaptation measures and risk management strategies to enhance resilience and sustainability.

Finally, the manager documents the findings of the data collection and impact assessment process, which may involve creating reports, maps, charts, and other visualizations to communicate key insights and trends related to normal and significant climate events. By documenting the historical context and consequences of climate variability and extreme weather events, the manager provides valuable information for decision-making and planning processes aimed at building climate resilience in natural and rural systems.

Detailing current and historical property and enterprise production involves a systematic process aimed at understanding the agricultural activities of the enterprise over time. Here's a breakdown of how the manager can accomplish this task:

Firstly, the manager identifies the specific types of data necessary to comprehensively detail both the current and historical aspects of property and enterprise production. This could encompass a wide range of information including crop yields, livestock performance metrics, land use practices, irrigation methods, fertilizer and pesticide application records, and other pertinent production-related data points.

Next, the manager embarks on gathering historical data pertaining to property and enterprise production activities. This involves sourcing information from various channels such as farm records, financial documents, agricultural surveys, government reports, and historical archives. The objective is to compile a robust dataset spanning multiple years or even decades to capture long-term trends and patterns in production.

In addition to historical data, the manager also collects up-to-date information on the current production activities of the enterprise. This includes data on crop yields, livestock inventory, harvest schedules, field management practices, and other relevant parameters. Various tools such as field surveys, interviews with farmers, and on-site observations may be employed to gather this current data.

Once the data is amassed, the manager conducts a comprehensive analysis to identify trends, patterns, and key performance indicators related to property and enterprise production. This analysis involves comparing current production levels with historical averages, assessing the impact of different management practices on productivity, and pinpointing areas of strength and weakness within the enterprise.

Following the analysis, the manager evaluates the enterprise's exposure to climate risks based on the production data. This entails identifying vulnerabilities such as sensitivity to temperature extremes, susceptibility to drought or water scarcity, and risks associated with pest and disease outbreaks. By understanding how climate variability and extreme weather events may impact production activities, the manager can develop strategies to mitigate these risks and enhance the enterprise's resilience.

During this evaluation, vulnerabilities that could potentially affect the enterprise's operations and productivity are identified. These vulnerabilities include sensitivity to temperature extremes, susceptibility to drought or water scarcity, and risks associated with pest and disease outbreaks.

Firstly, the manager assesses how the enterprise's crops, livestock, and infrastructure respond to extreme temperatures, both hot and cold. For instance, certain crops may be sensitive to heat stress, while livestock may suffer from heat-related health issues. Conversely, frost-sensitive crops may be vulnerable to damage during cold spells.

Secondly, the manager examines the enterprise's reliance on water resources and its capacity to endure periods of drought or limited water availability. This involves evaluating the efficiency of irrigation systems, water storage capacity, and the ability of crops and livestock to withstand water stress.

Thirdly, the manager considers the likelihood of pest infestations and disease outbreaks under various climatic conditions. Factors such as temperature, humidity, and rainfall patterns can influence the prevalence and severity of pests and diseases. The manager evaluates the enterprise's vulnerability to these risks and identifies potential strategies to mitigate them.

By comprehending how climate variability and extreme weather events could impact production activities through this evaluation process, the manager can develop tailored strategies to mitigate risks and enhance the enterprise's resilience. These strategies may involve implementing practices such as crop diversification, improving irrigation efficiency, adopting pest and disease management techniques, and investing in resilient crop varieties or livestock breeds. The overarching objective is to minimize the adverse impacts of climate risks on the enterprise while maximizing its capacity to adapt and thrive in a changing climate.

Lastly, armed with insights gleaned from analysing production data, the manager identifies adaptation strategies to help the enterprise navigate changing environmental conditions. These strategies may include implementing practices such as crop diversification, soil conservation measures, water-efficient irrigation techniques, improved livestock management practices, and the adoption of climate-resilient crop varieties. By proactively addressing climate risks, the enterprise can position itself to thrive in a changing climate while upholding sustainable production practices.

Reviewing short and long-term enterprise goals to ensure their alignment with climatic constraints involves a methodical assessment of the enterprise's objectives considering potential climate impacts. Initially, the manager scrutinizes the enterprise's short and long-term goals, encompassing targets related to production, market expansion, profitability, sustainability, and resilience, ensuring clarity regarding associated timelines and objectives.

Subsequently, the manager proceeds to evaluate the climatic constraints and uncertainties that could impede the attainment of these goals. This entails examining variables such as temperature fluctuations, precipitation trends, occurrences of extreme weather events, and the enduring repercussions of climate change on agricultural and land management practices.

Identifying potential impacts on the enterprise's operations, productivity, and profitability constitutes the next step. This involves contemplating how alterations in climate variables might influence crop yields, water availability, soil health, pest and disease pressures, and the overall viability of the enterprise's endeavours.

This process involves an examination of how variations in climate variables could affect different facets of the enterprise's functions and overall sustainability. These include:

- Crop Yields: The manager scrutinizes how shifts in climate variables such as temperature, precipitation, and humidity might influence the growth, development, and productivity of crops. For instance, warmer temperatures could hasten crop maturation but might also induce heat stress, while alterations in precipitation patterns could impact water availability for irrigation and crop nourishment.

- Water Availability: Assessing how changes in climate variables might impact water resources forms a crucial aspect of this process. This includes analysing alterations in precipitation patterns, the timing of snowmelt, and rates of evaporation, as well as the potential occurrence of droughts or water scarcity. Understanding these factors is vital for devising irrigation schedules, managing water resources efficiently, and ensuring an adequate water supply for crops, livestock, and other operational needs.

- Soil Health: The manager also considers how variations in climate variables could affect the health and fertility of the soil. For example, elevated temperatures and shifts in precipitation patterns may disrupt soil moisture levels and alter nutrient availability, consequently influencing plant growth and the activity of soil microbes. Evaluating these potential impacts helps determine if adjustments are needed in soil management practices to uphold or enhance soil health and productivity.

- Pest and Disease Pressures: Climate variables have a notable impact on the prevalence, distribution, and behaviour of pests and diseases that affect crops and livestock. Factors such as warmer temperatures, changes in precipitation patterns, and shifts in seasonal timing can all influence the life cycles, population dynamics, and transmission rates of pests and diseases. The manager considers how these changes could affect pest and disease pressures on the enterprise and formulates strategies to mitigate associated risks.

- Overall Viability: Lastly, the manager examines how the combined effects of alterations in climate variables may affect the enterprise's overall viability and resilience. This involves evaluating the collective influence of climate-related factors on production costs, market accessibility, profitability, and long-term sustainability. By comprehending these potential impacts, the manager can make informed decisions and implement adaptive measures to ensure the enterprise remains viable and competitive amid a changing climate.

The manager then undertakes an assessment of the feasibility and sustainability of existing business plans and practices in light of anticipated climate impacts. This evaluation scrutinizes whether current strategies possess adequate resilience to withstand climate-related adversities and uncertainties over both short and long-term horizons, while also considering the environmental, social, and economic sustainability of the enterprise's activities.

Based on the evaluation of goals and climatic constraints, the manager discerns opportunities for adaptation and innovation to bolster the enterprise's resilience. These opportunities may encompass adjustments to production practices, diversification of crops or livestock, investment in climate-resilient technologies and infrastructure, enhancement of water management strategies, and fortification of partnerships with stakeholders.

Finally, the manager integrates climate considerations into the enterprise's decision-making processes. This encompasses prioritizing climate resilience in strategic planning, allocating resources for adaptation measures, monitoring climate-related risks and opportunities, and periodically revisiting and refining goals and strategies to accommodate evolving climatic conditions.

Through this comprehensive review process, the manager can proactively navigate climate risks and leverage opportunities, thereby fostering the sustainability and resilience of the enterprise amidst a dynamic and evolving climatic landscape.

Sourcing, presenting, and updating climate and enterprise data according to enterprise requirements is a multifaceted process aimed at facilitating informed decision-making and proactive climate risk management strategies. The manager begins by identifying sources of relevant climate and enterprise data that align with the specific needs and objectives of the enterprise. These sources may include meteorological agencies, government databases, academic research, industry reports, internal records, and third-party data providers. The manager ensures that the selected data sources provide accurate, reliable, and up-to-date information that is pertinent to the enterprise's operations and decision-making processes.

Once the data is sourced, the manager presents it in a format that is suitable for decision-making and communication within the enterprise. This may involve organizing the data into clear and concise reports, dashboards, charts, graphs, maps, or other visualizations that effectively convey key insights and trends. The manager considers the preferences and information needs of various stakeholders within the enterprise and tailors the presentation format accordingly to ensure maximum comprehension and usability.

The manager ensures that the climate and enterprise data is regularly updated and reviewed to reflect changing climatic conditions and business priorities. This involves establishing mechanisms for collecting new data, monitoring ongoing trends, and revising existing datasets as needed. The manager stays informed about emerging climate-related developments, scientific research, and industry trends that may impact the enterprise and incorporates relevant updates into the data management process.

Managing and analysing climate and enterprise data requires effective data management and analytical skills. The manager employs data management tools, software applications, and analytical techniques to organize, process, and analyse large volumes of data efficiently. This may involve data cleaning, validation, transformation, statistical analysis, modelling, and interpretation to extract meaningful insights and identify actionable recommendations.

Finally, the manager utilizes strong communication skills to effectively communicate the findings and implications of the data analysis to key stakeholders within the enterprise. This may involve preparing presentations, reports, and briefings that clearly articulate the relevance of the data to decision-making processes and the potential implications for climate risk management strategies. The manager engages in collaborative discussions, solicits feedback, and fosters a culture of data-driven decision-making within the enterprise to ensure that climate and enterprise data are utilized effectively to support informed decision-making and proactive risk management efforts.

In summary, developing climate risk management strategies for an agricultural, horticultural, or land management enterprise requires proficiency in several key tasks. Firstly, the manager must obtain and interpret historical climate data from diverse sources such as meteorological agencies, government databases, and historical archives. This data is then analysed to understand past climate patterns and occurrences of natural disasters.

Next, the manager identifies weather and climate risk factors using the historical climate data gathered, assessing their potential impacts on the enterprise. This includes evaluating factors like the frequency, intensity, and duration of climatic events such as heatwaves, frosts, storms, or prolonged dry spells.

Additionally, the manager collects information on normal and significant climate events and their impact on natural and rural systems. This involves compiling data on climate patterns and events over time and assessing their implications for ecosystems, communities, and agricultural systems.

Moreover, the manager details current and historical property and enterprise production activities, gathering data on crop yields, livestock performance, land use practices, and resource management techniques. This data analysis provides insights into the enterprise's exposure to climate risks and its capacity to adapt.

Furthermore, the manager reviews short and long-term enterprise goals to ensure alignment with climatic constraints, evaluating the feasibility and sustainability of existing plans in light of potential climate impacts.

Lastly, the manager sources, presents, and updates climate and enterprise data according to enterprise requirements, utilizing effective data management, analysis, and communication skills to support informed decision-making and proactive risk management strategies. This involves sourcing relevant data, presenting it appropriately for decision-making, and ensuring regular updates to reflect changing climatic conditions and business priorities.

As a process and recording example, Figure 47 provides a sample Climate Risk Management Strategy Development Recording Template.

Climate Risk Management Strategy Development Recording Template
Date: [Date of recording]
Agribusiness Manager: [Name of manager]
Enterprise: [Name of agricultural, horticultural, or land management enterprise]

Task 1: Obtain and interpret historical climate data
- Sources utilized: [List of sources such as meteorological agencies, government databases, historical records, etc.]
- Data gathered: [Brief description of historical climate data collected]
- Analysis conducted: [Summary of analysis to understand past climate patterns, trends, and occurrences of natural disasters]

Task 2: Identify weather and climate risk factors
- Risk factors identified: [List of weather and climate risk factors assessed]
- Potential implications: [Analysis of the potential impacts on agricultural production and land management practices]

Task 3: Collect information on normal and significant climate events
- Data compiled: [Summary of data on typical climate patterns and significant climate events]
- Impact assessment: [Assessment of the impact of these events on natural ecosystems, rural communities, and agricultural production systems]

Task 4: Detail current and historical property and enterprise production
- Information gathered: [Description of current and historical production activities including crop yields, livestock performance, land use practices, etc.]
- Insights gained: [Summary of insights into the enterprise's exposure to climate risks and its capacity to adapt]

Task 5: Review short and long-term enterprise goals
- Goals evaluated: [Description of short and long-term goals reviewed]
- Assessment of feasibility: [Analysis of the feasibility and sustainability of existing business plans and practices in light of potential climate impacts]

Task 6: Source, present, and update climate and enterprise data
- Data sourcing: [Explanation of how relevant climate and enterprise data was sourced]
- Presentation format: [Description of the format used to present data for decision-making and communication within the enterprise]
- Update process: [Explanation of how data is regularly updated and reviewed to reflect changing climatic conditions and business priorities]

Comments/Notes: [Optional section for additional comments or notes related to the tasks performed or any other relevant information]

Figure 47: Sample Climate Risk Management Strategy Development Recording Template.

As a hypothetical example of using this process, the following outlines a climate risk management strategy for a vertical farming agribusiness.

Climate Risk Management Strategy Development Recording Template

Date: February 10, 20XX

Agribusiness Manager: Sarah Smith

Enterprise: GreenTech Vertical Farms

Task 1: Obtain and interpret historical climate data

- *Sources utilized: National Meteorological Service, Government Climate Database, Historical Records from Local Agricultural Research Institute*

- *Data gathered: Historical temperature and precipitation records for the past 20 years*

- *Analysis conducted: Analysis of past climate patterns reveals increasing temperature trends and sporadic precipitation fluctuations, with occasional extreme weather events like heatwaves and heavy rainfall.*

Task 2: Identify weather and climate risk factors

- *Risk factors identified: Increasing temperatures, unpredictable precipitation patterns, heatwaves, and heavy rainfall events*

- *Potential implications: Risks include reduced crop yields, water scarcity, heat stress on crops, and flooding in low-lying areas of the farm.*

Task 3: Collect information on normal and significant climate events

- *Data compiled: Records of typical temperature and precipitation patterns, as well as significant climate events such as heatwaves and heavy rainfall incidents.*

- *Impact assessment: Significant climate events have historically caused temporary disruptions in crop growth and led to minor infrastructure damage.*

Task 4: Detail current and historical property and enterprise production

- *Information gathered: Current and historical data on crop yields, plant health metrics, energy consumption, and resource management techniques.*

- *Insights gained: Analysis indicates consistent growth in crop yields over the past five years, but recent temperature increases are causing concerns about energy consumption for climate control systems.*

Task 5: Review short and long-term enterprise goals

- *Goals evaluated: Short-term goal to expand production by 20% within the next two years. Long-term goal to achieve carbon neutrality by 2030.*

- *Assessment of feasibility: The feasibility of expanding production is challenged by potential climate impacts on energy costs and water availability. Achieving carbon neutrality may require additional investments in renewable energy and sustainable practices.*

Task 6: Source, present, and update climate and enterprise data

- *Data sourcing: Climate data sourced from reliable meteorological agencies and internal production records. Enterprise data sourced from farm management software and energy consumption monitoring systems.*

- *Presentation format: Data presented in visual dashboards accessible to key stakeholders. Regular updates provided through monthly reports and quarterly meetings.*

- *Update process: Data updated in real-time through automated systems and reviewed quarterly to incorporate new climate projections and production trends.*

Comments/Notes: *Despite climate risks, opportunities exist to enhance resilience through investments in renewable energy and water-efficient technologies. Continuous monitoring and adaptation are essential for sustainable growth in the face of changing climate conditions.*

Chapter Nine

Identifying and Analysing Climate Risks and Opportunities

Assessing Seasonal Climate Forecasts, Identifying Climate Risks and Opportunities Specific to The Enterprise Site, and Analysing Their Impact on Production Using Qualitative and Quantitative Methods

To formulate effective climate risk management strategies for agricultural, horticultural, or land management enterprises, an agribusiness manager must undertake several key steps. Firstly, the manager would meticulously scrutinize weather forecasts and climate projections pertinent to the enterprise's geographical location. This comprehensive analysis encompasses various climatic parameters such as temperature fluctuations, precipitation levels, humidity trends, and the likelihood of extreme weather occurrences. By gaining insights into anticipated seasonal climate patterns, the manager can proactively prepare the enterprise for potential challenges and opportunities arising from climatic variability.

Subsequently, the manager embarks on a thorough assessment of climate-related risks and opportunities specific to the site and enterprise region. This involves identifying potential threats such as droughts, floods, heatwaves, and storms, which could adversely impact agricultural operations. Conversely, the manager also seeks out opportunities inherent in changing climatic conditions, such as extended growing seasons or the introduction of new crop varieties better suited to the evolving environment.

Furthermore, the manager conducts a detailed analysis of how different weather and climate factors may affect the enterprise's production processes and outcomes. By scrutinizing variables such as crop yields, soil moisture levels, pest and disease pressures, and livestock health, the manager gains a comprehensive understanding of the potential implications of climate risks on the enterprise's overall productivity and profitability.

Employing a blend of qualitative and quantitative methodologies, the manager evaluates climate risks and opportunities. Qualitative approaches, including risk assessments, scenario planning, and SWOT analysis, provide insights into the broader implications and contexts of climate-related challenges and opportunities. Meanwhile, quantitative techniques such as statistical modelling, climate data analysis, and economic modelling enable the manager to quantify the likelihood and severity of climate risks and identify potential avenues for adaptation and innovation.

By systematically addressing these critical aspects, the agribusiness manager can devise resilient and adaptive climate risk management strategies tailored to the specific needs and circumstances of the enterprise. These strategies are designed to enhance the enterprise's capacity to withstand and thrive amidst climatic variability and change, thereby ensuring its long-term sustainability and resilience in the face of evolving environmental conditions.

To analyse forecasted chances of seasonal climate for the enterprise region, the following steps are recommended:

- Access Weather Forecasts and Climate Projections: Begin by accessing reliable sources such as meteorological agencies, climate research institutions, and online platforms. These sources offer tailored weather forecasts and climate projections specific to the enterprise's location. Detailed information about temperature trends, precipitation levels, humidity variations, and the likelihood of extreme weather events over specific time frames can be obtained from these sources.

- Review Seasonal Climate Patterns: Study historical weather data and climate patterns for the enterprise region to comprehend typical seasonal variations in temperature, precipitation, and other meteorological factors. Identifying recurring trends and anomalies provides valuable insights into potential influences on seasonal climate conditions.

- Assess Forecasted Climate Variables: Analyse the forecasted climate variables for the upcoming season, including projected temperature ranges, expected precipitation amounts, humidity levels, and the likelihood of extreme weather events such as storms, droughts, or heatwaves. Paying attention to any deviations from historical norms or significant shifts in climate patterns is crucial.

- Consider Regional Influences: Take into account regional factors that may impact seasonal climate patterns, such as geographic features, proximity to water bodies, elevation changes, and urbanization effects. These factors contribute to variations in local microclimates and can influence weather and climate conditions.

- Evaluate Climate Model Outputs: Review climate model outputs and ensemble forecasts to gain insights into potential scenarios for the upcoming season. Comparing multiple models and assessing their agreement or divergence helps gauge the confidence level in forecasted climate projections.

- Anticipate Impacts on Enterprise Operations: Consider how forecasted seasonal climate patterns may affect various aspects of the enterprise's operations, including crop growth, livestock management, irrigation needs, pest and disease pressures, and overall productivity. Assessing potential risks and opportunities associated with projected climate conditions is essential.

- Develop Adaptive Strategies: Based on the analysis of forecasted seasonal climate data, develop adaptive strategies to mitigate risks and capitalize on opportunities. This may involve adjusting planting schedules, selecting crop varieties or breeds suited to projected climate conditions, implementing water conservation measures, upgrading infrastructure, and diversifying revenue streams to enhance resilience.

- Monitor Weather Updates: Continuously monitor weather updates and climate outlooks throughout the season to track any changes or deviations from initial forecasts. Staying informed about evolving weather patterns enables the adjustment of management practices to optimize outcomes for the enterprise.

To review seasonal climate patterns effectively, start by accessing historical weather data for the enterprise region from reliable sources like meteorological agencies, climate databases, or research institutions. This data encompasses records of temperature, precipitation, humidity, wind patterns, and other relevant meteorological factors spanning several years or decades. Once you've gathered the data, compile and organize it in a structured format, such as spreadsheets or databases, ensuring chronological organization and categorization by relevant time intervals like months or seasons for easy comparison and pattern identification.

Next, analyse historical temperature data to discern typical seasonal variations in temperature for the enterprise region. Look for recurring trends such as seasonal temperature changes, average temperature ranges, and fluctuations between day and night temperatures. Additionally, note any anomalies or extreme temperature events that deviate from expected seasonal norms. Following this, assess historical precipitation data to comprehend typical seasonal variations in rainfall or snowfall. Identify patterns such as wet and dry seasons, average precipitation amounts, and the frequency of extreme precipitation events such as heavy rainstorms or droughts, while also observing shifts in precipitation patterns over time.

Consideration of other relevant meteorological factors such as humidity, wind patterns, atmospheric pressure, and cloud cover is crucial. Analyse historical data for these factors to gain a comprehensive understanding of seasonal climate patterns and their potential influences on weather conditions. While analysing the data, pay close attention to recurring trends and patterns, such as seasonal cycles or consistent fluctuations in meteorological variables, and highlight any anomalies or irregularities that may indicate unusual weather events or climate phenomena.

Interpret the findings of the analysis to develop a comprehensive understanding of seasonal climate patterns in the enterprise region. Consider the implications of identified trends and anomalies for agricultural or land management practices, including potential risks and opportunities associated with seasonal weather variations. Finally, document the results of the analysis, including key findings, trends, anomalies, and potential implications for the enterprise. Communicate these results effectively to relevant stakeholders, such as management teams, agricultural professionals, or decision-makers, to inform strategic planning and decision-making processes.

By systematically reviewing historical weather data and climate patterns for the enterprise region, agribusiness managers can gain valuable insights into typical seasonal variations in temperature, precipitation, and other meteorological factors. Identifying recurring trends and anomalies provides essential information for understanding the influences on seasonal climate conditions and informing climate risk management strategies.

To assess forecasted climate variables effectively, begin by accessing reliable forecasted climate data for the upcoming season from trusted sources such as meteorological agencies, climate research institutions, or online platforms. This data typically encompasses projected temperature ranges, expected precipitation amounts, humidity levels, and forecasts for extreme weather events such as storms, droughts, or heatwaves.

Next, review the projected temperature ranges for the upcoming season to grasp anticipated variations in temperature. Analyse these forecasts in comparison to historical norms to detect any deviations or significant shifts in climate patterns. Pay close attention to forecasted trends, including potential increases or decreases in average temperatures, as well as the likelihood of temperature extremes.

Evaluate the forecasted precipitation amounts for the upcoming season to anticipate changes in rainfall or snowfall patterns. Compare the projected precipitation levels with historical averages to assess any deviations or notable variations. Consider the potential impact of forecasted precipitation on soil moisture, water availability, and agricultural practices.

Analyse the forecasted humidity levels for the upcoming season to comprehend atmospheric moisture conditions. Evaluate how changes in humidity may affect factors such as evaporation rates, plant transpiration, and overall environmental comfort. Take note of forecasted trends in humidity levels and consider their implications for agricultural operations.

Evaluate the likelihood of extreme weather events such as storms, droughts, or heatwaves based on forecasted data. Assess the probability of these events occurring and their potential impacts on the enterprise. Consider historical trends in extreme weather events and any changes in frequency or intensity indicated by the forecasts.

Compare the forecasted climate variables with historical data to identify any deviations or shifts from historical norms. Look for trends or patterns that differ from past climate conditions and consider the significance of these deviations for the enterprise region. Pay particular attention to any forecasted changes that may pose risks or opportunities for agricultural or land management practices.

Continuously monitor weather forecasts and climate outlooks throughout the season to stay informed about updates and revisions to forecasted climate variables. Be prepared to adapt strategies and management practices based on updated forecast information. By following these steps and carefully analysing forecasted climate variables, agribusiness managers can gain valuable insights into anticipated changes in temperature, precipitation, humidity, and the likelihood of extreme weather events. Identifying deviations from historical norms and monitoring forecast updates allows managers to proactively respond to climate-related risks and opportunities, ultimately optimizing decision-making and operational outcomes.

Considering regional influences effectively involves several steps. Firstly, it's essential to identify the relevant regional factors that could impact seasonal climate patterns in the enterprise area. These factors encompass geographic features like mountains, valleys, plains, and coastal areas, as well as the proximity to water bodies such as oceans, lakes, or rivers. Additionally, elevation changes and urbanization effects, including the presence of cities or industrial areas, are crucial considerations as they can significantly alter local climate conditions.

Understanding how geographic features influence weather patterns and climate conditions is vital. For instance, mountain ranges can influence temperature and precipitation by blocking or redirecting air masses, leading to variations in rainfall or snowfall patterns. Coastal areas may experience milder temperatures due to maritime influences, while inland areas may exhibit more extreme temperature fluctuations.

Assessing the proximity to water bodies is another critical step. Bodies of water can moderate temperature extremes, resulting in cooler summers and milder winters compared to inland areas. They also play a role in increasing humidity levels and influencing precipitation patterns through processes like evaporation and moisture transport.

Evaluating elevation changes within the region is essential for understanding their effects on climate. Higher elevations typically experience cooler temperatures and may receive more precipitation than low-lying areas. Elevation gradients can create microclimates with distinct climate conditions, such as cooler temperatures at higher elevations and warmer temperatures in valleys.

Considering the impact of urbanization on local climate conditions is also crucial. Urban areas often experience the urban heat island effect, where built-up infrastructure and human activities lead to higher temperatures compared to surrounding rural areas. Urbanization can also affect precipitation patterns and wind patterns through changes in surface albedo and land cover.

Analysing the combined effects of multiple regional factors is necessary to understand how they interact to influence seasonal climate patterns in the enterprise area. For example, coastal areas with high urbanization levels may experience unique climate conditions due to the combined influence of maritime effects and the urban heat island effect. Similarly, mountainous regions near water bodies may exhibit diverse microclimates based on elevation, proximity to the coast, and prevailing wind patterns.

Finally, integrating regional influences into climate analysis and forecasting efforts allows for more accurate assessments of seasonal climate patterns. Considering how these factors contribute to variations in local microclimates enables agribusiness managers to develop informed climate risk management strategies and adapt agricultural practices effectively to local climate conditions.

To effectively evaluate climate model outputs, several steps should be followed. Firstly, access climate model outputs from reputable sources, including climate research institutions, meteorological agencies, or online databases. These models simulate various aspects of the Earth's climate system, offering projections of future climate conditions based on different scenarios and input parameters.

Next, review ensemble forecasts, which involve running multiple simulations with slight variations in initial conditions or model parameters. These forecasts help capture uncertainty in climate projections, providing a range of possible outcomes for the upcoming season.

It's crucial to compare outputs from multiple climate models to assess their agreement or divergence regarding projected climate conditions. Each model may utilize different algorithms, assumptions, and input data, leading to variations in projections. Comparing multiple models allows for a more comprehensive understanding of potential scenarios and helps identify areas of consensus or disagreement among them.

Assessing consistency and robustness is essential. Look for consistent trends or patterns across different climate model outputs to gauge the confidence level in forecasted climate projections. Consistency among models increases confidence in projected outcomes, while significant discrepancies may indicate areas of uncertainty or limitations in predictive capabilities.

Evaluate the skill and performance of individual climate models based on their historical accuracy and ability to simulate past climate conditions. Models with a proven track record of accurately capturing observed climate variability and trends are generally more reliable for projecting future scenarios.

Quantifying uncertainty associated with climate model outputs is crucial. Examine the range of projected outcomes from ensemble forecasts and consider factors such as model sensitivity, internal variability, and external forcing. This helps communicate the level of confidence or uncertainty inherent in forecasted climate projections.

Lastly, seek expert guidance from climate scientists, researchers, or experts familiar with climate modelling techniques. Consulting experts can provide valuable insights into the strengths, limitations, and nuances of different climate models, helping interpret and contextualize forecasted climate projections effectively.

Climate modelling encompasses various techniques, each tailored to simulate and project future climate conditions, assess impacts, and inform decision-making processes. Global Climate Models (GCMs) simulate Earth's climate system in a three-dimensional grid, capturing atmospheric dynamics, ocean circulation, land interactions, and sea ice dynamics. They project long-term climate changes and assess the impacts of greenhouse gas emissions on a global scale.

Regional Climate Models (RCMs) operate at higher spatial resolutions, focusing on specific regions or areas of interest. Utilizing GCM outputs as boundary conditions, RCMs downscale climate projections to provide detailed insights into local climate variability and impacts on regional scales.

Earth System Models (ESMs) integrate various components of Earth's climate system, including the atmosphere, oceans, land surface, and biosphere. They simulate complex processes such as carbon cycling, vegetation dynamics, and biogeochemical cycles, offering a comprehensive understanding of climate change and its impacts.

Statistical Downscaling Models downscale coarse-resolution climate model outputs to finer spatial scales using statistical techniques. By establishing relationships between large-scale climate variables and local-scale observations, these models generate downscaled climate projections with improved spatial detail.

Integrated Assessment Models (IAMs) combine climate models with economic, social, and technological components to assess interactions between climate change and human activities. They evaluate mitigation and adaptation strategies, estimate socioeconomic impacts, and inform policy decisions related to climate policy and sustainable development.

Emulator Models are simplified representations of complex climate models that mimic their behaviour using statistical or machine learning techniques. They provide computationally efficient ways to generate large ensembles of climate simulations for uncertainty quantification, sensitivity analysis, and decision-making under uncertainty.

By following these steps and critically evaluating climate model outputs, agribusiness managers can gain insights into potential scenarios for the upcoming season, enabling them to make informed decisions to mitigate climate-related risks and capitalize on opportunities.

For identifying climate risks and opportunities for the site and enterprise region, the process is initiated by conducting a comprehensive risk assessment. This involves thoroughly evaluating potential climate-related threats, such as droughts, floods, heatwaves, storms, and other extreme weather events. Utilize historical climate data, local climate projections, and vulnerability assessments to pinpoint areas of potential risk.

Next, analyse vulnerabilities and exposure. Assess the enterprise's susceptibility to various climate risks by examining its exposure to climatic hazards and the potential repercussions of these hazards on operations, infrastructure, and resources. Identify critical assets, such as crops, livestock, buildings, and water sources, that may be particularly vulnerable to climate-related impacts.

Consideration of adaptation measures is crucial. Explore various adaptation strategies aimed at mitigating climate risks and leveraging opportunities. This may involve implementing water management techniques to address drought risk, fortifying infrastructure to withstand extreme weather events, diversifying crop or livestock portfolios to adapt to changing climate conditions, or exploring new market opportunities driven by shifts in climate suitability for certain crops or products.

Engage stakeholders throughout the process. Involve relevant parties, including farmers, agronomists, researchers, local communities, and government agencies, in the identification and assessment of climate risks and opportunities. Collaborate with experts and community members to gather insights, exchange knowledge, and craft tailored strategies aligned with the enterprise's specific needs and priorities.

Continuous monitoring of climate trends is essential. Stay vigilant regarding climate trends and projections to remain informed about evolving risks and opportunities. Keep abreast of updates from meteorological agencies, climate research institutions, and other reliable sources to monitor changes in temperature, precipitation, extreme weather events, and other climate variables affecting the enterprise region.

Lastly, integrate climate risk management considerations into decision-making processes. Incorporate climate resilience criteria into strategic planning, operational decisions, and investment priorities. Factor in climate-related risks and opportunities when making decisions regarding site selection, crop selection, infrastructure investments, and resource allocation. This integration enhances the enterprise's capacity to adapt to climate change and variability, ultimately bolstering resilience to climate-related challenges.

To determine the impact on production of various weather and climate risk factors, agribusiness managers need to pinpoint the relevant weather and climate risk factors that could influence agricultural production within the enterprise. These factors encompass temperature variations, precipitation levels, humidity, wind patterns, and the occurrence of extreme weather phenomena like droughts, floods, heatwaves, and storms.

Temperature variations, precipitation levels, humidity, wind patterns, and extreme weather phenomena such as droughts, floods, heatwaves, and storms are all significant climate risk factors that can profoundly impact agriculture. Fluctuations in temperature can affect crop growth, development, and yield, with extreme temperatures stressing crops and potentially leading to reduced yields or crop failure. Frost events can damage sensitive crops, while heatwaves can cause heat stress and diminish productivity. Moreover, temperature changes can disrupt the timing of planting and harvesting, influencing crop calendars and management practices.

Figure 48: Stud Murray Grey cows receiving supplementary feeding during a drought, Graman, NSW. Cgoodwin, CC BY-SA 3.0, via Wikimedia Commons.

Adequate and well-distributed rainfall is crucial for crop growth and soil moisture replenishment. Insufficient rainfall leading to drought conditions can stunt crop growth, reduce yields, and compromise crop quality. Conversely, excessive rainfall resulting in flooding or waterlogging can drown crops, promote soil erosion, and facilitate the spread of diseases. Variability in precipitation patterns can also impact planting schedules, irrigation needs, and crop selection, adding complexity to agricultural planning and management.

Humidity levels play a crucial role in agriculture, influencing plant transpiration rates, water uptake, and disease susceptibility. High humidity levels can create favourable conditions for fungal diseases, molds, and pests, increasing the risk of crop damage and yield losses. Conversely, low humidity levels can induce water stress in plants, particularly in arid regions, affecting growth and productivity.

Wind, depending on its speed, direction, and frequency, can impact agriculture in various ways. Strong winds can physically damage crops, disrupt pollination rates, and cause wind erosion, leading to soil degradation and nutrient loss. Wind patterns can also influence pest and disease spread, seed dispersal, and microclimate conditions within agricultural fields, posing additional challenges to farmers.

Figure 49: Floodwaters covering farmland more than two months after the initial flood event near Berrigan, New South Wales. Mattinbgn (talk · contribs), CC BY 3.0, via Wikimedia Commons.

Extreme weather events such as droughts, floods, heatwaves, and storms pose significant risks to agriculture. Droughts can parch soils, deplete water resources, and devastate crops, resulting in substantial yield losses and economic hardship for farmers. Floods can inundate fields, destroy crops, and cause soil erosion, leading to crop failures and infrastructure damage. Heatwaves can cause heat stress in crops and livestock, compromising yields and animal welfare. Similarly, storms, including hurricanes, tornadoes, and severe thunderstorms, can cause extensive damage to crops, infrastructure, and rural communities, disrupting agricultural operations and livelihoods.

Figure 50: Callery Pears and a brown, dried up lawn during a drought along Franklin Farm Road in the Franklin Farm section of Oak Hill, Fairfax County, Virginia. Famartin, CC BY-SA 4.0, via Wikimedia Commons.

Overall, these climate risk factors can have multifaceted and cascading impacts on agriculture, affecting crop yields, soil health, water availability, pest and disease dynamics, and overall farm profitability. Managing and adapting to these risks requires proactive planning, resilient agricultural practices, and effective risk mitigation strategies tailored to local climate conditions and agricultural systems.

Secondly, it's crucial to compile extensive data on agricultural production within the enterprise, including metrics such as crop yields, soil moisture levels, incidents of pest and disease outbreaks, and indicators of livestock health. This data serves as a foundation for evaluating the existing state of production and identifying vulnerabilities to climate risks.

Next, managers should analyse historical weather data alongside production records to discern past associations between weather/climate variables and agricultural outcomes. By recognizing patterns and trends, such as diminished crop yields during drought periods or heightened pest pressure following warm, humid conditions, they can better comprehend the interplay between climate dynamics and production outcomes.

Following this, conducting risk assessments becomes imperative to gauge the potential impact of various weather and climate risk factors on agricultural production. This involves evaluating the likelihood and severity of different climate-related hazards and their potential ramifications for crop yields, soil quality, pest and disease incidence, and livestock performance.

Subsequently, managers should leverage modelling and forecasting tools to project how future weather and climate scenarios might affect agricultural production. Utilizing crop models, soil moisture models, pest and disease models, and livestock production models enables them to forecast how changes in weather variables could influence productivity and profitability.

Moreover, it's essential to develop adaptive strategies based on the analysis of weather and climate risk factors and their potential repercussions on production. These strategies may entail implementing measures such as crop diversification, optimized irrigation practices, enhanced pest and disease monitoring, soil conservation initiatives, and adjustments in livestock management to bolster resilience against climate variability and change.

Lastly, continuous monitoring of weather conditions, production outcomes, and the efficacy of adaptive strategies is vital. Staying abreast of weather forecasts, climate outlooks, and emerging climate-related risks empowers managers to proactively refine management practices and optimize production in response to evolving environmental conditions.

By evaluating the impact of diverse weather and climate risk factors on agricultural production and implementing adaptive strategies accordingly, agribusiness managers can fortify the resilience and sustainability of their enterprises amidst the challenges posed by climate variability and change.

To effectively employ qualitative and quantitative techniques for analysing climate risks and opportunities in agricultural management, agribusiness managers should identify the most relevant methods according to the specific climate risks and opportunities their enterprise faces. This involves choosing between qualitative approaches like risk assessments, scenario planning, and SWOT analysis, which offer insights into broader contexts and stakeholder perspectives, and quantitative techniques such as statistical modelling, climate data analysis, and economic modelling, which provide more rigorous numerical assessments.

Next, it's essential to clearly define the objectives and scope of the analysis, specifying the climate-related risks and opportunities to be evaluated, the timeframe, and the relevant stakeholders. This clarity ensures that the analysis remains focused and aligned with the enterprise's needs and priorities.

Agribusiness managers should then gather qualitative and quantitative data from diverse sources, including climate data repositories, historical records, scientific literature, stakeholder surveys, and economic indicators. Ensuring the accuracy and relevance of the collected data is crucial for a robust analysis.

Once the data is assembled, they should apply qualitative methods like risk assessments, scenario planning, and SWOT analysis to identify and characterize climate risks and opportunities. Engaging stakeholders in these qualitative exercises can provide diverse perspectives and valuable insights.

Additionally, quantitative techniques such as statistical modelling, climate data analysis, and economic modelling should be employed to analyse historical climate data, quantify the likelihood and severity of specific risks, assess potential impacts on agricultural production, and identify adaptation strategies.

To develop a comprehensive understanding, it's important to integrate qualitative and quantitative insights from the analysis. Combining narratives from qualitative methods with quantitative data offers a holistic view of climate risks and opportunities, facilitating better-informed decision-making.

Finally, communicating the findings effectively to stakeholders, decision-makers, and other relevant parties is crucial. Presenting results clearly using visualizations, charts, and reports helps convey key messages and recommendations, fostering buy-in and informed decision-making regarding climate risk management strategies and adaptation measures.

By following these steps and integrating qualitative and quantitative techniques, agribusiness managers can effectively analyse climate risks and opportunities, bolster resilience to climate variability and change, and devise innovative adaptation strategies to navigate evolving environmental conditions.

SWOT analysis is a strategic planning tool used to identify and evaluate the Strengths, Weaknesses, Opportunities, and Threats facing an organization, project, or in this case, climate risks and opportunities in agricultural management. Here's how to conduct a SWOT analysis for climate risks and opportunities:

1. Identify Strengths (S): Strengths represent internal factors that give the agricultural enterprise an advantage in dealing with climate risks or leveraging opportunities. This could include aspects like robust infrastructure, access to technology, diversified crop portfolios, strong financial resources, or a skilled workforce.

2. Identify Weaknesses (W): Weaknesses refer to internal factors that may hinder the enterprise's ability to address climate risks or capitalize on opportunities. These could include factors like outdated infrastructure, limited access to capital, reliance on single crops, inadequate workforce training, or poor management practices.

3. Identify Opportunities (O): Opportunities are external factors in the broader environment that could be leveraged to mitigate

climate risks or enhance the enterprise's resilience. This might include factors like favourable government policies, technological innovations, market trends favouring climate-resilient crops, or partnerships with research institutions for climate adaptation.

4. Identify Threats (T): Threats represent external factors that could pose risks to the enterprise's operations or impede its ability to adapt to climate change. This could include factors like extreme weather events (droughts, floods, storms), changing regulatory environments, market volatility, pest and disease outbreaks exacerbated by climate change, or disruptions in supply chains.

Let's consider an example:

SWOT Analysis for a Small-scale Farming Enterprise:

- Strengths (S): The farm has a diversified crop portfolio, including both drought-resistant and flood-tolerant crops, which enhances resilience to climate variability. Additionally, the farm employs sustainable agricultural practices that improve soil health and water conservation, reducing vulnerability to climate-related risks.

- Weaknesses (W): The farm lacks access to advanced technology for weather forecasting and irrigation management, limiting its ability to respond proactively to climate risks. Furthermore, the farm relies heavily on seasonal labor, which may be disrupted by extreme weather events or changing climate conditions.

- Opportunities (O): There is growing consumer demand for organic and climate-resilient produce, presenting an opportunity for the farm to expand its market reach and increase profitability. Additionally, government subsidies and grants are available for implementing climate-smart agriculture practices, providing financial incentives for investing in resilience-building measures.

- Threats (T): The farm faces the threat of more frequent and intense extreme weather events, such as droughts and storms, which could damage crops and infrastructure, leading to significant economic losses. Furthermore, changing precipitation patterns and water scarcity pose risks to crop productivity and irrigation availability, exacerbating vulnerability to climate-related risks.

By conducting a SWOT analysis, the farm can gain insights into its internal strengths and weaknesses, as well as external opportunities and threats related to climate risks and opportunities. This information can inform strategic decision-making and help the farm develop resilience-building strategies to adapt to climate change effectively.

Scenario planning is a strategic tool that involves creating and analysing multiple plausible future scenarios to anticipate and prepare for uncertainties, including climate risks and opportunities. Here's how scenario planning can be used to analyse climate risks and opportunities, along with hypothetical examples:

1. Identify Key Drivers of Change: Begin by identifying the key drivers of climate change and variability that could impact the agricultural enterprise. These drivers may include factors such as greenhouse gas emissions, temperature trends, precipitation patterns, extreme weather events, technological advancements, policy changes, market dynamics, and societal shifts.

Example: In a scenario where global greenhouse gas emissions continue to rise unchecked, leading to more frequent and intense heatwaves and droughts, agricultural enterprises may face heightened risks of crop failures, water scarcity, and reduced yields. Conversely, in a scenario where there is a concerted global effort to mitigate climate change through carbon reduction measures, agricultural enterprises may benefit from improved climate resilience, enhanced soil health, and increased demand for sustainable agriculture practices.

1. Develop Plausible Future Scenarios: Based on the identified drivers of change, develop a range of plausible future scenarios that represent different climate trajectories, socio-economic conditions, and policy environments. These scenarios should encompass a spectrum of possibilities, from best-case to worst-case scenarios, to capture the full range of uncertainties and risks faced by the agricultural enterprise.

Example: Scenario 1: "Business as Usual" - Continued emissions growth leads to more frequent and severe extreme weather events, disrupting agricultural operations and supply chains. Scenario 2: "Green Transition" - Global efforts to mitigate climate change result in reduced emissions, leading to more stable weather patterns and increased demand for sustainable agriculture products. Scenario 3: "Adaptation Challenges" - Despite mitigation efforts, climate impacts intensify, requiring rapid adaptation measures and investment in resilient farming practices.

1. Assess Impacts and Implications: Evaluate the potential impacts and implications of each scenario on the agricultural enterprise, considering factors such as crop yields, water availability, soil health, pest and disease pressures, market demand, regulatory requirements, and financial viability. Assess the enterprise's vulnerability and resilience to each scenario and identify key adaptation strategies and opportunities.

Example: In Scenario 1, the agricultural enterprise may face increased risks of crop losses due to extreme weather events, prompting investments in climate-resilient crop varieties, irrigation infrastructure, and risk management strategies. In Scenario 2, the enterprise may capitalize on growing consumer demand for sustainable and organic products, leading to diversification of crop portfolios and adoption of regenerative agriculture practices. In Scenario 3, the enterprise may need to collaborate with other stakeholders to develop collective adaptation measures, such as water-sharing agreements or cooperative pest management programs.

1. Develop Adaptive Strategies: Based on the insights gained from scenario planning, develop adaptive strategies to mitigate climate risks and capitalize on opportunities identified in each scenario. These strategies should be flexible, robust, and tailored to the specific needs and priorities of the agricultural enterprise, taking into account the uncertainties and dynamics of the changing climate and business environment.

Example: Adaptive strategies may include investments in climate-resilient crop varieties, soil conservation practices, water-efficient irrigation systems, diversification of income sources, adoption of precision agriculture technologies, and participation in carbon sequestration initiatives. Additionally, building partnerships with local communities, research institutions, and government agencies can enhance the enterprise's capacity to adapt and thrive in a changing climate.

Quantitative techniques like statistical modelling, climate data analysis, and economic modelling are integral for analysing historical climate data, assessing risks, understanding potential impacts on agricultural production, and formulating adaptation strategies. Here's how each technique can be utilized:

Statistical Modelling:

- **Data Preparation:** Commence by collecting historical climate data encompassing variables like temperature, precipitation, humidity, and wind speed over a substantial time span. Organize this data meticulously to ensure its quality and comprehensiveness.

- **Trend Analysis:** Employ statistical methodologies such as time series analysis to detect trends and patterns in historical climate data. This involves scrutinizing changes over time, including long-term trends, seasonal fluctuations, and inter-annual variations.

- **Correlation Analysis:** Conduct correlation analysis to unveil relationships between climate variables and agricultural outcomes like crop yields, soil moisture levels, and livestock performance. This helps in discerning the primary climate factors impacting agricultural productivity.

- **Risk Assessment:** Utilize statistical models like probability distributions or regression analysis to quantify the probability of specific climate risks such as droughts, floods, or heatwaves. Estimating these probabilities based on historical trends aids in anticipating different risk scenarios.

Climate Data Analysis:

- **Climate Indices Calculation:** Compute climate indices like the Palmer Drought Severity Index (PDSI) or Standardized Precipitation Index (SPI) to characterize climate conditions and gauge their suitability for agricultural activities. These indices offer standardized metrics for climate variability assessment.

- **Extreme Event Analysis:** Analyse historical data to identify extreme weather events such as heatwaves, storms, or prolonged dry spells. Assessing the frequency, duration, and intensity of these events aids in evaluating their potential impact on agricultural production and infrastructure.

- **Spatial Analysis:** Conduct spatial analysis to discern climate condition variations across different regions or locations within the enterprise area. Identifying areas prone to specific climate risks enables tailored adaptation strategies.

Economic Modelling:
- **Cost-Benefit Analysis:** Employ economic modelling techniques like cost-benefit analysis to evaluate the financial implications of climate risks and adaptation strategies. This involves estimating costs associated with climate-related damages, productivity losses, and adaptation measures, and comparing them with potential benefits.

- **Market Analysis:** Assess the impact of climate variability on agricultural markets, supply chains, and commodity prices. Analysing changes in market demand, production costs, and input prices aids in understanding profitability for different agricultural products.

- **Investment Appraisal:** Utilize economic models to evaluate investment options for climate adaptation measures such as infrastructure upgrades or risk management strategies. Assessing the return on investment (ROI) and long-term financial viability under varying climate scenarios guides decision-making.

By employing these quantitative techniques, agribusiness managers can gain insights into historical climate data, quantify risks, assess impacts on agricultural production, and formulate adaptation strategies to bolster resilience in the face of climate variability and change.

Below are some hypothetical examples illustrating how each technique can be applied in the context of analysing climate risks and opportunities in agriculture:

Statistical Modelling:
1. Data Preparation: An agribusiness manager collects historical climate data spanning the past 50 years, including variables like temperature, precipitation, humidity, and wind speed, from meteorological stations across the region.

2. Trend Analysis: Using time series analysis, the manager identifies a significant increasing trend in average temperatures over the past three decades, accompanied by a decrease in precipitation during the summer months.

3. Correlation Analysis: By conducting correlation analysis, the manager discovers a strong negative correlation between soil moisture levels and rainfall patterns, indicating that decreased precipitation leads to drier soil conditions, adversely affecting crop yields.

4. Risk Assessment: Utilizing probability distributions, the manager calculates the probability of extreme weather events such as droughts or heatwaves occurring in the upcoming growing season based on historical trends, helping in anticipating potential risks to crop production.

Climate Data Analysis:
1. Climate Indices Calculation: The manager computes the Standardized Precipitation Index (SPI) for the region over the past 20 years, revealing that the area has experienced several periods of prolonged drought, with SPI values consistently below -1.5 during those times.

2. Extreme Event Analysis: Analysing historical data on extreme weather events, the manager identifies a significant increase in the frequency and intensity of heatwaves over the past decade, with temperatures exceeding 40°C for extended periods during the summer months.

3. Spatial Analysis: Through spatial analysis, the manager maps climate condition variations across different agricultural zones

within the region, identifying areas susceptible to specific risks like flooding in low-lying regions or soil erosion in areas with steep slopes.

Economic Modelling:

1. Cost-Benefit Analysis: Using cost-benefit analysis, the manager evaluates the financial implications of implementing a drip irrigation system to mitigate water scarcity during drought periods. The analysis compares the costs of installation and operation with the expected increase in crop yields and water savings, determining the system's economic viability.

2. Market Analysis: The manager conducts a market analysis to assess the impact of climate variability on the demand and prices of agricultural products. They observe that during periods of drought, the prices of water-intensive crops like rice and corn increase significantly due to reduced supply, impacting the profitability of farmers.

3. Investment Appraisal: Utilizing economic models, the manager evaluates different investment options for climate adaptation, such as investing in resilient crop varieties or implementing soil conservation practices. They analyse the potential returns on investment and the long-term benefits of each adaptation measure under different climate scenarios to inform decision-making.

Assessing the Significance of Climate Variability and Significant Events, Outlining Tactics to Manage Various Risks and Opportunities Stemming from Climate Variability, And Identifying Contingency Options for the Enterprise and Business

From the perspective of developing climate risk management strategies for an agricultural, horticultural, or land management enterprise, an agribusiness manager would approach the tasks initially by understanding the importance of climate variability and significant climate events is instrumental in assessing their impact on agricultural operations. The agribusiness manager should:

- Assess Historical Data: Analyse historical climate data to identify trends, patterns, and significant events that have affected the enterprise in the past. This includes examining temperature fluctuations, precipitation levels, extreme weather events, and their impacts on crop yields, soil health, and overall productivity.

- Consider Future Projections: Evaluate climate projections and forecasts to anticipate future trends and potential changes in climate variability. This involves studying climate models, ensemble forecasts, and other scientific assessments to understand how climate patterns may evolve over time and their implications for the enterprise.

- Conduct Risk Assessment: Conduct a comprehensive risk assessment to prioritize climate-related risks and opportunities based on their likelihood and potential impact on the enterprise. This involves identifying vulnerabilities, assessing exposure to climate hazards, and determining the significance of different climate events for the business.

Developing tactics to address various climate variability risks and opportunities requires a multifaceted approach tailored to the specific needs and circumstances of the enterprise. The agribusiness manager should:

- Implement Diversification Strategies: Encourage crop diversification and rotation to mitigate the risks associated with climate variability. This involves selecting resilient crop varieties, adjusting planting schedules, and diversifying revenue streams to spread risk across different products or markets (Biró et al., 2021; Duan et al., 2021; Kassa, 2021; Mohapatra et al., 2022; Turyasingura & Chavula, 2022; Vadlamudi, 2020).

- Invest in Resilient Infrastructure: Upgrade infrastructure and implement technologies that enhance resilience to climate-related hazards. This may include investing in irrigation systems, water storage facilities, greenhouses, or protective structures to safeguard crops and livestock from extreme weather events (Gairhe et al., 2021; Oostendorp et al., 2019; Subedi & Poudel, 2020; Viola & Mendes, 2022).

- Adopt Sustainable Practices: Promote sustainable land management practices that improve soil health, water conservation, and biodiversity while mitigating climate risks. This includes practices such as conservation tillage, cover cropping, agroforestry, and integrated pest management to build resilience and enhance ecosystem services (Amare & Gacheno, 2021; Arjoo et al., 2022; Brenya et al., 2022; Rao, 2018).

- Engage in Climate-Smart Agriculture: Embrace climate-smart agriculture approaches that combine adaptation, mitigation, and sustainable development goals. This involves adopting practices such as precision agriculture, agroecology, and climate-resilient crop breeding to optimize resource use, reduce greenhouse gas emissions, and enhance productivity in a changing climate (Biró et al., 2021; Duan et al., 2021; Kassa, 2021; Mohapatra et al., 2022; Rao, 2018; Turyasingura & Chavula, 2022; Vadlamudi, 2020)..

Identifying contingency options helps the enterprise prepare for and respond to unexpected climate events or disruptions. The agribusiness manager should:

- Develop Emergency Response Plans: Establish protocols and procedures for responding to climate-related emergencies such as droughts, floods, storms, or pest outbreaks. This includes creating contingency plans for crop loss, livestock management, evacuation procedures, and business continuity measures to minimize disruptions.

- Build Partnerships and Networks: Foster collaborations with stakeholders, industry partners, and community organizations to enhance resilience and share resources during climate-related crises. This involves forming alliances with local authorities, agricultural associations, research institutions, and emergency response agencies to coordinate efforts and access support services.

- Invest in Insurance and Risk Transfer Mechanisms: Explore options for insurance coverage, risk pooling, or financial instruments that provide protection against climate-related losses. This includes purchasing crop insurance, livestock insurance, or weather derivatives to mitigate financial risks associated with adverse weather conditions and ensure business continuity in times of crisis.

By evaluating the importance of climate variability, outlining tactics to address different climate risks and opportunities, and identifying contingency options for the enterprise, the agribusiness manager can develop robust climate risk management strategies that enhance resilience, sustainability, and long-term viability in the face of a changing climate.

To evaluate the importance of climate variability and significant climate events, a structured approach involving several steps is essential. Begin by delving into historical climate data, scrutinizing past trends, patterns, and notable events that have influenced the enterprise. This entails examining variables like temperature fluctuations, precipitation levels, and occurrences of extreme weather events over a pertinent time frame. By dissecting historical data, one can glean insights into how climate variability has impacted crop yields, soil health, and overall productivity in the past.

Next, shift focus towards evaluating climate projections and forecasts to anticipate forthcoming trends and potential alterations in climate variability. This stage necessitates a deep dive into climate models, ensemble forecasts, and other scientific evaluations to gain a comprehensive understanding of how climate patterns might evolve over time. By contemplating future projections, individuals can better grasp the potential ramifications of climate change on agricultural operations and proactively prepare for necessary adaptation measures.

Finally, undertake a thorough risk assessment to prioritize climate-related risks and opportunities based on their likelihood and potential impact on the enterprise. This entails identifying vulnerabilities within the agricultural system, evaluating exposure to climate hazards such as droughts, floods, or heatwaves, and gauging the significance of different climate events for the business. Through meticulous risk assessment, stakeholders can better comprehend the specific challenges posed by climate variability and formulate targeted strategies to mitigate risks while seizing available opportunities.

Conducting a comprehensive risk assessment to prioritize climate-related risks and opportunities for an agricultural enterprise involves several key steps. The process begins with identifying vulnerabilities within the agricultural system. This entails a thorough examination of various aspects of the enterprise, encompassing crops, livestock, infrastructure, and human resources. Factors such as susceptibility to extreme weather events, soil quality, water availability, and reliance on specific inputs like irrigation or pesticides are carefully considered. By assessing vulnerabilities, areas of weakness that may be susceptible to climate impacts can be pinpointed.

Next, it is essential to assess the exposure of the agricultural enterprise to different climate hazards. This includes evaluating risks such as droughts, floods, heatwaves, storms, and changes in temperature or precipitation patterns. Geographic location, historical climate data, and local climate projections are taken into account to gauge the likelihood and severity of exposure to these hazards. The assessment also considers how climate hazards could directly or indirectly affect various aspects of agricultural operations, from crop production to infrastructure resilience.

Determining the significance of different climate events for the business is another crucial step. This involves analysing historical data to understand the frequency, duration, and intensity of past climate events and their consequences for crop yields, livestock health, soil erosion, and financial performance. Climate events are prioritized based on their likelihood of occurrence and their potential to disrupt or enhance agricultural activities.

Quantifying the risks and opportunities associated with different climate events comes next. This is achieved by assessing their likelihood and potential impact on the enterprise using quantitative methods such as probability analysis, scenario modelling, or economic valuation. Factors such as crop yield losses, infrastructure damage, input costs, market fluctuations, and potential benefits of climate adaptation measures are carefully considered during this step.

Based on the findings of the risk assessment, targeted strategies to mitigate risks and capitalize on opportunities are developed. Specific actions and interventions are identified to enhance the resilience of the agricultural enterprise to climate variability and change. These may include measures such as crop diversification, improved water management practices, soil conservation techniques, infrastructure upgrades, insurance coverage, and market diversification. Strategies are tailored to address the unique vulnerabilities and opportunities identified during the risk assessment process.

Finally, it is essential to continuously monitor and review the effectiveness of risk mitigation and adaptation strategies over time. Risk assessments are regularly updated based on new information, changing climate conditions, and evolving business priorities. Strategies are adjusted as needed to optimize resilience and capitalize on emerging opportunities in a dynamic climate environment.

As an example of this process implementation, to conduct a hypothetical comprehensive risk assessment for an agricultural enterprise involved in agroforestry, let's outline the steps and apply them to this specific context:

1. **Identify Vulnerabilities within the Agricultural System:**

   ○ **Crops and Trees:** Assess the susceptibility of crops and trees to extreme weather events such as storms, droughts, and heatwaves. Consider the diversity of species and their resilience to different climate conditions.

   ○ **Soil Health:** Evaluate soil quality and erosion risks, considering the impact of heavy rainfall and flooding on erosion rates and nutrient leaching.

   ○ **Water Availability:** Examine water sources for irrigation and assess their reliability under changing precipitation patterns and drought conditions.

   ○ **Infrastructure:** Identify vulnerabilities in infrastructure such as irrigation systems, greenhouses, and storage facilities to climate-related risks like flooding, wind damage, or temperature extremes.

2. **Assess Exposure to Climate Hazards:**

   ○ **Climate Hazards:** Evaluate the historical frequency and severity of climate hazards like droughts, floods, storms, and temperature fluctuations in the region where the agroforestry enterprise operates.

○ **Geographic Considerations:** Consider the location of the enterprise and its proximity to water bodies, floodplains, and areas prone to extreme weather events.

○ **Local Climate Projections:** Review climate projections and forecasts to anticipate future changes in temperature, precipitation patterns, and the frequency of extreme weather events.

3. **Determine the Significance of Different Climate Events:**

○ **Historical Analysis:** Analyse historical climate data to understand how past climate events have impacted crop yields, tree health, soil erosion, and financial performance.

○ **Risk Prioritization:** Prioritize climate events based on their potential to disrupt agroforestry operations and their likelihood of occurrence. Focus on events with significant consequences for crop and tree productivity, soil stability, and infrastructure integrity.

4. **Quantify Risks and Opportunities:**

○ **Probability Analysis:** Use probability analysis to quantify the likelihood of different climate risks occurring within a given time frame.

○ **Scenario Modelling:** Develop scenarios to simulate the potential impacts of climate events on crop and tree yields, soil erosion rates, water availability, and infrastructure damage.

○ **Economic Valuation:** Estimate the financial implications of climate risks, including potential crop losses, damage to infrastructure, increased input costs, and the cost of implementing adaptation measures.

5. **Develop Mitigation and Adaptation Strategies:**

○ **Crop and Tree Management:** Implement practices to enhance the resilience of crops and trees to climate variability, such as selecting drought-tolerant species, improving soil health through agroforestry techniques, and optimizing irrigation efficiency.

○ **Infrastructure Upgrades:** Invest in infrastructure upgrades to mitigate climate-related risks, such as improving drainage systems to reduce flood risk, reinforcing structures to withstand strong winds, and installing irrigation systems with water-saving features.

○ **Insurance Coverage:** Explore options for insurance coverage to mitigate financial risks associated with climate-related crop losses, property damage, and business interruptions.

○ **Market Diversification:** Diversify revenue streams by exploring alternative markets for agroforestry products, value-added products, and eco-tourism opportunities that may be less vulnerable to climate risks.

6. **Monitor and Review:**

○ **Continuous Monitoring:** Establish a system for monitoring climate conditions, crop and tree performance, soil health, and infrastructure integrity to detect changes and emerging risks.

○ **Regular Review:** Conduct regular reviews of risk assessments and adaptation strategies to incorporate new information, adjust plans in response to changing climate conditions, and optimize resilience over time.

By following these steps and tailoring them to the specific context of agroforestry, the agricultural enterprise can effectively prioritize climate-related risks and opportunities, develop targeted mitigation and adaptation strategies, and enhance resilience to climate variability and change.

Addressing a range of different climate variability risks and opportunities in agricultural operations requires a intricate approach tailored to the specific needs and circumstances of the enterprise. These can include:

Implementing Diversification Strategies:

- Crop Diversification: Encouraging the planting of a variety of crops with different growth characteristics and resilience to climate variability helps spread the risk of crop failure due to adverse weather conditions. By diversifying the crop portfolio, the enterprise becomes less vulnerable to the impact of specific climate events on individual crops.

Figure 51: A farm in Rio Arriba County diversifies their operation by raising herbs for specialized markets. Jeff Vanuga / Photo courtesy of USDA Natural Resources Conservation Service.

- Crop Rotation: Rotating crops is essential to improve soil health, break pest and disease cycles, and reduce reliance on specific inputs. Rotating crops with different nutrient requirements and growth habits can mitigate soil erosion and enhance overall resilience by optimizing resource utilization and minimizing the buildup of pests and diseases.

Figure 52: Crop rotation near Aldunie Yellow and green striped hillside. Stanley Howe / Crop rotation near Aldunie.

- Revenue Stream Diversification: Diversifying revenue streams involves exploring alternative markets for agricultural products, value-added processing, agritourism, or other income-generating activities. By expanding revenue sources beyond traditional crop sales, the enterprise can reduce its dependence on weather-sensitive commodities and tap into more stable income streams.

Investing in Resilient Infrastructure:

- Irrigation Systems: Investing in efficient irrigation systems, such as drip irrigation or micro-sprinklers, ensures water availability during dry periods and reduces reliance on unpredictable rainfall. By optimizing water management, the enterprise can maintain consistent crop yields and minimize the impact of droughts or water scarcity.

Figure 53: Irrigation system at McCall's Pumpkin Patch in Moriarty, New Mexico. Daniel Schwen, CC BY-SA 4.0, via Wikimedia Commons.

- Water Storage Facilities: Building or expanding water storage facilities, such as ponds, tanks, or reservoirs, allows the enterprise to capture and store rainwater for irrigation during periods of water scarcity. Adequate water storage enhances resilience to climate variability by providing a reliable water source for agricultural activities.

- Greenhouses: Constructing greenhouses or high tunnels provides controlled environments for crop production, protecting plants from extreme temperatures, pests, and diseases. Greenhouse cultivation enables year-round production and minimizes the risk of weather-related crop losses.

Figure 54: Strawberry greenhouse, Aquitaine. Cjp24, CC BY-SA 3.0, via Wikimedia Commons.

- Protective Structures: Installing windbreaks, shade structures, or hail nets protects crops and livestock from extreme weather events, such as strong winds, hailstorms, or excessive sunlight. These structures reduce the impact of adverse weather conditions on agricultural productivity and infrastructure.

Adopting Sustainable Practices:
- Conservation Tillage: Implementing conservation tillage practices minimizes soil disturbance, reduces erosion, and improves soil structure and water retention. Conservation tillage enhances soil health and resilience to climate variability while promoting sustainable land management practices.

Figure 55: Minimum tillage plow used on the Steven Thomas Farm in Cochran County near Morton, Texas. USDA NRCS Texas, Public domain, via Wikimedia Commons.

Cover Cropping: Planting cover crops during fallow periods protects soil from erosion, enhances soil fertility, suppresses weeds, and improves water infiltration. Cover cropping improves soil health, reduces nutrient runoff, and mitigates the impact of climate variability on soil moisture and nutrient availability.

Figure 56: A cover crop of Tillage Radish. Ethanstuckey, CC BY-SA 4.0, via Wikimedia Commons.

- Agroforestry: Integrating trees or shrubs into agricultural landscapes provides multiple benefits, including soil stabilization, windbreaks, shade, biodiversity enhancement, and additional income streams from timber or non-timber forest products. Agroforestry systems enhance ecosystem resilience and contribute to climate change mitigation and adaptation.

- Integrated Pest Management (IPM): Implementing IPM strategies manages pests and diseases using a combination of cultural, biological, and chemical control methods while minimizing environmental impacts. IPM reduces reliance on synthetic pesticides, conserves natural enemies of pests, and promotes ecological balance in agricultural ecosystems.

Engaging in Climate-Smart Agriculture:

- Precision Agriculture: Using precision agriculture technologies optimizes resource use, minimizes environmental impacts, and improves crop yields and profitability. Precision agriculture tools, such as GPS-guided machinery and remote sensing, enable precise monitoring and management of crop inputs based on site-specific conditions and real-time data.

- Agroecology: Applying principles of agroecology designs resilient farming systems that mimic natural ecosystems, enhance biodiversity, improve soil health, and reduce reliance on external inputs. Agroecological practices promote ecosystem resilience, enhance soil fertility, and mitigate the impact of climate variability on agricultural productivity.

- Climate-Resilient Crop Breeding: Promoting the development and adoption of climate-resilient crop varieties through plant breeding programs focuses on traits such as drought tolerance, heat tolerance, disease resistance, and nutrient efficiency. Climate-resilient crop varieties enhance the resilience of agricultural systems to climate variability and contribute to food security and sustainability.

By considering these tactics, the agribusiness manager can develop a comprehensive strategy to address a range of climate variability risks and opportunities, enhancing the resilience, sustainability, and profitability of the agricultural enterprise.

Identifying contingency options for agricultural enterprises is paramount to prepare for and respond to unexpected climate events or disruptions and developing emergency response plans is a critical step for agricultural enterprises to mitigate the impact of climate-related emergencies on their operations. The agribusiness manager plays a pivotal role in this process, overseeing the assessment of vulnerabilities and the creation of protocols to address various scenarios. Here's an expanded explanation of how to develop emergency response plans:

The first step in developing an emergency response plan is to assess vulnerabilities within the agricultural system. This involves identifying potential climate-related emergencies such as droughts, floods, storms, or pest outbreaks. The agribusiness manager must carefully evaluate the susceptibility of crops, livestock, infrastructure, and human resources to these emergencies. By understanding the specific vulnerabilities, the enterprise can better prepare for and respond to potential threats.

Once vulnerabilities are identified, the next step is to create detailed emergency response plans with specific protocols for different scenarios. These plans outline the actions and responsibilities of personnel during emergencies and establish procedures for crop protection, livestock management, evacuation, and business continuity. Protocols should be comprehensive, addressing various aspects of emergency response to minimize disruptions and ensure the safety of both personnel and assets.

Training and education are essential components of effective emergency response plans. The agribusiness manager should provide comprehensive training to employees on how to implement emergency protocols effectively. This includes conducting drills and simulations to familiarize personnel with their roles and responsibilities during emergencies. By investing in training, the enterprise can ensure that all personnel are prepared to respond swiftly and effectively to any climate-related emergency.

Communication is key during emergencies, and establishing reliable communication channels is crucial for disseminating alerts, instructions, and updates. The agribusiness manager should set up communication systems that enable efficient communication with employees, stakeholders, and relevant authorities. These channels should be accessible and resilient, capable of functioning even during power outages or other disruptions. By ensuring effective communication, the enterprise can coordinate its response efforts and minimize the impact of emergencies.

Developing partnerships and networks is a crucial aspect of building resilience and enhancing resource-sharing during climate-related crises within agricultural enterprises. The agribusiness manager plays a pivotal role in this endeavour, facilitating collaboration with various stakeholders, industry partners, and community organizations. Here's an expanded explanation of how to build partnerships and networks:

The first step in building partnerships is to identify potential collaborators who can provide support and resources during emergencies. These may include local authorities, agricultural associations, research institutions, and emergency response agencies. By identifying key partners, the agribusiness manager can lay the foundation for effective collaboration and resource-sharing during times of crisis.

Once potential partners are identified, the next step is to establish collaborative relationships through regular communication and mutual assistance agreements. Building strong relationships with partners involves fostering open dialogue, sharing information, and exploring opportunities for joint planning efforts. By collaborating on preparedness activities, resource-sharing arrangements, and coordinated response strategies, partners can maximize their effectiveness in addressing climate-related emergencies.

Coordination is essential for ensuring that response efforts are well-integrated and streamlined during emergencies. The agribusiness manager should coordinate efforts with partners to develop integrated emergency response plans and protocols. This includes clarifying roles, responsibilities, and decision-making processes to ensure smooth coordination and communication during crises. By working together effectively, partners can optimize their collective response and minimize duplication of efforts.

In addition to formal partnerships, engaging with the local community is crucial for building resilience and raising awareness about climate-related risks and preparedness measures. The agribusiness manager should actively engage with community members to educate them about climate-related risks, emergency preparedness, and response strategies. This may involve organizing training workshops, conducting drills, and facilitating community-based monitoring initiatives. By encouraging community participation in preparedness activities, the enterprise can strengthen its overall resilience and foster a sense of collective responsibility for emergency response.

Investing in insurance and risk transfer mechanisms is a fundamental strategy for safeguarding agricultural enterprises against climate-related losses and ensuring business continuity during times of crisis. The agribusiness manager plays a critical role in this process, overseeing the assessment of insurance needs, researching available options, purchasing adequate coverage, and regularly reviewing and updating insurance policies. Here's an expanded explanation of how to invest in insurance and risk transfer mechanisms:

The first step in investing in insurance is to assess the specific needs of the enterprise in light of its exposure to climate risks, financial resources, and risk tolerance. This involves evaluating factors such as the value of crops and livestock, production volumes, market prices, and potential losses due to climate-related events. By conducting a thorough assessment, the agribusiness manager can determine the appropriate level of insurance coverage required to mitigate risks effectively.

Once the insurance needs are assessed, the next step is to research available insurance options tailored to agricultural enterprises. This may include exploring various types of insurance such as crop insurance, livestock insurance, weather derivatives, and business interruption insurance. The agribusiness manager should compare coverage options, premiums, deductibles, and terms and conditions offered by different insurers to identify the most suitable insurance products for the enterprise's specific needs and budget.

After identifying suitable insurance products, the agribusiness manager should purchase adequate coverage to protect against climate-related losses and ensure business continuity in emergencies. This involves carefully considering coverage limits, policy exclusions, claim procedures, and premium affordability when selecting insurance policies. By obtaining sufficient coverage, the enterprise can mitigate financial risks associated with climate-related events and minimize the impact on operations and profitability.

It is essential to regularly review and update insurance policies to reflect changes in the enterprise's operations, assets, and risk exposure. The agribusiness manager should review coverage levels, policy terms, and insured values annually or as needed to ensure that insurance coverage remains adequate and up-to-date. By staying informed about changes in the business environment and adjusting insurance policies accordingly, the enterprise can maintain effective risk management practices and protect against emerging threats.

On the basis of the discussion above, the following emerges a procedure that the agribusiness manager can develop robust climate risk management strategies tailored to the enterprise's specific needs and circumstances. This approach enhances resilience, sustainability, and long-term viability in the face of climate variability and change.

Climate Risk Management Strategies Development Procedure
Developing Emergency Response Plans:

1. **Assess Vulnerabilities:** Start by analysing historical climate data to understand past trends and events impacting the enterprise. Evaluate variables such as temperature fluctuations, precipitation levels, and extreme weather occurrences, and their effects on crops, soil health, and overall productivity.

2. **Consider Future Projections:** Evaluate climate forecasts and models to anticipate future trends in climate variability. Study projections to grasp potential changes over time and their implications for agricultural operations.

3. **Conduct Risk Assessment:** Perform a comprehensive risk assessment to prioritize climate-related risks and opportunities. Identify vulnerabilities, assess exposure to climate hazards, and determine their significance for the business.

Developing Tactics for Addressing Climate Variability Risks:

1. **Implement Diversification Strategies:** Encourage crop diversification, rotation, and revenue stream diversification to spread risks. Select resilient crop varieties, adjust planting schedules, and explore alternative markets to mitigate climate-related impacts.

2. **Invest in Resilient Infrastructure:** Upgrade infrastructure with technologies like irrigation systems, water storage facilities, and protective structures. These investments safeguard crops and livestock from extreme weather events and enhance overall resilience.

3. **Adopt Sustainable Practices:** Promote sustainable land management practices such as conservation tillage, cover cropping, and agroforestry. These practices improve soil health, conserve water, and mitigate climate risks while enhancing ecosystem services.

4. **Engage in Climate-Smart Agriculture:** Embrace climate-smart agriculture approaches like precision farming, agroecology, and climate-resilient crop breeding. These methods optimize resource use, reduce emissions, and enhance productivity in a changing climate.

Identifying Contingency Options:

1. Develop Emergency Response Plans: Establish protocols for responding to climate-related emergencies like droughts, floods, or pest outbreaks. Create contingency plans for crop loss, evacuation procedures, and business continuity measures.

2. Build Partnerships and Networks: Foster collaborations with stakeholders, industry partners, and community organizations. Form alliances with local authorities, research institutions, and emergency response agencies to coordinate efforts and access support services.

3. Invest in Insurance and Risk Transfer Mechanisms: Explore insurance coverage options such as crop insurance, livestock insurance, or weather derivatives. Purchase adequate coverage to mitigate financial risks and ensure business continuity during crises. Regularly review and update insurance policies to reflect changes in operations and risk exposure.

Chapter Ten

Prepare Climate Risk Management Strategies

Developing and Documenting Climate Risk Management Strategies

Analysing climate variability and seasonal climate forecasts requires a systematic approach to gather data, utilize climate models, and assess implications for agricultural operations. First, gathering data involves identifying reliable sources such as government meteorological agencies or research institutions and collecting information on key climate variables like temperature, precipitation, and extreme weather events over an extended period. This data, organized systematically using spreadsheets or database software, provides a foundation for analysis.

Next, utilizing climate models entails accessing reputable sources for climate models and seasonal forecasts and familiarizing oneself with modelling principles. Inputting relevant parameters into these models allows for the simulation of future climate scenarios specific to the region or area of interest. Analysing the outputs of these models, including projected changes in temperature, precipitation patterns, and extreme weather events, requires careful consideration of uncertainties and limitations to account for variability effectively.

Assessing the implications of climate variability involves reviewing historical data and model projections to understand potential impacts on agricultural operations. This assessment considers factors such as crop growth, water availability, soil health, and pest dynamics, evaluating the vulnerability of different crops, livestock breeds, and infrastructure components. Insights gained from this analysis inform decision-making and risk management strategies, identifying priority areas for intervention and adaptation measures.

Finally, documenting findings and conclusions is essential for sharing insights with relevant stakeholders to facilitate informed decision-making and planning processes. Noting key observations, trends, and potential challenges associated with climate variability ensures that strategies are developed with a comprehensive understanding of the climate context. Through this process, agricultural enterprises can better prepare for and adapt to the challenges posed by climate variability, enhancing resilience and sustainability in the face of changing environmental conditions.

The findings could be recorded in a format similar to the template following.

Title: Analysing Climate Variability and Seasonal Climate Forecasts

Introduction: Analysing climate variability and seasonal climate forecasts is crucial for agricultural operations to adapt and prepare for changing environmental conditions. This document outlines a systematic approach to gathering data, utilizing climate models, assessing implications, and documenting findings to enhance resilience and sustainability in the face of climate variability.

1. Gathering Data:

Objective: Identify reliable sources and collect historical climate data.

Steps:

1.1. Identify reliable sources such as government meteorological agencies or research institutions.

1.2. Collect data on key climate variables (temperature, precipitation, extreme weather events) over an extended period.

1.3. Organize data systematically using spreadsheets or database software.

2. *Utilizing Climate Models:*

Objective: *Access reputable sources, understand modelling principles, and analyse outputs.*

Steps:

2.1. Access reputable sources for climate models and seasonal forecasts.

2.2. Familiarize yourself with modelling principles and input relevant parameters. 2.3. Analyse outputs, considering uncertainties and limitations.

3. *Assessing Implications:*

Objective: *Review historical data and model projections to understand potential impacts.*

Steps:

3.1. Review historical data and model projections.

3.2. Assess implications on crop growth, water availability, soil health, and pest dynamics.

3.3. Evaluate vulnerability of crops, livestock, and infrastructure components.

4. *Documenting Findings and Conclusions:*

Objective: *Share insights with stakeholders for informed decision-making.*

Steps:

4.1. Document key observations, trends, and potential challenges.

4.2. Share findings with relevant stakeholders. 4.3. Develop strategies based on insights gained from analysis.

Conclusion: *Analysing climate variability and seasonal forecasts is essential for agricultural enterprises to adapt and thrive in a changing climate. By following a systematic approach to gather data, utilize climate models, assess implications, and document findings, enterprises can enhance resilience and sustainability, ensuring informed decision-making and effective planning processes.*

[End of Document Template]

This template provides a structured framework for agricultural enterprises to analyse climate variability and seasonal climate forecasts, guiding them through the process from data collection to decision-making. Adjustments can be made to tailor the template to specific needs and circumstances of the enterprise.

Predicting the impact of climate variability is crucial for agricultural enterprises to evaluate risks and devise effective strategies for adaptation and resilience. To achieve this, follow a structured approach to anticipate the effects of climate variability on the environment, property value, and equity of the enterprise.

Conducting a comprehensive assessment is key to quantifying potential impacts. Begin by identifying assets vulnerable to climate-related hazards, such as extreme weather events, droughts, or temperature fluctuations. Evaluate the potential financial losses associated with these hazards, considering factors like crop damage, infrastructure damage, and increased operational costs. Additionally, assess the overall vulnerability of the enterprise to climate variability, taking into account its geographic location, exposure to climate hazards, and existing resilience measures.

The objective here is to simulate the effects of various climate scenarios on the enterprise. Define a range of climate scenarios based on historical data, climate projections, and expert insights. Assess the likelihood and severity of potential impacts associated with each scenario, considering changes in temperature, precipitation patterns, and the frequency of extreme weather events. Utilize modelling techniques to simulate the effects of these scenarios on key variables such as crop yields, soil health, water availability, and property damage.

Estimating potential losses and assessing the financial viability of strategies is critical in this step. Incorporate predicted impacts of climate variability into financial models, accounting for both short-term and long-term effects. Estimate potential financial losses associated with climate-related hazards, including direct costs like crop losses and property damage, as well as indirect costs such as reduced productivity and increased insurance premiums. Evaluate the financial viability of different risk management options by considering the projected costs and benefits of adaptation measures. Furthermore, consider the long-term implications for property value and equity when evaluating risk management options, ensuring alignment with the enterprise's overall financial goals and objectives.

To effectively identify business strategies, identify and prioritize business strategies aimed at mitigating the major climate risk factors identified through analysis. This involves diversifying crops to reduce reliance on a single crop and spread risk, implementing irrigation systems, water conservation techniques, and drought-resistant crop varieties to enhance water efficiency and mitigate drought risk, and adopting practices such as conservation tillage, cover cropping, and agroforestry to improve soil health, reduce erosion, and enhance resilience to extreme weather events.

Develop adaptation strategies to help the enterprise adapt to changing climate conditions. This entails upgrading infrastructure, such as irrigation systems, barns, and storage facilities, to withstand extreme weather events and ensure operational continuity, embracing techniques like precision agriculture, integrated pest management, and agroecology to optimize resource use, enhance productivity, and minimize climate-related risks, and diversifying income sources by adding value to agricultural products, offering agritourism experiences, or engaging in renewable energy production to reduce reliance on climate-sensitive activities.

Integrate consideration of insurance options into business strategies to mitigate financial losses from climate-related events. This involves evaluating different insurance products such as crop insurance, livestock insurance, and business interruption insurance based on the enterprise's specific needs and risk profile, selecting appropriate coverage that provides adequate protection against potential losses from climate variability, considering factors like coverage limits, deductibles, and premium costs, and incorporating insurance coverage into broader risk management plans, alongside other strategies for risk mitigation and adaptation, to ensure comprehensive protection against climate-related risks.

To prepare financial forecasts effectively, initiate the process with a comprehensive cost-benefit analysis of all identified strategies. Evaluate both the costs and benefits associated with each option, taking into account the financial implications of implementing risk management and adaptation measures. This entails estimating upfront investments in infrastructure, technology, or training, alongside potential returns on investment such as increased productivity, reduced losses, or enhanced market competitiveness.

To initiate the process with a comprehensive cost-benefit analysis of agricultural strategies, several steps should be followed. Firstly, identify specific agricultural strategies relevant to your enterprise and its context. For instance, options like implementing drip irrigation systems, adopting cover cropping, or investing in greenhouse technology can be considered to address various agricultural challenges.

Next, estimate the costs associated with each strategy, encompassing expenses such as equipment purchase, materials, technology, installation, labour, and training. For example, installing a drip irrigation system would involve costs for drip lines, emitters, pumps, filters, and labour for installation.

Subsequently, assess the potential benefits linked with the implementation of each strategy. This involves considering factors like increased productivity, reduced losses, and enhanced market competitiveness. For instance, cover cropping may improve soil fertility and moisture retention, resulting in higher yields and better crop quality.

Quantify the financial impacts of each strategy by comparing estimated costs and benefits. Utilize financial metrics such as net present value (NPV), internal rate of return (IRR), or payback period to evaluate the financial viability of each option. For instance, if the NPV of investing in greenhouse technology is positive, it suggests that the expected benefits outweigh the upfront investment costs, indicating financial feasibility.

Calculating the Net Present Value (NPV) involves discounting the future cash flows generated by an investment to their present value, considering the time value of money. Here's how to calculate NPV with an example relevant to a dairy farm:

Example: Let's consider a dairy farm that is contemplating investing in new equipment for automated milking. The initial investment for the equipment is $100,000. Over the next five years, the investment is expected to generate additional annual cash flows of $30,000, $35,000, $40,000, $45,000, and $50,000, respectively.

Assumptions: Discount Rate (r) = 8% per annum

Step 1: Calculate the Present Value (PV) of each cash flow:

PV1 = Cash flow Year 1 / (1 + r)^1 = $30,000 / (1 + 0.08)^1 = $30,000 / 1.08 ≈ $27,778.

Similarly, PV2 ≈ $32,407 PV3 ≈ $34,568 PV4 ≈ $36,739 PV5 ≈ $38,920

Step 2: Calculate the NPV by summing up the present values of all cash flows and subtracting the initial investment:

NPV = PV1 + PV2 + PV3 + PV4 + PV5 - Initial Investment ≈ $27,778 + $32,407 + $34,568 + $36,739 + $38,920 - $100,000 ≈ $170,412 - $100,000 ≈ $70,412

Interpretation: With a discount rate of 8% per annum, the NPV of investing in the automated milking equipment for the dairy farm is approximately $70,412. This positive NPV indicates that the investment is expected to generate returns higher than the cost of capital, making it financially viable.

Note: The NPV rule states that if NPV > 0, the investment should be undertaken as it adds value to the business. If NPV < 0, the investment is not advisable as it would result in a loss. If NPV = 0, the investment neither adds nor detracts value, and further analysis may be needed.

Calculating the Internal Rate of Return (IRR) involves finding the discount rate that makes the net present value (NPV) of the cash flows from an investment equal to zero. Here's how to calculate IRR with an example relevant to a dairy farm:

Example: Let's continue with the example of the dairy farm considering an investment in new equipment for automated milking. The initial investment is $100,000, and over the next five years, the investment is expected to generate additional annual cash flows of $30,000, $35,000, $40,000, $45,000, and $50,000, respectively.

Assumptions: Cash flows: Year 1 = $30,000, Year 2 = $35,000, Year 3 = $40,000, Year 4 = $45,000, Year 5 = $50,000 Initial Investment = $100,000

Step 1: Set up the NPV equation:

NPV = 0 = Cash Flow Year 1 / (1 + IRR)^1 + Cash Flow Year 2 / (1 + IRR)^2 + ... + Cash Flow Year 5 / (1 + IRR)^5 - Initial Investment

Step 2: Use trial and error or software/excel functions to find the IRR:

Trial and error method: Start with a guess for the IRR (e.g., 10%). Calculate the NPV using the guessed IRR. If NPV is close to zero, stop; otherwise, adjust the guess and repeat until NPV is close to zero.

Using software/excel functions: Many spreadsheet software like Excel have built-in functions to calculate IRR. You can use the IRR function and input the cash flows to find the IRR directly.

For our example, let's use the trial and error method:

Guess IRR = 10%

NPV = $30,000 / (1 + 0.10)^1 + $35,000 / (1 + 0.10)^2 + $40,000 / (1 + 0.10)^3 + $45,000 / (1 + 0.10)^4 + $50,000 / (1 + 0.10)^5 - $100,000

NPV ≈ $30,000 / 1.10 + $35,000 / 1.10^2 + $40,000 / 1.10^3 + $45,000 / 1.10^4 + $50,000 / 1.10^5 - $100,000 NPV ≈ $27,273 + $28,512 + $29,211 + $28,367 + $27,144 - $100,000 NPV ≈ $40,507

Since the NPV is not close to zero, we adjust the guess.

Guess IRR = 12%

NPV ≈ $30,000 / 1.12 + $35,000 / 1.12^2 + $40,000 / 1.12^3 + $45,000 / 1.12^4 + $50,000 / 1.12^5 - $100,000 NPV ≈ $26,786 + $28,240 + $29,179 + $28,272 + $27,177 - $100,000 NPV ≈ $39,654

Since the NPV is still not close to zero, we continue adjusting the guess until we find the IRR that results in NPV ≈ 0.

Continue this process until the NPV is close to zero. In practice, you can use software or Excel functions to find the IRR more efficiently.

Interpreting the results of the Net Present Value (NPV) and Internal Rate of Return (IRR) calculations is crucial for making informed investment decisions. Here's how to interpret the results of the two examples provided:

1. Net Present Value (NPV):

  ◦ NPV measures the present value of all cash inflows and outflows associated with an investment, discounted at a specified rate (usually the cost of capital).

- If NPV is positive: The investment generates more cash inflows than outflows and is considered financially attractive. In other words, the project adds value to the business and increases wealth.

- If NPV is negative: The investment generates less cash inflows than outflows and may not be financially viable. It indicates that the project is likely to result in a loss and may not be worthwhile.

- In the dairy farm example: A positive NPV suggests that the investment in new equipment for automated milking adds value to the dairy farm. It implies that the present value of the expected cash inflows (returns) exceeds the initial investment cost, making the investment financially attractive.

2. Internal Rate of Return (IRR):

- IRR represents the discount rate at which the NPV of an investment equals zero. It is the rate of return at which the investment breaks even.

- If IRR is greater than the cost of capital: The investment is considered financially viable and generates returns higher than the cost of capital. In other words, the project earns more than the expected rate of return, making it attractive.

- If IRR is less than the cost of capital: The investment may not meet the required rate of return and may not be financially viable. It suggests that the project's returns are lower than expected, making it less attractive.

- In the dairy farm example: A higher IRR indicates that the investment in new equipment for automated milking generates a higher rate of return compared to the cost of capital. It implies that the project is financially viable and offers a return greater than the dairy farm's required rate of return.

In summary, positive NPV and higher IRR values indicate that the investment is financially attractive and likely to enhance the wealth of the business. However, it's essential to consider other factors such as risk, opportunity cost, and strategic alignment before making investment decisions.

As a further example. let's consider a farm considering undertaking vertical farming of arugula, butterhead lettuce, red leaf lettuce, and romaine lettuce:

1. Net Present Value (NPV) Calculation Example:

- Assume the initial investment for setting up the vertical farming system, including equipment, lighting, irrigation systems, and infrastructure, is $100,000.

- Estimate the expected annual cash inflows from selling arugula, butterhead lettuce, red leaf lettuce, and romaine lettuce grown through vertical farming. Let's say the projected annual cash inflows are $40,000.

- Determine the discount rate, typically the farm's cost of capital or the desired rate of return on investment. For this example, let's use a discount rate of 10%.

- Use the NPV formula to calculate the present value of the expected cash inflows: $NPV = \Sigma(CF_t / (1 + r)^t)$ - Initial Investment Where CF_t = Cash flow in year t, r = Discount rate, t = Time period

- Substituting the values: NPV = $40,000 / (1 + 0.10)^1 + $40,000 / (1 + 0.10)^2 + ... - $100,000

- Calculate NPV to determine if the investment is financially viable: NPV = $40,000 / 1.10 + $40,000 / 1.10^2 + ... - $100,000 NPV ≈ $40,000 / 1.10 + $40,000 / 1.21 + ... - $100,000 NPV ≈ $36,364 + $33,059 + ... - $100,000 NPV ≈ $69,423 - $100,000 NPV ≈ -$30,577

○ Interpretation: The negative NPV (-$30,577) suggests that the initial investment in vertical farming of arugula, butterhead lettuce, red leaf lettuce, and romaine lettuce may not be financially viable at the assumed discount rate of 10%. Further analysis and adjustments to the assumptions or investment plan may be necessary to achieve a positive NPV.

2. Internal Rate of Return (IRR) Calculation Example:

○ Use the same initial investment amount of $100,000 and projected annual cash inflows of $40,000 from the vertical farming operation.

○ Calculate the IRR, which is the discount rate that makes the NPV of the investment equal to zero.

○ Substitute the cash flows and the initial investment into the IRR formula and solve for the rate.

○ Solve for the rate using financial software, spreadsheet functions, or trial and error.

○ Let's assume the calculated IRR is 8%.

○ Interpretation: An IRR of 8% indicates that the investment in vertical farming of arugula, butterhead lettuce, red leaf lettuce, and romaine lettuce generates a return of 8%, which exceeds the farm's cost of capital or required rate of return. Therefore, the investment is financially viable at an IRR of 8%. However, it's essential to consider other factors such as market demand, operational costs, and risks associated with vertical farming before making a final investment decision.

Furthermore, consider non-monetary factors such as environmental sustainability, social impacts, and regulatory compliance when evaluating strategies. Sustainable farming practices not only benefit the environment but also enhance the enterprise's reputation and resilience to regulatory changes.

Continuing with the vertical farming example, when evaluating strategies for vertical farming of arugula, butterhead lettuce, red leaf lettuce, and romaine lettuce, several non-monetary factors should be considered:

1. Environmental Sustainability:

○ Resource Efficiency: Assess the efficiency of resource utilization, such as water and energy, in the vertical farming system. Evaluate technologies like hydroponics or aeroponics, which can significantly reduce water usage compared to traditional soil-based farming.

○ Reduction of Environmental Footprint: Consider the potential reduction in land use and deforestation associated with vertical farming, as it allows for high-density crop production in indoor or controlled environments without the need for large expanses of land.

○ Minimization of Pesticide Use: Vertical farming can offer better control over pests and diseases, potentially reducing the reliance on chemical pesticides and promoting more sustainable pest management practices.

2. Social Impacts:

○ Job Creation: Assess the potential for job creation within the local community through the establishment of vertical farming operations, including roles in production, maintenance, distribution, and marketing.

○ Access to Fresh Produce: Consider the impact on local food security and access to fresh, locally grown produce, particularly in urban areas where access to traditional farmland may be limited.

○ Community Engagement: Evaluate opportunities for community engagement and education around sustainable agriculture, healthy eating, and food sovereignty through partnerships, workshops, or educational programs.

3. Regulatory Compliance:

   ◦ Zoning and Land Use Regulations: Ensure compliance with local zoning regulations and land use policies governing the establishment and operation of vertical farming facilities, including building codes, permits, and environmental impact assessments.

   ◦ Food Safety Standards: Adhere to food safety regulations and standards to ensure the quality and safety of produce grown in vertical farming systems. Implement measures such as Good Agricultural Practices (GAP) and Hazard Analysis and Critical Control Points (HACCP) to minimize foodborne hazards.

   ◦ Environmental Regulations: Comply with environmental regulations related to waste management, water usage, and air quality to minimize the environmental impact of vertical farming operations and maintain ecological balance.

Considering these non-monetary factors alongside financial considerations can help ensure that strategies for vertical farming of arugula, butterhead lettuce, red leaf lettuce, and romaine lettuce are sustainable, socially responsible, and compliant with regulatory requirements, ultimately contributing to the long-term success and resilience of the enterprise.

Lastly, based on the results of the cost-benefit analysis, prioritize and select agricultural strategies that offer the highest net benefits and align with the enterprise's goals and objectives. Evaluate the trade-offs between costs, benefits, and risks associated with each option to make informed decisions that maximize the overall value and resilience of the agricultural operation.

Following the assessment of costs and benefits, proceed to develop budgets for implementing and maintaining the chosen strategies, adhering to enterprise guidelines. Allocate resources judiciously to ensure the financial sustainability of the enterprise while effectively managing climate-related risks. Consider various factors, including capital expenditure, operational expenses, and ongoing maintenance costs, to create budgets that align with the enterprise's financial objectives.

To effectively develop budgets for implementing and maintaining chosen strategies aimed at mitigating climate-related risks in agriculture, the following steps can be followed with specific examples:

In the case of a dairy farm considering the implementation of a strategy to enhance water management through the installation of a drip irrigation system, it's crucial to evaluate the associated costs and benefits. The costs may encompass expenses related to purchasing drip lines, emitters, pumps, filters, and labour for installation. On the other hand, the benefits could include reductions in water usage, improvements in crop yield, and enhancements in soil health due to the more efficient irrigation provided by the system.

Resource allocation should be done judiciously, taking into account the estimated costs of procuring and installing the drip irrigation system. It's essential to ensure that the allocated budget adequately covers all necessary components and labour expenses while also allowing for unforeseen costs. Prioritizing investments based on potential returns and the urgency of addressing climate-related risks can further enhance the effectiveness of resource allocation.

Apart from the initial capital expenditure for acquiring and installing the drip irrigation system, various ongoing operational expenses need consideration. These may include energy costs for operating the system, expenses for maintenance and replacement of worn-out components, as well as labour costs for monitoring and repairing the system. Allocating funds for periodic maintenance and potential upgrades is essential to ensure the long-term sustainability and effectiveness of the irrigation system.

It's imperative to ensure that the developed budgets align with the financial objectives of the dairy farm. These objectives may include maintaining profitability, enhancing operational efficiency, and bolstering resilience to climate-related risks. By evaluating the potential impact of the proposed budget on the overall financial performance of the farm, adjustments can be made as necessary to achieve the desired financial outcomes while effectively managing climate-related risks. A resulting sample budget follows:

Sample Budget for Implementing Drip Irrigation System:
   1. **Initial Investment:**

- Drip Lines: $2,500

- Emitters: $1,200

- Pumps: $3,000

- Filters: $800

- Labor for Installation: $1,500

- Total Initial Investment: $9,000

2. **Ongoing Operational Expenses (Annual):**

- Energy Costs for System Operation: $500

- Maintenance and Replacement of Components: $1,000

- Labor Costs for Monitoring and Repairs: $800

- Total Annual Operational Expenses: $2,300

3. **Reserve for Unforeseen Costs (10% of Initial Investment):**

- Unforeseen Costs Reserve: $900

4. **Total Budget for Year 1:**

- Initial Investment: $9,000

- Ongoing Operational Expenses: $2,300

- Unforeseen Costs Reserve: $900

- Total Budget for Year 1: $12,200

5. **Annual Budget for Subsequent Years (excluding Initial Investment):**

- Ongoing Operational Expenses: $2,300

- Unforeseen Costs Reserve: $900

- Total Annual Budget for Subsequent Years: $3,200

6. **Alignment with Financial Objectives:**

- Ensure profitability through improved crop yield and reduced water usage.

- Enhance operational efficiency by optimizing water management practices.

- Bolster resilience to climate-related risks by mitigating the impact of droughts and water scarcity.

This sample budget outlines the initial investment and annual expenses required for implementing and maintaining a drip irrigation system on the dairy farm. By aligning with the farm's financial objectives and considering potential benefits such as improved crop yield and water efficiency, the budget aims to ensure the effective management of climate-related risks while maintaining financial sustainability.

And for our other example related to vertical farming of lettuce, the budget may look something like:

Sample Budget for Vertical Farming of Arugula, Butterhead Lettuce, Red Leaf Lettuce, and Romaine Lettuce:

1. **Infrastructure Setup:**

   ◦ Vertical Farming Equipment (Vertical towers, LED grow lights, irrigation system): $20,000

   ◦ Indoor Farming Facility Renovation (Insulation, lighting installation, ventilation): $15,000

   ◦ Total Infrastructure Setup Cost: $35,000

2. **Seeds and Planting Materials:**

   ◦ Arugula Seeds: $500

   ◦ Butterhead Lettuce Seeds: $600

   ◦ Red Leaf Lettuce Seeds: $600

   ◦ Romaine Lettuce Seeds: $600

   ◦ Total Seeds and Planting Materials Cost: $2,300

3. **Operating Expenses (Annual):**

   ◦ Electricity for LED Grow Lights: $2,500

   ◦ Water and Nutrient Solutions: $1,200

   ◦ Labor (Harvesting, planting, maintenance): $12,000

   ◦ Miscellaneous Supplies (pH testers, trays, etc.): $800

   ◦ Total Annual Operating Expenses: $16,500

4. **Marketing and Distribution:**

   ◦ Packaging Materials: $1,000

   ◦ Marketing Campaigns (Online advertising, flyers): $2,000

   ◦ Distribution Costs (Transportation, delivery): $3,000

   ◦ Total Marketing and Distribution Costs: $6,000

5. **Reserve for Unforeseen Costs (10% of Total Expenses):**

   ◦ Unforeseen Costs Reserve: $6,580

6. **Total Budget for Year 1:**

- Infrastructure Setup: $35,000

- Seeds and Planting Materials: $2,300

- Operating Expenses: $16,500

- Marketing and Distribution: $6,000

- Unforeseen Costs Reserve: $6,580

- Total Budget for Year 1: $66,380

7. **Annual Budget for Subsequent Years (excluding Infrastructure Setup):**

- Seeds and Planting Materials: $2,300

- Operating Expenses: $16,500

- Marketing and Distribution: $6,000

- Unforeseen Costs Reserve: $6,580

- Total Annual Budget for Subsequent Years: $31,380

8. **Alignment with Financial Objectives:**

- Ensure profitability through the sale of high-value, locally grown specialty greens.

- Enhance operational efficiency through vertical farming techniques that optimize space and resource utilization.

- Bolster resilience to climate-related risks by growing crops indoors, independent of outdoor weather conditions.

Finally, perform sensitivity analysis to gauge the impact of different climate scenarios on financial forecasts. This entails evaluating the resilience of proposed strategies to changes in climate conditions and pinpointing potential areas of vulnerability. Analyse how variations in temperature, precipitation, or extreme weather events may influence the financial performance of the enterprise, and adjust forecasts accordingly. By integrating sensitivity analysis into financial planning, you can proactively address uncertainties and make well-informed decisions to safeguard the financial well-being of the enterprise.

To perform sensitivity analysis, firstly, identify a range of climate scenarios by leveraging historical data, climate projections, and expert insights. Consider variations in temperature, precipitation patterns, and the frequency and intensity of extreme weather events to encompass a comprehensive set of potential outcomes.

Next, evaluate the resilience of proposed strategies to changes in climate conditions. Assess how each strategy may perform under different climate scenarios, taking into account factors such as crop yield, water availability, input costs, and market demand. This analysis helps in understanding the potential effectiveness of each strategy in varying environmental conditions.

Identify potential areas of vulnerability within the enterprise's operations and financial performance. This includes identifying crops or livestock particularly sensitive to certain climate conditions, infrastructure prone to damage from extreme weather events, or increased operational costs due to changes in resource availability.

Analyse how variations in temperature, precipitation, or extreme weather events may impact the financial performance of the enterprise. Quantify the potential impact on revenue, expenses, and overall profitability under different climate scenarios to understand the magnitude of potential risks and opportunities.

Adjust financial forecasts based on the insights gained from sensitivity analysis. Incorporate the potential financial implications of different climate scenarios into budgeting, risk management, and decision-making processes. This involves adjusting revenue projections, cost estimates, and investment plans to account for climate-related risks and opportunities.

Integrate sensitivity analysis into financial planning processes to proactively address uncertainties and mitigate risks. Use the findings to make well-informed decisions that safeguard the financial well-being of the enterprise. Consider developing contingency plans or alternative strategies to mitigate potential adverse effects of climate variability on financial performance.

By performing sensitivity analysis and integrating the findings into financial planning, enterprises can better prepare for and adapt to the challenges posed by climate variability. This proactive approach enables them to make informed decisions that enhance resilience and sustainability in the face of changing environmental conditions.

Let's consider the example of the vertical farm that cultivates arugula, butterhead lettuce, red leaf lettuce, and romaine lettuce. Here's how sensitivity analysis can be applied to assess the financial impact of different climate scenarios on the farm's profitability:

1. Identify Climate Scenarios:

   ◦ Climate Scenario 1: Normal weather conditions with average temperatures and regular precipitation.

   ◦ Climate Scenario 2: Increased temperatures and reduced precipitation, leading to water stress.

   ◦ Climate Scenario 3: Extreme weather events such as heavy rainfall or hailstorms, causing damage to crops and infrastructure.

   ◦ Climate Scenario 4: Mild temperatures and optimal precipitation, resulting in ideal growing conditions.

2. Evaluate Resilience of Strategies:

   ◦ Assess how each crop variety responds to changes in temperature, water availability, and light conditions in different climate scenarios.

   ◦ Consider the adaptability of the vertical farming system to varying environmental conditions, such as the efficiency of irrigation systems and the resilience of vertical growing structures.

3. Pinpoint Areas of Vulnerability:

   ◦ Identify crops that are more susceptible to specific climate conditions. For example, arugula and butterhead lettuce may be more sensitive to heat stress, while romaine lettuce and red leaf lettuce may tolerate it better.

   ◦ Assess the vulnerability of infrastructure components such as vertical growing towers and irrigation systems to extreme weather events like storms or high winds.

4. Analyse Financial Impact:

   ◦ Quantify the potential impact of each climate scenario on crop yields, production costs, and revenue generation.

   ◦ Estimate the financial losses incurred due to crop damage or reduced yields in adverse climate scenarios and compare them to the financial gains in favourable conditions.

5. Adjust Forecasts Accordingly:

   ◦ Modify financial forecasts and budget allocations based on the insights gained from sensitivity analysis.

   ◦ Allocate resources towards implementing adaptive measures such as additional climate control equipment, crop diversification, or infrastructure upgrades to mitigate risks identified in vulnerable areas.

6. Integrate Sensitivity Analysis into Financial Planning:

○ Incorporate the findings of sensitivity analysis into the farm's financial planning processes, including risk management strategies and investment decisions.

○ Develop contingency plans to address potential losses and capitalize on opportunities arising from climate variability, ensuring the farm's long-term financial sustainability.

And as a hypothetical example:
Hypothetical Sensitivity Analysis for Vertical Farming of Arugula, Butterhead Lettuce, Red Leaf Lettuce, and Romaine Lettuce:

1. Identify Climate Scenarios:

○ Climate Scenario 1: Normal Weather Conditions

• Average temperatures and regular precipitation.

○ Climate Scenario 2: Increased Temperatures and Reduced Precipitation

• Higher temperatures and decreased rainfall leading to water stress.

○ Climate Scenario 3: Extreme Weather Events

• Heavy rainfall or hailstorms causing damage to crops and infrastructure.

○ Climate Scenario 4: Mild Temperatures and Optimal Precipitation

• Ideal growing conditions with moderate temperatures and adequate rainfall.

2. Evaluate Resilience of Strategies:

○ Assessing Crop Varieties:

• Arugula and Butterhead Lettuce: More sensitive to heat stress, requiring efficient cooling systems during warmer periods.

• Red Leaf Lettuce and Romaine Lettuce: Tolerate heat better but may require additional irrigation during dry spells.

○ Vertical Farming System:

• Efficiency of irrigation systems: Assess the capacity to adjust watering schedules and water distribution to mitigate water stress.

• Resilience of vertical growing structures: Evaluate the ability to withstand strong winds and heavy rainfall without compromising crop integrity.

3. Pinpoint Areas of Vulnerability:

○ Crop Sensitivity:

• Arugula and Butterhead Lettuce: Vulnerable to heat stress and reduced water availability.

- Red Leaf Lettuce and Romaine Lettuce: More resilient to heat but susceptible to damage from extreme weather events.

  - Infrastructure Vulnerability:

    - Vertical Growing Towers: Potential damage from strong winds or heavy rainfall, necessitating reinforcement or protective measures.

    - Irrigation Systems: Risk of malfunction or clogging during periods of increased sedimentation or debris carried by heavy rainfall.

4. Analyse Financial Impact:

  - Impact on Crop Yields and Production Costs:

    - Scenario 2: Decreased yields and increased water usage leading to higher production costs.

    - Scenario 3: Crop damage resulting in losses due to reduced harvestable yield and increased labour costs for recovery efforts.

  - Revenue Generation:

    - Scenario 4: Optimal conditions leading to higher yields and increased revenue generation.

  - Comparative Analysis:

    - Assess the net financial impact of each scenario by comparing losses incurred in adverse conditions with gains in favourable conditions.

5. Adjust Forecasts Accordingly:

  - Allocate Resources:

    - Increase budget allocation for climate-resilient infrastructure upgrades, such as reinforced growing structures and advanced irrigation systems.

  - Implement Adaptive Measures:

    - Invest in climate control equipment like cooling systems and shading structures to mitigate heat stress.

    - Diversify crop selection to include more heat-tolerant varieties and explore alternative irrigation sources to cope with water stress.

6. Integrate Sensitivity Analysis into Financial Planning:

  - Risk Management Strategies:

    - Develop contingency plans for crop loss insurance or emergency funds to mitigate financial risks associated with extreme weather events.

  - Investment Decisions:

    - Prioritize investments in climate-resilient infrastructure and adaptive technologies to ensure long-term viability and profitability of the vertical farming operation.

- Long-Term Financial Sustainability:

  - Incorporate sensitivity analysis findings into long-term financial planning to anticipate and address climate-related uncertainties, ensuring the farm's resilience and sustainability over time.

Reviewing and selecting strategies for agricultural enterprises involves a systematic process to assess proposed options and determine their suitability in achieving enterprise goals and objectives.

Engaging stakeholders is a critical step in the review process as it ensures that the perspectives and interests of all relevant parties are considered. Stakeholders may include investors, employees, suppliers, customers, regulatory authorities, and community members. By gathering feedback and input from diverse stakeholders, agricultural enterprises can gain valuable insights into the potential impacts and implications of proposed strategies. This consultation helps to ensure alignment with enterprise goals, values, and priorities. For example, investors may prioritize financial returns, while employees may prioritize job security and workplace safety. Community members may be concerned about environmental sustainability and social responsibility. By soliciting input from stakeholders, agricultural enterprises can identify potential concerns, address any conflicts of interest, and build consensus around the selected strategies.

Conducting a risk-return analysis is essential for evaluating the trade-offs between risk mitigation and potential returns associated with each strategy. This analysis involves assessing the likelihood of success, estimating expected financial outcomes, and evaluating the level of risk tolerance. Strategies with higher potential returns may also carry higher levels of risk, while those with lower risk profiles may offer more modest returns. Agricultural enterprises must weigh these factors carefully to determine the most appropriate strategy for their specific circumstances. For example, investing in high-yield crop varieties may offer the potential for increased profitability but may also be more susceptible to market fluctuations or climate risks. On the other hand, diversifying crop production or implementing sustainable farming practices may reduce risk exposure but may yield lower returns in the short term. By conducting a comprehensive risk-return analysis, agricultural enterprises can make informed decisions that optimize the balance between risk and reward.

As a sample Risk-Return Analysis for Vertical Farming of Arugula, Butterhead Lettuce, Red Leaf Lettuce, and Romaine Lettuce, this could include:

Strategy: Implementing Vertical Farming for Arugula, Butterhead Lettuce, Red Leaf Lettuce, and Romaine Lettuce

Risk Assessment:

- Market Risk: The demand for leafy greens may fluctuate due to factors such as changing consumer preferences, competition from traditional farming methods, and economic conditions.

- Production Risk: Vertical farming may be susceptible to technical challenges such as equipment failure, pest infestations, disease outbreaks, or crop failures.

- Climate Risk: Extreme weather events or climate variability may affect indoor farming conditions, leading to fluctuations in crop yields or quality.

Potential Returns:

- Market Opportunities: Vertical farming offers the potential to produce high-quality, locally grown leafy greens year-round, tapping into growing consumer demand for fresh and sustainable produce.

- Operational Efficiency: Vertical farming can optimize space utilization, water usage, and resource efficiency, leading to higher yields and reduced production costs.

- Premium Pricing: The unique selling proposition of locally grown, pesticide-free leafy greens may command premium prices in niche markets such as upscale restaurants, specialty grocers, and farmers' markets.

Risk-Return Analysis:

- High Market Potential: The growing trend towards healthy eating and sustainable food production presents a favourable market environment for vertical farming of leafy greens.

- Moderate Production Risks: While vertical farming offers greater control over growing conditions, there are still risks associated with crop management, equipment reliability, and labour availability.

- Climate Resilience: Indoor farming mitigates some climate-related risks, but infrastructure vulnerabilities and energy costs may impact profitability.

Decision Making:

- Based on the risk-return analysis, the vertical farming of arugula, butterhead lettuce, red leaf lettuce, and romaine lettuce presents an attractive opportunity with moderate to high potential returns.

- However, careful consideration must be given to managing production risks, optimizing operational efficiency, and adapting to market dynamics to ensure the success of the venture.

- By implementing best practices in crop management, investing in reliable technology, and establishing strategic partnerships, the enterprise can maximize returns while mitigating potential risks associated with vertical farming.

A similar process can be applied to the dairy farm example implementing drip irrigation system:

Sample Risk-Return Analysis for Implementing Drip Irrigation System in a Dairy Farm:

1. Strategy: Implementing Drip Irrigation System for Water Management

Risk Assessment:

- Investment Risk: There is a risk associated with the initial investment required for purchasing and installing the drip irrigation system, including the cost of equipment, installation labour, and system maintenance.

- Crop Yield Risk: Drip irrigation systems may not guarantee a significant increase in crop yields if not implemented effectively. Factors such as improper system design, maintenance issues, or inadequate water distribution may lead to suboptimal results.

- Operational Risk: There is a risk of operational disruptions due to system failures, clogging of drip lines, or damage to components caused by factors such as pests, rodents, or extreme weather events.

Potential Returns:

- Water Savings: Drip irrigation systems offer efficient water usage by delivering water directly to the root zone of crops, minimizing evaporation and runoff compared to traditional irrigation methods. This can result in significant water savings and reduced water costs for the dairy farm.

- Increased Crop Yield: Properly managed drip irrigation systems ensure consistent and uniform water distribution, optimizing soil moisture levels and promoting healthy plant growth. This can lead to increased crop yields, improved crop quality, and enhanced overall productivity.

- Operational Efficiency: Drip irrigation systems require less labour and management compared to conventional irrigation methods, resulting in operational efficiency gains for the dairy farm. This may translate into cost savings, reduced labour expenses, and improved resource allocation.

Risk-Return Analysis:

- Moderate Investment Risk: While there is an initial investment required for purchasing and installing the drip irrigation system, the potential water savings and crop yield improvements justify the upfront costs over the long term.

- Moderate to High Crop Yield Risk: The effectiveness of the drip irrigation system in increasing crop yields depends on various factors such as proper system design, maintenance, and crop management practices. However, with proper implementation and management, the potential returns in terms of increased productivity outweigh the associated risks.

- Low Operational Risk: Drip irrigation systems are generally reliable and require minimal maintenance if properly installed and managed. While there is a risk of operational disruptions, proactive maintenance and monitoring can mitigate these risks and ensure consistent performance of the system.

Decision Making:
- Based on the risk-return analysis, implementing a drip irrigation system presents a viable opportunity for the dairy farm to improve water management, increase crop yields, and enhance operational efficiency.

- While there are risks associated with the initial investment and crop yield outcomes, the potential water savings, productivity gains, and operational benefits outweigh the associated risks.

- By carefully selecting and implementing a drip irrigation system, providing staff training, and establishing maintenance protocols, the dairy farm can maximize returns while mitigating risks associated with water management.

Once the review and analysis process is complete, agricultural enterprises must review preferred production, enterprise, or alternative strategies and select options appropriate for their specific context. This decision-making process involves weighing the findings of the stakeholder consultation and risk-return analysis to identify the most suitable strategies. Priority should be given to strategies that offer the best balance of risk reduction and financial viability while also aligning with enterprise goals and objectives. Agricultural enterprises may need to make trade-offs and compromises to select the most optimal strategy given the constraints and uncertainties they face. For example, an enterprise may prioritize strategies that enhance environmental sustainability and social responsibility, even if they entail higher upfront costs or lower immediate returns. By carefully evaluating the potential benefits and drawbacks of each option, agricultural enterprises can make well-informed decisions that support their long-term success and resilience.

Documenting and presenting strategies is crucial to ensure clarity, transparency, and alignment among stakeholders regarding approaches chosen to manage risks associated with variable climate. When documenting strategies, it's important to provide a comprehensive overview that includes the following components:
- Rationale and Objectives: Begin by articulating the reasoning behind each chosen strategy and delineating the specific objectives it intends to accomplish. These objectives may encompass risk mitigation, resilience enhancement, or leveraging opportunities arising from variable climate conditions.

- Key Assumptions: Clearly delineate the fundamental assumptions that underpin each strategy. These assumptions may relate to climate projections, market dynamics, or advancements in technology. Acknowledge any uncertainties or constraints associated with these assumptions to provide a realistic assessment of the strategy's feasibility.

- Expected Outcomes: Define the anticipated outcomes resulting from the implementation of each strategy, encompassing both short-term and long-term impacts on the enterprise. Whenever possible, quantify these outcomes, such as projected cost savings, yield enhancements, or reductions in climate-induced losses. This quantification aids in evaluating the strategy's effectiveness.

- Implementation Plan: Detail the steps involved in executing each strategy, including timelines, responsible parties, and resource requirements. This entails outlining any necessary infrastructure upgrades, technological investments, or training initiatives needed to support implementation effectively.

- Monitoring and Evaluation: Establish a set of metrics and indicators to gauge the performance of each strategy over time. Develop a framework for regular evaluation to track progress towards objectives and identify areas requiring refinement or enhancement. This monitoring and evaluation process ensures that strategies remain aligned with goals and allows for adjustments as needed to optimize outcomes.

When presenting strategies to stakeholders, effective communication is key to ensuring understanding and alignment. This requires:

- Clear Communication: Craft a concise and visually engaging presentation that effectively communicates the documented strategies to stakeholders. Utilize clear language and visual aids to simplify complex concepts and present data in an easily digestible format. Clarity in communication helps stakeholders grasp the essence of the strategies and their implications.

- Highlight Benefits: Accentuate the potential benefits associated with the selected strategies to stakeholders. Illustrate how these strategies can lead to improved resilience, enhanced productivity, or reduced financial risks for the enterprise. Provide concrete examples and case studies wherever possible to demonstrate the positive impact of similar strategies in real-world scenarios.

- Address Concerns: Proactively anticipate and address any concerns or inquiries raised by stakeholders regarding the proposed strategies. Foster transparency by openly acknowledging potential challenges or risks associated with the strategies. Clearly articulate how these challenges will be managed or mitigated to alleviate stakeholders' apprehensions and build confidence in the proposed approaches.

- Seek Feedback: Encourage stakeholders to actively engage in the presentation by inviting their feedback and input on the documented strategies. Create an environment conducive to open dialogue and collaboration, where stakeholders feel empowered to share their perspectives and suggestions for improvement. Incorporating stakeholder feedback fosters a sense of ownership and buy-in, ensuring greater support for the proposed strategies.

Continuous monitoring and review are essential aspects of strategy implementation to ensure effectiveness and adaptability over time. Begin by setting up a structured framework for continuous monitoring and review of the implemented strategies. This includes establishing regular reporting mechanisms and scheduling review meetings to assess progress and performance. Define clear roles and responsibilities for stakeholders involved in monitoring progress and collecting relevant data, ensuring accountability and coordination.

Remain flexible and responsive to changes in climate conditions, emerging risks, or new information that may impact the effectiveness of implemented strategies. Regularly evaluate the performance of strategies against predetermined metrics and indicators, and be prepared to make necessary adjustments to optimize outcomes. This may involve refining implementation plans, reallocating resources, or introducing new measures to address evolving challenges and opportunities.

Foster ongoing communication and engagement with stakeholders throughout the monitoring and review process. Provide regular updates on progress, successes, and challenges encountered, maintaining transparency and accountability. Actively solicit feedback and input from stakeholders to gather insights, identify issues, and collaboratively explore solutions. By involving stakeholders in the monitoring and review process, you can harness their expertise and support to drive continuous improvement and ensure alignment with overarching goals and objectives.

In summary, developing climate risk management strategies for implementing vertical farming for arugula, butterhead lettuce, red leaf lettuce, and romaine lettuce involves a comprehensive understanding of various aspects related to weather patterns, climate variability, and their impacts on business activities. Here's a detailed summary covering each aspect:

1. Types and Sources of Data Used to Record Weather Patterns: Utilize data from various sources such as meteorological stations, satellite imagery, and weather sensors to record weather patterns. Historical weather data can provide insights into temperature fluctuations, precipitation levels, and extreme weather events relevant to the region.

2. Qualitative and Quantitative Techniques to Analyse Risks: Employ qualitative techniques such as risk assessments and SWOT analyses, along with quantitative methods like statistical modelling and scenario analysis, to assess climate-related risks. These

techniques help identify potential vulnerabilities and prioritize strategies for mitigating risks.

3. Impact of Weather and Climate on Business Activities: Understand how weather and climate variability can affect crop growth, water availability, pest infestations, and market demand for agricultural products. Analyse how these factors impact production costs, revenue streams, and overall profitability of vertical farming operations.

4. Difference Between Weather and Climate: Distinguish between weather, which refers to short-term atmospheric conditions like temperature and precipitation, and climate, which represents long-term patterns and trends in weather over a specific region. Recognize that climate variability encompasses fluctuations in weather patterns over time.

5. Causes of General Patterns of Weather and Climate Over Australia: Consider factors such as atmospheric circulation patterns, ocean currents, and geographical features that influence weather and climate patterns over Australia. Understand how phenomena like El Niño and La Niña events can impact rainfall and temperature distributions across the continent.

6. Climate Variability and Climate Change Impacts for the Local Region: Assess the local region's vulnerability to climate variability and climate change impacts, including shifts in temperature regimes, changes in precipitation patterns, and increased frequency of extreme weather events. Recognize how these changes may affect crop yields, water availability, and resource management practices.

7. Property and Enterprise Management Decisions Affected by Climatic Variability: Evaluate property and enterprise management decisions, such as crop selection, irrigation scheduling, and infrastructure investments, in light of current and predicted climatic variability. Implement adaptive measures to mitigate risks and capitalize on opportunities presented by variable climate conditions.

8. Recognition of Climate Risks and Opportunities: Identify potential climate risks, such as droughts, heatwaves, and flooding, that may threaten vertical farming operations. Explore opportunities for innovation, such as implementing climate-resilient crop varieties or adopting sustainable water management practices, to enhance the resilience of the enterprise.

9. Seasonal Climate Forecasting Systems and Related Indicators: Utilize seasonal climate forecasting systems and relevant indicators, such as sea surface temperatures and atmospheric pressure patterns, to anticipate seasonal climate trends and plan farming activities accordingly. Incorporate forecasts into decision-making processes to optimize resource allocation and minimize weather-related risks.

10. Contingency Planning Including Natural Disaster Planning for Site: Develop contingency plans, including natural disaster preparedness protocols, to respond effectively to climate-related emergencies such as floods, storms, or wildfires. Establish evacuation procedures, emergency communication channels, and backup systems to safeguard personnel and assets during crises.

11. Definition of Risk Management: Define risk management as the systematic process of identifying, assessing, and prioritizing risks, followed by the coordinated and economical application of resources to minimize, monitor, and control the probability and/or impact of adverse events. Implement risk management strategies tailored to specific climate-related threats faced by vertical farming operations.

12. Potential Impacts of Climate Change on Land and Natural Resource Management: Anticipate potential impacts of climate change on land and natural resource management, including changes in soil fertility, water availability, and ecosystem dynamics. Develop adaptive strategies to mitigate these impacts and ensure the sustainability of agricultural practices in the face of changing environmental conditions.

13. Strategic Options and Planning in Response to Climate Variability: Formulate strategic options and contingency plans to address

different climate scenarios, including normal, drier, or wetter than normal seasons. Implement flexible farming techniques, such as hydroponics or vertical farming, that can adapt to varying weather conditions and optimize resource use efficiency.

14. Calculation of Financial Returns for Different Strategic Options: Calculate financial returns for different strategic options using metrics such as return on investment (ROI), net present value (NPV), and internal rate of return (IRR). Consider factors such as capital investments, operating costs, and revenue projections to assess the profitability and viability of each option.

15. Computer Applications to Access, Record, Analyse, and Model Data: Utilize computer applications and data management systems to access, record, analyse, and model weather and climate data. Implement Geographic Information Systems (GIS), weather forecasting software, and crop simulation models to support decision-making processes and optimize resource management practices.

16. Principles for Decision-Making Based on Variable Climate and Seasonal Climate Forecasts: Adopt principles for decision-making that account for variable climate conditions and seasonal climate forecasts. Emphasize flexibility, adaptability, and resilience in decision-making processes, and prioritize strategies that can withstand uncertainty and variability in weather patterns. Utilize probabilistic forecasting techniques and scenario planning to evaluate alternative courses of action and make informed choices in dynamic environmental conditions.

By integrating these considerations into the development of climate risk management strategies for vertical farming, enterprises can effectively mitigate risks, capitalize on opportunities, and enhance the resilience and sustainability of agricultural operations in the face of climate variability and change.

Large-Scale Farming Context

Large-scale farming, also known as industrial or commercial farming, is a form of agricultural production that occurs on a large land area and involves the use of modern machinery, technology, and management practices (Mugera & Langemeier, 2011). This type of farming is characterized by the cultivation of crops or the raising of livestock on a much larger scale compared to small-scale or subsistence farming operations (Mugera & Langemeier, 2011). Large-scale farming operations often focus on maximizing efficiency and productivity to meet the demands of a growing population and global markets (Paul et al., 2004). To achieve this, farmers may utilize advanced irrigation systems, mechanized equipment, genetically modified crops, and agrochemical inputs such as fertilizers and pesticides to optimize yields and minimize production costs (Paul et al., 2004).

Large-scale farming can involve the cultivation of a wide range of crops or the raising of various types of livestock, including grains, fruits, vegetables, poultry, cattle, and aquaculture (Liu et al., 2019). These operations are often characterized by specialized monoculture production systems, where one or a few crops dominate the farming landscape (Liu et al., 2019). Large-scale farming plays a significant role in modern agriculture, contributing to food security, economic development, and employment opportunities in many regions around the world (Stevens, 2023).

However, large-scale farming also raises concerns about environmental sustainability, resource depletion, and biodiversity loss due to its intensive nature and reliance on synthetic inputs (Jadidi et al., 2021). The use of agrochemical inputs such as fertilizers and pesticides can have detrimental effects on the environment, and the monoculture production systems may lead to a loss of biodiversity (Jadidi et al., 2021). Balancing the benefits and challenges of large-scale farming is an ongoing concern for policymakers, farmers, and stakeholders involved in agricultural production and food systems (Xu et al., 2022).

In the context of large-scale farming, studies have been conducted to assess the efficiency and productivity of farming operations. For example, research has investigated the relative technical performance of grain farms and observed that the major source of inefficiency was scale inefficiency, particularly for large farms (Mugera & Langemeier, 2011). Additionally, there have been studies on the environmental efficiency of dairy farming in China, which found regional differences in environmental efficiency (Isgin et al., 2020). Furthermore, the decision to become a large-scale farmer has been found to be significantly influenced by household human capital, cooperative membership, marketing channels, land-transfer contracts, and government policies (Yu et al., 2019).

Large-scale farming operations rely on various types and sources of data to record weather patterns. This includes data from meteorological stations, satellite imagery, and weather sensors, which provide insights into temperature fluctuations, precipitation levels, and extreme weather events relevant to the region. Historical weather data is particularly valuable in understanding long-term trends.

To assess climate-related risks, both qualitative and quantitative techniques are employed. Qualitative methods such as risk assessments and SWOT analyses help in understanding vulnerabilities, while quantitative methods like statistical modelling and scenario analysis provide deeper insights into potential risks associated with climate variability.

Understanding the impact of weather and climate variability on business activities is crucial for large-scale farming operations. Factors such as crop growth, water availability, pest infestations, and market demand significantly influence production costs, revenue streams, and overall profitability.

Distinguishing between short-term weather conditions and long-term climate trends is essential for effective planning and decision-making. Large-scale farming operations need to consider atmospheric circulation patterns, ocean currents, and geographical features that influence weather and climate patterns, including phenomena like El Niño and La Niña events.

Assessing the vulnerability of the local region to climate variability and climate change impacts is crucial for large-scale farming. This involves understanding potential shifts in temperature regimes, changes in precipitation patterns, and increased frequency of extreme weather events and their implications for crop yields, livestock health, and resource management practices.

Property and enterprise management decisions are heavily influenced by climatic variability. Evaluating crop selection, irrigation scheduling, livestock management practices, and infrastructure investments in light of current and predicted climatic variability is essential for sustainable farming practices.

Identifying potential climate risks such as droughts, heatwaves, and flooding, and exploring opportunities for innovation to enhance the resilience of large-scale farming operations, is crucial for long-term success.

Utilizing seasonal climate forecasting systems and relevant indicators aids in planning farming activities, optimizing resource allocation, and minimizing weather-related risks.

Developing contingency plans, including natural disaster preparedness protocols, to respond effectively to climate-related emergencies is essential for safeguarding personnel, livestock, and assets during crises.

Implementing risk management strategies tailored to specific climate-related threats is necessary for minimizing adverse events and ensuring business continuity.

Anticipating potential impacts of climate change on land fertility, water availability, livestock health, and ecosystem dynamics is crucial. Developing adaptive strategies to mitigate these impacts ensures the sustainability of agricultural practices in the long term.

Formulating strategic options and contingency plans to address different climate scenarios, and implementing flexible farming techniques that can adapt to varying weather conditions, are essential for resilience.

Calculating financial returns for different strategic options considering factors such as capital investments, operating costs, market demand, and weather-related risks is necessary to assess profitability and viability.

Utilizing computer applications and data management systems to access, record, analyse, and model weather and climate data supports decision-making processes, optimizes resource management practices, and enhances operational efficiency.

Adopting principles for decision-making that account for variable climate conditions and seasonal climate forecasts is essential for effective management of large-scale farming operations. Flexibility, adaptability, and resilience should be emphasized to navigate dynamic environmental conditions successfully, utilizing probabilistic forecasting techniques and scenario planning to evaluate alternative courses of action and make informed choices.

Let's apply these principles to a hypothetical large-scale grain farm in the United States:

Scenario: **Jones Grain Farm**

Jones Grain Farm is a large-scale operation spanning thousands of acres in the Midwest, primarily focused on producing corn and soybeans. The farm leverages advanced technology and data-driven approaches to optimize production and mitigate risks associated with weather variability.

1. Data Collection: Jones Grain Farm utilizes data from meteorological stations, satellite imagery, and on-site weather sensors to monitor weather patterns. Historical weather data is analysed to understand long-term trends in temperature, precipitation, and extreme weather events relevant to the region.

2. Risk Assessment: The farm conducts both qualitative and quantitative assessments of climate-related risks. Qualitative methods such as SWOT analyses identify vulnerabilities such as susceptibility to droughts or flooding, while quantitative techniques like statistical modelling assess the potential impact of climate variability on crop yields and profitability.

3. Impact on Business Activities: Understanding the impact of weather and climate variability is crucial for Jones Grain Farm. Fluctuations in temperature and precipitation directly affect crop growth and yield potential, influencing production costs, revenue streams, and overall profitability.

4. Distinguishing Weather and Climate Trends: Jones Grain Farm distinguishes between short-term weather conditions and long-term climate trends. They consider atmospheric circulation patterns, ocean currents, and phenomena like El Niño and La Niña events to make informed decisions regarding planting schedules and crop rotations.

5. Assessing Vulnerability: The farm assesses its vulnerability to climate variability and climate change impacts. They analyse potential shifts in temperature regimes, changes in precipitation patterns, and increased frequency of extreme weather events to develop adaptive strategies for crop management and resource allocation.

6. Property and Enterprise Management: Jones Grain Farm evaluates crop selection, irrigation scheduling, and infrastructure investments in light of climatic variability. They implement adaptive measures such as drought-resistant crop varieties and precision irrigation systems to mitigate risks and optimize resource use efficiency.

7. Identifying Climate Risks and Opportunities: The farm identifies climate risks such as droughts and explores opportunities for innovation, such as adopting conservation tillage practices or investing in renewable energy sources, to enhance resilience and long-term success.

8. Utilizing Seasonal Climate Forecasts: Jones Grain Farm utilizes seasonal climate forecasting systems to anticipate weather trends and plan farming activities accordingly. They adjust planting dates and input usage based on forecasted weather conditions to minimize weather-related risks.

9. Contingency Planning: The farm develops contingency plans to respond to climate-related emergencies. This includes protocols for drought management, flood mitigation, and crop insurance coverage to safeguard personnel, assets, and financial stability during crises.

10. Risk Management Strategies: Jones Grain Farm implements risk management strategies tailored to climate-related threats, such as diversifying crop portfolios or investing in irrigation infrastructure to mitigate water scarcity risks.

11. Anticipating Climate Change Impacts: The farm anticipates potential impacts of climate change on land fertility and water availability. They develop adaptive strategies such as soil conservation practices and water-efficient irrigation techniques to ensure the sustainability of agricultural practices in the face of changing environmental conditions.

12. Strategic Planning: Jones Grain Farm formulates strategic options and contingency plans to address different climate scenarios. They invest in research and development to identify new crop varieties resilient to climate variability and explore opportunities

for diversification to reduce dependence on weather-sensitive crops.

13. Financial Analysis: The farm calculates financial returns for different strategic options considering factors such as capital investments, operating costs, and weather-related risks. They assess the profitability and viability of alternative courses of action to optimize resource allocation and maximize returns.

14. Utilizing Computer Applications: Jones Grain Farm utilizes computer applications and data management systems to access, record, analyse, and model weather and climate data. This supports decision-making processes and enhances operational efficiency by optimizing resource management practices.

15. Principles for Decision-Making: The farm adopts principles for decision-making that account for variable climate conditions and seasonal climate forecasts. They prioritize flexibility, adaptability, and resilience in their management approach, utilizing probabilistic forecasting techniques and scenario planning to make informed choices in dynamic environmental conditions.

And applying these principles to a hypothetical cattle ranch in Australia:

Scenario: **Outback Cattle Ranch**

Outback Cattle Ranch is a large-scale operation situated in the arid regions of Western Australia. The ranch covers vast expanses of land and specializes in breeding and raising cattle for meat production. The management of Outback Cattle Ranch is keen on implementing advanced techniques to adapt to the challenging environmental conditions and mitigate risks associated with climate variability.

1. Data Collection: Outback Cattle Ranch relies on meteorological stations, satellite imagery, and weather sensors to gather data on weather patterns. Historical weather data is particularly valuable in understanding long-term trends in temperature, rainfall, and extreme weather events, which are critical for managing cattle health and productivity.

2. Risk Assessment: The ranch conducts both qualitative and quantitative risk assessments to evaluate climate-related risks. Qualitative methods such as SWOT analyses identify vulnerabilities such as water scarcity or heat stress on cattle, while quantitative methods like statistical modelling assess the potential impact of climate variability on cattle production and profitability.

3. Impact on Business Activities: Understanding the impact of weather and climate variability is crucial for Outback Cattle Ranch. Fluctuations in temperature and rainfall directly affect pasture growth, water availability, and cattle health, thereby influencing production costs, revenue streams, and overall profitability.

4. Distinguishing Weather and Climate Trends: The ranch distinguishes between short-term weather conditions and long-term climate trends. They consider atmospheric circulation patterns, ocean currents, and phenomena like El Niño and La Niña events to make informed decisions regarding pasture management, breeding schedules, and infrastructure investments.

5. Assessing Vulnerability: Outback Cattle Ranch assesses its vulnerability to climate variability and climate change impacts. They analyse potential shifts in temperature regimes, changes in rainfall patterns, and increased frequency of extreme weather events to develop adaptive strategies for cattle management and resource allocation.

6. Property and Enterprise Management: The ranch evaluates livestock management practices, water infrastructure investments, and breeding programs in light of climatic variability. They implement adaptive measures such as rotational grazing, drought-resistant cattle breeds, and water storage facilities to mitigate risks and optimize resource use efficiency.

7. Identifying Climate Risks and Opportunities: Outback Cattle Ranch identifies climate risks such as prolonged droughts and explores opportunities for innovation, such as implementing sustainable grazing practices or investing in alternative feed sources, to enhance resilience and long-term success.

8. Utilizing Seasonal Climate Forecasts: The ranch utilizes seasonal climate forecasting systems to anticipate weather trends and plan cattle management activities accordingly. They adjust mustering schedules, supplementary feeding regimes, and breeding decisions based on forecasted weather conditions to minimize weather-related risks.

9. Contingency Planning: Outback Cattle Ranch develops contingency plans to respond to climate-related emergencies such as bushfires or heatwaves. This includes protocols for emergency mustering, provision of supplementary feed and water, and relocation of cattle to safer areas to safeguard livestock and assets during crises.

10. Risk Management Strategies: The ranch implements risk management strategies tailored to climate-related threats, such as diversifying grazing areas or investing in shade infrastructure to mitigate heat stress risks on cattle.

11. Anticipating Climate Change Impacts: Outback Cattle Ranch anticipates potential impacts of climate change on pasture productivity and water availability. They develop adaptive strategies such as soil conservation practices, reforestation efforts, and water recycling systems to ensure the sustainability of cattle production in the face of changing environmental conditions.

12. Strategic Planning: The ranch formulates strategic options and contingency plans to address different climate scenarios. They invest in research and development to identify cattle breeds resilient to climate variability and explore opportunities for diversification into value-added products to reduce dependence on weather-sensitive markets.

13. Financial Analysis: Outback Cattle Ranch calculates financial returns for different strategic options considering factors such as capital investments, operating costs, and weather-related risks. They assess the profitability and viability of alternative courses of action to optimize resource allocation and maximize returns.

14. Utilizing Computer Applications: The ranch utilizes computer applications and data management systems to access, record, analyse, and model weather and climate data. This supports decision-making processes and enhances operational efficiency by optimizing resource management practices and improving cattle health and productivity.

15. Principles for Decision-Making: Outback Cattle Ranch adopts principles for decision-making that account for variable climate conditions and seasonal climate forecasts. They prioritize flexibility, adaptability, and resilience in their management approach, utilizing probabilistic forecasting techniques and scenario planning to make informed choices in dynamic environmental conditions.

And finally, applying the principles to a sugarcane plantation in Brazil:

Scenario: **Tropical Sugarcane Plantation in Brazil**

Brazil is one of the world's largest producers of sugarcane, with vast plantations spanning across the tropical regions of the country. A typical example is the Tropical Sugarcane Plantation, located in the state of São Paulo. This plantation covers thousands of hectares and is dedicated to the cultivation of sugarcane for sugar and ethanol production.

1. Data Collection: The Tropical Sugarcane Plantation utilizes data from meteorological stations, satellite imagery, and weather sensors to monitor weather patterns. Historical weather data helps in understanding long-term trends in temperature, rainfall, and extreme weather events, which are crucial for optimizing sugarcane growth and harvest.

2. Risk Assessment: The plantation conducts both qualitative and quantitative risk assessments to evaluate climate-related risks. Qualitative methods such as SWOT analyses identify vulnerabilities such as water scarcity or pest outbreaks, while quantitative techniques like statistical modelling assess the potential impact of climate variability on sugarcane yields and profitability.

3. Impact on Business Activities: Understanding the impact of weather and climate variability is crucial for the Tropical Sugarcane

Plantation. Factors such as temperature, rainfall, and humidity directly affect sugarcane growth, pest infestations, and harvest quality, thereby influencing production costs, revenue streams, and overall profitability.

4. Distinguishing Weather and Climate Trends: The plantation distinguishes between short-term weather conditions and long-term climate trends. They consider atmospheric circulation patterns, ocean currents, and phenomena like El Niño and La Niña events to make informed decisions regarding planting schedules, irrigation management, and pest control strategies.

5. Assessing Vulnerability: The Tropical Sugarcane Plantation assesses its vulnerability to climate variability and climate change impacts. They analyze potential shifts in temperature regimes, changes in rainfall patterns, and increased frequency of extreme weather events to develop adaptive strategies for sugarcane cultivation and resource management.

6. Property and Enterprise Management: The plantation evaluates irrigation systems, pest control measures, and harvest schedules in light of climatic variability. They implement adaptive measures such as drip irrigation, biological pest control, and mechanized harvesting to mitigate risks and optimize sugarcane yield and quality.

7. Identifying Climate Risks and Opportunities: The Tropical Sugarcane Plantation identifies climate risks such as droughts, flooding, and hurricanes, and explores opportunities for innovation to enhance resilience. This includes investing in drought-resistant sugarcane varieties, improving water storage infrastructure, and diversifying into value-added products like biofuels.

8. Utilizing Seasonal Climate Forecasts: The plantation utilizes seasonal climate forecasting systems to anticipate weather trends and plan farming activities accordingly. They adjust planting dates, irrigation schedules, and fertilizer applications based on forecasted weather conditions to minimize weather-related risks and optimize sugarcane productivity.

9. Contingency Planning: The Tropical Sugarcane Plantation develops contingency plans to respond to climate-related emergencies such as floods or cyclones. This includes protocols for emergency drainage, crop insurance coverage, and evacuation procedures to safeguard personnel, equipment, and infrastructure during crises.

10. Risk Management Strategies: The plantation implements risk management strategies tailored to specific climate-related threats, such as investing in weather-resistant infrastructure or diversifying income streams to reduce dependence on sugarcane production alone.

11. Anticipating Climate Change Impacts: The Tropical Sugarcane Plantation anticipates potential impacts of climate change on sugarcane yields and quality. They develop adaptive strategies such as soil conservation practices, carbon sequestration initiatives, and renewable energy production to ensure the sustainability of sugarcane cultivation in the face of changing environmental conditions.

12. Strategic Planning: The plantation formulates strategic options and contingency plans to address different climate scenarios. They invest in research and development to identify sugarcane varieties resilient to climate variability and explore opportunities for vertical integration and market diversification to reduce risks and maximize returns.

13. Financial Analysis: The Tropical Sugarcane Plantation calculates financial returns for different strategic options considering factors such as capital investments, operating costs, and weather-related risks. They assess the profitability and viability of alternative courses of action to optimize resource allocation and maximize returns in the dynamic agricultural market.

14. Utilizing Computer Applications: The plantation utilizes computer applications and data management systems to access, record, analyze, and model weather and climate data. This supports decision-making processes and enhances operational efficiency by optimizing resource management practices and improving sugarcane yield and quality.

15. Principles for Decision-Making: The Tropical Sugarcane Plantation adopts principles for decision-making that account for variable climate conditions and seasonal climate forecasts. They prioritize flexibility, adaptability, and resilience in their management approach, utilizing probabilistic forecasting techniques and scenario planning to make informed choices in dynamic environmental conditions and ensure long-term sustainability.

The application of principles across Jones Grain Farm, Outback Cattle Ranch, and the Tropical Sugarcane Plantation demonstrates both similarities and differences in their approaches to managing climate-related risks and optimizing agricultural operations.

Firstly, each operation relies on various sources of data, including meteorological stations, satellite imagery, and weather sensors, to record weather patterns. While Jones Grain Farm and the Tropical Sugarcane Plantation focus on general weather data, the Outback Cattle Ranch also considers pasture conditions and water availability specific to livestock management.

Regarding risk assessment, all three operations conduct qualitative and quantitative assessments, albeit with different focuses. Jones Grain Farm evaluates risks associated with grain production, while the Outback Cattle Ranch assesses risks related to livestock health and pasture management. Similarly, the Tropical Sugarcane Plantation evaluates risks pertinent to sugarcane growth, such as droughts and pest infestations.

In terms of the impact on business activities, each operation considers how weather and climate variability affect crop yields, livestock health, and overall profitability. However, the specific implications differ: Jones Grain Farm focuses on crop growth and revenue streams, the Outback Cattle Ranch on cattle health and market demand for beef, and the Tropical Sugarcane Plantation on sugarcane yield and quality.

All three operations recognize the importance of distinguishing short-term weather conditions from long-term climate trends and incorporate this understanding into decision-making processes. They also assess vulnerability to climate variability and climate change, developing adaptive strategies to mitigate risks associated with shifts in temperature, precipitation, and extreme weather events.

Property and enterprise management strategies vary based on the nature of each operation. While Jones Grain Farm focuses on crop selection and irrigation management, the Outback Cattle Ranch emphasizes livestock management practices and water infrastructure. Meanwhile, the Tropical Sugarcane Plantation prioritizes irrigation systems, pest control measures, and soil conservation practices tailored to sugarcane cultivation.

Despite these differences, all operations identify climate risks and explore opportunities for innovation to enhance resilience. They utilize seasonal climate forecasts to plan farming activities and develop contingency plans to respond to climate-related emergencies effectively. Additionally, they adopt tailored risk management strategies and anticipate potential impacts of climate change, incorporating principles for decision-making that prioritize flexibility, adaptability, and resilience.

Although Jones Grain Farm, Outback Cattle Ranch, and the Tropical Sugarcane Plantation share common principles in managing climate-related risks, their specific focus areas and strategies reflect the unique characteristics and challenges of each agricultural operation.

The approach taken by Vertical Farming for Arugula, Butterhead Lettuce, Red Leaf Lettuce, and Romaine Lettuce in managing climate-related risks and optimizing agricultural operations differs significantly from traditional large-scale farming practices such as those employed by Jones Grain Farm, Outback Cattle Ranch, and the Tropical Sugarcane Plantation.

Vertical farming operations for leafy greens like arugula and lettuce typically occur in controlled indoor environments, where climate variables are tightly regulated. Here's a comparison and contrast of their approach with the principles applied by traditional farming operations:

Data Collection: Vertical farming relies on sophisticated climate control systems and sensors within indoor facilities to monitor and adjust environmental conditions such as temperature, humidity, and light levels. While traditional farms utilize meteorological stations and satellite imagery, vertical farms focus more on real-time monitoring and management within their controlled environments.

Risk Assessment: Vertical farming involves assessing risks associated with indoor climate control systems, such as equipment malfunctions or power outages, which could affect crop growth. While both vertical and traditional farms conduct risk assessments, the focus and nature of the risks differ due to the distinct farming environments.

Impact on Business Activities: Vertical farms consider how indoor climate variability affects crop growth and production costs. Unlike traditional farms, they do not face the same challenges related to natural weather patterns but must manage energy consumption and operational costs associated with maintaining controlled environments.

Vulnerability Assessment: Vertical farms focus on vulnerabilities related to indoor climate control systems and technological infrastructure. Traditional farms assess vulnerability to natural weather events and climate change impacts on outdoor crops and livestock.

Property and Enterprise Management: Vertical farms prioritize the management of indoor growing systems, including hydroponic or aeroponic setups and artificial lighting. Traditional farms manage outdoor land, livestock, and water resources, with a focus on crop rotation, pasture management, and irrigation.

Climate Risk Identification and Innovation: Vertical farms innovate by developing advanced indoor growing technologies to enhance crop productivity and resource efficiency. Traditional farms may innovate by adopting climate-resilient crop varieties or implementing sustainable land management practices.

Utilization of Climate Forecasts: While traditional farms rely on seasonal climate forecasts for planning outdoor farming activities, vertical farms focus on optimizing indoor environmental conditions year-round based on real-time data and crop growth requirements.

Contingency Planning: Vertical farms develop contingency plans for addressing equipment failures or power outages to minimize crop loss. Traditional farms develop contingency plans for responding to natural disasters such as droughts, floods, or wildfires.

Risk Management Strategies: Both vertical and traditional farms implement risk management strategies tailored to their specific farming environments, addressing climate-related threats accordingly.

Anticipation of Climate Change Impacts: Vertical farms may anticipate potential impacts of climate change on energy costs or technological infrastructure but are less directly affected by outdoor climate variability. Traditional farms anticipate impacts on crop yields, water availability, and livestock health due to changing weather patterns.

While both vertical and traditional farming operations share some principles in managing climate-related risks, the specific approaches and focus areas differ significantly due to the distinct farming environments and technological requirements of vertical farming for leafy greens.

The comparison underscores how the principles of managing climate-related risks and optimizing agricultural operations are context-dependent and vary significantly based on the type and scale of the farm. Vertical farming operations for leafy greens, with their controlled indoor environments and sophisticated technology, require different approaches compared to traditional large-scale farming operations that rely on outdoor cultivation methods and natural weather patterns. Understanding the specific context of each farm type is crucial for effectively implementing strategies to mitigate climate-related risks and ensure agricultural sustainability.

Part 5 - Developing Workplace Policy and Procedures for Environment and Sustainability

This section examines the practical steps involved in implementing environmental sustainability initiatives. It serves as a comprehensive guide, providing insights into navigating through various stages effectively.

The process begins with the identification of legislative, regulatory, and industry requirements relevant to environmental sustainability. This step ensures that initiatives are in compliance with established standards and aligned with legal obligations.

Next, the scope of environment and sustainability policy is defined, outlining the objectives and parameters within which initiatives will operate. This serves as a foundational framework for guiding decision-making and action.

Strategies for gathering information from diverse sources are discussed to facilitate informed decision-making and policy development. This includes methods for collecting data and insights necessary for effective planning and implementation.

The importance of stakeholder engagement is emphasized as a critical component of the policy development process. This involves identifying and consulting with stakeholders to ensure inclusivity, garner support, and align initiatives with organizational goals and values.

Appropriate strategies for minimizing resource use, employing life cycle management approaches, and reducing hazardous material use are integrated into the policy framework. These strategies aim to enhance sustainability and mitigate environmental impacts.

Methods for making recommendations for policy options based on effectiveness, timeframes, and cost are explored. This is followed by the development of policies reflecting the organization's commitment to environmental sustainability.

Techniques for selecting suitable implementation methods and promoting workplace environment and sustainability policy to key personnel and stakeholders are outlined. Effective communication and promotion are essential for generating awareness and fostering buy-in.

Strategies for informing work teams of policy changes, developing workplace environment and sustainability procedures, and establishing recordkeeping systems are addressed. Clear communication and robust recordkeeping are vital for ensuring accountability and tracking progress.

The process of investigating policy outcomes, monitoring records to identify trends requiring remedial action, and promoting continuous improvement of performance is discussed. This iterative approach allows for adaptive management and optimization of initiatives over time.

Procedures for modifying policies and procedures to incorporate improvements based on outcomes and providing feedback to key personnel and stakeholders are elucidated. This feedback loop ensures that policies remain responsive to changing circumstances and stakeholder needs.

Overall, this section provides a roadmap for effectively implementing and managing environment and sustainability policies within organizations.

Chapter Eleven

Developing Workplace Environment and Sustainability Policy

Environment and Sustainability Policies and Procedures

A sustainability policy serves as an internal document that articulates a company's stance on managing and enhancing environmental, social, and governance (ESG) issues. Typically, these policies amalgamate guidelines, procedures, rules, and expectations for both employees and the organization as a whole. Consequently, the policy acts as both a proactive framework for enhancing performance and a reactive guide for addressing sustainability-related challenges as they arise (Caush, 2024).

Prioritizing the development of a sustainability policy is crucial for several compelling reasons (Caush, 2024):

1. Legal Requirements: In numerous countries globally, Environmental, Social, and Governance (ESG) disclosures are legally mandated. With approximately 25 countries enforcing these disclosures, it's highly likely that your company falls under such regulatory requirements. A sustainability policy ensures compliance with these laws and regulations, mitigating the risk of legal repercussions.

2. Performance Improvement: A sustainability policy acts as a pivotal tool for setting and achieving ESG-related goals. By clearly delineating specific procedures and expectations, companies can ensure alignment and consistency in their sustainability efforts. This proactive approach enhances the company's performance in addressing environmental, social, and governance issues, leading to long-term sustainability and growth.

3. Risk Management: ESG issues pose significant risks to businesses, ranging from reputational damage to financial losses. A comprehensive sustainability policy enables companies to identify, manage, and mitigate these risks effectively. By implementing proactive measures and robust protocols outlined in the policy, companies can minimize their exposure to various risks associated with environmental, social, and governance factors.

4. Investor Appeal: Investors increasingly scrutinize ESG factors when making investment decisions. Companies with robust sustainability policies demonstrate their commitment to responsible business practices and long-term sustainability. Consequently, such companies may be more appealing to potential investors, leading to increased investment opportunities, enhanced shareholder confidence, and improved access to capital.

When drafting a sustainability policy, adhering to several best practices is essential to ensure its effectiveness and successful implementation. Firstly, seeking input from key stakeholders is crucial. Engaging with senior management, employees, investors, customers, and suppliers allows for gathering diverse perspectives and ensuring buy-in throughout the organization.

Next, it's vital to define your company's approach to sustainability. This involves articulating your organization's values, principles, and goals regarding Environmental, Social, and Governance (ESG) issues. Synthesizing stakeholder input, legal requirements, and industry standards helps create a comprehensive framework for sustainability (Caush, 2024).

Identifying specific goals and objectives is another critical step. Establishing SMART (specific, measurable, achievable, relevant, and time-bound) goals aligned with your company's sustainability approach provides clarity and direction for the policy's implementation.

Clear delineation of procedures and responsibilities is essential to ensure everyone understands their role in implementing the policy and achieving the established goals.

Developing a timeline for implementation is crucial for effective planning and execution. Establishing milestones and deadlines allows for periodic assessment and adjustment, ensuring progress towards sustainability objectives.

Regular review and updating of the policy are necessary to ensure its relevance and effectiveness over time. Considering changes in industry trends, legal requirements, and company objectives helps keep the policy aligned with evolving circumstances.

Communication of the policy to all employees is vital for fostering understanding and compliance. Disseminating the policy and emphasizing the importance of adherence and the consequences of non-compliance promotes organizational alignment.

Comprehensive training for employees on the policy is essential for successful implementation. Providing instruction on key aspects of the policy, its objectives, procedures, and compliance requirements ensures that everyone is equipped to contribute effectively to sustainability efforts.

Execution of the policy involves implementing procedures and monitoring compliance. Effective enforcement ensures that sustainability practices are integrated into daily operations across the organization.

Lastly, evaluating the effectiveness of the policy is crucial for assessing progress and identifying areas for improvement. Utilizing qualitative and quantitative measures, such as employee feedback, compliance rates, costs and benefits analysis, and achievement of goals and objectives, helps gauge the policy's impact and inform future decision-making (Caush, 2024).

An environmental policy serves as a written declaration, typically endorsed by senior management, outlining a business's objectives and principles concerning the management of its environmental impacts and aspects across its operations. Having an environmental policy is paramount, especially when seeking to implement an environmental management standard such as ISO 14001. It is also crucial for businesses currently engaged with, or intending to engage with, larger organizations, or those aiming to demonstrate to customers and other stakeholders their commitment to responsible environmental management practices (Chamber of Commerce of Metropolitan Montreal, 2023a).

An environmental policy lays the groundwork for environmental improvements within a business, as defined by senior management, by establishing key objectives and principles. Embracing an environmental policy can yield substantial benefits for a business (Chamber of Commerce of Metropolitan Montreal, 2023a):

- Ensuring legal compliance

- Enhancing employee awareness of environmental roles and responsibilities

- Improving cost management

- Mitigating incidents that may lead to liability

- Conserving raw materials and energy

- Enhancing monitoring of environmental impacts

- Optimizing process efficiency

However, the benefits extend beyond internal operations. By showcasing a commitment to environmental management, businesses can cultivate positive relationships with external stakeholders, including investors, insurers, customers, suppliers, regulators, and the local community. This can, in turn, enhance corporate reputation and yield financial benefits such as increased investments, customer sales, and market share.

It is important to note that merely having an environmental policy in place may not yield these benefits. Implementing an environmental management system (EMS) necessitates the systematic delivery of the policy in a strategic manner. Certification of the EMS by external bodies can substantiate to customers, investors, regulators, and other stakeholders that the environmental claims outlined in the policy are credible, reliable, and independently verified (Chamber of Commerce of Metropolitan Montreal, 2023a). In cases where a formal EMS is not established, it is advisable to at least apply some of its principles to ensure the effectiveness of the policy. This may involve assessing the environmental impact of the business, establishing relevant key performance indicators, setting objectives and targets, and regularly reviewing these components.

While there is no standardized format for drafting an environmental policy, careful planning is essential to maximize its effectiveness. Securing buy-in from management is critical, emphasizing key benefits such as cost reduction, improved risk management, and marketing.

Once management commitment is obtained, assessing the current state of environmental management within the business is advisable. This could entail compiling an environmental history, assessing impacts, and identifying risks.

Tailoring the environmental policy to reflect the unique characteristics and culture of the business is vital. Reviewing policies from other businesses can provide inspiration, but copying them should be avoided. Adherence to certain guidelines can enhance the effectiveness of the policy (Chamber of Commerce of Metropolitan Montreal, 2023a):

- Keeping the statement concise
- Ensuring readability and comprehension
- Ensuring realism, achievability, and relevance
- Demonstrating commitment and obtaining endorsements
- Making the policy readily accessible, such as on the company website
- Ensuring dissemination to new employees and suppliers

While there is no standardized content for an environmental policy, certain themes are typically addressed. The policy should be customized to align with the activities, priorities, and concerns most relevant to the business.

Prior to drafting the policy, conducting an assessment of the business's environmental impact and potential risks is advisable. Various assessment techniques can be employed to identify key issues pertinent to the business. The content of the policy should be informed by the results of this assessment, focusing on key criteria such as (Chamber of Commerce of Metropolitan Montreal, 2023a):

- Business mission and operational information
- Commitment to continual improvement
- Management of significant environmental impacts
- Expectations regarding external parties
- Compliance with relevant environmental legislation
- Employee education and training
- Progress monitoring and review

Additionally, addressing issues specific to the business, such as transportation, waste minimization, and energy efficiency, can enhance the policy's relevance and effectiveness.

Integration with other policies on health and safety, quality management, corporate social responsibility, or sustainability may also be considered to ensure coherence across organizational objectives (Chamber of Commerce of Metropolitan Montreal, 2023a).

The distinction between an Environment policy and procedures and a Sustainability policy and procedures predominantly revolves around their scope and emphasis.

Environment policies and procedures are chiefly concerned with addressing environmental matters and considerations within an organization. This encompasses elements such as pollution prevention, waste management, resource conservation, and adherence to environmental regulations.

The primary focus of environment policies and procedures is to minimize the adverse effects of an organization's operations on the natural environment. They often delineate specific actions and protocols aimed at ensuring environmental compliance and fostering responsible environmental stewardship.

On the other hand, Sustainability policies and procedures encompass a broader spectrum of factors that extend beyond purely environmental concerns. While environmental sustainability remains a pivotal aspect, sustainability policies also encompass social and economic dimensions. This includes aspects such as social equity, community engagement, ethical sourcing, economic viability, and long-term resilience.

Sustainability policies and procedures are designed to ensure that an organization's activities contribute positively to environmental, social, and economic well-being, both in the short and long term. They typically involve strategies aimed at achieving a harmonious balance between environmental protection, social responsibility, and economic prosperity, often guided by principles such as the triple bottom line (people, planet, profit).

Where Environment policy and procedures are primarily centred on addressing environmental issues and ensuring compliance, Sustainability policy and procedures adopt a more holistic approach, encompassing environmental, social, and economic dimensions with the overarching goal of achieving sustainable development over the long term.

Environment policy and procedures and Sustainability policy and procedures can be combined into an Environment and Sustainability policy and procedure. This amalgamation allows organizations to adopt a more comprehensive and integrated approach to managing their environmental, social, and economic impacts.

The combined policy and procedures address a broader range of concerns, including environmental, social, and economic aspects. This encompasses elements such as pollution prevention, waste management, resource conservation, social equity, community engagement, ethical sourcing, economic viability, and long-term resilience.

The primary focus is on promoting sustainable practices that not only minimize environmental harm but also foster social responsibility and economic viability. The policy and procedures aim to achieve a balance between environmental protection, social equity, and economic prosperity, guided by principles such as the triple bottom line (people, planet, profit).

Combining Environment policy and procedures with Sustainability policy and procedures into an Environment and Sustainability policy and procedure provides a more holistic framework for organizations to manage their operations and ensure they align with sustainable development goals. This integrated approach enables organizations to address environmental, social, and economic challenges in a coordinated manner, leading to more effective and impactful sustainability initiatives.

Developing workplace policies and procedures for environment and sustainability in agribusiness requires a number of steps as follows:

- Identify Legislative, Regulatory, and Industry Requirements: Begin by conducting comprehensive research to identify pertinent laws, regulations, and industry standards concerning environmental sustainability. This entails reviewing environmental protection laws, agricultural regulations, and specific guidelines applicable to the agribusiness sector. Pay close attention to requirements regarding resource utilization, waste management, pollution prevention, and other environmental considerations.

- Define the Scope of the Environment and Sustainability Policy: Once the legislative and regulatory requirements are identified, define the scope of the environment and sustainability policy. Outline the policy's objectives, goals, and areas of focus, taking into account factors such as resource conservation, waste reduction, energy efficiency, and sustainable practices throughout the agribusiness operations.

- Gather Information from a Range of Sources to Plan and Develop Policy: Collect information from diverse sources to inform the development of the environment and sustainability policy. This may involve consulting scientific research, industry reports, best practices from other organizations, and seeking input from internal stakeholders such as employees, managers, and environmental experts.

- Identify and Consult with Stakeholders as a Key Component of the Policy Development Process: Engage stakeholders throughout the policy development process to ensure their perspectives and concerns are considered. Identify key stakeholders, including employees, suppliers, government agencies, community groups, and industry associations, and seek their input through meetings, surveys, and consultations.

- Include Appropriate Strategies in Policy at All Stages of Work for Minimizing Resource Use and Employing Life Cycle Management Approaches: Integrate strategies into the policy to minimize resource use and adopt life cycle management approaches across all stages of work in the agribusiness. This may entail implementing practices such as efficient water and energy use, waste reduction and recycling, sustainable land management, and responsible sourcing of inputs.

- Identify and Incorporate Strategies to Reduce Hazardous Material Use Where Applicable: Identify areas where hazardous materials are utilized in agribusiness operations and develop strategies to mitigate their use. This might involve substituting hazardous materials with safer alternatives, implementing proper handling and storage procedures, and providing employee training on safe practices.

- Make Recommendations for Policy Options Based on Likely Effectiveness, Timeframes, and Cost: Evaluate different policy options based on their potential effectiveness in achieving environmental sustainability goals, as well as their timeframes for implementation and associated costs. Consider factors such as technological feasibility, regulatory compliance, resource availability, and potential benefits to the agribusiness.

- Develop Policy That Reflects the Organization's Commitment to Environment and Sustainability Work Practices: Craft a comprehensive environment and sustainability policy document that embodies the organization's dedication to environmental stewardship and sustainable business practices. Clearly articulate the organization's values, goals, and commitments to environmental sustainability, and outline specific policies and procedures to be implemented.

- Select Appropriate Methods of Implementation: Choose suitable methods for implementing the environment and sustainability policy within the organization. This may involve establishing procedures, assigning responsibilities, providing training and resources, setting targets and milestones, and establishing mechanisms for monitoring, reporting, and continuous improvement. Consider organizational structure, resources, and culture when determining the most effective implementation methods.

Legislative, Regulatory, and Industry Requirements

Identifying legislative, regulatory, and industry requirements concerning environmental sustainability is a crucial undertaking for any organization, particularly those in sectors with significant environmental impacts like agribusiness. Begin by conducting comprehensive research to pinpoint relevant laws, regulations, and industry standards pertaining to environmental sustainability. This research involves accessing various sources such as government websites, regulatory agencies, industry associations, and specialized publications. It's imperative to examine both national and local regulations since environmental laws may vary depending on jurisdiction.

Key attention should be given to environmental protection laws, which serve as the cornerstone of environmental regulations. These laws typically cover a wide array of environmental issues, including air and water quality, biodiversity conservation, hazardous waste management, and ecosystem protection. Examples include the Clean Air Act, Clean Water Act, Endangered Species Act, and Resource Conservation and Recovery Act (RCRA). The Clean Air Act, Clean Water Act, Endangered Species Act, and Resource Conservation and Recovery Act (RCRA) are all environmental laws from the United States. They were enacted by the U.S. Congress to address various environmental issues and protect natural resources within the country.

As other examples, in Canada, the Environmental Protection Act is a comprehensive law designed to prevent pollution, safeguard the environment, and conserve natural resources. Covering a broad spectrum of environmental concerns, this act addresses issues such as air and water quality standards, waste management practices, and the assessment of environmental impacts for proposed projects.

Malaysia's Environmental Quality Act is aimed at regulating activities that have the potential to cause pollution or degrade the environment. This legislation establishes stringent standards for air and water quality, waste disposal methods, and requires environmental impact assessments for development projects to mitigate adverse effects.

Within the European Union, the Water Framework Directive is a pivotal piece of legislation focused on attaining and maintaining high water quality standards across member states. This directive mandates effective monitoring and management of water resources, pollution prevention measures, and initiatives to restore degraded aquatic ecosystems.

Australia's Environmental Protection and Biodiversity Conservation Act serves as a cornerstone for environmental preservation and biodiversity conservation efforts within the country. It regulates activities that may impact the environment, including land clearing, management of wildlife trade, and oversight of development projects in ecologically sensitive areas.

Japan's Environmental Conservation Law addresses a wide array of environmental concerns, emphasizing pollution prevention, waste management, and the conservation of natural habitats and biodiversity. This legislation establishes stringent standards for environmental protection and mandates businesses to implement measures for pollution control and environmental stewardship.

These examples underscore the diverse approaches countries take in addressing environmental challenges, reflecting the unique ecological landscapes, priorities, and regulatory frameworks within their respective regions.

Additionally, specific agricultural regulations are pertinent within the context of agribusiness. These regulations may address agricultural practices, land use, pesticide usage, water management, and animal welfare. Examples include regulations set by agricultural departments, food safety agencies, and animal welfare organizations.

Furthermore, it's essential to review industry-specific guidelines and standards applicable to the agribusiness sector. These guidelines may be established by industry associations, trade organizations, or certification bodies. For instance, organizations like the Sustainable Agriculture Network (SAN) provide standards for sustainable agricultural practices.

The Sustainable Agriculture Network (SAN) serves as a global network headquartered in Costa Rica, dedicated to fostering sustainable agricultural practices on a worldwide scale. Through collaboration with a diverse array of stakeholders, SAN endeavours to promote agricultural methods that prioritize environmental sustainability, social responsibility, and economic viability. It's noteworthy that SAN's standards and certification schemes are not confined to Costa Rica or any specific region; instead, they are implemented internationally, reflecting the organization's commitment to global sustainability initiatives.

Furthermore, various organizations around the world share SAN's mission and work diligently to promote sustainable agriculture practices in their respective countries. One such example is the Rainforest Alliance, headquartered in the United States, which operates globally to safeguard biodiversity and uphold sustainable livelihoods by advocating for environmentally responsible agricultural practices.

In the United Kingdom, the Soil Association champions the cause of organic farming and sustainable land management. By advocating for practices that enhance soil health, preserve biodiversity, and bolster ecosystem resilience, the Soil Association contributes significantly to sustainable agriculture efforts within the UK.

Similarly, Australian Organic, based in Australia, focuses on promoting organic farming practices and certifying organic products. By endorsing organic agriculture, Australian Organic contributes to environmental sustainability and supports consumer health.

In Germany, Fairtrade International stands out for its commitment to empowering small-scale farmers and promoting sustainable agriculture. Through initiatives aimed at ensuring fair prices, decent working conditions, and environmental stewardship, Fairtrade International plays a pivotal role in advancing sustainability in agriculture.

Lastly, Organic Farmers & Growers, also based in the UK, specializes in certifying organic products and assisting farmers transitioning to organic farming methods. By facilitating the adoption of sustainable practices and promoting environmental conservation, Organic Farmers & Growers contributes to the global effort to foster sustainable agriculture practices.

Special emphasis should be placed on key focus areas within environmental sustainability, such as resource use, waste management, pollution prevention, and other environmental considerations relevant to agribusiness operations. This encompasses understanding requirements related to water usage, soil conservation, energy efficiency, greenhouse gas emissions, chemical management, and biodiversity conservation.

Identify requirements associated with resource use, including regulations governing water usage, land management practices, and energy consumption. Understanding any restrictions, permits, or reporting obligations associated with these resources is crucial to ensure compliance.

Review waste management regulations to comprehend requirements for handling, storing, transporting, treating, and disposing of agricultural waste, including animal waste, crop residues, and packaging materials. Compliance with waste disposal regulations is critical to minimize environmental impacts and avoid potential legal liabilities.

Finally, investigate pollution prevention measures mandated by regulatory authorities to mitigate environmental pollution stemming from agribusiness activities. This may encompass regulations on air emissions, water pollution, soil contamination, and chemical spills. Implementing pollution prevention measures not only ensures compliance but also promotes sustainable business practices.

By adhering to these steps and conducting thorough research, agribusinesses can effectively identify and comprehend the legislative, regulatory, and industry requirements concerning environmental sustainability. This knowledge lays the groundwork for developing effective environmental management strategies and ensuring compliance with relevant laws and regulations.

Policy and Procedure Development

Defining the scope of an environment and sustainability policy is a critical step in the development of a comprehensive framework that guides an organization's commitment to environmental stewardship and sustainable business practices. This process entails a thorough examination of the legislative and regulatory requirements relevant to the organization's operations, followed by a meticulous delineation of the policy's objectives, goals, and areas of focus.

Once the legislative and regulatory requirements pertaining to environmental sustainability are identified, organizations must proceed to define the scope of their environment and sustainability policy. This involves outlining the overarching objectives and goals that the policy aims to achieve, which serve as guiding principles for the organization's environmental initiatives. These objectives typically encompass a wide range of environmental considerations, such as resource conservation, waste reduction, energy efficiency, and sustainable practices across all facets of agribusiness operations.

Resource conservation stands as a fundamental aspect of any environment and sustainability policy, emphasizing the prudent and responsible utilization of natural resources to minimize environmental impact. This may entail strategies to optimize water usage, reduce reliance on finite resources, and adopt sustainable land management practices that preserve ecosystems and biodiversity. By outlining specific objectives related to resource conservation, organizations can prioritize efforts to minimize their ecological footprint and promote long-term sustainability.

Waste reduction represents another critical component of the policy's scope, focusing on minimizing the generation of waste and maximizing resource efficiency throughout agribusiness operations. This may involve implementing practices to reduce packaging waste, recycle agricultural by-products, and optimize production processes to minimize waste generation. By setting clear goals for waste reduction, organizations can enhance operational efficiency, reduce environmental pollution, and contribute to the circular economy by promoting resource reuse and recycling.

Energy efficiency emerges as a key priority within the scope of the environment and sustainability policy, underscoring the importance of minimizing energy consumption and optimizing energy use to mitigate greenhouse gas emissions and combat climate change. Organizations may set objectives to improve energy efficiency in agricultural processes, invest in renewable energy sources, and implement energy-saving technologies to reduce their carbon footprint. By prioritizing energy efficiency, organizations can not only reduce operational costs but also enhance environmental performance and contribute to climate resilience.

Moreover, sustainable practices encompass a broad spectrum of initiatives aimed at fostering environmentally responsible behaviour and promoting sustainable development within the agribusiness sector. This may include adopting sustainable agricultural practices, promoting biodiversity conservation, and ensuring ethical sourcing of inputs to minimize environmental and social impacts. By integrating sustainability principles into all aspects of agribusiness operations, organizations can uphold their commitment to environmental stewardship, social responsibility, and economic viability.

Defining the scope of an environment and sustainability policy involves a comprehensive assessment of legislative and regulatory requirements, followed by the delineation of objectives, goals, and areas of focus that guide the organization's environmental initiatives. By prioritizing resource conservation, waste reduction, energy efficiency, and sustainable practices, organizations can establish a robust framework for promoting environmental sustainability and fostering responsible business practices within the agribusiness sector.

The following are examples of the scope of an environmental and sustainability policy:

1. **Resource Conservation:** The policy aims to optimize the use of natural resources such as water, land, and energy to minimize environmental impact. Specific objectives may include reducing water usage through efficient irrigation methods, implementing sustainable land management practices to preserve ecosystems and biodiversity, and transitioning to renewable energy sources to decrease reliance on fossil fuels.

2. **Waste Reduction:** Another key focus area is waste reduction, with the policy targeting the minimization of waste generation and the promotion of resource efficiency throughout agribusiness operations. Objectives within this scope may include reducing packaging waste by adopting eco-friendly packaging materials, implementing recycling programs for agricultural by-products such as crop residues and animal waste, and optimizing production processes to minimize waste generation at source.

3. **Energy Efficiency:** The policy prioritizes energy efficiency as a means of mitigating greenhouse gas emissions and combating climate change. Objectives in this area may include improving energy efficiency in agricultural processes by upgrading equipment and machinery to energy-efficient models, investing in renewable energy sources such as solar and wind power to meet energy needs, and implementing energy-saving technologies such as LED lighting and automated systems to reduce energy consumption.

4. **Sustainable Practices:** Sustainable practices encompass a wide range of initiatives aimed at fostering environmentally responsible behaviour and promoting sustainable development within the agribusiness sector. This includes adopting sustainable agricultural practices such as crop rotation and organic farming methods to minimize chemical inputs and soil degradation, promoting biodiversity conservation through habitat restoration and wildlife-friendly farming practices, and ensuring ethical sourcing of inputs such as seeds and fertilizers to minimize environmental and social impacts along the supply chain.

Overall, the scope of the environment and sustainability policy encompasses objectives, goals, and areas of focus related to resource conservation, waste reduction, energy efficiency, and sustainable practices within agribusiness operations. By prioritizing these aspects, organizations can establish a robust framework for promoting environmental sustainability, fostering responsible business practices, and contributing to long-term economic viability within the agribusiness sector.

As a further example of the scope of an environmental and sustainability policy, the following scope could apply to a previously used example of a hypothetical vertical farming operation specializing in arugula, butterhead lettuce, red leaf lettuce, and romaine lettuce:

1. **Resource Conservation:** Our environmental and sustainability policy prioritizes the efficient use of resources to minimize environmental impact. We aim to optimize water usage by implementing advanced irrigation systems, such as drip irrigation or hydroponics, to reduce water consumption per unit of produce grown. Additionally, we commit to sustainable land management practices, such as soilless farming techniques utilized in vertical farming, which minimize land usage while maximizing crop yield.

2. **Waste Reduction:** We are dedicated to minimizing waste generation throughout our vertical farming operations. Our policy includes initiatives to reduce packaging waste by utilizing eco-friendly packaging materials that are biodegradable or recyclable. We also implement closed-loop systems to recycle organic waste generated during the cultivation process, such as crop residues and unused plant matter, back into nutrient-rich compost for soil enrichment or bioenergy production.

3. **Energy Efficiency:** Energy efficiency is a key focus area in our sustainability efforts. We invest in energy-efficient LED lighting systems and utilize renewable energy sources, such as solar panels or wind turbines, to power our vertical farming facilities. Additionally, we employ smart automation and control systems to optimize energy usage, including temperature control, ventilation, and lighting schedules, to minimize electricity consumption and reduce our carbon footprint.

4. **Sustainable Practices:** Our environmental and sustainability policy encompasses a range of sustainable farming practices aimed at promoting ecosystem health and biodiversity conservation. We prioritize organic farming methods, free from synthetic pesticides and fertilizers, to minimize chemical runoff and soil contamination. We also implement integrated pest management (IPM) strategies to control pests and diseases using natural predators and biological controls, reducing reliance on chemical pesticides.

5. **Ethical Sourcing and Social Responsibility:** We are committed to ethical sourcing practices and social responsibility throughout our supply chain. We partner with local suppliers and farmers who share our values of environmental sustainability and fair labour practices. Additionally, we prioritize community engagement and education initiatives to promote environmental awareness and support local communities where our vertical farming facilities are located.

By defining the scope of our environmental and sustainability policy in these areas, we aim to uphold our commitment to environmental stewardship, promote sustainable agriculture practices, and contribute to the well-being of our planet and local communities.

Gathering information from various sources is pivotal in the planning and development of an environment and sustainability policy. Initiate the process by conducting an extensive review of scientific research pertinent to environmental sustainability within your industry or sector. This involves delving into peer-reviewed journals, academic publications, and scientific databases to procure data and insights on environmental issues, trends, and potential solutions. Scientific research serves as a robust foundation of evidence-based knowledge that can significantly inform policy development and decision-making processes.

Dive into industry-specific reports, studies, and publications to garner insights into the environmental challenges and opportunities prevalent within your sector. These reports often furnish valuable data on environmental trends, regulatory advancements, emerging technologies, and best practices embraced by leading organizations. By scrutinizing industry reports, you can identify sector-specific risks, benchmarks, and areas necessitating improvement, thereby shaping the development of your environment and sustainability policy.

Harness the experiences and successes of other organizations by examining their environmental policies, initiatives, and sustainability practices. Seek out case studies, success stories, and benchmarking studies that spotlight effective strategies for environmental management, resource conservation, and sustainable business practices. Analysing best practices from other organizations provides invaluable inspiration and guidance for refining your environment and sustainability policy.

Engage with internal stakeholders, including employees, managers, and environmental experts within your organization, to glean insights, perspectives, and recommendations pertinent to policy development. Utilize various methods such as interviews, focus groups, or surveys to solicit feedback on environmental priorities, challenges, and opportunities. Internal stakeholders possess firsthand knowledge of operational processes, organizational culture, and environmental concerns, which can significantly inform the crafting of targeted and effective policies.

Collaborate with external stakeholders, including customers, suppliers, regulators, and community groups, to gather diverse perspectives and input on environmental sustainability issues. Foster dialogue through consultation forums, stakeholder meetings, or other engagement mechanisms to grasp stakeholder expectations, concerns, and priorities concerning environmental management. External stakeholders offer valuable insights into the broader social, economic, and environmental context, facilitating the alignment of your environment and sustainability policy with external expectations and stakeholder interests.

Engaging stakeholders throughout the policy development process is essential to ensure that their perspectives and concerns are properly addressed. Firstly, identify key stakeholders who have a vested interest or may be impacted by the environment and sustainability policy. These stakeholders may include employees, suppliers, government agencies, community groups, industry associations, customers, investors, and regulatory bodies. Recognizing the diversity of stakeholder groups is crucial as each may offer unique insights and perspectives on environmental sustainability initiatives.

Conduct a thorough stakeholder analysis to understand the interests, influence, and significance of each stakeholder group. This analysis helps prioritize stakeholder engagement efforts and tailor communication strategies to effectively engage with each group. Factors such as the level of influence, expertise, proximity to the organization, and potential impact on the policy development process should be considered.

Initiate the engagement process by reaching out to key stakeholders through various communication channels. This may involve holding stakeholder meetings, workshops, focus groups, webinars, or consultation sessions to facilitate dialogue and exchange of ideas. Additionally, utilize surveys, questionnaires, and feedback mechanisms to gather input from stakeholders who may not be able to participate in face-to-face meetings.

Encourage stakeholders to provide input, feedback, and suggestions throughout the policy development process. Actively listen to their perspectives, concerns, and recommendations, ensuring that their voices are heard and respected. Creating a supportive and inclusive environment where stakeholders feel comfortable expressing their views is essential for meaningful contribution to the policy formulation process.

Address the concerns, priorities, and expectations raised by stakeholders during the consultation process. Analyse the feedback received and incorporate relevant insights into the policy development framework. Strive to address stakeholders' concerns and accommodate their interests while maintaining the integrity and effectiveness of the policy.

Maintain transparency and open communication with stakeholders by providing regular updates on the policy development process. Keeping stakeholders informed about key milestones, decisions, and developments, and soliciting additional feedback as needed, builds trust and credibility, enhancing stakeholder engagement and buy-in for the finalized policy.

Cultivate positive relationships and partnerships with stakeholders to foster ongoing collaboration and support for environmental sustainability initiatives. Recognize and acknowledge stakeholders' contributions and involvement in the policy development process, exploring opportunities for future collaboration on implementation, monitoring, and evaluation of the policy.

Incorporating strategies into a policy to minimize resource use and employ life cycle management approaches throughout all stages of work in agribusiness is paramount for promoting sustainability and mitigating environmental impact. Initiate the process by conducting a comprehensive assessment of the agribusiness operations to pinpoint opportunities for resource minimization. This entails scrutinizing processes, inputs, and outputs across all production stages, from cultivation to distribution, to identify where resources like water, energy, and raw materials can be conserved or optimized.

Based on the assessment findings, establish specific objectives and targets for resource minimization within the policy framework. These objectives should be SMART (specific, measurable, achievable, relevant, and time-bound) and in alignment with the organization's overarching goals. For instance, objectives could revolve around reducing water consumption by a specific percentage or optimizing energy use to lower greenhouse gas emissions.

Incorporate strategies for efficient water and energy use into the policy across all stages of work. This may encompass investing in water-efficient irrigation systems, adopting energy-saving technologies, and implementing maintenance protocols to curtail energy wastage. Develop protocols for monitoring and managing water and energy consumption to ensure adherence to targets.

Integrate waste reduction and recycling practices into the policy to minimize waste generation and facilitate resource recovery. Measures such as composting organic waste, recycling packaging materials, and embracing circular economy principles should be implemented. Set targets for waste diversion and recycling rates to monitor progress over time.

Address sustainable land management practices within the policy framework to safeguard soil health, biodiversity, and ecosystem resilience. This involves implementing conservation tillage techniques, employing cover crops, and practicing crop rotation to prevent soil erosion and degradation. Incorporate soil testing and monitoring protocols to evaluate soil health and guide land management decisions.

Ensure responsible sourcing of inputs such as seeds, fertilizers, and pesticides by including provisions within the policy framework. Encourage the use of sustainable agricultural inputs that minimize environmental impact and uphold social responsibility. Establish criteria for evaluating suppliers based on sustainability credentials like environmental certifications and adherence to fair labour practices.

Embed life cycle management approaches into the policy to assess and mitigate environmental impacts throughout the entire life cycle of products and processes. Consider the environmental implications of production, distribution, use, and disposal stages, and identify opportunities for enhancement. Conduct life cycle assessments (LCAs) to quantify environmental impacts and inform decision-making.

In the context of a hypothetical vertical farming operation specializing in arugula, butterhead lettuce, red leaf lettuce, and romaine lettuce, embedding life cycle management approaches into the policy involves assessing and mitigating environmental impacts across all stages of the products' life cycles. Let's break down how this might apply to each stage:

1. **Production Stage**:

   ○ Assessing Environmental Impacts: Evaluate the environmental implications of the production process, including factors such as water and energy consumption, use of fertilizers and pesticides, and greenhouse gas emissions associated with indoor farming operations.

   ○ Example: Conducting a life cycle assessment (LCA) to quantify the environmental impacts of different cultivation methods (e.g., hydroponics vs. aeroponics) and identify areas where resource efficiency can be improved.

2. **Distribution Stage**:

   ○ Assessing Environmental Impacts: Consider the environmental impacts associated with transportation, packaging materials, and storage practices during the distribution of the harvested lettuce products.

   ○ Example: Analysing the carbon footprint of transportation methods (e.g., trucking vs. local distribution) and exploring opportunities to reduce emissions through efficient route planning or utilizing renewable energy sources for transportation.

3. **Use Stage**:

   ○ Assessing Environmental Impacts: Evaluate how consumers use the lettuce products, including factors such as food waste generation, packaging disposal, and energy consumption during food preparation.

   ○ Example: Conducting surveys or interviews to understand consumer behaviour regarding lettuce storage and consumption habits, then implementing educational campaigns to promote optimal storage practices and reduce food waste.

4. **Disposal Stage**:

- ○ Assessing Environmental Impacts: Consider the environmental consequences of disposing of lettuce waste, including organic matter decomposition and packaging disposal.

- ○ Example: Implementing composting programs for organic lettuce waste to divert it from landfills and reduce methane emissions, or using biodegradable packaging materials to minimize environmental impact.

By conducting life cycle assessments (LCAs) at each stage of the products' life cycles, the vertical farming operation can quantify environmental impacts such as carbon emissions, water usage, and waste generation. These assessments provide valuable data to inform decision-making and identify opportunities for improvement, such as optimizing production methods, reducing transportation emissions, minimizing food waste, and implementing sustainable packaging solutions. Ultimately, integrating life cycle management approaches into the policy helps the vertical farming operation minimize its environmental footprint and promote sustainability across all stages of its operations.

Provide employees with adequate training and capacity building opportunities to effectively implement the strategies outlined in the policy. Offer training programs on resource conservation practices, waste reduction techniques, and sustainable land management principles to empower employees with the requisite knowledge and skills.

Establish monitoring and evaluation mechanisms to track progress towards resource minimization targets and identify areas for improvement. Collect pertinent data on resource consumption, waste generation, and environmental performance indicators to assess policy effectiveness. Utilize this information to make informed decisions, tweak strategies as necessary, and continually enhance performance over time.

Incorporating strategies to diminish hazardous material usage in agribusiness operations is pivotal for mitigating environmental and health hazards. This includes:

- Commence with a Hazardous Material Inventory: Conduct an exhaustive inventory encompassing all hazardous materials employed in agribusiness operations, ranging from pesticides and herbicides to fertilizers and cleaning agents. Thorough documentation of the quantity, type, and purpose of each hazardous material is imperative.

- Evaluate Risks and Impacts: Assess the potential risks and environmental ramifications associated with hazardous material utilization. Factors like toxicity, persistence, bioaccumulation, and groundwater contamination potential should be carefully weighed. Identify areas where these materials pose the most significant threats to human health, wildlife, soil quality, water resources, and ecosystem integrity.

- Explore Safer Alternatives: Undertake research to pinpoint safer alternatives to hazardous materials presently utilized. Seek out environmentally friendly pesticides, organic fertilizers, and non-toxic cleaning products that deliver comparable outcomes without jeopardizing human health or the environment. Consider factors such as efficacy, cost-effectiveness, and regulatory compliance in the selection process.

- Prioritize Substitution and Reduction: Develop strategies aimed at substituting hazardous materials with safer alternatives wherever feasible. Prioritize the elimination or reduction of high-risk substances and emphasize practices that curtail reliance on hazardous materials. Integrated pest management (IPM) techniques, for instance, can diminish the need for chemical pesticides by integrating biological controls, crop rotation, and habitat manipulation.

- Implement Stringent Handling and Storage Protocols: Establish rigorous protocols governing the handling, storage, and disposal of hazardous materials to mitigate the risk of accidents, spills, and contamination. Offer comprehensive training to employees on safe handling practices, proper utilization of personal protective equipment (PPE), emergency response protocols, and appropriate disposal methods. Employ engineering controls such as ventilation systems, spill containment measures, and secure storage facilities to pre-empt accidental releases.

- Invest in Training and Education: Allocate resources to comprehensive training programs designed to educate employees about the hazards associated with hazardous materials and foster awareness of safer alternatives and best practices. Regular refresher training sessions should be offered to reinforce safe handling procedures and ensure regulatory compliance. Cultivate an environment conducive to open communication and encourage the reporting of safety concerns or incidents to facilitate continuous improvement.

- Monitor and Evaluate Performance: Institute robust monitoring and evaluation mechanisms to track progress in reducing hazardous material use and gauge the effectiveness of implemented strategies. Key performance indicators such as hazardous material volume, incidents of exposure or spills, and adherence to safety protocols should be closely monitored. Regular audits and inspections must be conducted to pinpoint areas for improvement and address any emerging risks or issues promptly.

- Pursue Continuous Improvement: Foster a culture centred on continuous improvement by actively soliciting feedback from employees, stakeholders, and regulatory authorities. Promote innovation and collaboration to identify fresh opportunities for reducing hazardous material use and enhancing environmental sustainability. Stay abreast of technological advancements, regulatory requirements, and best practices to adapt and refine strategies over time.

When making recommendations for policy options, a comprehensive evaluation encompassing effectiveness, timeframes, and costs is imperative. Here's a detailed guide on how to navigate this process:

Firstly, evaluate effectiveness by assessing the potential impact of each policy option on environmental sustainability goals. Analyse factors such as resource conservation, waste reduction, and energy efficiency, drawing insights from research studies and best practices. Consider scalability and alignment with organizational objectives.

Next, assess timeframes associated with implementation, considering the planning, approval, and monitoring phases. Prioritize options capable of delivering timely results without compromising long-term sustainability objectives. Some measures may offer immediate benefits, while others require longer-term investments.

Analysing costs is crucial. Evaluate upfront investment requirements, ongoing operational expenses, and potential cost savings or benefits. Assess the financial viability of each option and calculate the return on investment (ROI) concerning environmental benefits achieved. Seek cost-effective solutions that maximize impact while minimizing financial strain.

Evaluate technological feasibility by considering available technologies, infrastructure requirements, and compatibility with existing systems. Assess the maturity of technologies and potential barriers to adoption such as complexity or interoperability issues. Ensure alignment with organizational capabilities and resources.

Regulatory compliance is paramount. Consider the implications of each policy option on relevant laws, regulations, and industry standards. Assess the risks associated with non-compliance and prioritize options that align with regulatory requirements while advancing environmental sustainability. Engage with regulatory authorities for guidance and support.

Evaluate resource availability including financial, human, and technological resources required for implementation. Assess the organization's capacity to mobilize resources internally or secure external funding. Identify constraints and explore strategies such as partnerships or phased implementation to overcome limitations.

Consider the potential benefits of each policy option to the agribusiness including operational efficiency, risk reduction, and market competitiveness. Evaluate opportunities for innovation, differentiation, and value creation through sustainable practices. Assess broader societal and environmental benefits including community well-being and climate resilience.

Lastly, prioritize options based on effectiveness, timeframes, and costs. Consider trade-offs and synergies between different options, aiming to identify strategies offering the greatest overall value and impact. Engage stakeholders throughout the decision-making process to ensure alignment with organizational goals and priorities.

By systematically evaluating policy options across these dimensions, agribusinesses can make well-informed recommendations that advance environmental sustainability while considering practical constraints and opportunities for implementation.

Developing a policy that reflects an organization's commitment to environmental stewardship and sustainable business practices is a multifaceted endeavour requiring careful consideration and strategic planning. Begin by clearly defining the organization's values and goals regarding environmental sustainability. This involves articulating the organization's overarching commitment to environmental stewardship and its vision for integrating sustainability into all aspects of its operations. These values and goals serve as the foundation upon which the policy will be built, guiding decision-making and actions.

Engage with internal and external stakeholders to gather input and perspectives on environmental sustainability. This includes employees, management, customers, suppliers, regulatory agencies, and community groups. Solicit feedback on their expectations, concerns, and priorities related to environmental practices. This stakeholder engagement process ensures that the policy reflects diverse perspectives and fosters buy-in and support from key stakeholders.

Assess the organization's environmental footprint and identify key areas where improvements can be made. This may include resource conservation, waste reduction, energy efficiency, pollution prevention, and sustainable sourcing practices. Prioritize areas for action based on their significance to the organization's operations and their potential impact on environmental sustainability.

Based on the organization's values, goals, and identified areas for action, develop specific policies and procedures to be implemented. These policies should outline clear guidelines, protocols, and standards for environmental practices across all aspects of the organization's operations. Address key areas such as water and energy use, waste management, chemical handling, land management, and emissions reduction.

Ensure that the policy aligns with relevant legal and regulatory requirements governing environmental sustainability. This may include compliance with environmental protection laws, industry regulations, and certification standards. Incorporate provisions that demonstrate the organization's commitment to meeting or exceeding these requirements and maintaining regulatory compliance.

Establish measurable targets and objectives to track progress towards environmental sustainability goals. Define specific metrics, such as reduction targets for greenhouse gas emissions, water usage, or waste generation, and set timelines for achieving these targets. Incorporate mechanisms for monitoring, reporting, and reviewing progress on a regular basis to ensure accountability and transparency.

Implement training programs to educate employees about the organization's environment and sustainability policy, as well as their roles and responsibilities in implementing it. Offer training on relevant environmental practices, procedures, and compliance requirements. Build organizational capacity by empowering employees with the knowledge and skills needed to contribute to environmental sustainability efforts.

Communicate the environment and sustainability policy to all stakeholders, both internally and externally. Develop communication strategies to raise awareness of the policy's objectives, expectations, and benefits. Emphasize the importance of environmental stewardship and encourage stakeholders to actively participate in implementing the policy. Promote the organization's commitment to sustainability through internal communications, external marketing efforts, and public relations initiatives.

Establish mechanisms for monitoring, evaluating, and continuously improving the effectiveness of the environment and sustainability policy. Implement regular audits, assessments, and performance reviews to track progress towards goals and identify areas for improvement. Solicit feedback from stakeholders and incorporate lessons learned into policy revisions and updates. Foster a culture of continuous improvement and innovation to drive ongoing progress towards environmental sustainability.

A sample Environment and Sustainability Policy follows.

Environment and Sustainability Policy

1. Introduction

Our organization is committed to environmental stewardship and sustainable business practices. This policy outlines our values, goals, and procedures to integrate sustainability into all aspects of our operations.

2. Organizational Values and Goals

We value environmental sustainability and are committed to minimizing our environmental footprint. Our goals include reducing resource consumption, waste generation, and greenhouse gas emissions while promoting sustainable practices throughout our operations.

3. Stakeholder Engagement Process

We engage with internal and external stakeholders to gather input and perspectives on environmental sustainability. Stakeholders include employees, management, customers, suppliers, regulatory agencies, and community groups. Feedback from stakeholders informs our environmental policies and practices.

4. Key Areas for Action

We assess our environmental footprint and prioritize areas for action, including resource conservation, waste reduction, energy efficiency, pollution prevention, and sustainable sourcing practices.

5. Specific Policies and Procedures

We develop specific policies and procedures to be implemented across all aspects of our operations. These policies address water and energy use, waste management, chemical handling, land management, emissions reduction, and other key areas of environmental concern.

6. Legal and Regulatory Compliance

Our policies align with relevant legal and regulatory requirements governing environmental sustainability. We ensure compliance with environmental protection laws, industry regulations, and certification standards.

7. Measurable Targets and Objectives

We establish measurable targets and objectives to track progress towards environmental sustainability goals. Metrics include reduction targets for greenhouse gas emissions, water usage, waste generation, and other key performance indicators.

8. Training and Capacity Building

We implement training programs to educate employees about our environment and sustainability policy. Training covers relevant environmental practices, procedures, and compliance requirements to empower employees to contribute to our sustainability efforts.

9. Communication and Promotion

We communicate our environment and sustainability policy to all stakeholders through internal and external channels. We raise awareness of our objectives, expectations, and benefits and encourage stakeholders to actively participate in implementing the policy.

10. Monitoring, Evaluation, and Continuous Improvement

We establish mechanisms for monitoring, evaluating, and continuously improving the effectiveness of our environment and sustainability policy. Regular audits, assessments, and performance reviews track progress towards goals and identify areas for improvement. Feedback from stakeholders informs policy revisions and updates to drive ongoing progress towards environmental sustainability.

Implementing the environment and sustainability policy within an organization necessitates the selection of suitable methods to ensure the effective realization of the policy's objectives. The following provides a breakdown of how to choose and execute these methods.

Firstly, it's crucial to establish clear and detailed procedures delineating step-by-step instructions for implementing the environment and sustainability policy. These procedures should comprehensively cover various aspects such as resource conservation, waste management, energy efficiency, and sustainable sourcing practices. By providing employees with clear guidance, these procedures ensure tasks are executed in alignment with the policy.

Secondly, clearly defining roles and responsibilities for implementing the environment and sustainability policy is essential. Assigning specific tasks to individuals or teams within the organization fosters accountability for achieving objectives. This clarity in responsibilities ensures everyone comprehends their role in policy implementation, preventing confusion or overlap.

Offering training programs and resources is the next step. Educating employees about the environment and sustainability policy equips them with the knowledge and skills necessary for effective implementation. Training should encompass environmental practices, procedures, compliance requirements, and the utilization of tools or technologies supporting sustainability efforts. Adequate resources ensure employees have the necessary tools, equipment, and support to fulfill their responsibilities.

Setting measurable targets, goals, and milestones is crucial for tracking progress towards achieving environmental sustainability objectives. Defining specific metrics and timelines for targets related to resource conservation, waste reduction, energy efficiency, and other key performance indicators provides a roadmap for implementation and facilitates progress monitoring over time.

Establishing mechanisms for monitoring, reporting, and continuous improvement is imperative. These mechanisms track progress, identify areas for enhancement, and make necessary adjustments to the implementation process. Regular monitoring and reporting procedures track key performance indicators and assess compliance with the policy. Processes for collecting feedback from stakeholders, evaluating performance, and identifying opportunities for improvement foster a culture of continuous improvement.

Finally, considering the organization's structure, resources, and culture is vital when selecting implementation methods. Factors such as organizational size, availability of financial and human resources, and the organizational culture regarding sustainability should be taken into account. Choosing methods compatible with the organization's capabilities, constraints, and values ensures successful implementation.

Chapter Twelve

Communicating Workplace Environment and Sustainability Policy

From the perspective of developing workplace policies and procedures for environment and sustainability in agribusiness, effectively promoting these policies to key personnel and stakeholders is essential. Promoting workplace policies and procedures for environment and sustainability to key personnel and stakeholders is crucial for several reasons.

Firstly, it ensures understanding and compliance. By effectively promoting these policies, key personnel and stakeholders grasp their importance, rationale, and implications for daily operations. Clear communication clarifies expectations and responsibilities, reducing misunderstandings and facilitating compliance.

Secondly, effective promotion drives cultural change. Highlighting the significance of sustainable practices raises awareness and fosters a culture of environmental stewardship and responsibility among employees and stakeholders, contributing to long-term sustainability efforts.

Moreover, promoting these policies gains buy-in and support from key stakeholders. When stakeholders understand the value and benefits of sustainability initiatives, including senior management, employees, suppliers, customers, regulatory agencies, and community groups, they are more likely to actively support and participate in their implementation.

Furthermore, clear communication and promotion encourage engagement and participation. Involving employees and stakeholders in the development and implementation of these policies harnesses their ideas, expertise, and commitment to achieving sustainability goals.

Additionally, informed and engaged personnel and stakeholders lead to improved performance and impact. By aligning behaviours and practices with sustainability objectives, organizations can achieve tangible environmental benefits and contribute significantly to long-term sustainability.

Moreover, effective promotion maintains transparency and accountability within the organization. Openly communicating about sustainability goals, progress, and challenges demonstrates a commitment to transparency and accountability to stakeholders.

Lastly, promoting environment and sustainability policies effectively helps manage reputation and mitigate risks associated with environmental and social issues. A proactive approach to sustainability enhances reputation, builds trust with stakeholders, and reduces the likelihood of reputational damage due to environmental incidents or controversies.

Identifying key personnel and stakeholders is a fundamental step in the successful implementation of environment and sustainability policies within an agribusiness. These individuals and groups play pivotal roles in shaping and executing initiatives that drive environmental stewardship and sustainable practices. The process of identifying them involves a systematic approach to recognizing those who have a direct or indirect influence on policy implementation and outcomes.

Firstly, senior management stands as a cornerstone in driving organizational priorities, including sustainability initiatives. Their support and commitment are vital for allocating resources, setting strategic direction, and fostering a culture of sustainability throughout the organization. Identifying senior management ensures their involvement from the outset, facilitating decision-making and resource allocation processes.

Secondly, department heads and supervisors are key conduits for cascading policies and procedures down to frontline employees. Their understanding of operational challenges, resources, and constraints makes them instrumental in translating overarching sustainability goals into actionable plans at the departmental level. Identifying them enables targeted communication and engagement strategies tailored to their specific roles and responsibilities.

Frontline employees represent the workforce directly involved in day-to-day operations, making their buy-in and participation essential for effective policy implementation. Their firsthand insights and experiences can inform the identification of operational inefficiencies, opportunities for improvement, and potential barriers to adoption. Engaging frontline employees ensures their alignment with sustainability objectives and fosters a sense of ownership and accountability for implementing policies on the ground.

Suppliers and customers are external stakeholders whose practices and preferences can significantly impact an agribusiness's environmental footprint. Identifying them allows for collaboration and dialogue on sustainable sourcing, supply chain transparency, and product stewardship initiatives. Engaging suppliers and customers in discussions about sustainability expectations and requirements fosters mutually beneficial relationships and drives positive environmental outcomes across the value chain.

Regulatory agencies play a critical role in setting environmental standards, enforcing regulations, and monitoring compliance within the agribusiness sector. Identifying relevant regulatory bodies ensures proactive engagement and alignment with legal requirements, thereby minimizing compliance risks and ensuring adherence to environmental laws and regulations.

Community groups represent local stakeholders whose interests and concerns may intersect with the agribusiness's operations. Engaging with community groups fosters dialogue, builds trust, and strengthens social license to operate. Identifying community groups enables the agribusiness to address community needs, mitigate potential conflicts, and contribute positively to the socio-economic development of the local area.

Identifying key personnel and stakeholders is essential for effective implementation of environment and sustainability policies within an agribusiness. By understanding the roles, interests, and influence of these individuals and groups, organizations can tailor communication strategies, foster collaboration, and mobilize support to drive meaningful change towards environmental stewardship and sustainable practices.

Developing a communication plan is a crucial component of promoting workplace environment and sustainability policies to key personnel and stakeholders within an organization. A well-designed communication plan ensures that relevant information is effectively disseminated, stakeholders are engaged, and organizational objectives are communicated clearly and consistently.

Firstly, the communication plan should outline the key messages to be conveyed regarding the environment and sustainability policies. These messages should articulate the organization's commitment to sustainability, the rationale behind the policies, and the expected benefits for both the organization and its stakeholders. Clarity and consistency in messaging are essential to ensure that all stakeholders understand the purpose and significance of the policies.

Next, the communication plan should identify the target audience, including key personnel and stakeholders who have a vested interest in the implementation and outcomes of the environment and sustainability policies. This may include senior management, department heads, frontline employees, suppliers, customers, regulatory agencies, and community groups. Tailoring communication strategies to each audience segment ensures that messages resonate with their specific needs, interests, and concerns.

The communication plan should specify the communication channels to be utilized for promoting the environment and sustainability policies. These channels may include email newsletters, intranet announcements, staff meetings, training sessions, one-on-one discussions, posters, bulletin boards, and social media platforms. Leveraging a combination of channels allows for maximum reach and engagement, catering to diverse communication preferences and accessibility needs among stakeholders.

Determining the frequency and timing of communication is another important aspect of the communication plan. Regular and timely communication ensures that stakeholders are kept informed of policy updates, progress, and upcoming initiatives. Consider factors such as the urgency of the message, the availability of stakeholders, and the cadence of existing communication channels when determining the frequency and timing of communication.

In addition to formal communication channels, the communication plan should also incorporate opportunities for interactive dialogue and feedback. This may include hosting town hall meetings, focus groups, workshops, or webinars where stakeholders can ask questions, share feedback, and participate in discussions about the environment and sustainability policies. Creating a two-way communication flow fosters engagement, collaboration, and a sense of ownership among stakeholders.

Finally, the communication plan should outline mechanisms for monitoring and evaluating the effectiveness of communication efforts. This may involve tracking metrics such as message reach, engagement levels, feedback received, and stakeholder perceptions of the policies. Regularly reviewing and adjusting the communication plan based on feedback and performance data ensures continuous improvement and maximizes the impact of communication efforts.

A sample communication plan follows.

Communication Plan: Environment and Sustainability Policy

Objective: *The objective of this communication plan is to effectively promote and communicate our organization's Environment and Sustainability Policy to key personnel and stakeholders, fostering understanding, engagement, and support for our sustainability initiatives.*

1. Target Audience:

- *Internal: All employees, including senior management, department heads, supervisors, and frontline staff.*

- *External: Customers, suppliers, regulatory agencies, community groups, and other relevant stakeholders.*

2. Key Messages:

- *Our commitment to environmental stewardship and sustainable business practices.*

- *The importance of integrating sustainability into all aspects of our operations.*

- *The specific values, goals, and procedures outlined in our Environment and Sustainability Policy.*

- *The role of stakeholders in contributing to and benefiting from our sustainability efforts.*

3. Communication Channels:

- *Internal:*

  - *Email newsletters: Regular updates and announcements sent to all employees.*

  - *Intranet announcements: Posting the policy document and related materials on our company intranet.*

  - *Staff meetings: Presenting the policy and facilitating discussions during team meetings.*

  - *Training sessions: Conducting training sessions to educate employees about the policy and its implementation.*

- *External:*

  - *Customer communications: Including information about our sustainability initiatives in newsletters, invoices, and other customer-facing communications.*

  - *Supplier communications: Sharing the policy document and expectations with our suppliers and seeking their support.*

  - *Community engagement: Participating in community events and initiatives to raise awareness of our sustainability efforts.*

4. Frequency and Timing:

- *Internal communications: Regular updates and reminders about the policy will be communicated through email newsletters and intranet announcements on a monthly basis.*

- *Staff meetings: The policy will be discussed and reinforced during quarterly staff meetings to ensure alignment and understanding across departments.*

- *Training sessions: Initial training sessions will be conducted upon the release of the policy, followed by annual refresher sessions to reinforce key concepts and updates.*

5. Interactive Dialogue and Feedback:

- *Town hall meetings: Hosting quarterly town hall meetings to provide employees with the opportunity to ask questions, share feedback, and engage in discussions about the policy and its implementation.*

- *Feedback mechanisms: Establishing an anonymous feedback system to encourage employees to provide input, suggestions, and concerns related to the policy.*

6. Monitoring and Evaluation:

- *Tracking engagement metrics: Monitoring email open rates, intranet traffic, and attendance at staff meetings and training sessions to gauge employee engagement with the policy.*

- *Soliciting feedback: Conducting periodic surveys or focus groups to gather feedback from employees and stakeholders on their understanding, perception, and satisfaction with the policy communication efforts.*

7. Responsible Parties:

- *Internal Communications Team: Responsible for developing and disseminating internal communications materials, coordinating staff meetings, and organizing training sessions.*

- *Sustainability Committee: Responsible for overseeing the implementation of the policy, monitoring progress, and soliciting feedback from stakeholders.*

8. Timeline:

- *Initial rollout: Policy communication efforts will commence immediately upon the release of the policy document.*

- *Ongoing: Communication efforts will be ongoing, with regular updates and reinforcement scheduled throughout the year.*

9. Evaluation Criteria:

- *Employee awareness and understanding of the policy.*

- *Stakeholder engagement and support for sustainability initiatives.*

- *Feedback and suggestions received from employees and stakeholders.*

- *Progress towards achieving sustainability goals outlined in the policy.*

10. Continuous Improvement:

- *Regular review and refinement of communication strategies based on feedback and performance data.*

- *Collaboration with stakeholders to identify new opportunities for engagement and improvement in policy communication efforts.*

By following this communication plan, we aim to ensure that our Environment and Sustainability Policy is effectively communicated, understood, and embraced by all key personnel and stakeholders, driving meaningful change towards environmental stewardship and sustainable practices within our organization and beyond.

Crafting clear and compelling messaging is crucial when communicating environment and sustainability policies to key stakeholders. Begin by understanding the needs, interests, and concerns of your audience. This includes employees, management, customers, suppliers, regulatory agencies, and community groups. Tailor your messaging to address their specific interests and priorities related to sustainability.

Clearly articulate the importance of environment and sustainability policies and their alignment with the organization's values, goals, and objectives. Emphasize how these policies contribute to the organization's overall mission and vision, reinforcing the commitment to environmental stewardship and corporate social responsibility.

Highlight the benefits of adopting sustainable practices, both for the organization and its stakeholders. This may include cost savings through resource efficiency and waste reduction, regulatory compliance, reputation enhancement, and contribution to environmental conservation and social well-being. Use concrete examples and case studies to illustrate the tangible benefits of sustainability initiatives.

Communicate in clear, straightforward language that is accessible to all stakeholders, regardless of their level of expertise or familiarity with sustainability concepts. Avoid jargon and technical terms, opting instead for plain language that resonates with a broad audience. Use visuals, infographics, and storytelling techniques to make complex concepts easier to understand and memorable.

Instead of dwelling on the challenges or negative impacts of unsustainable practices, focus on the solutions and opportunities presented by environment and sustainability policies. Highlight innovative approaches, best practices, and success stories that demonstrate the positive outcomes achievable through sustainable business practices.

Acknowledge and address any concerns or scepticism that stakeholders may have about environment and sustainability policies. Provide evidence-based information and transparent communication to build trust and credibility. Invite dialogue and feedback to address questions, clarify misconceptions, and demonstrate a commitment to openness and accountability.

Close your messaging with a clear call to action, encouraging stakeholders to actively engage in supporting and implementing environment and sustainability policies. This may include inviting participation in training sessions, suggesting ways to incorporate sustainability into daily practices, or encouraging feedback and suggestions for improvement.

By crafting clear and compelling messaging that emphasizes the importance, benefits, and alignment of environment and sustainability policies with organizational values and goals, organizations can effectively engage stakeholders and inspire action towards a more sustainable future. As an example, here are examples of clear and compelling messaging for each section of the Environment and Sustainability Policy:

1. **Introduction:**

   ○ "Welcome to our commitment to environmental stewardship! At [Organization Name], we are dedicated to integrating sustainability into every aspect of our operations. This policy serves as our roadmap towards a greener future."

2. **Organizational Values and Goals:**

   ○ "At [Organization Name], environmental sustainability is at the core of everything we do. We strive to minimize our environmental footprint by reducing resource consumption, waste generation, and greenhouse gas emissions. Our goal is to promote sustainable practices throughout our operations, ensuring a brighter and more sustainable future for all."

3. **Stakeholder Engagement Process:**

   ○ "Your voice matters! We actively engage with our employees, management, customers, suppliers, regulatory agencies, and community groups to gather input and perspectives on environmental sustainability. Your feedback informs our policies and practices, ensuring that we make informed decisions that benefit our stakeholders and the planet."

4. **Key Areas for Action:**

   ○ "We're taking action! By assessing our environmental footprint, we've identified key areas where we can make a difference. From resource conservation to waste reduction, energy efficiency, pollution prevention, and sustainable sourcing practices, we're committed to making a positive impact on the environment."

5. Specific Policies and Procedures:

- "Our policies mean business! We've developed specific policies and procedures to be implemented across all aspects of our operations. From water and energy use to waste management, chemical handling, land management, and emissions reduction, we're setting the standard for environmental responsibility."

6. Legal and Regulatory Compliance:

- "We're in compliance! Our policies align with relevant legal and regulatory requirements governing environmental sustainability. By ensuring compliance with environmental protection laws, industry regulations, and certification standards, we're demonstrating our commitment to responsible stewardship."

7. Measurable Targets and Objectives:

- "Tracking our progress! We've established measurable targets and objectives to monitor our journey towards environmental sustainability. From reducing greenhouse gas emissions to minimizing water usage and waste generation, we're setting clear goals to guide our efforts."

8. Training and Capacity Building:

- "Empowering our team! We're investing in training programs to educate our employees about our environment and sustainability policy. By providing relevant environmental practices, procedures, and compliance requirements, we're empowering our team to contribute to our sustainability efforts."

9. Communication and Promotion:

- "Join us on the journey! We're communicating our environment and sustainability policy to all stakeholders through internal and external channels. By raising awareness of our objectives, expectations, and benefits, we're inviting everyone to actively participate in implementing the policy."

10. Monitoring, Evaluation, and Continuous Improvement:

- "Continuous improvement is our motto! We've established mechanisms for monitoring, evaluating, and continuously improving the effectiveness of our environment and sustainability policy. Through regular audits, assessments, and performance reviews, we're tracking our progress and driving ongoing improvements."

Engaging leadership support is a critical step in promoting workplace environment and sustainability policies within an organization. Start by clearly communicating the rationale and benefits of workplace environment and sustainability policies to organizational leadership. Explain how these policies align with the organization's mission, values, and long-term goals. Emphasize the potential benefits, such as cost savings, regulatory compliance, enhanced reputation, and improved employee morale. Highlighting the business case for sustainability can help leaders understand the importance of these policies in driving organizational success.

Equip senior management with the information and resources they need to effectively promote workplace environment and sustainability policies to key personnel and stakeholders. Provide them with talking points, presentations, case studies, and examples of successful sustainability initiatives. Offer access to relevant research, industry best practices, and tools for measuring and monitoring sustainability performance. By arming leaders with credible and compelling information, they can confidently advocate for sustainability within the organization.

Create opportunities for senior management to actively engage in promoting workplace environment and sustainability policies. Encourage them to lead by example and demonstrate their commitment to sustainability through their actions and decisions. Provide platforms for leaders to communicate their support for sustainability initiatives, such as town hall meetings, employee forums, and internal communications channels. By visibly championing sustainability, leaders can inspire and motivate employees to embrace environmental stewardship.

Ensure that workplace environment and sustainability policies are aligned with broader organizational goals and objectives. Emphasize how sustainability initiatives contribute to achieving strategic priorities, such as operational efficiency, risk management, innovation, and growth. By demonstrating the link between sustainability and organizational success, leaders are more likely to recognize the value and importance of these policies in driving overall performance.

Foster open dialogue and collaboration between leadership and stakeholders on workplace environment and sustainability policies. Encourage leaders to seek feedback and input from employees, customers, suppliers, and other key stakeholders to ensure that policies are relevant, practical, and impactful. By involving stakeholders in the decision-making process, leaders can build trust, foster engagement, and strengthen the organization's commitment to sustainability.

Regularly monitor and evaluate the progress of workplace environment and sustainability initiatives and celebrate successes along the way. Provide leaders with updates on key performance indicators, milestones achieved, and positive outcomes resulting from sustainability efforts. Recognize and reward individuals and teams for their contributions to sustainability, reinforcing the importance of these policies and fostering a culture of continuous improvement.

Offering comprehensive training sessions and resources is paramount for ensuring that key personnel and stakeholders grasp and effectively implement workplace environment and sustainability policies. Begin by tailoring training programs to suit the specific needs and roles of key personnel and stakeholders. Develop training materials that cover the workplace environment and sustainability policies, elucidating their rationale and implications for daily operations. Ensure that the training content is easily accessible, engaging, and pertinent to the intended audience.

During training sessions, clearly convey the objectives of the workplace environment and sustainability policies. Explain the significance of these policies for the organization, its employees, and the wider community. Emphasize the potential benefits of adopting sustainable practices, such as cost savings, regulatory compliance, reputation enhancement, and environmental stewardship.

Offer practical guidance on how employees can integrate sustainable practices into their work responsibilities. Present examples and case studies illustrating how sustainable practices can be applied in various contexts. Provide tips and suggestions for reducing resource consumption, minimizing waste, conserving energy, and fostering environmental responsibility in daily tasks and decision-making.

Encourage employees to actively engage in achieving environmental objectives by integrating sustainable practices into their work routines. Cultivate a culture of environmental stewardship and responsibility by empowering employees to identify improvement opportunities and implement sustainable solutions. Encourage collaboration, innovation, and knowledge-sharing among employees to drive continuous improvement in environmental performance.

Provide continuous support to employees by granting access to resources, tools, and assistance to aid them in implementing workplace environment and sustainability policies effectively. Appoint designated points of contact or resource persons capable of addressing inquiries, offering guidance, and providing support as needed. Establish platforms for employees to exchange best practices, learn from one another's experiences, and collaborate on sustainability initiatives.

Evaluate the effectiveness of training programs by gathering feedback from participants and assessing their comprehension of workplace environment and sustainability policies. Utilize surveys, quizzes, and evaluations to collect feedback on the relevance, clarity, and usefulness of training content. Employ this feedback to pinpoint areas for improvement and refine training programs to better meet the needs of participants.

Continually enhance training efforts by incorporating feedback from participants, monitoring changes in employee behaviour and performance, and staying informed about developments in environmental best practices. Regularly update training materials to reflect alterations in policies, procedures, and regulatory requirements. Through continuous improvement of training efforts, organizations can ensure that employees remain informed, engaged, and committed to achieving environmental objectives.

Providing training sessions and resources is indispensable for ensuring that key personnel and stakeholders comprehend workplace environment and sustainability policies, their rationale, and their implications for day-to-day operations.

Encouraging two-way communication within an organization is essential for fostering an environment of openness and transparency, particularly regarding environment and sustainability policies. Implement various channels through which key personnel and stakeholders can provide feedback, ask questions, and share ideas related to environment and sustainability policies. This can include online feedback forms, suggestion boxes placed strategically in the workplace, and dedicated email addresses or hotline numbers. Ensure that these channels are easily accessible and well-publicized to encourage participation.

Next, host regular town hall meetings or forums where employees and stakeholders can engage in open dialogue about environment and sustainability policies. These meetings provide a platform for sharing updates, addressing concerns, and soliciting input from participants. Encourage active participation by inviting diverse perspectives and fostering a respectful and inclusive atmosphere where all voices are heard.

Thirdly, organize focus groups consisting of representatives from different departments or stakeholder groups to delve deeper into specific environmental issues or policy-related topics. These smaller, more intimate settings allow for in-depth discussions, brainstorming sessions, and collaborative problem-solving. Encourage participants to share their experiences, insights, and suggestions for improvement.

Actively listen to concerns and suggestions raised by employees and stakeholders regarding environment and sustainability policies. Demonstrate empathy and understanding by acknowledging their perspectives and valuing their input. Avoid dismissing or ignoring feedback, even if it is critical or challenging, as this can undermine trust and discourage future engagement.

Actively incorporate feedback received from employees and stakeholders into the development and implementation of environment and sustainability policies. Analyse common themes, identify actionable insights, and prioritize areas for improvement based on stakeholder input. Demonstrate a willingness to adapt policies and procedures in response to feedback, demonstrating a commitment to continuous improvement.

Lastly, foster a culture of collaboration and cooperation where employees and stakeholders are encouraged to work together towards common environmental goals. Recognize and celebrate successes achieved through collaborative efforts, and encourage cross-departmental or cross-functional teamwork on sustainability initiatives. By promoting a culture of collaboration, organizations can harness the collective expertise and creativity of their workforce to address environmental challenges effectively.

Celebrating successes and milestones in the implementation of workplace environment and sustainability policies is also essential for reinforcing positive behaviour, motivating employees, and fostering a culture of environmental stewardship. Take the time to recognize and acknowledge achievements made in advancing environmental and sustainability initiatives within the organization. This could include meeting specific targets for resource conservation, reducing waste generation, implementing innovative sustainability practices, or achieving certification or recognition for environmental performance. Recognize both individual and team contributions to highlight the collective effort towards sustainability goals.

Share success stories and best practices that demonstrate the tangible impact of environment and sustainability policies. Highlight examples of successful initiatives, projects, or interventions that have resulted in measurable environmental benefits, cost savings, or improvements in operational efficiency. Use internal communication channels such as newsletters, bulletin boards, intranet platforms, or staff meetings to showcase these success stories and inspire others to follow suit.

Acknowledge the contributions of employees, teams, and departments who have played a role in implementing environment and sustainability policies. Express appreciation for their dedication, creativity, and hard work in driving sustainability initiatives forward. Consider implementing recognition programs, awards, or incentives to reward outstanding contributions and encourage continued engagement in sustainability efforts.

Moreover, foster a positive work environment, Celebrating successes and milestones creates a sense of pride, accomplishment, and camaraderie among employees, fostering a positive work environment conducive to sustainability. Celebrate achievements in a manner that is inclusive, respectful, and supportive of diversity. Encourage employees to share their experiences, lessons learned, and ideas for future sustainability initiatives, creating opportunities for peer learning and collaboration.

Use celebrations of successes and milestones as an opportunity to reaffirm the organization's commitment to environmental stewardship and sustainable business practices. Emphasize the importance of sustainability as a core value and strategic priority for the organization, and highlight the role of employees in driving positive change. Reinforce the connection between sustainability goals and the organization's mission, vision, and long-term success.

Lastly, set the stage for continuous improvement. Use celebrations of successes and milestones as a platform for setting new goals, targets, and aspirations for furthering environmental sustainability efforts. Encourage employees to reflect on past achievements and identify areas for improvement or expansion of sustainability initiatives. Set ambitious yet achievable targets for the future and rally employees around a shared vision of a more sustainable future.

Celebrating successes and milestones in the implementation of workplace environment and sustainability policies is essential for recognizing achievements, inspiring employees, and reinforcing organizational commitment to environmental stewardship.

Regularly reviewing and updating workplace environment and sustainability policies is crucial for ensuring their continued effectiveness and relevance in addressing evolving environmental challenges and opportunities.

Set up a regular schedule for reviewing workplace environment and sustainability policies, taking into account factors such as regulatory changes, technological advancements, industry best practices, and organizational priorities. Consider conducting reviews annually or biennially, or more frequently if significant changes occur in the operating environment.

Secondly, gather stakeholder input. Engage key personnel and stakeholders in the review process to gather their input, perspectives, and feedback on the effectiveness of existing policies. This could involve conducting surveys, focus groups, interviews, or workshops to solicit feedback on policy strengths, weaknesses, and areas for improvement. Ensure that stakeholders representing diverse interests and perspectives are included in the review process.

Thirdly, assess policy effectiveness. Evaluate the effectiveness of existing policies in achieving their intended objectives and outcomes. Assess policy performance against key performance indicators, targets, and objectives established during the policy development phase. Identify areas where policies have been successful in driving positive change and areas where improvements or adjustments are needed to address emerging challenges or opportunities.

Next, based on stakeholder input and policy effectiveness assessments, identify specific areas where workplace environment and sustainability policies can be strengthened or enhanced. This may involve updating policy language, revising procedures or protocols, incorporating new sustainability practices or technologies, or addressing gaps or deficiencies in existing policies. Prioritize areas for improvement based on their significance, feasibility, and potential impact on environmental sustainability.

Transparently communicate any policy changes or updates to key personnel and stakeholders in a timely and clear manner. Clearly explain the rationale behind the changes, the objectives they aim to achieve, and their implications for stakeholders' responsibilities and operations. Provide opportunities for stakeholders to ask questions, seek clarification, and provide feedback on the proposed changes.

Additionally, once policy changes have been communicated and agreed upon, implement the necessary updates across the organization. Disseminate updated policy documents, procedures, guidelines, or training materials to relevant stakeholders to ensure consistency and compliance with the revised policies. Provide support and resources to help employees understand and adapt to the changes effectively.

Continuously monitor and evaluate the impact of policy updates on environmental sustainability performance and organizational operations. Track key performance indicators, metrics, and targets to assess progress towards sustainability goals and objectives. Solicit feedback from stakeholders on the effectiveness of policy changes and their impact on day-to-day operations. Use this feedback to inform future policy revisions and updates.

Lastly, foster a culture of continuous improvement. Encourage a culture of continuous improvement and learning by fostering open dialogue, collaboration, and innovation around workplace environment and sustainability policies. Encourage employees to actively participate in the review process, share ideas for improvement, and contribute to the ongoing evolution of environmental sustainability initiatives. By regularly reviewing and updating policies, organizations can adapt to changing circumstances, stay ahead of emerging trends, and drive continuous progress towards environmental sustainability goals.

Chapter Thirteen

Implementing Workplace Environment and Sustainability Procedures

Developing and communicating workplace environment and sustainability procedures in agribusiness is an essential endeavour to ensure that environmental sustainability practices are seamlessly integrated into daily operations. Firstly, it is important to identify the specific procedures necessary to support environment and sustainability goals within the agribusiness context. This involves pinpointing procedures related to waste management, water and energy conservation, pesticide use, soil health management, biodiversity conservation, and sustainable sourcing practices, among others. Each procedure should delineate clear steps, guidelines, and best practices for executing tasks in an environmentally responsible manner.

Secondly, engage key personnel, stakeholders, and subject matter experts in the development of workplace environment and sustainability procedures. Gather input, feedback, and insights from individuals with relevant knowledge and expertise in agribusiness operations, environmental management, and sustainability practices. Incorporating their perspectives ensures that the procedures are practical, effective, and aligned with organizational goals.

Next, document the identified procedures in a clear, concise, and accessible format that is easily understandable for employees. Utilize simple language, visual aids, and step-by-step instructions to facilitate comprehension. Provide examples, illustrations, and case studies to demonstrate how the procedures should be implemented in real-world scenarios. Ensure that the procedures align with regulatory requirements, industry standards, and best practices.

Once the procedures are documented, communicate them to all relevant personnel and stakeholders within the agribusiness. Utilize various communication channels such as email announcements, staff meetings, training sessions, intranet postings, and signage in relevant work areas. Clearly explain the purpose, objectives, and expected outcomes of each procedure, along with employees' roles and responsibilities in implementing them.

Provide comprehensive training sessions and resources to educate employees about the workplace environment and sustainability procedures. Ensure that employees understand the rationale behind the procedures, their importance for environmental stewardship, and their implications for day-to-day operations. Offer practical guidance, demonstrations, and hands-on exercises to facilitate effective implementation.

Additionally, develop recordkeeping systems for tracking environment and sustainability work practices within the agribusiness. Determine the types of records and data that need to be collected, stored, and maintained to monitor environmental performance, compliance with procedures, and progress towards sustainability goals. Choose appropriate tools, software, or databases for recording and managing data, ensuring usability, reliability, and security.

Assign responsibility for maintaining the recordkeeping system to designated personnel within the agribusiness. Clearly define roles, responsibilities, and accountabilities for individuals responsible for collecting, entering, verifying, and updating records. Provide training and guidance to ensure that personnel understand their roles and obligations regarding recordkeeping. Establish procedures for reviewing, auditing, and validating records to ensure accuracy, completeness, and compliance with established protocols.

Identifying relevant procedures to support environment and sustainability goals within the agribusiness context is a foundational step in integrating sustainable practices into daily operations. This process involves a thorough examination of various aspects of the business to pinpoint areas where environmental improvements can be made.

One key area to consider is waste management. This includes procedures for reducing, reusing, and recycling waste generated during agricultural activities. Strategies may involve composting organic waste, implementing recycling programs for packaging materials, and properly disposing of hazardous substances.

Water and energy conservation are also critical aspects of sustainability in agribusiness. Procedures in this area may focus on efficient irrigation techniques, the use of renewable energy sources such as solar power, and minimizing energy consumption in facilities and equipment.

Pesticide use is another important consideration, given its potential impact on human health and the environment. Procedures may involve integrated pest management strategies to reduce reliance on chemical pesticides, proper storage and handling of pesticides to prevent contamination, and regular monitoring to ensure compliance with safety regulations.

Soil health management is essential for maintaining the long-term productivity and sustainability of agricultural land. Procedures in this area may include soil testing to assess nutrient levels, crop rotation to prevent soil depletion, and the use of cover crops to prevent erosion and improve soil structure.

Biodiversity conservation is increasingly recognized as vital for the health and resilience of agricultural ecosystems. Procedures may involve preserving natural habitats on farmland, implementing wildlife-friendly farming practices, and minimizing the use of pesticides and chemical fertilizers that can harm non-target species.

Sustainable sourcing practices are also integral to promoting environmental sustainability in agribusiness. Procedures may include sourcing inputs from certified sustainable suppliers, promoting fair labor practices throughout the supply chain, and reducing the carbon footprint associated with transportation and distribution.

Each procedure should outline clear steps, guidelines, and best practices for carrying out tasks in an environmentally responsible manner. This ensures consistency and clarity in implementation and helps employees understand their role in achieving environmental objectives. By identifying relevant procedures across these various areas, agribusinesses can develop a comprehensive framework for promoting sustainability and environmental stewardship in their operations.

Engaging key personnel, stakeholders, and subject matter experts in the development of workplace environment and sustainability procedures is essential for creating policies that are practical, effective, and aligned with organizational goals. This collaborative approach ensures that the diverse perspectives and expertise of stakeholders are taken into account, resulting in procedures that are well-informed and relevant to the specific needs of the agribusiness.

To begin the consultation process, it's important to identify and reach out to relevant stakeholders who have a vested interest in environmental sustainability within the organization. This may include employees from different departments, management representatives, suppliers, customers, regulatory agencies, and community groups. By involving a wide range of stakeholders, the organization can gather diverse insights and perspectives on environmental management and sustainability practices.

During the consultation process, stakeholders should be given the opportunity to provide input, feedback, and insights based on their knowledge and expertise. This can be done through various means such as meetings, workshops, focus groups, surveys, interviews, and online forums. Encourage stakeholders to share their experiences, concerns, priorities, and suggestions for improving environmental practices within the agribusiness.

Subject matter experts, including individuals with experience in agribusiness operations, environmental management, and sustainability practices, play a crucial role in providing technical guidance and expertise during the consultation process. Their insights can help ensure that the procedures developed are based on best practices and are feasible to implement in the agricultural context.

Throughout the consultation process, it's important to foster an environment of open communication and collaboration, where all stakeholders feel valued, respected, and heard. Actively listen to the perspectives and feedback provided by stakeholders, and demonstrate a willingness to incorporate their input into the development of workplace environment and sustainability procedures.

By incorporating the perspectives of key personnel, stakeholders, and subject matter experts, agribusinesses can develop procedures that are practical, effective, and aligned with organizational goals and objectives. This collaborative approach not only enhances the quality of the procedures but also fosters a sense of ownership and commitment among stakeholders towards achieving environmental sustainability within the organization.

Documenting procedures is a critical step in the development of workplace environment and sustainability policies, as it provides employees with clear guidance on how to implement these practices effectively. To ensure that procedures are accessible and comprehensible to all employees, it's essential to present them in a clear, concise, and user-friendly format. This involves using simple language, avoiding jargon or technical terms, and breaking down complex processes into easy-to-follow steps.

Visual aids such as flowcharts, diagrams, and infographics can be effective tools for enhancing understanding and retention of information. These visual representations help to illustrate the sequence of steps involved in carrying out procedures and can make complex concepts more digestible for employees. Additionally, incorporating images, icons, and color-coding can further enhance the clarity and accessibility of the documentation.

In addition to written instructions, providing examples, illustrations, and case studies can help employees understand how the procedures should be implemented in real-world scenarios. By presenting practical examples and demonstrating the desired outcomes of the procedures, employees can better grasp their relevance and importance in achieving environmental sustainability goals.

It's important to ensure that the documented procedures are consistent with regulatory requirements, industry standards, and best practices in environmental management. This may involve conducting thorough research and consulting relevant sources to ensure that the procedures align with legal obligations and industry norms. By adhering to established standards and guidelines, organizations can demonstrate their commitment to compliance and excellence in environmental stewardship.

Regular review and updates of the documented procedures are also necessary to ensure their ongoing relevance and effectiveness. As regulations, technologies, and best practices evolve over time, it's essential to keep the procedures up-to-date to reflect these changes. This may involve conducting periodic audits or assessments to identify areas for improvement and incorporating feedback from stakeholders to refine the procedures as needed.

Overall, documenting procedures in a clear, concise, and accessible format is essential for ensuring that employees understand and follow workplace environment and sustainability policies effectively. By providing clear guidance, visual aids, practical examples, and ensuring consistency with regulatory requirements, organizations can empower employees to adopt sustainable practices and contribute to environmental stewardship efforts within the organization.

The following is an example of a general environment and sustainability procedure:

INTRODUCTION

The Earth's environment faces significant stress from uncontrolled human activity, posing a threat to both our society and the mission of [Name of Organization]. Recognizing this challenge, [Name of Organization] acknowledges its responsibility to actively contribute to the preservation of environmental sustainability. This commitment extends across all levels of our operations – from our internal practices to our engagement within the broader community and Australian social discourse. At [Name of Organization], we aim to minimize our environmental impact while maximizing the efficient use of resources. To achieve this, we emphasize communication, awareness, and responsible environmental behaviour among our staff, volunteers, and users at every level. Furthermore, we are dedicated not only to meeting applicable legal requirements but also to minimizing risks and impacts through the development of robust environmental management systems.

PURPOSE

This Environmental Sustainability Policy serves as a framework to consolidate various accountabilities related to environmental responsibility within [Name of Organization]. Its purpose is to integrate principles of sustainable development into all our activities and to promote sound environmental practices across our operations.

POLICY

[Name of Organization] commits to minimizing its environmental impact through the following actions:

- *Providing a safe and healthful workplace.*

- *Cultivating an environmentally aware culture where responsibility is clearly assigned and understood.*

- *Being a responsible neighbour in our community.*

- *Conserving natural resources through reuse and recycling.*

- *Utilizing processes in our operations that do not harm the environment.*

- *Ensuring responsible energy use throughout the organization.*

- *Participating in environmental protection efforts.*

- *Leveraging our professional expertise to address environmental challenges.*

- *Continuously improving our environmental performance through rigorous audits and evaluations.*

- *Collaborating with suppliers who promote sound environmental practices.*

- *Enhancing awareness among employees, volunteers, and users to foster environmentally responsible behaviour.*

RESPONSIBILITIES

The Board is responsible for establishing and maintaining policies and procedures, while the CEO ensures their implementation.

PROCEDURES

1. *[Name of Organization] will develop clear guidelines for staff, volunteers, and users to adopt sound environmental practices, accompanied by adequate training to ensure compliance.*

2. *[Name of Organization] will promptly address and report incidents or conditions endangering health, safety, or the environment.*

3. *Efforts will be made to reuse, recycle, and purchase recycled materials.*

4. *Services and products will prioritize safety, energy efficiency, and environmental protection.*

5. *Operations will minimize materials and energy use, prevent pollution, and dispose of waste responsibly.*

6. *Energy conservation will be prioritized, with a preference for renewable energy sources.*

7. *Expertise will be utilized to advance environmentally sustainable techniques.*

8. *Biodiversity will be maintained and enhanced across landholdings.*

9. *Compliance with all applicable regulations and voluntary standards will be ensured.*

10. *Regular audits and self-assessments will measure progress against environmental goals.*

11. *Transparent communication will be maintained regarding environmental performance.*

12. *All personnel will be expected to adhere to this policy and report any concerns promptly.*

13. *Suppliers and contractors will be selected based on their environmental performance.*

14. *Investments will align with the environmental objectives outlined in this policy.*

RELATED DOCUMENTS
- *Purchasing Policy*

- *Code of Ethics*

Communicating workplace environment and sustainability procedures effectively is essential to ensure that all relevant personnel and stakeholders within the agribusiness are aware of and understand their roles and responsibilities in implementing these procedures. Commence by identifying the most appropriate communication channels to reach all relevant personnel and stakeholders within the agribusiness. This may include email announcements, staff meetings, training sessions, intranet postings, and signage in relevant work areas. By utilizing multiple channels, you can ensure widespread dissemination of information and reach individuals with different communication preferences and accessibility needs.

Develop clear and comprehensive messages that clearly explain the purpose, objectives, and expected outcomes of each workplace environment and sustainability procedure. Use simple language that is easy to understand and avoid technical jargon or overly complex terms. Clearly outline employees' roles and responsibilities in implementing the procedures to ensure clarity and understanding.

Contextualize the procedures by explaining their relevance to the agribusiness and its overall goals and objectives. Help employees understand how the procedures contribute to environmental sustainability, operational efficiency, regulatory compliance, and the organization's reputation. Providing context helps employees see the bigger picture and understand the importance of their roles in implementing the procedures.

Incorporate interactive and engaging methods to capture employees' attention and encourage active participation in the communication process. This could include interactive training sessions, Q&A sessions, demonstrations, and hands-on activities that allow employees to practice implementing the procedures in a simulated environment. Engaging employees in the communication process fosters better understanding and retention of information.

Ensure that the communication of workplace environment and sustainability procedures is accessible to all employees, including those with disabilities or language barriers. Provide materials in multiple formats, such as written documents, visual presentations, and audio recordings, to accommodate different learning styles and accessibility needs. Consider translating materials into multiple languages if necessary to reach employees who may not be fluent in the primary language of communication.

Encourage employees to provide feedback and ask questions about the workplace environment and sustainability procedures. Create opportunities for open dialogue and discussion where employees can seek clarification, share concerns, and provide input on how the procedures can be improved or implemented more effectively. Actively listening to employees' feedback demonstrates a commitment to transparency and responsiveness.

Reinforce the messages about workplace environment and sustainability procedures consistently over time to ensure that they remain top of mind for employees. Use regular reminders, updates, and reinforcement activities to reinforce key messages and encourage ongoing adherence to the procedures. Consistent communication helps embed the procedures into the organizational culture and promotes long-term behaviour change.

Providing training and support is crucial for ensuring that employees understand and effectively implement workplace environment and sustainability procedures. Start by designing comprehensive training programs that cover all aspects of the workplace environment and sustainability procedures. These programs should be tailored to the specific needs and roles of employees within the agribusiness. Develop training materials that explain the rationale behind the procedures, their importance for environmental stewardship, and their implications for day-to-day operations.

Clearly communicate the purpose and importance of the workplace environment and sustainability procedures during training sessions. Help employees understand how their actions contribute to environmental sustainability, operational efficiency, and the overall success of the organization. Emphasize the significance of their role in implementing the procedures and the positive impact it can have on the environment and the community.

Offer practical guidance, demonstrations, and hands-on exercises to help employees apply the procedures effectively in their work responsibilities. Use real-life examples and case studies to illustrate how the procedures should be implemented in different situations. Provide opportunities for employees to practice applying the procedures in a simulated or controlled environment to build their confidence and competence.

Offer ongoing support, assistance, and access to resources to help employees implement the procedures successfully. Establish designated points of contact or resource persons who can answer questions, provide guidance, and offer support as needed. Create forums for employees to share best practices, learn from each other's experiences, and collaborate on finding solutions to challenges.

Utilize various training methods to accommodate different learning styles and preferences. This could include classroom-style training sessions, online courses, workshops, seminars, and on-the-job training opportunities. Tailor the training methods to the needs and preferences of employees, ensuring that they are engaging, interactive, and effective in conveying the required information.

Monitor employees' progress in implementing the workplace environment and sustainability procedures and provide feedback on their performance. Offer constructive feedback and guidance to help employees improve their implementation of the procedures and address any challenges or issues that may arise. Recognize and celebrate employees' successes and achievements to motivate and inspire continued adherence to the procedures.

Encourage employees to engage in continuous learning and professional development related to environmental sustainability. Provide opportunities for further education, training, and certification in relevant areas to empower employees to become champions of sustainability within the organization. Foster a culture of continuous improvement and innovation by encouraging employees to explore new ideas and approaches to environmental stewardship.

Establishing recordkeeping systems is essential for tracking environment and sustainability work practices within an agribusiness. Here's a detailed guide on how to develop such systems effectively. Begin by identifying the types of records and data that need to be collected, stored, and maintained to monitor environmental performance, compliance with procedures, and progress towards sustainability goals. This may include data related to resource consumption, waste generation, energy usage, emissions, water usage, pesticide application, soil health, biodiversity, and sustainable sourcing practices.

Determine the most appropriate methods for collecting the required data. This may involve manual data entry, automated data collection systems, sensor technology, metering devices, monitoring equipment, sampling protocols, or other data collection methods. Consider factors such as accuracy, reliability, cost-effectiveness, and ease of implementation when selecting data collection methods.

Choose appropriate tools, software, or databases for recording and managing the collected data. Look for recordkeeping systems that are user-friendly, reliable, and secure. Consider factors such as ease of data entry, accessibility, scalability, compatibility with existing systems, data backup and recovery capabilities, and security features when selecting recordkeeping tools and software.

Develop clear procedures for collecting, storing, and maintaining records related to environment and sustainability work practices. Outline step-by-step instructions for data collection, entry, validation, storage, retrieval, and disposal. Define roles and responsibilities for individuals involved in recordkeeping processes and establish quality control measures to ensure data accuracy and integrity.

Implement data management protocols to ensure the integrity, confidentiality, and security of collected data. Establish data encryption, access controls, user authentication, and data backup procedures to safeguard sensitive information. Regularly review and update data management protocols to address emerging security threats and comply with relevant regulations and standards.

Offer training sessions and resources to educate personnel responsible for recordkeeping about data collection procedures, recordkeeping tools, and data management protocols. Ensure that personnel understand their roles and responsibilities regarding recordkeeping and provide ongoing support and assistance as needed. Foster a culture of accountability and compliance with recordkeeping procedures within the organization.

Monitor and evaluate recordkeeping practices regularly to ensure compliance with established procedures and identify areas for improvement. Conduct audits, reviews, and assessments of recordkeeping systems to assess data accuracy, completeness, and reliability. Use feedback from audits and evaluations to refine recordkeeping procedures, address deficiencies, and enhance overall data quality and integrity.

RICHARD SKIBA

Document recordkeeping processes, including data collection methods, recordkeeping tools, data management protocols, and training materials. Maintain detailed records of data sources, data entry procedures, data validation procedures, and data storage locations. Document any changes or updates made to recordkeeping processes and ensure that documentation is accessible to relevant personnel.

Documenting recordkeeping processes plays a pivotal role in guaranteeing the consistency, accuracy, and transparency necessary for effectively managing data related to environment and sustainability practices.

It is essential to document the methods utilized for data collection, encompassing various techniques such as manual data entry, sensor technology, or monitoring equipment. This documentation should include specifics on sampling protocols, the frequency of data collection, and the individuals responsible for carrying out these tasks. For instance, a documented data collection method could detail how soil moisture levels are measured using sensors installed in the field, with measurements conducted weekly by designated field technicians.

Secondly, document the tools or software employed for recording and managing data, such as spreadsheets, databases, or specialized software platforms. Provide comprehensive information on how to access and effectively utilize these tools. For instance, an example documentation could outline that environmental data is recorded and managed using a specific software platform, with authorized personnel granted access and provided training on data entry and retrieval processes.

Thirdly, establish clear protocols for managing data, covering areas like data validation procedures, storage, backup, and security measures. These protocols should outline steps to ensure data integrity, confidentiality, and accessibility. For example, a documented data management protocol might specify double-entry verification and cross-checking against field measurements for data validation, alongside details on centralized data storage with regular backups performed.

Next, document the training materials utilized to educate personnel responsible for recordkeeping about data collection procedures, recordkeeping tools, and data management protocols. This documentation should encompass presentations, manuals, and instructional videos, ensuring comprehensive training for all involved parties. For instance, a documented set of training materials might include a manual titled 'Environmental Data Management Guidelines,' offering detailed instructions on various aspects of data management.

Furthermore, document the sources of data utilized, ranging from field measurements to laboratory analyses or external databases. Specify the locations where data is stored, encompassing both physical and digital storage facilities. For example, documentation could detail that soil nutrient data is obtained from laboratory analyses conducted by specific laboratories, with raw data files stored in designated folders on shared network drives.

Lastly, maintain a log of any changes or updates made to recordkeeping processes, including revisions to data collection methods, tools, protocols, or training materials. This log should provide details on the reasons for the change and the individuals involved in implementing it. For instance, a change log might record an update to the frequency of soil moisture measurements, along with the rationale behind the change and the recommendation from relevant experts.

By meticulously documenting recordkeeping processes, agribusinesses can ensure transparency, accountability, and reliability in managing environmental data. This comprehensive documentation facilitates effective decision-making and ensures compliance with regulatory requirements, ultimately contributing to the organization's environmental sustainability goals.

Assigning responsibility for recordkeeping within an agribusiness is a critical step in ensuring the accuracy, reliability, and compliance of environmental data management processes. Here's a detailed explanation of how to effectively assign these responsibilities:

Firstly, designate specific personnel within the agribusiness who will be responsible for maintaining the recordkeeping system. These individuals should have the necessary skills, knowledge, and expertise to effectively manage environmental data. Consider factors such as job roles, departmental responsibilities, and relevant experience when selecting personnel for these roles.

Once personnel have been identified, clearly define their roles, responsibilities, and accountabilities regarding recordkeeping. This includes outlining tasks such as data collection, data entry, data verification, and record updating. Provide detailed job descriptions or role profiles to ensure that personnel understand their duties and obligations regarding recordkeeping.

To ensure that personnel are equipped to fulfill their recordkeeping responsibilities effectively, provide training and guidance on recordkeeping processes, procedures, and tools. This training should cover topics such as data collection methods, recordkeeping tools, data management protocols, and compliance requirements. Offer training sessions, workshops, or online resources to ensure that personnel have the necessary knowledge and skills to perform their recordkeeping duties competently.

Establish clear procedures for reviewing, auditing, and validating records to ensure accuracy, completeness, and compliance with established protocols. Define the frequency and scope of record reviews and audits, as well as the criteria for evaluating record accuracy and completeness. Implement quality control measures, such as double-checking data entries and cross-referencing records with source documents, to minimize errors and discrepancies.

Regularly monitor and evaluate the performance of personnel responsible for recordkeeping to identify any areas for improvement or additional support. Provide ongoing feedback, coaching, and mentoring to help personnel enhance their recordkeeping skills and practices. Encourage open communication and collaboration among recordkeeping personnel to foster a culture of continuous improvement and accountability.

By assigning clear responsibilities, providing adequate training and guidance, and implementing robust procedures for recordkeeping, agribusinesses can ensure the accuracy, reliability, and compliance of their environmental data management processes. This systematic approach helps mitigate risks, enhance decision-making, and support the organization's overall environmental sustainability goals.

Chapter Fourteen

Reviewing Workplace Environment and Sustainability Policy Implementation

I nvestigating the outcomes of policies, monitoring records, modifying policies and procedures, and documenting outcomes constitute essential steps in the continuous improvement process of workplace policies and procedures for environment and sustainability in agribusiness.

To investigate the outcomes of policy, the initial step involves assessing the efficacy of implemented policies and procedures to ensure their alignment with the targeted environmental and sustainability objectives. This evaluation aims to determine whether these policies are effectively fulfilling their intended purposes and to identify any areas where performance may be lacking or falling short of expectations.

Various methodologies should be employed to gather data and assess outcomes comprehensively. These methods include conducting audits, administering surveys, conducting interviews, and performing performance assessments. Through these avenues, both quantitative data, such as metrics related to energy consumption and waste generation, and qualitative feedback from stakeholders are gathered to provide a holistic understanding of policy effectiveness.

Following data collection, a detailed analysis should be conducted to discern the strengths, weaknesses, opportunities, and threats associated with policy implementation. This analysis involves scrutinizing the collected data to identify trends, patterns, and areas that require improvement. By identifying these aspects, organizations can derive insights to inform future actions and decision-making processes, ensuring that policies are continually refined to enhance their effectiveness and alignment with environmental and sustainability objectives.

To monitor records and identify trends, it is essential to regularly examine the data and information gathered through recordkeeping systems to pinpoint patterns and trends that may signal areas requiring corrective measures or enhancements. This process involves systematically reviewing records to track performance and assess whether objectives are being met effectively.

Key performance indicators (KPIs) and predefined benchmarks established in the policy serve as vital tools for monitoring progress and evaluating performance against predetermined targets. These indicators may encompass various metrics, including but not limited to energy consumption, waste generation, water usage, and adherence to environmental regulations. By aligning monitoring efforts with these indicators, organizations can effectively gauge their performance and progress towards environmental and sustainability goals.

A structured approach should be implemented for record monitoring, involving scheduled reviews and analyses to detect any deviations from expected outcomes. This systematic review process helps to identify emerging issues, recurring challenges, or shifts in performance trends that may warrant further investigation or corrective action. By remaining vigilant and proactive in monitoring records, organizations can promptly address any discrepancies and make informed decisions to optimize performance and ensure alignment with environmental objectives.

Upon reviewing the findings from policy investigations and record monitoring, it is imperative to identify opportunities for improvement and make adjustments to existing policies and procedures accordingly. This involves pinpointing areas where enhancements can be made to better align with organizational objectives and address any shortcomings in policy effectiveness.

A collaborative approach is essential in the modification process, necessitating the engagement of key stakeholders, subject matter experts, and personnel. By involving individuals with diverse perspectives and expertise, organizations can ensure that proposed changes are well-informed, practical, and in line with overarching goals. This collaborative effort fosters a sense of ownership and collective responsibility for the success of policy modifications.

When deliberating modifications to policy and procedures, it is essential to consider a range of factors, including emerging environmental challenges, advancements in technology, updates to regulations, and feedback from stakeholders. By taking these factors into account, organizations can make informed decisions that reflect current best practices and address evolving needs and priorities.

Following the identification of necessary modifications, it is crucial to update policy documents, procedural manuals, and training materials accordingly. Clear communication of these updates is essential to ensure that relevant stakeholders and personnel are aware of the changes and understand their implications. By providing transparent and accessible information, organizations can facilitate comprehension and promote adherence to revised policies and procedures.

To ensure a thorough understanding of performance and improvement efforts, it is essential to diligently document the outcomes stemming from policy assessments, record monitoring endeavours, and policy modifications. By maintaining detailed records, organizations can establish a comprehensive repository of information that chronicles their progress over time and informs future decision-making processes.

This documentation process involves compiling comprehensive reports, summaries, and analyses that encapsulate the findings from various assessments and evaluations. These documents should encapsulate trends, lessons learned, and areas requiring further attention or action. By incorporating both quantitative data and qualitative insights, organizations can provide a nuanced perspective on outcomes and facilitate a deeper understanding of the factors influencing performance.

Once outcomes have been documented, it is crucial to disseminate this information to key stakeholders and personnel. Constructive feedback should be delivered in a clear and transparent manner, allowing stakeholders to gain insights into policy outcomes and improvement initiatives. Reports, presentations, or other communication materials can be utilized to communicate progress, address challenges, and outline future directions.

Cultivating a culture of transparency, accountability, and continuous improvement is essential for the long-term success of environmental and sustainability initiatives. To achieve this, organizations should actively engage key stakeholders and personnel in the feedback loop. By fostering open dialogue, facilitating discussions, and encouraging collaboration, organizations can harness the collective insights and expertise of their stakeholders to drive ongoing improvement efforts forward.

Part 6 - Applying Agribusiness Management Practices

This book section outlines the process of determining the scope of a business plan and associated systems, emphasizing thorough planning and strategic alignment. It begins by highlighting the necessity of consulting with key personnel to identify the scope, ensuring alignment with stakeholders' needs and objectives. Accessing relevant information, including market trends and legal obligations, is crucial for informed decision-making during plan development. The section emphasizes aligning the plan with strategic goals and incorporating risk management strategies to safeguard operations.

In preparing the business plan, various factors such as resource requirements, input supply chain options, and marketing strategies are considered. Trial systems are tested before full implementation to assess their feasibility and effectiveness. Once finalized, the plan is documented and regularly reviewed to ensure relevance and alignment with evolving needs. Additionally, monitoring performance against the plan allows for the identification of strengths, weaknesses, and areas for improvement, guiding future development efforts.

Furthermore, the book section also focuses on determining the directions for the business by establishing long-term goals and strategies aligned with stakeholders' expectations. It involves conducting SWOT analysis to identify internal strengths and weaknesses and external opportunities and threats, informing strategic decision-making. Auditing natural resources and infrastructure, addressing legal requirements, and developing sustainable management strategies are integral components of this process.

Moreover, the section delves into determining production process requirements, emphasizing the importance of thorough planning and efficient management. It involves accessing information about products and market dynamics, analysing data from organizational records, and establishing robust production and monitoring systems. Monitoring and evaluating production processes allow for continuous improvement and informed decision-making based on performance data.

Lastly, the book section discusses evaluating the need for agricultural technology, emphasizing strategic planning and informed decision-making. It involves identifying tasks that could benefit from technology, assessing opportunities and limitations, and evaluating the cost-benefit ratio. Implementing technology to manage production, collecting and analysing data, and integrating technology to improve efficiency and sustainability are key aspects. Continuous evaluation ensures cost-effectiveness and alignment with operational goals, fostering agility and adaptation to technological advancements.

Chapter Fifteen

Developing and Reviewing a Business Plan

The Need for Business Planning

Planning is fundamental to the success of any business, regardless of its stage of development. Regularly reviewing the business plan ensures its alignment with evolving needs and circumstances (Chamber of Commerce of Metropolitan Montreal, 2023b). This ongoing assessment is crucial for identifying growth opportunities and ensuring strategic focus.

Once progress has been assessed and key growth areas identified, revisiting the business plan is necessary to chart the next phase of development. An up-to-date business plan serves as a roadmap guiding the business toward its objectives.

The significance of continuous business planning extends beyond securing investment. While investors often require a business plan as a prerequisite, its role extends throughout the lifespan of the business. It serves as a tool for resource allocation, aids in attracting funding, and outlines the strategic utilization of resources (Chamber of Commerce of Metropolitan Montreal, 2023b).

Continuous business planning facilitates monitoring of progress toward business objectives. It provides a means to assess the current position of the business and chart its desired growth trajectory. By setting clear targets and objectives, a business plan ensures that key priorities are managed effectively.

Maximizing the chances of success requires adopting a regular business planning cycle. This involves periodic meetings involving key stakeholders to assess performance against set targets and identify areas for improvement. Regular assessment, typically conducted every three to six months, enables timely adjustments and helps in demonstrating commitment to stakeholders.

A well-rounded business plan encompasses various elements essential for effective management and growth. It should provide a summary of the business's activities, development, and future trajectory. Critical components include marketing strategies, operational details, financial forecasts, and a summary of objectives and targets.

From an agribusiness perspective, determining the scope of a business plan involves careful consideration of various factors to ensure that the plan addresses the specific needs and goals of the enterprise. This includes:

1. Consultation with Key and Specialist Personnel:

   ○ Engage key personnel, such as management, agronomists, and financial advisors, to understand the current state of the business and its future aspirations.

   ○ Seek input from specialists in relevant areas, such as market analysts or legal advisors, to ensure comprehensive coverage of all aspects affecting the business.

2. Access Information for Business Plan Development:

- Gather relevant data and information from various sources, including market research reports, industry publications, and government statistics.

- Analyse potential business opportunities by assessing market trends, consumer preferences, and competitor activities to identify areas for growth and development.

3. Account for Trends and Seasonal Variations:

- Consider seasonal fluctuations and trends in the agricultural industry when developing the business plan.

- Incorporate strategies to mitigate the impact of seasonal variations on production, sales, and cash flow.

4. Account for Strategic Goals and Directions:

- Align the business plan with the strategic goals, targets, and directions of the enterprise, as outlined in the organization's mission and vision statements.

- Ensure that operational goals and targets are consistent with the overall strategic plan of the enterprise.

5. Identify Legal Obligations and Compliance Requirements:

- Identify and understand the legal obligations and compliance requirements relevant to the agribusiness, including environmental regulations, food safety standards, and labour laws.

- Incorporate measures to ensure compliance with applicable laws and regulations into the business plan.

Once the scope of the business plan is determined, the next steps involve preparing and documenting the plan, as well as reviewing and refining it on an ongoing basis to ensure its effectiveness and relevance to the enterprise's objectives.

Crafting a comprehensive business plan is a pivotal initial step for any agricultural endeavour, regardless of its scale or complexity. Establishing a robust business plan is essential because it enhances organization, serves as a guiding framework, and facilitates financial support (USDA, 2024a). Firstly, a well-developed business plan ensures that all pertinent details are considered, facilitating comprehensive organization and ensuring that necessary steps are taken. Acting as a roadmap, a business plan prompts thoughtful consideration of the reasons behind embarking on farming or ranching ventures and outlines future objectives. It serves as a reference point to evaluate progress towards achieving set goals. Additionally, a detailed business plan is often a prerequisite for securing financial assistance, such as loans from financial institutions or guarantees from commercial lenders. Lenders scrutinize business plans to gauge the feasibility of loan repayment (USDA, 2024a).

When building a farm business plan, various core elements must be addressed. Firstly, assessing operational details such as the type of products, the scale of operations, and the expertise of individuals involved lays the foundation for further business plan development (USDA, 2024a). Defining the mission, vision, and objectives of the enterprise establishes a framework for aligning operational goals with broader strategic aspirations. Determining the legal structure of the business and delineating management responsibilities is essential for effective decision-making and communication. Identifying target consumers, assessing demand for products, and formulating marketing strategies are crucial components, considering relevant regulations and licensing requirements (USDA, 2024a).

Moreover, evaluating existing resources and identifying additional requirements, such as land, equipment, and labour, supports business operations effectively. Assessing financing options, analysing current assets and liabilities, and forecasting income and expenses are vital aspects of financial planning to ensure viability and sustainability. Anticipating potential risks and challenges associated with business operations and developing strategies to mitigate them effectively are essential for risk management. Establishing metrics for evaluating the success of the business and monitoring performance against predetermined benchmarks is crucial for performance measurement (USDA, 2024a).

By comprehensively addressing these components, agribusinesses can develop robust business plans that serve as invaluable tools for guiding their operations and securing support for their endeavours.

For businesses with multiple departments or divisions, integrating individual business plans into a cohesive strategy document is essential (Chamber of Commerce of Metropolitan Montreal, 2023b). This ensures alignment with the overarching organizational goals and facilitates consistent progress toward shared objectives.

To effectively allocate resources throughout the organization, the business plan plays a pivotal role. It guides resource allocation to support the achievement of set objectives. Regular review ensures that the plan remains relevant and reflects the business's current position and future direction.

Setting SMART objectives and targets is imperative for translating strategic goals into actionable plans. Targets help in clarifying expectations and monitoring performance. Performance indicators such as sales figures, productivity benchmarks, and market share statistics aid in tracking progress toward set targets (Chamber of Commerce of Metropolitan Montreal, 2023b).

Regular review of the business plan is essential for ensuring ongoing relevance and effectiveness. The review process, conducted either continuously or at fixed intervals, involves assessing progress, analysing market dynamics, and adjusting strategies as necessary. This iterative approach enables businesses to adapt to changing circumstances and maintain a competitive edge (Chamber of Commerce of Metropolitan Montreal, 2023b).

To begin the process of assessing your business's performance and potential for growth, it's essential to evaluate your core activities, including the products or services you offer. Start by examining what makes your core activities successful and how they could be enhanced. Consider whether there are opportunities to introduce new or complementary products or services to meet evolving customer needs. By asking these questions, you can lay the groundwork for enhancing performance and profitability.

Addressing key questions about your products or services is crucial. Evaluate how well your offerings align with customer needs and which ones are achieving success. Identify any underperforming products or services and delve into the root causes of their challenges, considering factors like pricing, marketing, and customer service. Additionally, regularly reviewing costs and exploring ways to reduce expenses can contribute to improved efficiency and financial performance.

Efficiency is a vital aspect of business operations, particularly for new ventures navigating growth and development. While short-term flexibility can be advantageous, it's essential to balance this with a clear overall strategy to ensure that actions taken align with long-term objectives. Evaluate internal factors that may be hindering business progress and strategize solutions to address them effectively.

Assess various aspects of your business, including premises, facilities, information technology, personnel, and professional skills. Consider factors such as long-term property commitments, the modernity of equipment, and the effectiveness of IT systems in supporting business operations. Additionally, evaluate the skills and satisfaction levels of your workforce to ensure they align with business objectives.

Financial management is critical for business success, as poor financial practices can lead to failure. Regularly reviewing your financial position and updating your business plan can help ensure sound financial management. Assess cash flow, working capital needs, cost structures, borrowing arrangements, and plans for future growth to maintain financial stability and support business objectives.

Conducting a competitor analysis is essential to understanding market dynamics and identifying opportunities for growth. Gather information about competitors' offerings, pricing strategies, customer profiles, and competitive advantages. Utilize tools like SWOT analysis to assess your position relative to competitors and inform strategic decision-making.

Similarly, conducting a customer and market analysis allows you to revisit your target market and adjust marketing strategies accordingly. Consider changes in market trends, customer preferences, and competitive activity to refine your marketing plan and enhance business performance.

Regularly reviewing your business goals and strategic direction is essential for long-term success. Set aside time to evaluate where your business stands, where it's headed, and how it will get there. Develop work plans to implement new ideas and regularly review progress to ensure alignment with business objectives. Continuous improvement and adaptation to change are crucial for navigating today's dynamic business environment.

Finally, consider seeking expert input to support business growth and development. Skilled consultants, non-executive directors, or management consultants can provide valuable insights and guidance to strengthen your management structure and drive business growth. Collaborating with external experts can help you address challenges and seize opportunities more effectively.

Various business-analysis models offer valuable tools to enhance strategic thinking about your business. Among these, the SWOT analysis (Strengths, Weaknesses, Opportunities, Threats) stands out as particularly popular. This approach involves a comprehensive examination of internal strengths and weaknesses, as well as external opportunities and threats. By identifying these factors, businesses can strategize ways to leverage strengths, mitigate weaknesses, seize opportunities, and mitigate threats effectively.

It's crucial to recognize that opportunities may also pose threats, and threats can sometimes present opportunities. For instance, while new markets may offer growth opportunities, they could also intensify competition, posing a threat to existing market share. Conversely, a competitor's rapid growth may open new market segments, potentially benefiting your business. Employing a SWOT analysis offers a structured framework for evaluating business performance and future prospects, whether as part of routine reviews or in preparation for strategic decisions like fundraising or consulting engagements.

Another useful tool is the STEEPLE analysis, which delves into various external influences on a business, encompassing Social, Technological, Economic, Environmental, Political, Legal, and Ethical factors. Scenario planning aids in exploring multiple plausible future scenarios, enabling businesses to prepare for uncertainty effectively. Critical success factor analysis helps identify key areas crucial for achieving business objectives, while the Five Forces model examines factors influencing market and business development, including potential entrants, existing competitors, buyers, suppliers, and alternative products/services.

As a business owner or member of the management team, periodic strategic reviews are essential to assess business performance comprehensively. Key areas to evaluate include market performance and direction, product and service offerings, operational efficiency, financial health, and organizational structure and personnel management. Identifying internal issues that hinder business progress is particularly important during this review process.

After identifying areas needing attention, businesses can strategize next steps accordingly. Addressing issues promptly is critical to sustaining business viability, especially in its early stages. Once areas of improvement are addressed, businesses can focus on planning the next phase of growth. Developing a cohesive strategy tailored to the business's strengths and opportunities is key to navigating future growth effectively.

Careful planning and ongoing monitoring are essential to executing the chosen growth strategy successfully. By updating the business plan and tracking progress against predefined milestones, businesses can adapt to evolving circumstances and maximize their chances of success in achieving long-term objectives.

Strategic planning is a pivotal aspect of leadership within any organization, encompassing both strategic and operational functions. It falls upon organizational leaders to craft a realistic strategic direction and ensure its dissemination throughout the entire organization. This strategic direction must be grounded in an understanding of the organization's business environment, the influence of stakeholders, and the organization's resource capacity.

Integral to effective strategic planning is the alignment of business unit plans and individual responsibilities with the overarching strategic plan. The strategic plan, formulated after a thorough internal assessment and consideration of environmental trends, serves as a blueprint for resource allocation and procurement aimed at achieving the organization's desired strategic outcomes.

A strategic plan typically comprises goals or objectives to be attained within a defined timeframe. These plans are developed through a comprehensive internal process that involves analysing environmental trends and leveraging the organization's strengths. Subsequently, plans are cascaded throughout the organization, ensuring alignment from departmental plans down to individual objectives.

The values upheld by an organization play a foundational role in shaping the behaviour of its employees, both internally and externally. These values guide employees in their interactions and decision-making processes, thereby influencing the organization's public perception and its products or services' reception.

Moreover, aligning strategic plans and values with the evolving needs and expectations of customers and the broader community is paramount for sustained business success. Given that social values evolve over time, organizations must continuously review their values to ensure alignment with the evolving business environment.

A crucial aspect of strategic planning is the development of a vision that outlines the desired future state of the organization. Whether at the corporate, business unit, or brand level, a vision provides guidance for the organization's trajectory over a specified timeframe. However, the development of a vision often occurs at the senior management level without sufficient consultation with frontline employees, necessitating effective communication strategies to bridge this gap.

Following environmental and business analyses, crafting a mission statement serves as the foundational step for identifying the organization's purpose and desired trajectory. This statement articulates the organization's core values and provides a framework for individual business units to formulate their own plans. Importantly, a mission statement is not static and must evolve in response to changing market dynamics and organizational competencies.

Clear communication of goals and objectives is essential for frontline staff to understand their role in achieving organizational objectives. Goals outline what the organization aims to achieve, while objectives provide measurable outcomes aligned with these goals. By distilling broad goals into measurable objectives, organizations can effectively communicate their strategic priorities and empower employees to contribute to their attainment.

Defining the strategic direction of an organization marks the initial step in aligning and allocating resources to various groups and individuals within the organization. This process involves establishing the organization's objectives, values, and standards, which serve as guiding principles for its operations.

Once the Vision, Mission, Strategic Goals, Objectives, and Strategies of Action are formulated and structured within the organization, the organizational leader must ensure that the responsibility for achieving specific objectives is delegated to the most suitable work units (departments, sections, teams), and individuals. This delegation process often involves the use of Key Performance Indicators (KPIs) linked to Job Descriptions.

Job Descriptions outline the specific tasks and responsibilities associated with positions within the organization, including the required knowledge, skills, and qualifications. Additionally, they delineate levels of authority and reporting relationships. KPIs, on the other hand, establish performance standards that employees must meet to fulfill their work unit goals and objectives, thereby contributing to the organization's overall strategic direction.

The primary objective of this process is to ensure that all individuals and work groups within the organization are aligned with the organization's strategic goals. Ideally, strategic management should be fully integrated throughout all levels of the organization, facilitating alignment and coordination of activities towards the organization's strategic direction.

Linking work teams/groups and individual goals and activities to strategic plans offers several potential advantages, including alignment towards common goals, reduced likelihood of divergence from organizational goals, efficient resource utilization, improved workflow integration, enhanced employee understanding of their contributions to the organization's strategic plan, and increased accountability and motivation.

Furthermore, aligning with the external environment is crucial for organizations, as the public represents their potential customer base. Listening to public needs and concerns, both general community issues and those specific to the organization, is essential for developing a positive public image that influences consumer behaviour and brand loyalty.

With the advent of sophisticated technology and globalization, organizations face new challenges and opportunities in interacting with their external environment. While technology enables immediate and widespread communication, it also exposes organizations to global scrutiny, requiring leaders to navigate diverse public expectations across various communities and countries where they operate. Despite the benefits, organizations must address the challenges posed by emerging technologies and globalization to effectively manage their external environment and maintain a positive public image.

Emphasizing Business Growth

As your business progresses beyond its initial stages, shifting focus towards identifying avenues for growth becomes imperative to ensure sustainability. Evaluating growth can be gauged through various key indicators such as turnover, market share, profits, sales, and staff numbers. However, determining which metric provides the most accurate insight into your business performance relies on factors such as your business type and its current stage of development.

For instance, a retail establishment might boast high sales figures, but if profit margins on inventory are narrow, this could lead to diminished profitability, thus compromising the business's long-term viability. Generally, a balanced approach encompassing both sales and profits serves as an effective measure of growth.

Even if your current performance seems satisfactory, continuously seeking avenues for development is crucial. Failing to do so risks allowing competitors to seize opportunities for growth and erode your market share, potentially weakening your position significantly. Initiating growth efforts often begins with consolidating existing markets, ensuring your business operates efficiently, and maintaining a strong customer base to sustain cash flow—a vital aspect during this phase.

The success of any growth strategy hinges on timing. Answering critical questions such as whether your business can handle expansion without strain, if existing resources and systems can support simultaneous business operations and expansion efforts, and how to mitigate disruptions to current performance due to new initiatives is essential. Adjustments like augmenting staffing, refining production processes, or outsourcing tasks might be necessary to facilitate growth seamlessly.

Expanding market share involves capturing customers from competitors or attracting new ones, necessitating a comprehensive understanding of your customer base and competitors. Addressing questions like who your existing customers are, identifying untapped market segments, and assessing competitors' strengths can provide insights into gaining a larger market share.

Many businesses opt for growth through diversification, although this approach entails risks due to resource constraints. Diversification can manifest in various forms, including introducing new products or services to existing customers, targeting new markets with existing products, or venturing into entirely new products and markets.

Collaborating with other businesses offers advantages such as pooling resources, sharing expertise, and accessing larger markets. Partnerships and joint ventures enable businesses to tap into complementary skills and expand market reach, while mergers and acquisitions facilitate growth by combining forces and resources.

Selecting the most suitable growth strategy necessitates a thorough analysis of your business's current performance, aligning with its strengths, and addressing weaknesses. Ensuring feasibility, practicality, availability of funding, and potential profitability are crucial factors to consider before embarking on a growth strategy. Additionally, planning and measuring progress are vital to track and steer growth effectively.

When embarking on business growth initiatives, it's crucial to meticulously outline all associated costs and compare them against projected profits. Realism and practicality are key when setting growth objectives to ensure that financing development won't jeopardize funding for core activities.

Return on Investment (ROI) serves as a popular method for assessing whether projected figures align with targets. Utilizing the ROI formula helps determine the percentage of return over a specified period, typically three years, a timeframe commonly employed by expanding businesses.

To calculate ROI, the total investment amount is considered alongside anticipated increased sales and resulting net profit, expressed as a percentage of the investment. For instance, envision a business planning to introduce a new product line requiring a $200,000 investment in development, plant, marketing, and promotion. It foresees generating $400,000 in sales and $40,000 in net profit annually.

Testing ROI with various sales scenarios is advisable to account for potential factors like development challenges, delays, or marketing issues that could affect early-stage sales. Adjusting calculations to accommodate annual inflation may also be prudent.

Sound financial planning forms the cornerstone of any growth strategy. Begin by determining the necessary investment, its timing, and availability. A detailed cash flow forecast is indispensable, considering that expenses typically escalate sooner and faster than revenues. Sufficient funds must be allocated to sustain core business operations, with a surplus for unforeseen delays in project fruition.

Comprehensive forecasts covering sales, working capital, and funding sources, including seed funding and potential secondary funding, should be developed. Apart from bank loans, businesses seeking capital investment can explore equity capital from owners, friends, or family, venture capital, business angels, or government grants and loans.

Equity finance involves investing in a business without direct repayment, often sourced from personal funds or external investors acquiring a stake in the business. Venture capital typically targets early-stage businesses, often entailing management involvement. Business angels are private investors offering financial support and valuable business expertise, often in exchange for a minority or majority ownership stake.

Initiating Your Business Plan

Before delving into crafting your business plan, it's imperative to conduct a thorough self-assessment to determine the viability of your business idea. This analysis aids in foreseeing potential challenges and devising strategies to overcome them.

Evaluate the feasibility of your concept. Investigate the demand for your products or services and assess the sustainability of the market. Research your competitors to gauge market competitiveness. Key questions to ponder include:

- What products/services will you offer?

- Is your idea economically viable?

- How will you safeguard your concepts?

- Does a market exist for your products/services?

- What skills are essential?

- Who are your competitors?

- What sets you apart in the market?

- Do you possess the financial resources to launch a business?

Evaluate your readiness to embark on entrepreneurship. Beyond working for oneself, running a small business necessitates managerial acumen, industry proficiency, technical expertise, financial prowess, and a long-term vision for growth and success. Reflect on:

- Why are you initiating a business?

- What are your business and personal objectives?

- What skills do you possess?

- What income level do you aim to achieve?

- What are the pros and cons of entrepreneurship? Business Planning Phase

Upon completing research into your business idea's feasibility, you're prepared to draft your business plan. This blueprint provides guidance, ensures adherence to objectives, and is often a prerequisite for securing financial backing. Depending on your business type, the plan may include the following sections:

- Business Summary: A concise overview post-finalization of the plan.

- About Your Business: Details encompassing business structure, registrations, location, staff, and products/services.

- Market Analysis: Insights into industry, customer, and competitor landscapes, alongside marketing targets and strategies.

- Future Outlook: Vision statement, business goals, and significant milestones.

- Financial Projections: Financing mechanisms, costings, and revenue forecasts. Ongoing Review

Business planning is an iterative process. As your business evolves, strategies outlined in the plan must adapt to ensure alignment with goals. Maintaining an up-to-date plan aids in staying focused and prepared for future contingencies.

A business plan unveils your business's operational framework and future trajectory. Consequently, discretion is crucial when sharing the plan to prevent competitors from gaining insights. Establishing agreements with third parties regarding plan distribution may suffice for some, while others may opt for confidentiality agreements to safeguard innovative practices or products/services.

Numerous government services offer support for business planning, initiation, and growth through advice, workshops, seminars, networking events, and mentorship opportunities. Leveraging these resources can enhance your understanding of business dynamics and bolster your chances of success.

You should also:

- Conduct thorough research before commencing your plan.

- Clarify the purpose and target audience of your plan.

- Prioritize relevant sections, ignoring those that don't apply initially.

- Utilize italicized prompts for guidance but disregard irrelevant questions.

- Seek professional assistance if needed.

- Clearly differentiate between actual and expected figures.

- Craft your summary last, ensuring brevity and persuasiveness.

- Review your plan meticulously to maintain a professional image.

Apart from the sections outlined above, consider including a title page, supporting documentation, and details on distribution and protection measures.

Regularly review and update your plan to reflect business changes. Exercise caution when distributing and protecting your plan to safeguard proprietary information and innovations.

Developing Performance Objectives and Measures for Business Plan, in Consultation with Relevant Stakeholders

You may be contemplating various avenues for business expansion, whether it involves creating new products, refining business processes, or simply staying motivated to progress. Regardless of your choice, incorporating strategies to foster business growth into your business plan can propel you towards your objectives and uncover innovative pathways to success.

To explore inspiration and opportunities for business growth, consider the following approaches: Looking at case studies of other businesses to gather fresh ideas; Acquiring new skills that can enhance your business operations; Staying updated by attending business or industry-related workshops and events; Networking with other business owners or industry professionals to exchange ideas; and. Reviewing your business goals and plans to identify areas for improvement.

If you're wondering how to grow your business, here are some ideas and opportunities to consider:

- Attending business events, such as small business workshops and seminars, which cover various topics including planning, financial management, and innovation.

- Engaging in networking events to expand your business connections and glean insights from other professionals.

- Utilizing government services that offer advice, workshops, seminars, and networking events, and can even provide mentorship or coaching opportunities.

- Participating in mentoring or coaching programs to gain a deeper understanding of business processes and acquire essential skills for growth.

- Conducting market research by collecting information from diverse sources to refine your business and marketing plans, and identify new growth opportunities.

- Benchmarking against other businesses to learn different or improved ways to run your business, and establish networks for skill-sharing and staying updated on industry trends.

- Regularly reviewing and updating your business plan and marketing strategy to ensure alignment with market trends and objectives.

- Assessing your business's financial performance and procedures to identify areas for improvement and enhance efficiency.

- Exploring opportunities to access government tenders and contracts, and considering collective bargaining as a strategy for negotiation.

- Implementing an innovative business culture to generate new ideas and adapt to market trends, facilitating business growth and resilience.

- Considering importing or exporting goods or services to access new markets and diversify revenue streams, while ensuring compliance with regulations.

- Expanding your business online to reach a wider audience, increase brand awareness, and facilitate customer engagement.

If you seek guidance and support for your business endeavours, mentoring services offer valuable resources. Mentors, with their extensive business experience and knowledge, can provide tailored advice and guidance to help you navigate challenges and achieve your business goals, whether you're starting out or expanding your business. Whether delivered one-on-one or in group settings, mentoring can enrich your entrepreneurial journey and expand your business support network.

Developing performance objectives and measures requires careful consultation with key stakeholders to ensure alignment with organizational goals and strategies. Employees are encouraged to set their individual objectives in line with departmental and organizational objectives, considering both internal and external factors that may impact the achievement of these goals.

Setting organizational objectives can be challenging due to the multitude of variables involved. Best practices for setting higher-level objectives include conducting a SWOT analysis to identify strengths, weaknesses, opportunities, and threats. Additionally, considering future industry trends and utilizing methodologies such as the Hedgehog Concept by Jim Collins can help define long-term objectives and vision.

A strategic approach involves envisioning where the organization aims to be in the future, such as in 3 months, 1 year, and 5 years, and aligning objectives accordingly. Utilizing the SMART model—Specific, Measurable, Attainable, Relevant, and Time-bound—ensures that objectives are well-defined and achievable.

Understanding the contributions of various stakeholders is essential in distinguishing between organizational and departmental goals. Engaging employees in the objective-setting process can provide valuable insights and enhance organizational strategies.

Performance measurement is crucial for assessing the effectiveness of organizational activities and ensuring alignment with objectives. This involves evaluating both individual employee performance and organizational performance through metrics such as effectiveness, cost-effectiveness, impact, and best practices.

Employee evaluations should be conducted regularly to provide feedback and identify areas for improvement. Measurement tools, such as establishing targets and assessing outcomes, help gauge performance against objectives.

Organizational evaluations focus on outcomes related to the organization's public purpose, considering factors such as input, output, process, and benchmarks. Key metrics for organizational evaluation include margins, growth, market share, and customer satisfaction/retention, providing insights into overall performance and competitiveness.

Identifying Financial, Human and Physical Resource Requirements for the Business

A resource encompasses any physical or virtual entity with limited availability, essential for the effective functioning of an individual or organization. These resources, which include money, materials, staff, and other assets, are categorized into six broad types: financial, physical, human, technological, reputation, and organizational. Each type plays a vital role in exploiting opportunities, operating a business, and delivering products and services to customers. However, the specific resources required vary depending on the nature, type, and size of the business.

Accessing resources, especially for start-up businesses, can be challenging. It's crucial to identify your business's resource requirements early on and devise strategies for obtaining them. While many resources are acquired over time, during the start-up phase, critical resources typically include capital, knowledge, workspace, and necessary supplies or equipment. The limitations of available resources can significantly impact the nature and scope of your business. For instance, many entrepreneurs initially operate service-based businesses from home due to resource constraints but expand as they acquire more resources.

Strategic planning necessitates careful consideration of resources. Before finalizing and implementing a strategic plan, leaders must ensure sufficient resources are available for each activity at every stage of the plan's execution. Conducting a resource forecast is essential, analysing planned activities, stages, and objectives to identify any discrepancies between required and available resources. Prioritizing resources is crucial, considering factors beyond human resources, such as financial, physical, and systems resources, both internal and external.

Organizational infrastructure, systems, policies, procedures, location, and supplier quality are vital considerations in resource planning. Adequate front-line and support function resources, effective management, and employee involvement are necessary for operational success. Maintaining positive relationships with external stakeholders is crucial, as they provide valuable support for strategic direction. Intangible resources, including goodwill, reputation, and brands, must be managed effectively to support strategic objectives.

Leadership plays a pivotal role in ensuring appropriate resource allocation and guiding the organization toward success. Effective leadership involves leading the organization in line with strategic goals and adapting leadership styles as needed throughout the strategic journey. Ultimately, successful resource planning and allocation are essential for realizing strategic objectives and organizational success.

Developing the Business Plan

Title page of your plan

Regardless of whether you are writing a business, marketing or emergency management plan, here are some suggestions on what to include in the title page of your plan.

Business plan summary

- Insert business logo - Adding a logo gives a more professional image.

- Name and overview of the plan- Is it a business plan, a marketing plan or an Emergency Management and Recovery Plan?

- Your name - Enter the business owner's name. Enter multiple names if there are multiple owners.

- Your title - The titles of the business owner(s) listed above. For example: Owner/Manager.

- Business name - Enter your business name as registered in your state/territory.

- Main business address - Enter your main business address. This can be your home address if you are a home-based business or your head office if you have more than one location.

- Business Registration Details – As applicable to your locality.

- Prepared - The date you finished preparing your plan. This may also be important when making note of a date for revision

- Revision history table - Detail the changes made to your plan: (version number, changes made, person responsible and date updated).

- Communication strategy table - Details of who & how often you will communicate your plan: (manager/staff, type of communication, person responsible and frequency).

- Table of contents - Remember to update the table of contents as you write your plan.

Your business summary should be completed last, be no longer than a page and should focus on why your business is going to be successful. Your answers below should briefly summarise your more detailed answers provided throughout the main body of your business plan.

The business

Under the business summary section enter your:

- Business name as registered If you haven't registered your business name, add your proposed business name.

- Business structureEnter whether you're a sole trader, partnership, trust or company.

- Any relevant business registration number(s)If you're a business and have registered for an ABN enter it here.

- Main business locationEnter your main business location such as your city/town. Briefly describe the location and space occupied/required.

- Date the business was establishedThis is the date you started trading. Whether it was the date you opened your doors or the date your purchased business opened its doors.

- List of business ownersList the names of all business owners.

- Relevant owner experienceBriefly outline your experience and/or years in the industry and any major achievements/awards.

- Products/servicesBriefly list the products/services you're selling and the anticipated demand for your products/services. These answers should briefly summarise your answers under the main products/services section.

The future

Under the future summary section enter your:

- Vision statementYour vision statement should briefly outline your future plan for the business. It should also clearly state your overall goals. To keep it brief you may prefer to use dot points in your summary.

- Goals/objectivesList your short and long term goals and the activities you'll undertake to meet them.

The market

Under the market summary section enter your:

- Target marketBriefly explain who your customers are and why they'd buy your products/services over your competitors. This section should briefly summarise your answers under the main Market section.

- Marketing strategyBriefly explain how you plan to enter the market, how you intend to attract customers and how and why this strategy will work. This section should briefly summarise your answers under the main Market section.

The finances

Under the finances summary section briefly outline your sales forecast or how much profit you intend on making in a particular timeframe. Include how much money you'll need up-front, where you'll source these funds, what portion of funds you'll be seeking from other sources and how much of your own money you'll be contributing towards the business.

This section can be quite brief in the summary. It should give the reader a quick idea of your current financial position, where you're heading financially and how much you need to get there. You can provide more detail in the main Finances section of your business plan.

Business details section

Business information

Under the business information section of your business plan you can include the following information:

- Business purpose – Briefly explain the main purpose of your business, including products/services.

- Business size – Enter the number of employees and or estimated/actual annual turnover of your business.

- Operating history – Enter how long your business has been in operation and the progress you've had to date, including any major milestones achieved.

Registration details

Under the registration details section of your business plan you can include the following information:

- Business name - Enter your business name as registered in your state/territory. If you haven't registered your business name, add your proposed business name.

- Trading names - Enter your registered trading name or names. Your trading name is the name you'll use to trade under. It's the name that appears on your marketing and advertising material. If you have multiple trading names, list them all and explain the purpose of each name.

- Date registered - Enter the date that appears on your business name registration.

- Locations registered – List each state in which you've registered your business name. If your business is located in more than one state, you'll need to register your business name in each state.

- Business structure - Enter whether you're a sole trader, partnership, trust or company.

- Registered Business Number (ABN) - If you're a business and have registered for an ABN, enter it here.

- Registered Company Number (ACN) - Only fill this in if you're a company.

- Domain names - List any domain names you have registered and plan to use in the business. If you're planning on setting up a website for your business, you may also like to register a domain name to match your registered business name.

- Licences and permits - List all the licences or permits you have registered. These will vary depending on your type of business and location. Some examples include a tradesperson certificate, travel agents licence, or kerbside café permit.

Business premises

Under the business premises section of your business plan you can include the following information:

Organisation details

- Business location - Enter your main business location and include details of the location and space occupied/required, the size of the space you occupy/require, the city or town, and the proximity of your location to landmarks/main areas. If you have a retail business, you can also include details of where you're located in relation to other shops, and what the retail traffic is like.

- Buy/lease - If you've purchased a business premises or are currently leasing, briefly outline the arrangements. These can include purchase price, length of the lease agreement, rental cost or lease terms. If you're still looking for a lease or property, outline your commercial lease or purchase requirements and any utilities/facilities required.

Organisation chart

Under this section of your business plan, attach your current organisation chart. An organisation chart is a visual way of representing your business structure. Generally the roles are represented in order of seniority with lines between them representing relationships, responsibilities or ranks of individuals in the organisation. If you're still recruiting staff, you may wish to include a proposed structure and label it appropriately. For example, you can add 'Vacant' to the positions that are currently vacant or you could label the whole chart as a 'Proposed Organisation Chart'.

Example of an organisation chart

Mr J Citizen (owner manager) is on the first rung of the organisation and has two managers that report directly to him, Mr Chris Brantley (Marketing Manager) and Mrs Cherie Laws (Office Manager). Below this second level rung of the organisation there are four additional staff including, Mr John Blue (Events Coordinator) and Ms Fran Reid (Salesperson) who both report to the Marketing Manager, and Mrs Eileen Lawry (receptionist) and Mr Jo Stevens (Operations) who both report to the Office Manager.

Management and ownership

Under the management and ownership section of your business plan you can include the following information:

- Names of owners - List the names of all business owners.

- Details of management & ownership - Detail who'll be running the business (owners or a Chief Executive Officer) and what involvement the owners will have. If it's a partnership, briefly outline for each partner: their percentage share, role in the business, the strengths of each partner and whether you have a partnership agreement/contract in place.

- Experience - Detail what experience the business owner or owners have, why people should invest in you, and how many years you've owned or run a business. Also list any previous businesses owned/managed, any major achievements/awards and any other relevant experience. Don't forget to attach your resume or resumes to the back of your plan.

Key personnel

Under the key personnel section of your business plan you can include the following information.

Current staff

List your current staff. For each job enter the following:

- Job Title - E.g. Marketing/Sales Manager

- Name - E.g. Mr Chris Cole

- Expected staff turnover - E.g. 12-18 Months

- Skills or strengths - E.g. Relevant qualifications in Sales/Marketing. At least 5 years experience in the industry. Award in marketing excellence.

You may also like to attach a copy of their resume to the back of your plan.

Required staff

List your required staff. For each vacant position list the following:

- Job Title - E.g. Office Manager

- Quantity - E.g. 1 position

- Expected staff turnover - E.g. 2-3 Years

- Skills necessary - E.g. Relevant qualifications in Office Management. At least 2 years experience.

- Date required - E.g. Month/Year

Recruitment options

Detail how you will attract required staff. Whether it's advertising in the local paper, online advertising and/or training current staff members.

Training programs

List any training programs you'll be organising in the event you cannot find the required skills. Detail whether these are in-house or external providers. Also include any training you'll undertake as the business owner/manager to keep your skills current.

Skill retention strategies

Detail what procedural documentation you'll provide to ensure the skills of your staff are maintained, how you'll implement an appropriate allocation of responsibilities, and how these responsibilities will be documented and communicated to staff. You can also detail any internal processes you'll implement to regularly check that the current skills of staff members are still appropriate for the business.

Finances section

The finances section of your business or emergency management plan may includes the following information. This list is based on our free templates and downloads to help you get started in the planning process.

Innovation section

- Current creditors tableList all current creditors and any arrangements you have made during the recovery period: (Creditor name, contact details, special arrangement details, period of special arrangement and amount to pay).

- Current debtors tableList all current debtors you have contacted, their agreed payment amount and date: (Debtor name, contact details, details, agreed payment date and amount to receive).

- Government funding tableList all government funding you have applied for and the expected amount: (Program name, contact details, funding details, date of application and amount to receive).

- Expected cash flow tableThis section contains a list of suggested incoming and outgoing cash items your business may have. When completing this section in your Emergency Management plan, this would including your expected incoming and outgoing cash items when recovering from an emergency or disaster type scenario. Don't forget to include the years and months relevant to your situation. Our free plan and template provided includes a cash flow table with the ability to calculate the figures.

Research and development (R&D)/innovation activities

Detail any R&D activities you'll implement to encourage innovation in your business. Include any financial and/or staff resources you plan to allocate.

Intellectual property strategy

Detail how you plan to protect your innovations. List any current trade marks, patents, designs you have registered and if you have any confidentiality agreements in place.

Protecting your innovations can include registering for intellectual property protection, ensuring your staff sign a confidentiality agreement, and generally ensuring your competition does not find out what you are developing.

Legal considerations section

Under the legal considerations section, list all the relevant legislation that will have some impact on the running of your business. For example: consumer law, business law, or specific legislation to your industry.

List the legislation in order of the most impact on your business. You should also detail what you will do or have already done to ensure you comply. You may also like to include details about your disclosure and general obligations.

Insurance and risk management section

Insurance

Under the insurance section of your business plan you can include the following information:

- Workers compensation - Provide details of any workers compensation insurance. This is mandatory if you have employees.

- Public liability insurance - Provide details of any public liability insurance. This covers you for third party death or injury.

- Professional indemnity - Provide details of any professional indemnity insurance. This covers you for legal action taken out as a result of your professional advice.

- Product liability - Provide details of any product liability insurance. This covers you for legal action taken out as a result of injury, damage or death from your product.

- Business assets - Provide details of any insurance you have taken out for your business assets in the event of a fire, burglary, or damage. This insurance covers things like buildings, contents and motor vehicles.

- Business revenue - Provide details if you have insured your business in the event of business interruption. This covers you when you cannot trade because of a particular event and are unable to make money.

Risk management

Under the risk management section of your business plan, list all of the potential risks (in order of likelihood) that could impact your business. For each risk include:

Sustainability plan section

- Risk - Describe the risk and the potential impact to your business

- Likelihood - Rate the likelihood of this risk happening (either highly unlikely, unlikely, likely, or highly likely)

- Impact - Rate the level of impact it may have on your business (high, medium or low)

- Strategy - Detail your strategies for minimising/mitigating each potential risk.

Under the sustainability plan section of your business plan you can include the following information:

Operations section

- Environmental/resource impacts - Describe the impact your business could potentially have on the environment. For example, a particular manufacturing process may contribute negatively on the local water supply. To help you answer this question, you could consider conducting an environmental audit.

- Community impact & engagement – Describe how your environmental impact affects the local community. Detail how you can engage the community in minimising your impact.

- Risks/constraints - List any risks/constraints to your business resulting from any environmental impacts.

- Strategies – Detail any strategies you will implement to minimise/mitigate your environmental impact and any risks to your business. Detail also if you plan on conducting an environmental audit or use an Environmental management system. Consider how you can reduce your energy costs through better managing your energy usage.

- Action plan - List your key sustainability/environmental milestones and include the following details for each:

  - Sustainability milestone - Detail the milestone you are trying to achieve. For example, reduce water consumption.

  - Target - Detail the target you are trying to achieve. For example, reduce water consumption by 60%.

  - Target Date - Enter the date and year you expect to reach each target.

Under the operations section of your business plan you can include the following information:

- Production process – Detail the process involved in producing your products or services. This process will vary depending on your business. Cover any manufacturing processes, the people involved, any third parties involved, and details on how you deliver the product/service to your customers.

- Suppliers - List your main suppliers and detail what they supply to your business. Also explain how you will maintain a good relationship with them.

- Plant & equipment - List your current plant and equipment purchases. These can include vehicles, computer equipment, phones and fax machines. For each item include a description of the plant or equipment (make and model if applicable), purchase date, purchase price and running costs. If you have not purchased all of your equipment yet, you can include a separate table and include an expected purchase date.

- Inventory - List your current inventory items. For each item include:

  - a brief description of the inventory item (including make/model, name or reference number)

  - unit price

  - quantity you have in stock

  - total cost or value of the inventory item (Calculate unit price multiplied by quantity).

If you have a substantial inventory, you may prefer to attach a full inventory list to the back of your business plan.

Products or services section

- Technology (Software) – List the technology you require to run your business. For example: website, point of sale software or accounting package. For each technology solution, detail the main purpose, whether they will be off-the-shelf or purpose built and the estimated cost of each. Although this section focuses on software (as your hardware will be listed above in your plant and equipment table), for more clarity you may also wish to give a brief description of hardware including servers, specialised technology and computer equipment required.

- Trading hours – List your normal trading hours. Detail your expected peak trading times and which times you expect to be more profitable. Also include how this will change over different seasons and how your trading hours will accommodate these changes.

- Communication channels - Explain how your customers can get in contact with you. These channels can include: telephone (landline/mobile), post box, shopfront, email, fax, internet blog or social media channel.

- Payment types accepted - List which payment types you will accept. For example, cash, credit, cheque, gift cards, Paypal.

- Credit policy - Detail the terms of your credit policy for customers/suppliers. Include the length of your credit period, and your collection strategies/procedures. Also detail what credit your business receives and the terms that apply.

- Warranties & refunds - If you manufacture certain goods, detail your warranty terms and your refund/exchange policy for these goods.

- Quality control - Describe your quality control process. Detail any checks or balances you have in place to ensure the product or service you offer is produced to the same standard of quality. Also list the steps you take to meet product safety standards.

- Memberships & affiliations – Provide details of any business memberships or affiliations such as with an industry association or club.

Under the products/services section of your business plan you can include the following information:

Balance sheet

- Details of products/services - List each product/service your business currently offers.

- Market position – Detail where your products/services fit in the market and whether they are high-end, competitive or a low cost alternative to the products/services offered by your competitors. Also detail how this compares to your competitors.

- Unique selling position - Detail how your products/services will succeed in the market where others may have failed. Also include what gives your products/services the edge.

- Anticipated demand - Detail the anticipated quantity of products/services your customers are likely to purchase. For example, detail the quantity of products/services an individual customer will buy in 6 months and 12 months. If you have previous sales data, calculating your inventory turnover may help you anticipate product demand.

- Pricing strategy – If you have a particular pricing strategy, provide details on how your strategy will work and your reasons for choosing this strategy over other alternatives. To help you develop your pricing strategy, you could refer to your customer research, market position, anticipated demand and costs/expenses to get an idea. Whatever strategy you use when setting your price, you will also need to adhere to any relevant fair trading legislation or codes of conduct.

- Value to customer – Explain how your customers view your products/services. Whether they are seen as a necessity, luxury or something in between.

- Growth potential – Detail the anticipated percentage growth of your products/services in the future and explain what you expect will drive this growth. To help you determine this growth potential, you could refer to your research to get an idea of any industry/regional growth that could affect your business in a positive way.

A balance sheet is a snapshot of your business on a particular date. It lists all of your business's assets and liabilities and works out your net assets. A balance sheet can also help you work out your working capital (money needed to fund day-to-day operations) and business liquidity (how quickly you are able to pay your current debts), which can give you a good indication of the financial health of your business.

On your balance sheet, list all your balance sheet items with the dollar amount for the next three years. For each year list the items and total the figures under the headings Total assets and Total liabilities. Use the outline below as your starting point for your balance sheet for each year.

Balance sheet per year

- Current assets

  ◦ Cash

  ◦ Petty cash

  ◦ Inventory

  ◦ Pre-paid expenses

- Fixed assets

  ◦ Leasehold

  ◦ Property & land

  ◦ Renovations/improvements

  ◦ Furniture & fitout

  ◦ Vehicles

  ◦ Equipment/tools

  ◦ Computer equipment

- Total assets (Add up all current assets and fixed assets)

- Current/short-term liabilities

  ◦ Credit cards payable

  ◦ Accounts payable

  ◦ Interest payable

  ◦ Accrued wages

  ◦ Income tax

- Long-term liabilities

  ◦ Loans

- Total liabilities (Add up all short-term and long-term liabilities)

- Net assets (Calculate Total assets minus Total liabilities)

Whether you've already started or intending to start you'll need to fill in actual or estimated figures against each item. If using estimated costs, you'll need to label them clearly. When preparing a balance sheet, ensure you also clearly state whether your figures are GST inclusive or exclusive.

Break-even analysis

The break-even analysis calculates the point where your business has reached a zero balance i.e. when your income covers your expenses exactly. Before you can calculate your break-even point, complete the following details:

- Timeframe (e.g. monthly/yearly)

- Average price of each product/service sold

- Average cost of each product/service to make/deliver

- Fixed costs for the month/year

Once you have your figures above, you can work out your break even by completing the calculations below:

Cash flow statement

- Percentage of price that is profit - Calculate (Average price of each product/service sold minus Average cost of each product/service to make/deliver) divided by Average price of each product/service sold.

- Number of units sold needed to break-even - Calculate Fixed costs for the month/year divided by (Average price of each product/service sold minus Average cost of each product/service to make/deliver).

- Total sales needed to break-even - Calculate Number of units sold needed to break-even multiplied by Average price of each product/service sold.

A cash flow statement can be one of the most important tools in managing your finances. It tracks all the money flowing in and out of your business and can reveal payment cycles or seasonal trends that require additional cash to cover payments. This cycle or pattern can help you plan ahead and make sure you always have money to cover your payments.

On your cash flow statement, list all your incoming and outgoing cash items with the dollar amount for the next 12 months. For each month list the items and total the figures under the headings Cash incoming and Cash outgoing. Use the outline below as your starting point for your cash flow statement for each month:

- Opening balance (in the first month this will be your opening bank balance. In subsequent months this figure will be the closing balance from the previous month)

- Cash incoming

  ○ Sales

  ○ Asset sales

  ○ Debtor receipts

  ○ Other income

- Total incoming (Add up all cash incoming items above)

- Cash outgoing

    ○ Purchases (Stock etc)

    ○ Accountant fees

    ○ Solicitor fees

    ○ Advertising & marketing

    ○ Bank fees & charges

    ○ Interest paid

    ○ Credit card fees

    ○ Utilities (electricity, gas, water)

    ○ Telephone

    ○ Lease/loan payments

    ○ Rent & rates

    ○ Motor vehicle expenses

    ○ Repairs & maintenance

    ○ Stationery & printing

    ○ Membership & affiliation fees

    ○ Licensing

    ○ Insurance

    ○ Superannuation

    ○ Income tax

    ○ Wages

- Total outgoing (Add up all cash outgoing items above)

- Monthly cash balance (Calculate Total incoming minus Total outgoing)

- Closing balance (Calculate Opening balance plus Total incoming minus Total outgoing)

Whether you've already started or intending to start, you'll need to fill in actual or estimated figures against each item. If using estimated costs, you'll need to label them clearly. When preparing a cash flow statement, ensure you also clearly state whether your figures are GST inclusive or exclusive.

Profit and loss statement

A profit and loss or income statement lists your sales and expenses and is generally recorded on a monthly, quarterly or yearly basis. It tells you how much real profit you're making or losing. A profit and loss statement can help you develop sales targets and an appropriate sales price for goods/services using tools like the break-even, profit margin and mark up calculators.

On your profit and loss statement, list all your sales and expense items with the dollar amount for the next three years. For each year list the items and total the figures under the headings Sales and Expenses.

Use the below as starting point for your profit and loss statement for each year.

Profit and loss statement for each year

- Sales

  - Total sales

  - Cost of goods sold

  - Gross profit/net sales (Calculate total sales minus cost of goods sold minus any other expenses related to the production of a good or service)

- Expenses

  - Accountant fees

  - Advertising & marketing

  - Bank fees & charges

  - Bank interest

  - Credit card fees

  - Utilities (electricity, gas, water)

  - Telephone

  - Lease/loan payments

  - Rent & rates

  - Motor vehicle expenses

  - Repairs & maintenance

  - Stationery & printing

  - Insurance

  - Superannuation

  - Income tax

  - Wages

- Total expenses (Total all of your expenses above)

- Net profit (Calculate Gross profit/net sales minus Total expenses)

Whether you have already started or intending to start, you'll need to fill in actual or estimated figures against each item. If using estimated costs, you'll need to label them clearly. When preparing a profit and loss, ensure you also clearly state whether your figures are GST inclusive or exclusive.

Start-up costing of your business

If you're thinking of starting a business, a start-up costing sheet can help you determine how much money you need to start. It can also help you find the right amount of finance and determine how much money you need to invest from other sources.

How much will it cost to start a business?

Your start-up costing sheet can be prepared well before you start your business and can give you a more realistic idea of what it will cost.

Use the items below as a starting point to create your start-up costing sheet, listing all your current or expected start-up costs with each dollar amount. You will need to research each item to determine the cost for your particular business circumstances.

Once you have all of your items listed, total the amounts under your group headings.

Conducting this activity as part of your business planning process will help give you an idea of how much it will cost to start your business. Not all start up costs listed below will be relevant to your business, as these costs will depend on your business structure and type.

We've also included links to our business topic information to help you understand more about the possible costs listed.

Business start-up cost template

These are examples of different items that may incur a cost when starting your business. Do your research for each to determine the costs for your personal circumstances.

- Market research, including conducting primary research or gather research from secondary sources about your market, customers and competitors.

- Registrations, including registering for an ABN and Business name registration

- Licences and permits

- Domain names registration

- Intellectual Property (IP) rights protection, including Trade marks/designs/patents

- Vehicle registration

- Membership fees

- Accountant fees

- Solicitor fees

- Rental lease cost (Rent advance/deposit)

- Utility connections & bonds (Electricity, gas, water)

- Phone connection

- Internet connection

- Computer software

- Training

- Wages

- Stock/raw materials or inventory

- Insurance

  - Building & contents

  - Vehicle

  - Public liability

  - Professional indemnity

  - Product liability

  - Workers compensation

- Business assets

- Business revenue

- Printing

- Stationery and office supplies

- Marketing and advertising

- Online business and website set up

- Total start-up costs (add up all your start-up costs)

Equipment or capital costs

- Business purchase price

- Franchise fees

- Start-up capital

- Plant and equipment

- Vehicles

- Computer equipment

- Computer software

- Phones

- Fax machine

- Security system

- Office equipment

- Furniture

- Shop fit-out

- Total equipment or capital costs (add up all your equipment or capital costs)

Whether you have already started or intending to start you'll need to fill in actual or estimated figures against each cost item. If using estimated costs, you'll need to label them clearly. When preparing a start-up costing sheet, ensure you also clearly state whether your figures are tax inclusive or exclusive

.

Communicating Business Plan to all Relevant Stakeholders

Effective communication is a vital skill for organizational leaders, both internally and externally. It entails transmitting information in a manner that ensures the receiver comprehends it with the same meaning intended by the sender. Timeliness is also crucial, as delayed information may lose its relevance. Within an organization, effective two-way communication is necessary for conveying directives and information to and from the leader. Externally, leaders interact with government agencies, customers, industry bodies, and the media, representing the organization in various contexts.

Understanding the communication process is essential for effective communication, encompassing encoding, transmission, reception, and decoding of messages. Robbins et al. discuss organizational communication, emphasizing formal and informal channels, directional flows, and the role of technology. Different situations require different communication techniques, and avoiding industry-specific jargon is crucial for clarity. While communication technology offers benefits, such as accessibility and speed, it also poses challenges like information overload and the need for constant updates.

Organizational leaders must handle problems professionally and empathetically, utilizing interpersonal, listening, and feedback skills. Empathy facilitates understanding others' perspectives, enhancing problem-solving outcomes. Effective communication can also prevent conflicts and incidents from occurring. Promptly investigating and communicating incidents using appropriate mediums and skills is essential for resolving issues and maintaining the organization's reputation.

Ultimately, the leader's image influences stakeholders' perception of the organization's management and products. Therefore, effective communication is not only about transmitting information but also about managing the organization's reputation and relationships.

Ongoing Review Processes

Monitoring and reviewing a business plan is essential to ensure that the company stays on track towards its objectives and adapts to changes in its environment. This process involves systematically assessing the plan's implementation, identifying any deviations or challenges, and making necessary adjustments to enhance performance and achieve desired outcomes.

The first step in monitoring and reviewing a business plan is to establish clear performance indicators or key performance indicators (KPIs) that align with the organization's strategic goals and objectives. These indicators should be specific, measurable, achievable, relevant, and time-bound (SMART). Examples of KPIs include sales revenue, customer satisfaction scores, profit margins, market share, and employee productivity metrics. These indicators serve as benchmarks for evaluating the plan's progress and effectiveness.

Once KPIs are defined, regular monitoring of performance against these indicators is necessary. This can involve setting up a schedule for data collection and analysis, which may vary depending on the nature of the business and the frequency of key activities. For example, sales figures may be reviewed monthly, while customer feedback surveys may be conducted quarterly. It's important to have a designated individual or team responsible for overseeing the monitoring process to ensure consistency and accuracy.

During the monitoring phase, it's crucial to compare actual performance against the targets or benchmarks established in the business plan. This allows for early identification of any deviations or areas where performance is falling short of expectations. Additionally, it's important to analyze the underlying factors contributing to these deviations, whether they're internal issues such as operational inefficiencies or external factors like changes in market conditions or customer preferences.

Once deviations are identified, the next step is to conduct a thorough review to understand the root causes and determine appropriate corrective actions. This may involve gathering additional data, conducting performance reviews with key stakeholders, and engaging in strategic discussions to explore potential solutions. It's important to involve relevant stakeholders in this review process to gain diverse perspectives and ensure buy-in for any proposed changes.

Based on the findings of the review, adjustments may need to be made to the business plan to address identified issues and improve performance. This could involve revising specific objectives, updating strategies and tactics, reallocating resources, or implementing new initiatives to capitalize on emerging opportunities. The revised plan should be communicated effectively to all stakeholders to ensure alignment and clarity regarding expectations moving forward.

In addition to regular monitoring and review, it's important to incorporate a feedback loop into the process to capture lessons learned and continuously improve the business planning process. This could involve soliciting feedback from employees, customers, suppliers, and other stakeholders, as well as conducting post-mortem reviews after major projects or initiatives. By fostering a culture of continuous improvement and learning, businesses can adapt more effectively to changing circumstances and drive long-term success.

Sample Business Plan

In order to exemplify the processes outlined above, a sample business plan for the hypothetical vertical farming of arugula, butterhead lettuce, red leaf lettuce, and romaine lettuce follows.

Title Page
Vertical Farming Solutions Business Plan
- Agricultural Innovations for Sustainable Urban Agriculture

Prepared by: *John Doe Owner/Manager*
Business Name: *Vertical Farming Solutions*
Main Business Address: *123 Main Street, Cityville, State, ZIP*
Business Registration Details:
- *Business Number: 1234567890 (As applicable)*

Prepared: *February 13, 2024*
Revision History Table:

Version	Changes Made	Person Responsible	Date Updated
1.0	Initial Draft	John Doe	02/13/2024
1.1	Incorporated Feedback from Stakeholders	John Doe	02/20/2024
1.2	Updated Financial Projections	John Doe	03/05/2024

Communication Strategy Table:

Recipient	Type of Communication	Person Responsible	Frequency
Manager/Staff	Monthly Updates	John Doe	Monthly
Investors	Quarterly Reports	John Doe	Quarterly

| Stakeholders | Annual Meeting | John Doe | Annually |

Table of Contents

Business Plan Summary

The Business: Vertical Farming Solutions is dedicated to revolutionizing urban agriculture by employing innovative vertical farming techniques. Specializing in the cultivation of arugula, butterhead lettuce, red leaf lettuce, and romaine lettuce, we aim to provide high-quality, locally grown produce to consumers while promoting sustainability and minimizing environmental impact.

The Future: Our vision is to become a leader in urban agriculture, providing fresh, nutritious produce to communities across the region. Our short-term goals include establishing our brand presence and expanding our customer base, while our long-term objectives involve scaling our operations and diversifying our product offerings.

The Market: Our target market consists of health-conscious consumers, urban dwellers, restaurants, and retailers seeking high-quality, locally sourced produce. By leveraging innovative marketing strategies and partnerships, we aim to differentiate ourselves from competitors and capture market share.

The Finances: We project steady revenue growth, driven by increasing demand for locally grown produce and our commitment to operational excellence. With a focus on efficiency and cost management, we anticipate achieving sustainable profitability and long-term success in the market.

Business Details Section

Business Information:

- *Business Purpose:* Vertical Farming Solutions aims to revolutionize urban agriculture by employing innovative vertical farming techniques to cultivate high-quality, locally grown produce.

- *Business Size:* Estimated annual turnover of $500,000 in the first year of operation, with plans for expansion.

- *Operating History:* Founded in 2023, Vertical Farming Solutions has achieved significant milestones in developing our vertical farming infrastructure and securing partnerships with key stakeholders.

Registration Details:

- *Business Name:* Vertical Farming Solutions

- *Business Structure:* Limited Liability Company (LLC)

- *Business Number:* 1234567890

- *Date Registered:* January 1, 2023

- *Business Location:* Cityville, State

- *Trading Names:* N/A

- *Locations Registered:* State

- *Domain Names:* www.verticalfarmingsolutions.com

- *Licenses and Permits:* All necessary licenses and permits obtained for agricultural operations.

Business Premises:

- *Business Location:* 123 Main Street, Cityville, State, ZIP

- *Buy/Lease:* Leased premises with a 5-year agreement, renewable.

Organisation Details

Organisation Chart: (To be attached separately)

Management & Ownership:

- *Owners:* John Doe (100% ownership)

- *Management:* John Doe (Owner/Manager)

- *Experience:* John Doe has over 10 years of experience in agriculture and business management, with a proven track record of success in sustainable farming practices.

Key Personnel:

- *Current Staff: John Doe (Owner/Manager)*

- *Required Staff: Office Manager (1 position)*

- *Recruitment Options: Online advertising, recruitment agencies.*

- *Training Programs: In-house training programs for staff development.*

- *Skill Retention Strategies: Regular performance evaluations and training updates.*

Finances Section

(To be filled with relevant financial data, including current creditors, debtors, government funding, and cash flow projections.)

Innovation Section

Research & Development (R&D)/Innovation Activities: *Vertical Farming Solutions is committed to continuous innovation in agricultural practices. We will allocate resources for ongoing R&D activities to enhance our vertical farming techniques, optimize crop yields, and develop new varieties of arugula, butterhead lettuce, red leaf lettuce, and romaine lettuce. These activities will be led by our experienced team of agricultural scientists and supported by dedicated financial investments to ensure the success of our innovation initiatives.*

Intellectual Property Strategy: *To protect our innovations, Vertical Farming Solutions will implement a comprehensive intellectual property strategy. This includes obtaining patents for proprietary farming technologies and processes, registering trademarks for our brand identity, and securing confidentiality agreements with key stakeholders. Currently, we have trademark registrations for our brand name and are in the process of filing patents for our unique vertical farming systems. We will continue to monitor and update our intellectual property portfolio to safeguard our competitive advantage in the market.*

Legal Considerations Section

Legal Considerations: *Vertical Farming Solutions will adhere to all relevant legislation governing agricultural operations, business practices, and environmental protection. Key legal considerations include compliance with consumer protection laws, agricultural regulations, and environmental conservation policies. We will also ensure compliance with industry-specific standards and certifications to maintain the highest level of quality and safety in our operations.*

Insurance and Risk Management Section

Insurance:

- *Workers Compensation: Vertical Farming Solutions has obtained workers' compensation insurance to cover any work-related injuries or illnesses suffered by our employees.*

- *Public Liability Insurance: We have secured public liability insurance to protect against third-party claims for injury or property damage arising from our farming activities.*

- *Professional Indemnity: Our business is covered by professional indemnity insurance to mitigate risks associated with professional advice or services provided to clients.*

- *Product Liability: We have product liability insurance to safeguard against legal action arising from product-related injuries or damages.*

- *Business Assets: Vertical Farming Solutions has comprehensive insurance coverage for our business assets, including buildings, equipment, and vehicles, to mitigate financial losses in the event of fire, theft, or damage.*

- *Business Revenue: We have business interruption insurance to provide financial protection in case of disruptions to our operations that result in loss of revenue.*

Risk Management:

- *Risk: Crop Failure Due to Environmental Factors*

  ○ *Likelihood: Likely*

  ○ *Impact: High*

  ○ *Strategy: Implement climate control systems in our vertical farms and diversify crop varieties to minimize the impact of adverse environmental conditions.*

Sustainability Plan Section

Environmental/Resource Impacts: *Vertical Farming Solutions recognizes the potential environmental impacts of our farming operations, including water usage, energy consumption, and waste generation. We will conduct regular environmental audits to assess our resource usage and identify opportunities for improvement.*

Community Impact & Engagement: *We are committed to engaging with the local community to address concerns and promote sustainable farming practices. We will organize educational workshops, open-house events, and community outreach programs to raise awareness about urban agriculture and encourage community involvement in our initiatives.*

Risks/Constraints: *Potential risks to our business resulting from environmental impacts include regulatory compliance issues, reputational damage, and operational disruptions. We will proactively mitigate these risks by implementing sustainable farming practices and maintaining open communication with regulatory authorities and community stakeholders.*

Strategies: *To minimize our environmental impact and mitigate business risks, Vertical Farming Solutions will:*

- *Invest in energy-efficient farming technologies*

- *Implement water conservation measures*

- *Recycle organic waste for composting*

- *Partner with local organizations for environmental conservation initiatives*

Action Plan:

- *Sustainability Milestone: Reduce Water Consumption*

  ○ *Target: Reduce water consumption by 30%*

  ○ *Target Date: December 31, 2025*

Operations Section

Production Process: *Our production process involves seeding, germination, cultivation, and harvesting of arugula, butterhead lettuce, red leaf lettuce, and romaine lettuce in our vertical farming systems. We employ hydroponic and aeroponic techniques to optimize plant growth and minimize resource usage. Our team of skilled workers is responsible for monitoring crop health, maintaining equipment, and ensuring product quality throughout the production cycle.*

Suppliers: *We source seeds, nutrients, and equipment from reputable suppliers with a proven track record of quality and reliability. Maintaining strong relationships with our suppliers is essential to ensure timely delivery of materials and minimize production disruptions.*

Plant & Equipment: *Our current plant and equipment inventory include LED grow lights, nutrient delivery systems, climate control units, and harvesting tools. We regularly maintain and upgrade our equipment to ensure optimal performance and efficiency in our farming operations.*

Inventory: *We maintain inventory of seeds, nutrients, and packaging materials to support our production activities. Our inventory management system tracks stock levels, facilitates ordering, and minimizes wastage through efficient inventory control practices.*

Technology (Software): *We utilize specialized software for crop monitoring, inventory management, and sales tracking. Our custom-built software integrates data from our farming operations to provide real-time insights and optimize decision-making processes.*

Trading Hours: *Our trading hours are Monday to Friday, 8:00 AM to 6:00 PM. We adjust our trading hours seasonally to accommodate peak demand periods and ensure optimal customer service.*

Communication Channels: *Customers can contact us via phone, email, or social media channels for inquiries, orders, and support. We maintain active communication channels to engage with customers and address their needs promptly.*

Payment Types Accepted: *We accept cash, credit/debit cards, and electronic payments for purchases. Payment terms for bulk orders are negotiable based on customer requirements.*

Credit Policy: *Our credit policy offers flexible payment terms for corporate clients and wholesale buyers. We conduct credit checks and establish credit limits to manage credit risks effectively.*

Warranties & Refunds: *We provide warranties for our products against defects in materials and workmanship. Refunds or exchanges are offered for damaged or unsatisfactory products, subject to our refund policy.*

Quality Control: *We implement rigorous quality control measures to ensure product consistency and safety. Our quality assurance team conducts regular inspections and tests to uphold industry standards and meet customer expectations.*

Memberships & Affiliations: *Vertical Farming Solutions is a member of the Urban Agriculture Association and actively participates in industry events and initiatives to promote sustainable farming practices and community engagement.*

Chapter Sixteen

Developing a Farm Plan

Determining Business Direction and Farm Planning

Managing a farm effectively requires meticulous planning that encompasses both the physical development of the property and the growth of the farm business itself. Such planning is often long-term, spanning years or even decades, to ensure sustainable growth and resilience in the face of challenges. Various aspects need to be considered in this planning process, including facility upgrades, operational expansions, strategic shifts, major maintenance tasks, staff training, holiday schedules, and contingency measures to mitigate risks such as economic downturns, droughts, or illnesses.

A Whole Farm Plan is a comprehensive approach to farm planning that considers natural resources, economic factors, and environmental sustainability. It involves assessing the farm's potential and limitations, designing farm enterprises and management practices to align with these factors, and integrating environmental considerations into overall farm management. This planning process aims to improve farm efficiency, guide sustainable business practices, and protect sensitive environmental areas while maximizing the farm's potential for profitability.

The benefits of developing a Whole Farm Plan are both immediate and long-term. In the short term, it provides a thorough assessment of current resources and helps facilitate changes in farm enterprises or preparations for extreme weather events. In the long term, it enhances overall farm management, integrates environmental planning into property management, identifies opportunities for improvement, and prepares for future challenges such as climate change. Additionally, it fosters a holistic understanding of the farm's capabilities and limitations, allowing for better decision-making and long-term sustainability.

Short-term operations may involve buying and developing a farm for immediate gains, which requires short-term planning focused on rapid improvements and eventual resale. However, for most farmers, long-term planning is essential, considering that they often live on the same property for decades or even generations. Farm business structures, such as partnerships or family companies, need clear terms and conditions to avoid conflicts and ensure long-term viability.

Quality Management Systems involve documenting and implementing procedures to ensure consistency and compliance with standards. Whole Farm Planning takes a holistic approach to farm planning, considering all farm assets and aspirations over multiple generations. Conservation is a crucial aspect, emphasizing the protection and rehabilitation of natural resources on the farm.

Diversification is increasingly important for farms facing financial pressures, offering multiple income sources to mitigate risks. Possibilities for diversification include accommodation, tourism, agroforestry, alternative animals or crops, and value-added processing of farm produce. With creativity, hard work, and strategic marketing, diversification can bring additional income while leveraging existing farm assets. It's essential to choose activities aligned with personal interests and market demand for successful diversification endeavours.

Preparing a comprehensive farm plan requires careful consideration of various factors, which may differ based on the location, size, and scale of the proposed development. The level of detail required to demonstrate sound management varies, influencing whether a partial or complete farm plan is necessary to support the application process. Understanding the zoning regulations and requirements is crucial, especially in rural areas, where different zones have specific guidelines governing land use and development.

A Whole Farm Plan (WFP) should encompass a continuous improvement process involving planning, implementation, evaluation, and review. It begins with identifying existing conditions on the farm, such as topography, natural features, built structures, soil types, and land uses. Future conditions should also be outlined, including proposed land use changes, new infrastructure, and farming practices.

Core Components of a Whole Farm Plan

1. Farm Planning Process: This core process outlines the management steps for delivering a farm planning service, emphasizing a continuous improvement approach based on landholder vision, values, and purpose.

2. Client Focus and Integrated Services: This process addresses landholders' needs and regional priorities, establishing networks with relevant service providers to ensure comprehensive support.

3. Risk Management: Identifying and mitigating on-site and off-site impacts related to property, business, and personnel management is essential to minimize potential risks.

4. Land Capability and Soils: Understanding land classes and soil properties helps determine suitable management practices for different enterprises while considering climate change effects.

5. Water: Assessing water requirements, sources, quality, and sustainable management options is crucial, particularly in irrigated areas where efficient water application and drainage control are vital.

6. Biodiversity: Identifying native flora and fauna species and implementing management strategies to protect and enhance biodiversity contribute to sustainable farming practices.

7. Biosecurity: Recognizing biosecurity issues, assessing their impacts, and implementing appropriate management measures help prevent and control potential threats to farm operations and the environment.

A comprehensive Whole Farm Plan integrates these core components to ensure effective management, sustainability, and compliance with regulatory requirements while supporting the long-term viability of the farm.

Establishing long-term directions and goals for a farm business involves a comprehensive assessment of various factors, particularly the values, expectations, and goals of stakeholders. This process begins by engaging with all relevant stakeholders, including owners, managers, employees, and possibly even community members or regulatory bodies. Through interviews, surveys, or workshops, stakeholders' values, expectations, and aspirations for the farm business are identified and analysed.

Once stakeholder values and expectations are understood, the next step is to align them with the overarching vision for the farm. This vision should encapsulate the long-term direction and goals of the business, taking into account not only financial objectives but also environmental and social considerations. For example, stakeholders may prioritize sustainable land management practices or community engagement initiatives alongside profitability targets.

After establishing the long-term direction and vision, it's essential to conduct a SWOT (Strengths, Weaknesses, Opportunities, Threats) analysis to assess the internal and external factors that may impact the achievement of these goals. This analysis involves evaluating the farm's internal strengths and weaknesses, such as its resources, capabilities, and operational efficiency, as well as external opportunities and threats arising from factors like market trends, regulatory changes, or environmental risks.

Identifying business and personal strengths involves recognizing the areas where the farm excels, such as skilled labour, access to quality land or water resources, or strong customer relationships. Conversely, weaknesses may include limited access to capital, outdated infrastructure, or inefficiencies in production processes. Personal strengths and weaknesses refer to the attributes and skills of individuals involved in the farm business, including owners, managers, and employees.

Similarly, identifying opportunities involves recognizing potential avenues for growth, diversification, or innovation within the farm business. This could include emerging market trends, technological advancements, or opportunities for collaboration with other stakeholders. On the other hand, threats encompass external factors that pose risks to the farm's success, such as economic downturns, adverse weather events, or changes in regulatory requirements.

Once the SWOT analysis is complete, strategies can be developed to address the identified strengths, weaknesses, opportunities, and threats in a manner consistent with the farm's vision and long-term goals. These strategies should leverage the farm's strengths to capitalize on opportunities while mitigating weaknesses and minimizing the impact of threats. For example, if a farm's SWOT analysis reveals a lack of access to capital as a weakness, one strategy could involve seeking grants or loans to invest in infrastructure upgrades or diversification initiatives.

Documenting these strategies is crucial for ensuring clarity, accountability, and alignment throughout the organization. A written strategic plan should outline specific actions, timelines, responsibilities, and performance indicators for each strategy. Regular review and monitoring of progress against the plan allow for adjustments as needed to stay on track towards achieving the farm's long-term directions and goals.

Setting the strategic direction of any farm enterprise, whether it's focused on beef production or any other agricultural activity, holds significant importance. A clear strategic direction provides a roadmap for the future, guiding decisions and actions to achieve both financial and non-financial objectives. Whether this plan is formalized in writing or exists informally, it serves as a tool for clarifying goals, accessing shared objectives within the business, and adapting to changing circumstances.

In essence, strategic planning enables farm managers to navigate uncertainties and risks inherent in agricultural operations while striving to achieve sustainable profitability. By outlining clear goals, pathways, and steps to achieve them, strategic planning helps in managing resources effectively. This includes time, management skills, and financial investments, which are often limited on farms. Regular review of the strategic plan allows for adjustments in response to changing market conditions, climate variations, and other factors affecting the business environment.

For a beef enterprise, setting a strategic direction involves delineating objectives, identifying pathways for profitability, and ensuring sustainability over the long term. Key elements of a sound strategic plan include outlining goals, improving profitability while managing climate variability, and providing flexibility to capitalize on favourable conditions.

Moreover, strategic planning facilitates communication and decision-making within the enterprise, especially when multiple stakeholders are involved in ownership or management. It enables stakeholders to align their efforts and investments towards common goals, avoiding wasted resources on distractions and enhancing job satisfaction through clear progress tracking.

In practical terms, the strategic direction of a farm enterprise can be shaped by various tools and analyses. One such tool is the enterprise profitability tree, which highlights key areas impacting productivity and profitability. Assessing the current position of the business involves reviewing farm resources, financial performance, natural resource management, marketing strategies, and risk management practices.

Analysing farm business performance entails evaluating both physical and financial aspects over a period of time to identify trends and areas for improvement. Calculating the cost of production provides insights into efficiency and financial risk, while understanding feed supply and demand helps optimize resource utilization.

Interpreting farm business analysis involves comparing performance against industry benchmarks and conducting SWOT analyses to identify strengths, weaknesses, opportunities, and threats. Managing risks is integral to strategic planning, requiring measures such as maintaining a low cost structure, diversifying income streams, building reserves, and implementing effective management systems.

Continual monitoring and review of business performance, cash flow, profitability, and risk assessments ensure ongoing alignment with the strategic direction. This iterative process of planning, execution, and evaluation forms the foundation for effective farm management and long-term success.

Establishing business goals and objectives is a fundamental step in the ongoing planning process for any farm. This process begins with the development of a mission statement and the identification of goals that encompass various aspects of the business. The mission statement serves as a clear declaration of purpose, guiding decision-making in alignment with the farm's values and objectives. It should succinctly describe the target market, products or services offered, and what sets the farm apart from others.

When setting goals and objectives, it's essential to ensure they are specific, measurable, achievable, realistic, and time-bound (SMART). These goals should cover personal, social, environmental, and financial dimensions of the business and be revisited regularly to maintain focus. Prioritization of goals is necessary, considering the levels of planning involved—strategic decisions for long-term direction, tactical decisions for medium-term steps, and operational decisions for day-to-day activities.

Tools and guidelines are available to aid in the process of setting goals and objectives. These resources provide a structured approach to analysing the farm's various aspects, including business structure, financial management, production management, marketing, staff management, and risk management. Professional assistance may also be sought to facilitate the development of formal business plans.

Measuring progress towards these goals requires establishing concrete criteria and regularly assessing performance against key performance indicators (KPIs). The frequency of measurement depends on the specific targets, with operational measures often requiring weekly assessment and overall farm business analysis conducted annually.

After establishing goals, it's crucial to explore options for improvement and evaluate their benefits and feasibility. This involves considering various strategies, from simple modifications to complex changes, and assessing their potential impact on profitability, resource management, and lifestyle. Different analytical techniques, such as gross margin analysis, partial budgets, and discounted cash flow analysis, can help compare these options effectively.

Risk management is an integral part of planning any new enterprise or implementing changes. Sensitivity analysis and scenario planning should be conducted to assess the potential impact of various factors, such as market fluctuations and production risks. It's essential to manage the risks associated with transitioning enterprises to avoid destabilizing the farm business.

Constraints, including considerations of quality of life and environmental concerns, should also be factored into decision-making. While some benefits may be difficult to quantify, it's essential to weigh the pros and cons carefully when evaluating options.

Regular reviews and assessments are necessary to ensure that strategies are yielding the expected results and to make adjustments as needed. By following these guidelines and continually monitoring progress, farms can effectively establish and work towards achieving their business goals and objectives.

The farm planning process encompasses various steps aimed at effectively managing the property and achieving long-term goals. Initially, property management plans (PMPs) can range from basic maps outlining key actions to comprehensive reports covering all aspects of farm business and natural resource management. The level of detail in a PMP depends on the time invested in its development and its alignment with the property owner's vision.

The first step in farm planning involves clarifying the property owner's vision for the land and their lifestyle aspirations. This entails envisioning the property's potential over the long term and identifying necessary infrastructure and natural resources to realize this vision. As understanding of the property's characteristics grows, adjustments to the vision may be necessary to better align with its capacity.

Surveying the land constitutes the second step, providing insights into the property's layout, physical features, and existing conditions. Mapping the property allows for the identification of built infrastructure, natural assets, and potential risks. It also involves consulting local planning schemes to ensure alignment with land use regulations.

Gathering information about the property, including topography, soil types, and climate, is crucial for understanding its capability to support various land uses. This information informs decisions about land management practices and the placement of different activities to optimize productivity and sustainability.

Setting goals and objectives follows, focusing on natural resource management, lifestyle preferences, financial considerations, and production targets. These goals should be reflective of the property owner's vision and identify priority areas for action.

The next step involves creating action plans tailored to achieve the defined objectives. Each action plan should be specific, achievable, and aligned with time and financial constraints. Prioritizing objectives based on their feasibility and impact helps in resource allocation and maximizing outcomes.

Budgeting and cash flow analysis are essential to ensure the financial viability of the property management plan. This involves estimating costs associated with property maintenance, implementing action plans, and exploring new income opportunities. Developing a cash flow chart helps in understanding funding sources and managing expenses effectively.

Finally, turning the property management plan into reality requires flexibility and regular review. Property owners should stay informed about best practices and be open to adapting their plans as needed. Regular review of the PMP ensures alignment with evolving goals and changing circumstances, thereby maximizing the likelihood of achieving the desired property vision.

Conducting a SWOT analysis for farm planning involves a systematic evaluation of both internal strengths and weaknesses within the farm operation, as well as external opportunities and threats present in the broader agricultural environment. This comprehensive assessment enables farmers to gain valuable insights into their farm's current position in the market and develop informed strategies to enhance competitiveness, sustainability, and resilience in an ever-changing agricultural landscape.

In identifying strengths (S), farmers should consider what sets their farm apart from others in the industry. This could encompass various factors such as fertile soil, access to water sources, skilled labor, modern equipment, diversified crops or livestock, strong brand reputation, or efficient operational processes. By evaluating internal resources and capabilities that contribute to the farm's competitive advantage, farmers can better leverage these strengths to optimize performance and profitability.

On the other hand, recognizing weaknesses (W) involves reflecting on areas where the farm may be lacking or underperforming. This could include outdated machinery, limited financial resources, poor infrastructure, inadequate workforce training, vulnerability to weather conditions, or inefficient management practices. By honestly identifying internal constraints or challenges, farmers can take proactive measures to address these weaknesses and enhance the farm's overall competitiveness in the market.

Exploring opportunities (O) entails identifying external factors or trends in the agricultural industry that could benefit the farm. These opportunities may arise from emerging market demands for specific products, technological advancements in farming equipment or techniques, favourable government policies or subsidies, expansion into new markets, or collaborations with other farms or agribusinesses. By capitalizing on these opportunities, farmers can strategically grow revenue, improve efficiency, and enhance sustainability.

Assessing threats (T) involves evaluating external factors that may pose risks or challenges to the farm's operations or profitability. These threats could stem from market fluctuations, changes in consumer preferences, increased competition from larger farms or agribusiness conglomerates, environmental concerns such as climate change or water scarcity, regulatory changes, or disruptions in the supply chain. By anticipating potential threats and developing strategies to mitigate or adapt to these risks, farmers can safeguard their farm's viability and sustainability.

Following the analysis, farmers should prioritize key areas for improvement or action and develop strategies that leverage strengths and opportunities while addressing weaknesses and mitigating threats. This may involve investing in technology upgrades, diversifying crops or livestock, improving marketing efforts, enhancing sustainability practices, or seeking financial assistance or partnerships. Establishing clear goals and action plans based on the SWOT analysis will guide decision-making and resource allocation in farm planning and management.

The farm business direction is determined from the SWOT analysis by synthesizing the findings to identify strategic priorities and charting a course of action that capitalizes on strengths, addresses weaknesses, seizes opportunities, and mitigates threats. Here's how the SWOT analysis informs the farm business direction:

1. Leveraging Strengths: The analysis identifies the farm's internal strengths, such as fertile soil, skilled labour, or modern equipment. These strengths can be leveraged to capitalize on market opportunities, differentiate the farm from competitors, and enhance overall performance. For example, if the farm has a reputation for producing high-quality organic crops, the business direction may focus on expanding the organic product line to meet growing consumer demand.

2. Addressing Weaknesses: Weaknesses identified in the SWOT analysis, such as outdated machinery or inadequate infrastructure, present areas for improvement. The farm business direction should include strategies to address these weaknesses and strengthen the farm's competitive position. For instance, if the farm lacks sufficient workforce training, the direction may involve investing in employee development programs to enhance skills and efficiency.

3. Seizing Opportunities: Opportunities identified in the external environment, such as emerging market trends or favourable government policies, offer avenues for growth and expansion. The farm business direction should prioritize strategies to capitalize on these opportunities and maximize returns. For example, if there is increasing demand for sustainable farming practices, the direction may involve transitioning to organic farming methods to meet consumer preferences and access premium markets.

4. Mitigating Threats: Threats identified in the SWOT analysis, such as market fluctuations or environmental risks, pose potential challenges to the farm's success. The business direction should include proactive measures to mitigate these threats and safeguard the farm's viability. For instance, if the farm is vulnerable to weather-related risks, the direction may involve diversifying crops or investing in irrigation systems to reduce dependence on rainfall.

The farm business direction derived from the SWOT analysis should be strategic, actionable, and aligned with the farm's long-term goals and objectives. By leveraging strengths, addressing weaknesses, seizing opportunities, and mitigating threats, farmers can navigate uncertainties in the agricultural landscape and position their businesses for sustainable growth and success. The following provides an example of the process for a hypothetical agribusiness with livestock and crops.

Example of Hypothetical Business SWOT Analysis for Agribusiness:
Strengths (S):

1. ***Fertile Land***: *The agribusiness owns large tracts of fertile land suitable for both crops and livestock grazing.*

2. ***Diversified Operations***: *The business has both livestock (cattle and sheep) and crop (corn and soybeans) operations, providing multiple revenue streams.*

3. ***Modern Equipment***: *The agribusiness has invested in state-of-the-art machinery and equipment for planting, harvesting, and livestock management.*

4. ***Skilled Workforce***: *The business employs experienced farmers, agronomists, and veterinarians who contribute to efficient operations and high-quality production.*

5. ***Established Market Presence***: *The agribusiness has a strong reputation in the local market for producing high-quality crops and livestock products.*

Weaknesses (W):

1. ***Seasonal Dependence***: *The business is vulnerable to seasonal fluctuations in crop yields and market prices, impacting revenue stability.*

2. ***Limited Water Resources***: *The availability of water for irrigation is limited, leading to dependency on rainfall and potential yield variability.*

3. ***Aging Infrastructure***: *Some farm buildings and facilities are outdated and require maintenance or renovation.*

4. ***Market Dependency***: *The agribusiness relies heavily on a few key buyers for its livestock and crop products, posing risks in terms of market volatility and bargaining power.*

5. ***Inefficient Supply Chain***: *Inefficiencies in the supply chain lead to higher transportation costs and longer delivery times for inputs and products.*

Opportunities (O):

1. ***Growing Demand for Organic Products***: *There is increasing consumer demand for organic crops and grass-fed livestock*

products, presenting an opportunity for premium pricing and market differentiation.

2. ***Technological Advancements***: *Adoption of precision agriculture technologies, such as drones and IoT sensors, can improve farm productivity, resource efficiency, and decision-making.*

3. ***Value-Added Products***: *The agribusiness can explore opportunities to develop value-added products, such as branded meat products or specialty crops, to capture higher margins and expand market reach.*

4. ***Export Markets***: *Exploring export markets for livestock and crop products can diversify revenue streams and reduce dependency on local market conditions.*

5. ***Government Subsidies***: *Government subsidies and incentives for sustainable farming practices or renewable energy production can provide financial support and cost-saving opportunities.*

Threats (T):

1. ***Market Price Volatility***: *Fluctuations in commodity prices and market demand pose risks to revenue and profitability.*

2. ***Climate Change***: *Increasing frequency of extreme weather events, such as droughts or floods, can impact crop yields and livestock health.*

3. ***Regulatory Changes***: *Changes in government regulations related to agriculture, such as environmental standards or trade policies, may require compliance costs or affect market access.*

4. ***Competitive Pressure***: *Competition from larger agribusiness conglomerates or international producers can erode market share and pricing power.*

5. ***Pandemics and Disease Outbreaks***: *Outbreaks of livestock diseases or pandemics, such as foot-and-mouth disease or zoonotic infections, can disrupt operations and market access.*

Business Direction Derived from SWOT Analysis:

1. ***Leveraging Strengths***: *Capitalize on fertile land and modern equipment to optimize crop yields and livestock productivity. Invest in workforce training to enhance operational efficiency and product quality.*

2. ***Addressing Weaknesses***: *Upgrade irrigation infrastructure to mitigate water scarcity risks and improve crop resilience. Renovate aging facilities to ensure operational reliability and compliance with safety standards. Diversify customer base to reduce dependency on a few key buyers and strengthen market position.*

3. ***Seizing Opportunities***: *Expand organic farming practices and grass-fed livestock production to meet growing consumer demand for premium products. Embrace precision agriculture technologies to enhance resource management and maximize productivity. Explore value-added product development to capture higher margins and tap into niche markets.*

4. ***Mitigating Threats***: *Implement risk management strategies to hedge against market price volatility and climate-related risks. Stay abreast of regulatory changes and adapt operational practices accordingly to ensure compliance and market competitiveness. Enhance biosecurity measures to mitigate the risk of disease outbreaks and protect livestock health.*

Overall, the business direction for the agribusiness should focus on optimizing operations, diversifying revenue streams, embracing innovation, and mitigating risks to enhance competitiveness, sustainability, and resilience in the dynamic agricultural landscape.

To outline the portability of the approach, the following example is a revised SWOT for the hypothetical business allowing for possible diversification of operations to include vertical farming of arugula, butterhead lettuce, red leaf lettuce, and romaine lettuce.

Revised SWOT Analysis for Hypothetical Agribusiness with Vertical Farming Expansion:

Strengths (S):

1. **Fertile Land**: *The agribusiness owns large tracts of fertile land suitable for both traditional farming and vertical farming operations.*

2. **Diversified Operations**: *With existing livestock (cattle and sheep) and crop (corn and soybeans) operations, the business already has multiple revenue streams and operational expertise.*

3. **Modern Equipment**: *The agribusiness has invested in state-of-the-art machinery and equipment for traditional farming, which can be repurposed or complemented for vertical farming operations.*

4. **Skilled Workforce**: *The business employs experienced farmers, agronomists, and technicians who can adapt to the requirements of vertical farming with appropriate training.*

5. **Established Market Presence**: *The agribusiness has a strong reputation in the local market for producing high-quality crops and livestock products, providing a solid foundation for introducing new vertical farming products.*

Weaknesses (W):

1. **High Initial Investment**: *Establishing vertical farming infrastructure requires significant upfront investment in equipment, technology, and facility construction, which may strain financial resources.*

2. **Technical Expertise**: *Operating vertical farming systems demands specialized knowledge and skills in hydroponics, lighting, and climate control, potentially requiring additional training or hiring qualified personnel.*

3. **Market Uncertainty**: *The demand for specialty greens from vertical farming may be uncertain, necessitating market research and strategic marketing efforts to establish and sustain consumer demand.*

4. **Space Limitations**: *Limited space availability within existing facilities or on-farm locations may constrain the scalability of vertical farming operations and production capacity.*

5. **Operational Complexity**: *Managing both traditional farming and vertical farming operations simultaneously may increase operational complexity and management challenges.*

Opportunities (O):

1. **Expanding Market Reach**: *Vertical farming enables the agribusiness to access new market segments, including urban areas with high demand for locally grown produce and upscale restaurants and grocery stores seeking premium specialty greens.*

2. **Premium Pricing**: *Fresh, locally grown specialty greens from vertical farming can command premium prices in the market, offering opportunities for higher profit margins and revenue growth.*

3. **Product Innovation**: *Continuous experimentation and innovation in vertical farming techniques allow for the development of new varieties of arugula, butterhead lettuce, red leaf lettuce, and romaine lettuce tailored to consumer preferences and market trends.*

4. **Partnerships and Collaborations**: *Collaborating with restaurants, chefs, and food retailers to supply fresh, locally grown specialty greens can enhance brand visibility and market penetration.*

5. **Sustainability Credentials**: *Vertical farming aligns with consumer preferences for sustainable farming practices, offering the agribusiness an opportunity to enhance its sustainability credentials and appeal to environmentally conscious consumers.*

Threats (T):

1. ***Competitive Landscape****: Increased adoption of vertical farming by other agribusinesses or new market entrants may intensify competition and pressure prices, affecting profitability.*

2. ***Technological Risks****: Technical malfunctions or disruptions in vertical farming systems, such as equipment failures or climate control issues, can lead to production downtime and yield losses.*

3. ***Supply Chain Vulnerabilities****: Dependency on external suppliers for inputs such as seeds, nutrients, and equipment for vertical farming may expose the agribusiness to supply chain disruptions and price fluctuations.*

4. ***Regulatory Challenges****: Compliance with regulations governing food safety, labeling, and organic certification for vertical farming products may impose additional administrative burdens and costs.*

5. ***Consumer Preferences****: Shifting consumer preferences or market trends away from specialty greens produced through vertical farming could impact demand and market acceptance of the agribusiness's products.*

Business Direction: *Given the opportunities presented by vertical farming expansion and the strengths that can be leveraged, the agribusiness should pursue the following strategic directions:*

1. ***Investment in Vertical Farming Infrastructure****: Allocate resources to establish vertical farming facilities equipped with advanced technology and infrastructure to support the cultivation of arugula, butterhead lettuce, red leaf lettuce, and romaine lettuce.*

2. ***Market Research and Branding****: Conduct thorough market research to identify target consumer segments and preferences for specialty greens, and develop a compelling brand identity and marketing strategy to differentiate the agribusiness's vertical farming products.*

3. ***Partnerships and Distribution Channels****: Forge strategic partnerships with local restaurants, supermarkets, and food service providers to secure distribution channels and market access for vertical farming products.*

4. ***Sustainability and Quality Assurance****: Emphasize sustainable farming practices and quality assurance measures in vertical farming operations to uphold product integrity, meet regulatory standards, and enhance consumer trust and loyalty.*

5. ***Continuous Innovation and Adaptation****: Foster a culture of innovation and agility to respond to evolving market trends, technological advancements, and consumer preferences in the vertical farming sector.*

By capitalizing on its strengths, addressing weaknesses, and seizing opportunities while mitigating threats, the agribusiness can successfully navigate the complexities of vertical farming expansion and position itself for long-term growth and competitiveness in the agricultural industry.

Auditing Natural Resources and Infrastructure of Property

Surveying the land and recording its physical characteristics, natural resources, and soil characteristics is a foundational step in developing a comprehensive farm plan that integrates natural resource management with business objectives and production plans. The process includes:

1. Surveying the Land: Conduct a thorough survey of the entire property, taking note of its size, boundaries, and any significant landmarks or features. This survey should include both physical inspections of the land and the collection of relevant data

through maps, satellite imagery, and other sources.

2. Recording Physical Characteristics: Document the topography, elevation, slope, and drainage patterns of the land. Identify any water bodies, such as rivers, streams, ponds, or wetlands, as well as any man-made structures like roads, buildings, fences, or irrigation systems.

3. Assessing Natural Resources: Evaluate the availability and quality of natural resources on the property, including water sources, vegetation cover, wildlife habitats, and biodiversity. Consider factors such as soil fertility, erosion potential, and microclimate variations across different areas of the farm.

4. Producing a Soil Map: Create a detailed soil map of the property, delineating different soil types, textures, and fertility levels. Use classification terminology such as the USDA soil classification system or local soil survey standards to categorize and label each soil type accurately.

5. Identifying Water Resources: Identify and assess the condition of water resources on the property, including surface water sources like rivers, lakes, or ponds, as well as groundwater aquifers or wells. Evaluate water quality, availability, and potential risks such as pollution or over-extraction.

6. Determining Land Capability: Assess the land capability and suitability for various agricultural or land use purposes based on soil characteristics, slope, drainage, and other factors. Classify different land classes or zones based on their productivity, constraints, and management requirements.

7. Mapping Natural Features and Infrastructure: Create a detailed property map that overlays natural property features, such as vegetation types, water bodies, and topographic variations, with existing infrastructure like buildings, roads, fences, and irrigation systems. Note the condition and functionality of each infrastructure element.

8. Identifying Areas at Risk of Soil Degradation: Identify and prioritize areas of the property that are at risk of soil degradation due to factors such as erosion, compaction, salinity, or nutrient depletion. Implement measures to mitigate these risks, such as erosion control practices, soil conservation techniques, or improved land management practices.

9. Classifying Native Vegetation: Classify and map the native vegetation communities present on the property, distinguishing between different vegetation types, habitats, and ecological zones. Assess the condition and health of native vegetation, identifying any areas of degradation or fragmentation.

10. Identifying Rare or Endangered Species: Conduct surveys and assessments to identify any rare, threatened, or endangered species of flora or fauna present on the property. Document their habitats, population sizes, and conservation status, and develop strategies to protect and enhance their habitats as appropriate.

By systematically surveying and recording these physical characteristics, natural resources, and ecological features, farmers can gain valuable insights into the land's potential and constraints. This information forms the basis for developing land management plans that optimize productivity, protect environmental values, and promote long-term sustainability.

Surveying the land is an essential step in understanding the characteristics and layout of a property. It involves conducting a comprehensive assessment of the entire area, taking into account its size, boundaries, and notable landmarks or features. This process provides valuable information that serves as the foundation for various land management decisions.

To begin, it is crucial to conduct physical inspections of the land to visually assess its terrain, vegetation, soil types, and any existing infrastructure. This on-site examination allows for a firsthand understanding of the property's condition and potential opportunities or challenges.

In addition to physical inspections, gathering relevant data through maps, satellite imagery, and other sources enhances the surveying process. Maps provide valuable insights into the property's boundaries, neighbouring land uses, and topographic features. Satellite imagery offers a bird's-eye view of the landscape, highlighting patterns, vegetation cover, and land use practices. These data sources complement on-site inspections by providing a broader perspective and helping to identify areas of interest or concern.

During the survey, attention should be given to documenting any significant landmarks or features that may impact land management decisions. This could include natural elements such as rivers, hills, or forests, as well as man-made structures like buildings, roads, or fences. Understanding the location and characteristics of these features is essential for planning and implementing appropriate land use strategies.

Overall, conducting a thorough survey of the land involves a combination of physical inspections and data collection from various sources. By comprehensively assessing the property's size, boundaries, and notable features, landowners and managers can make informed decisions that optimize the use and conservation of the land.

Land capability refers to the inherent capacity of a piece of land to sustainably support specific agricultural or land use activities within its natural limits. If land is utilized beyond its capability, it inevitably leads to degradation and loss of productivity.

This concept is rooted in the understanding that each component of the land possesses its unique capacity to provide ecosystem services. The primary services, crucial for maintaining soil and land health, must be preserved, while the remaining capacity can be utilized for various human enterprises. Assessing land capability allows for the harmonization of production and conservation objectives.

For land use to be ecologically sustainable, it must not compromise the primary ecosystem services. The health of soil and land is contingent upon maintaining these essential services. Conversely, any land use practices that undermine these services lead to land degradation.

The availability of ecosystem services is influenced by factors such as land component features, topography, and climate. While it's challenging to quantify these services in absolute terms, they closely correlate with our understanding of land and soil quality. Moreover, seasonal and climatic variations further complicate the assessment.

The sustainable utilization of resources from the land is directly linked to the surplus capacity of ecosystem services beyond what is essential for basic ecosystem functioning. Any exploitation of goods and services that disrupts basic ecosystem activity results in the deterioration of land health.

It's important to note that land capability assessment solely focuses on ecological sustainability and does not consider social or economic factors. After evaluating land capability, it's necessary to integrate social and economic constraints to determine land suitability for specific purposes, leading to a comprehensive land suitability assessment.

Land capability and surveying the land are interconnected processes that contribute to understanding the potential uses and limitations of a piece of land. Surveying the land involves assessing its physical characteristics, boundaries, and features through both on-site inspections and data collection from maps, satellite imagery, and other sources. This process provides valuable information about the size, layout, and existing conditions of the property.

Land capability, on the other hand, refers to the inherent ability of a piece of land to sustainably support specific land uses over time. It takes into account factors such as soil type, slope, drainage, and climate, which influence the land's suitability for various purposes.

The information gathered during the surveying process directly informs the assessment of land capability. By conducting physical inspections and collecting data on soil types, topography, vegetation, and other relevant factors, landowners and managers can evaluate the land's potential for different uses, such as agriculture, forestry, or conservation.

For example, surveying the land may reveal that a property has steep slopes prone to erosion, limiting its suitability for intensive agriculture but making it suitable for forestry or conservation purposes. Similarly, data collected during the surveying process, such as soil maps or climate records, can help assess the land's capability for specific crops or livestock grazing.

In this way, surveying the land provides essential information that feeds into the assessment of land capability, enabling landowners and managers to make informed decisions about land use planning, management, and conservation. By understanding the physical characteristics and limitations of the land, they can develop strategies to maximize its productivity while minimizing environmental degradation and ensuring long-term sustainability.

Soil degradation, a significant concern in farm planning, refers to the deterioration of soil quality due to improper land use or inadequate management practices, particularly in agricultural, industrial, or urban contexts. Understanding the factors leading to soil degradation is crucial as soils are fundamental to terrestrial life and ecosystem health, making their preservation essential for overall well-being.

The deterioration potential of soil, influenced by various factors such as land use, climate, and slope, underscores the importance of sustainable and efficient agricultural practices. Predicting degradation susceptibility involves assessing inherent land deterioration hazards, independent of specific land use or management inputs. Moderate to high susceptibility indicates accelerated degradation under inappropriate practices.

Various criteria, including soil depth, texture, salt content, pH, and structure, aid in predicting susceptibility to specific degradation processes such as water erosion, wind erosion, salination, compaction, and contamination. Soil degradation manifests in physical, chemical, and biological declines, including loss of organic matter, decline in fertility, erosion, salinity changes, and contamination by pollutants.

Figure 57: Gully erosion The low sun highlights were heavy water flow has carved a gully in the surface of the field.
Alan Murray-Rust / Gully erosion. CC BY 4.0, via Wikimedia Commons.

Gully erosion, a prevalent form of soil erosion, results from flowing surface water, impacting agricultural land, infrastructure, and water quality. Sparse ground cover, concentrated runoff, unstable soils, intense rainfall, and unfavourable catchment shapes contribute to gully erosion. Preventative measures involve maintaining adequate ground cover and stabilizing erosion-prone areas through vegetation, earthworks, or other structures.

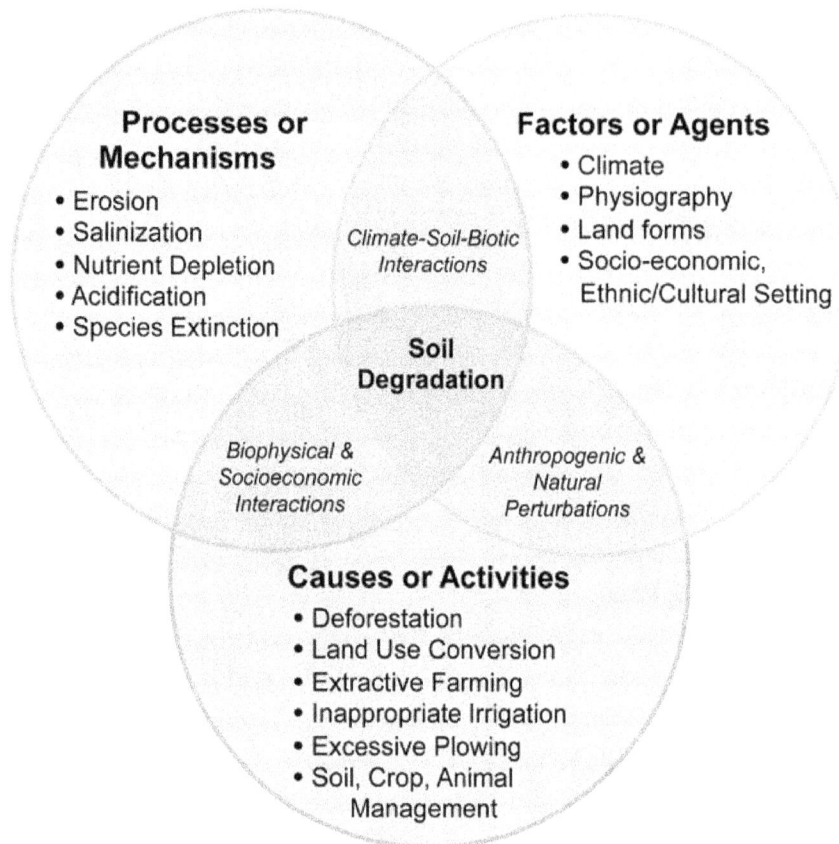

Figure 58: Soil degradation venn diagram, shows some attributing factors, causes, and effects. Rattan Lal, CC BY 4.0, via Wikimedia Commons.

Wind erosion, driven by natural processes and exacerbated by human activities, poses economic, environmental, and health risks. Activities like land clearing, over-grazing, and cropping reduce ground cover, leaving soil vulnerable to wind erosion. Drought exacerbates wind erosion by reducing vegetation growth and soil moisture. Mitigation strategies include maintaining vegetation cover and minimizing soil disturbance.

Figure 59: Wind Erosion. Wind Erosion by Anne Burgess, CC BY-SA 2.0 , via Wikimedia Commons,

Recording physical characteristics of a piece of land is a fundamental step in effective land management and planning. To begin, it's essential to document various aspects of the land's topography, including its elevation, slope, and drainage patterns. This information provides valuable insights into how water flows across the landscape and how different areas may be affected by erosion or waterlogging.

Elevation data allows for the visualization of the land's highs and lows, aiding in decision-making regarding land use and infrastructure placement. Slope analysis helps identify areas prone to erosion or runoff, guiding the implementation of erosion control measures and suitable farming practices. Understanding drainage patterns is crucial for optimizing irrigation systems and mitigating waterlogging issues that can harm crops or livestock.

Additionally, identifying water bodies such as rivers, streams, ponds, or wetlands is essential for assessing water resources and potential water-related risks. These features influence the availability of water for irrigation, livestock watering, and other agricultural activities. Furthermore, knowledge of water bodies helps protect sensitive habitats and ensures compliance with environmental regulations.

Man-made structures, including roads, buildings, fences, and irrigation systems, should also be documented. Roads provide access to different parts of the land and may influence traffic patterns or noise levels. Buildings, such as barns or storage facilities, determine the available infrastructure for agricultural operations. Fences delineate property boundaries and may impact livestock management practices. Irrigation systems, whether natural or man-made, affect water distribution and crop irrigation strategies.

By recording these physical characteristics comprehensively, land managers can make informed decisions about land use planning, conservation efforts, and infrastructure development. This data serves as a foundation for designing sustainable farming practices, minimizing environmental impacts, and maximizing the productivity and resilience of the land.

Assessing natural resources is a crucial aspect of farm management, as it provides valuable insights into the land's capacity to support agricultural activities sustainably. One of the primary considerations is evaluating the availability and quality of water sources on the property. This includes assessing the reliability of groundwater wells, surface water sources such as streams or ponds, and the potential for rainwater harvesting. Understanding water availability is essential for irrigation planning, livestock watering, and ensuring overall farm resilience during periods of drought or water scarcity.

Another important aspect of assessing natural resources is evaluating vegetation cover and wildlife habitats. This involves identifying different types of vegetation present on the land, ranging from natural forests to cultivated crops or pasturelands. Understanding vegetation cover helps in assessing land productivity, biodiversity, and potential for wildlife habitat restoration or enhancement. By preserving or enhancing natural habitats, farmers can promote ecological balance, pest control, and pollination services, contributing to overall farm sustainability.

Furthermore, assessing biodiversity is essential for maintaining ecosystem health and resilience. This involves identifying the variety of plant and animal species present on the farm and understanding their roles within the ecosystem. Biodiversity assessments help farmers recognize the importance of conserving native species, managing invasive species, and creating habitats that support beneficial wildlife. A diverse ecosystem enhances soil health, nutrient cycling, and pest regulation, reducing the need for chemical inputs and enhancing overall farm productivity.

Consideration of soil fertility and erosion potential is also critical in assessing natural resources. Soil fertility assessments involve analysing soil composition, nutrient levels, pH, and organic matter content. Understanding soil fertility guides farmers in implementing appropriate soil management practices, such as crop rotation, cover cropping, and organic amendments, to maintain or enhance soil productivity over time. Evaluating erosion potential helps identify areas prone to soil erosion and implement erosion control measures, such as contour ploughing, terracing, or vegetative buffer strips, to protect soil integrity and prevent nutrient runoff into water bodies.

Lastly, assessing microclimate variations across different areas of the farm is essential for optimizing agricultural production and managing climate-related risks. Microclimate factors include temperature variations, humidity levels, wind patterns, and frost occurrence. Understanding microclimate variations helps farmers select suitable crop varieties, timing planting and harvesting activities, and implementing climate adaptation strategies, such as installing windbreaks or shading structures, to mitigate weather-related risks and optimize farm productivity. Overall, assessing natural resources provides farmers with valuable information for making informed decisions about land management practices, conservation efforts, and sustainable agricultural development.

Assessing microclimate variations across different areas of the farm involves understanding the localized climatic conditions that can significantly influence agricultural production and management decisions. This can include:

1. Identify Monitoring Locations: Divide the farm into distinct zones or areas based on factors such as topography, elevation, vegetation cover, and land use. Select representative locations within each zone to monitor microclimate variations effectively.

2. Select Microclimate Parameters: Determine the specific climatic parameters to measure, such as temperature, humidity, wind speed and direction, solar radiation, and precipitation. These parameters provide insights into how environmental conditions vary across different areas of the farm.

3. Deploy Monitoring Instruments: Install weather monitoring instruments at selected locations to collect data on the chosen microclimate parameters. Common instruments include thermometers, hygrometers, anemometers, pyranometers, and rain gauges. Ensure that the instruments are calibrated and positioned correctly to obtain accurate measurements.

4. Collect Data: Continuously collect microclimate data over an extended period, preferably throughout the year, to capture seasonal variations and long-term trends. Record data at regular intervals, such as hourly or daily, to capture diurnal and daily fluctuations in climatic conditions.

5. Analyse Data: Analyse the collected data to identify patterns and trends in microclimate variations across different areas of the farm. Look for correlations between microclimate parameters and factors such as topography, land cover, proximity to water bodies, and exposure to prevailing winds.

6. Interpret Results: Interpret the findings to understand how microclimate variations impact agricultural activities and crop performance. Identify microclimatic hotspots or cold spots, areas prone to frost damage, zones with high wind exposure, and locations with moisture or heat stress.

7. Implement Management Strategies: Based on the microclimate assessment results, develop tailored management strategies to optimize agricultural production and mitigate climate-related risks. For example, in frost-prone areas, consider installing frost

protection measures such as wind machines, sprinkler irrigation, or thermal covers. In windy zones, plant windbreaks or adjust planting densities to minimize wind damage to crops.

8. Monitor and Adapt: Continuously monitor microclimate variations and evaluate the effectiveness of implemented management strategies. Adjust farming practices and mitigation measures as needed based on ongoing observations and changing climatic conditions.

Salinity, the accumulation of salt in land and water, poses significant challenges across various types of farms. While some regions naturally harbor saline conditions, human activities can exacerbate salt levels. The interconnection between salinity and water dynamics, particularly influenced by climate variability, underscores the complexity of managing salinity issues prone to climatic extremes. Salinity often coexists with other natural resource challenges, including soil and water quality decline, erosion, and loss of native vegetation.

The sources of salt accumulation in soil are diverse, ranging from rainfall carrying airborne salts to weathering of rocks and aeolian deposits redistributing salt across landscapes. The impacts of salinity are multifaceted, affecting agricultural productivity, wetland health, river ecosystems, infrastructure integrity, and human health. Economic costs associated with salinity management are substantial, with estimates indicating significant expenses incurred by both agricultural production and broader societal sectors.

Various salts contribute to salinity. Different types of salinity, such as dryland, irrigation, urban, industrial, and river salinity, necessitate tailored management approaches. The scale and response times of salinity issues vary, with local, intermediate, and regional groundwater systems exhibiting differing rates of problem occurrence and rehabilitation.

Another related concern is acid sulphate soils, which contain iron sulphides and can release acid when disturbed or exposed to air. Acid sulphate soils pose risks to built structures, ecosystems, and human health. Past drainage and flood-mitigation practices have contributed to current acid sulphate soil problems, particularly in coastal regions. Mapping and risk assessment tools aid in identifying areas susceptible to acid sulphate soils, informing land management decisions and development planning.

Preventing salinity and mitigating acid sulphate soil impacts require concerted efforts across agricultural, infrastructure, construction, and extraction sectors. Best practices include maintaining appropriate drainage systems, adhering to regulatory guidelines, and implementing remediation measures tailored to local conditions. Research and ongoing monitoring support evolving strategies for effectively managing salinity and acid sulphate soil issues, promoting sustainable land use and environmental stewardship.

Producing a soil map is a crucial step in farm planning as it provides valuable insights into the spatial distribution of soil characteristics across the property. Here's how to create a detailed soil map:

1. Surveying the Property: Begin by conducting a comprehensive survey of the entire property to gather information on soil characteristics. This involves physically inspecting different areas of the farm to observe variations in soil colour, texture, structure, and drainage patterns. Take note of any observable differences in vegetation growth, as soil types often influence plant distribution.

2. Collecting Soil Samples: Once the property has been surveyed, collect soil samples from various locations representative of different soil types and landscape features. Use a soil auger or coring tool to collect samples at multiple depths, typically ranging from surface soil to subsoil layers. Collect samples systematically, ensuring adequate coverage of the entire farm area.

3. Analysing Soil Properties: Take the collected soil samples to a laboratory for detailed analysis of soil properties such as texture, pH, organic matter content, nutrient levels, and compaction. The laboratory analysis provides quantitative data on soil characteristics, which are essential for accurately delineating soil types and fertility levels.

4. Creating the Soil Map: Utilize the survey data and laboratory analysis results to create a detailed soil map of the property. Use GIS (Geographic Information System) software or traditional mapping techniques to delineate different soil types, textures, and fertility levels across the farm. Classify soils according to established soil classification systems, such as the USDA soil taxonomy or local soil survey standards, using appropriate terminology and symbols.

5. Labelling and Legend: Clearly label each soil type on the map and provide a legend that explains the classification scheme used and the corresponding soil properties. Include information on soil texture (e.g., sandy, loamy, clayey), drainage characteristics, fertility status, and any soil limitations or management considerations.

6. Mapping Soil Variability: Pay close attention to variations in soil properties within the farm, such as differences in soil texture, depth, and fertility between fields or within individual fields. Use contour lines or colour shading to depict areas of soil variability, highlighting zones with distinct soil characteristics or management requirements.

7. Interpreting the Soil Map: Once the soil map is complete, interpret the findings to understand how soil variability influences agricultural productivity and management decisions. Identify areas with optimal soil conditions for specific crops or livestock production and areas that may require soil amendments or conservation practices to improve soil fertility or mitigate soil-related constraints.

8. Integrating Soil Information: Integrate the soil map with other spatial data layers, such as topography, land use, and vegetation cover, to gain a comprehensive understanding of the farm's physical environment. Use the soil map as a valuable tool for guiding land use planning, crop selection, irrigation management, soil conservation practices, and overall farm management strategies.

Producing a detailed soil map allows farmers to make informed decisions based on the spatial variability of soil properties, ultimately optimizing agricultural productivity, sustainability, and resource management on the farm.

Identifying water resources on a property is a crucial aspect of farm planning, given their significant impact on agricultural operations and sustainability. Here's a step-by-step guide on how to identify and assess water resources:

To begin, conduct a comprehensive survey of the property to locate all potential water sources. This entails physically inspecting the landscape to identify surface water features like rivers, streams, lakes, ponds, or wetlands. Additionally, identify existing groundwater sources such as wells, springs, or aquifers.

Once water resources are identified, the next step is to assess their quality. Collect water samples from surface water bodies and groundwater sources for laboratory analysis. Parameters such as pH, turbidity, dissolved oxygen, nutrient levels, and presence of contaminants or pollutants should be evaluated against relevant regulatory standards to ensure suitability for agricultural use.

Determine the quantity and reliability of water available from different sources on the property. Monitor water levels in surface water bodies and groundwater wells over time to assess seasonal variations and long-term trends. Consider factors such as precipitation patterns, hydrological cycles, and potential impacts of climate change on water availability.

Identify potential risks or threats to water resources, such as pollution from agricultural runoff, industrial discharge, or urban development, as well as over-extraction of groundwater. Evaluate the vulnerability of water sources to contamination, depletion, or degradation, and assess their potential impacts on agricultural productivity and environmental sustainability.

Create a map of water resources on the property, delineating the location and characteristics of surface water features and groundwater sources. Use GIS software or traditional mapping techniques to depict water bodies, wells, and aquifers, including relevant information such as water quality data, flow rates, and well depths.

Based on the assessment of water resources, develop comprehensive water management strategies to optimize water use efficiency, minimize risks, and ensure sustainable practices. Implement measures such as irrigation scheduling, water conservation techniques, soil moisture monitoring, and groundwater recharge practices to enhance water security and resilience.

Establish a monitoring program to regularly assess the condition of water resources and track changes over time. Monitor water quality parameters, water levels, and usage patterns to identify emerging issues or potential threats early and take proactive measures to address them. Routine maintenance of water infrastructure such as wells, pumps, and irrigation systems is essential to ensure reliable access to water resources.

Determining land capability is an important step in effective land management and agricultural planning, as it involves assessing the suitability of the land for various agricultural or land use purposes. Here's how to undertake this process:

Begin by conducting a thorough assessment of the land's characteristics, including soil properties, slope, drainage patterns, and other relevant factors. Soil characteristics such as texture, structure, pH levels, nutrient content, and organic matter play a crucial role in determining land capability. Evaluate these parameters through soil testing and analysis to understand the land's inherent qualities and limitations.

Consider the topography of the land, including its slope and elevation, as these factors influence water drainage, erosion potential, and accessibility. Steep slopes may be suitable for certain types of crops or grazing but may require erosion control measures to mitigate soil loss. Flat or gently sloping areas may be more suitable for mechanized farming or irrigation but could be prone to waterlogging without proper drainage.

Assess the land's drainage patterns to identify areas with poor drainage or susceptibility to waterlogging. Wet or poorly drained soils may be unsuitable for certain crops or livestock and may require land improvement techniques such as drainage tile installation or soil amendment to enhance drainage and productivity.

Classify different land classes or zones based on their productivity, constraints, and management requirements. This classification helps in identifying the most suitable land uses and management practices for each area of the property. Land classes may range from highly productive agricultural land with minimal constraints to marginal areas with significant limitations or environmental sensitivity.

Evaluate the land's capability for various agricultural activities, such as crop production, livestock grazing, forestry, or conservation purposes, based on its inherent qualities and limitations. Consider factors such as soil fertility, moisture retention, erosion risk, climate suitability, and proximity to markets or infrastructure when determining land use options.

Develop a comprehensive land capability map or zoning plan that delineates different land classes or zones across the property. This map should highlight areas of high productivity, moderate suitability, and low potential, along with any specific management recommendations or restrictions for each zone.

Use the land capability assessment to inform land use planning decisions, such as crop selection, pasture management, conservation practices, and infrastructure development. Tailor management practices to suit the specific characteristics and constraints of each land class, aiming to maximize productivity while minimizing environmental impacts and sustainability risks.

Regularly monitor and review the land capability assessment to adapt management strategies as conditions change over time. Periodic soil testing, erosion monitoring, and vegetation surveys help in assessing the effectiveness of management practices and identifying areas for improvement or optimization.

Mapping natural features and infrastructure is an essential aspect of farm planning and land management, as it provides a comprehensive visual representation of the property's landscape and existing assets. Here's how to create a detailed property map that overlays natural property features and infrastructure:

Begin by conducting a thorough survey of the property to identify and document natural features such as vegetation types, water bodies, topographic variations, and other significant landscape elements. Use aerial imagery, satellite imagery, or ground surveys to capture detailed information about the property's natural characteristics.

Identify and delineate vegetation types across the property, including forests, grasslands, wetlands, riparian zones, and other ecological communities. Note the distribution, extent, and composition of each vegetation type to understand habitat diversity, wildlife corridors, and ecological functions on the property.

Map water bodies such as rivers, streams, lakes, ponds, wetlands, and drainage channels to understand the hydrological network and water flow patterns across the landscape. Note the size, shape, orientation, and connectivity of water bodies, as well as any seasonal variations or fluctuations in water levels.

Overlay topographic variations on the property map to depict elevation contours, slopes, ridges, valleys, and other terrain features. Use contour lines or digital elevation models (DEMs) to visualize the land's relief and understand how elevation influences water drainage, soil erosion, and microclimate variations.

Identify existing infrastructure elements on the property, including buildings, roads, fences, irrigation systems, wells, barns, sheds, and other man-made structures. Note the location, size, condition, and functionality of each infrastructure element to assess its role in farm operations and land management.

Create a detailed property map that integrates natural features and infrastructure layers, using Geographic Information System (GIS) software or traditional mapping techniques. Overlay vegetation, water bodies, and topographic data with infrastructure layers to visualize spatial relationships and interactions between natural and built elements.

Annotate the property map with additional information such as property boundaries, land use zones, property access points, utility lines, soil types, and other relevant details. Use symbols, legends, and color-coding to enhance readability and convey key information effectively.

Regularly update and maintain the property map to reflect changes in the landscape, such as land use modifications, infrastructure upgrades, or natural disturbances. Incorporate new data sources, aerial imagery, or survey data to ensure the map remains accurate and up-to-date over time.

Identifying areas at risk of soil degradation is another crucial for sustainable land management and agricultural productivity. Here's how to effectively identify and address soil degradation risks on a property:

Begin by conducting a comprehensive assessment of the property's soil condition, taking into account factors such as erosion potential, compaction, salinity levels, nutrient content, and organic matter. Utilize soil testing methods, including soil sampling and laboratory analysis, to quantify soil parameters and identify areas of concern.

Identify and prioritize areas of the property that are particularly vulnerable to soil degradation based on factors such as slope, soil type, land use practices, and proximity to water bodies. Focus on areas with high erosion potential, excessive compaction, elevated salinity levels, nutrient deficiencies, or other indicators of soil degradation.

Implement measures to mitigate soil degradation risks in vulnerable areas, such as erosion control practices, soil conservation techniques, or improved land management practices. These measures may include contour ploughing, terracing, cover cropping, crop rotation, reduced tillage, mulching, agroforestry, or the establishment of riparian buffers.

Promote soil health and resilience through sustainable land management practices that enhance soil structure, fertility, and biological activity. Encourage the adoption of soil conservation practices that minimize soil disturbance, improve water infiltration, and protect soil from erosion, compaction, and degradation.

Educate landowners, farmers, and agricultural practitioners about the importance of soil conservation and sustainable land management practices. Provide training, resources, and technical assistance to support the implementation of soil conservation measures and promote responsible stewardship of natural resources.

Monitor and evaluate the effectiveness of soil conservation measures over time to assess their impact on soil health, erosion control, and agricultural productivity. Conduct regular soil testing and field assessments to track changes in soil condition, identify emerging risks, and adjust management practices as needed.

Collaborate with relevant stakeholders, including government agencies, conservation organizations, agricultural extension services, and local communities, to coordinate efforts and leverage resources for soil conservation and land management initiatives. Engage in participatory approaches that involve stakeholders in decision-making processes and foster collective action towards soil conservation goals.

Classifying native vegetation is a fundamental step in understanding and managing the ecological diversity and health of a property. Here's how to effectively classify native vegetation:

- Begin by conducting a thorough survey of the property to identify and map the various native vegetation communities present. This involves documenting different vegetation types, habitats, and ecological zones, as well as distinguishing between dominant plant species, forest types, grasslands, wetlands, and other distinctive vegetation communities.

- Utilize mapping tools, such as GIS (Geographic Information Systems) software or aerial imagery, to delineate and classify native vegetation communities across the property. Create detailed vegetation maps that accurately depict the spatial distribution and composition of different vegetation types and habitats.

- Assess the condition and health of native vegetation communities, taking into account factors such as species diversity, structural complexity, canopy cover, regeneration capacity, and overall ecosystem integrity. Evaluate the presence of invasive species, habitat fragmentation, degradation, or loss of biodiversity within native vegetation areas.

- Identify and prioritize areas of native vegetation that are particularly sensitive or valuable for conservation and restoration efforts. Focus on protecting high-quality habitats, critical wildlife corridors, riparian zones, and ecologically significant areas that provide important ecosystem services and support biodiversity.

- Monitor and track changes in native vegetation over time to assess trends in vegetation composition, structure, and health. Conduct regular field surveys, vegetation assessments, and biodiversity monitoring to document any shifts or disturbances in native vegetation communities.

- Develop management strategies and conservation plans to protect, restore, and enhance native vegetation on the property. Implement measures such as habitat restoration, invasive species control, reforestation, revegetation, prescribed burning, and land stewardship practices to promote the health and resilience of native ecosystems.

- Engage stakeholders, including landowners, conservation organizations, government agencies, indigenous communities, and local residents, in collaborative efforts to conserve and manage native vegetation resources. Foster partnerships and community involvement in conservation initiatives to promote shared stewardship of natural landscapes and biodiversity.

Identifying rare or endangered species on a property is crucial for effective conservation and biodiversity management. The following outlines how to conduct surveys and assessments to identify these species:

- Begin by researching the local flora and fauna species known to inhabit the region where the property is located. Consult relevant databases, species lists, conservation organizations, and government agencies to gather information on rare, threatened, or endangered species that may occur in the area.

- Conduct field surveys and habitat assessments to systematically search for rare or endangered species on the property. Utilize various survey techniques such as visual observations, wildlife cameras, trapping, acoustic monitoring, and habitat sampling to detect the presence of target species.

- Document the occurrence of rare or endangered species, recording important information such as species identity, abundance, distribution, habitat preferences, and ecological requirements. Collect data on population sizes, breeding habitats, foraging areas, and any specific threats or vulnerabilities facing these species.

- Assess the conservation status of rare or endangered species using criteria established by conservation agencies or scientific organizations. Evaluate factors such as population trends, habitat loss or degradation, genetic diversity, and susceptibility to threats to determine the species' level of risk and urgency for conservation action.

- Develop strategies and management plans to protect and enhance the habitats of rare or endangered species on the property. Implement measures such as habitat restoration, vegetation management, invasive species control, predator exclusion, and landscape connectivity to support the conservation of these species.

- Collaborate with conservation partners, wildlife experts, government agencies, and local stakeholders to coordinate conservation efforts and share information on rare or endangered species conservation. Engage in community outreach, education, and awareness-raising initiatives to promote public support and involvement in species conservation programs.

- Monitor and assess the effectiveness of conservation measures implemented to protect rare or endangered species over time. Conduct regular surveys, population monitoring, and habitat assessments to track changes in species abundance, distribution, and habitat quality, and adjust management strategies as needed to address emerging threats or conservation challenges.

By identifying and conserving rare or endangered species on the property, land managers can contribute to the preservation of biodiversity, ecosystem resilience, and the long-term sustainability of natural ecosystems. Prioritizing the protection of these species helps maintain ecological balance, ecosystem services, and the integrity of natural habitats for future generations.

Address Legal Requirements that Impact on Management of Property

Identifying legislation, regulations, and codes of practice relevant to activities on the property is essential for developing a farm plan that aligns with legal requirements and ensures compliance with environmental and agricultural standards. Integrating legal considerations into business management plans helps mitigate risks, protect natural resources, and maintain the sustainability of farm operations.

The process begins with research and identification of applicable laws governing farming activities in the region. This involves examining national, state/provincial, and local regulations pertaining to agriculture, environmental protection, water management, land use planning, wildlife conservation, and workplace safety. Seeking advice from legal experts, agricultural extension officers, environmental agencies, and industry associations is crucial to understanding specific legal requirements and obligations.

A comprehensive list of all relevant laws, regulations, and codes of practice impacting farm operations should be compiled and organized systematically based on different areas of activity such as land management, water usage, waste management, pesticide use, animal welfare, and occupational health and safety. Understanding the legal obligations imposed by each law or regulation is imperative, including key compliance requirements such as obtaining permits or licenses, maintaining records, submitting reports, and implementing best management practices.

Conducting a thorough assessment of potential risks and liabilities associated with non-compliance or violation of relevant laws and regulations is essential. This involves considering financial penalties, legal sanctions, loss of permits or certifications, reputational damage, and environmental harm. Integrating legal requirements into farm planning processes ensures alignment with business objectives, production plans, and resource management strategies, reflecting a commitment to sustainable and responsible farming practices.

Developing action plans and protocols to address specific legal requirements identified during the planning process is crucial. Assigning responsibilities, setting deadlines, and establishing monitoring and reporting mechanisms ensure ongoing compliance with relevant laws and regulations. Staying informed about changes or updates to existing laws and regulations is vital, as is maintaining accurate records of compliance efforts.

Seeking professional advice from lawyers, consultants, or regulatory experts with expertise in agricultural law and environmental regulations is recommended when faced with complex legal issues. Their guidance can help ensure that farm operations remain compliant and legally sound. By systematically identifying, understanding, and addressing legal requirements within farm planning processes, farmers can proactively manage risks, uphold regulatory compliance, and safeguard the long-term viability and sustainability of their operations.

Developing Management Strategies to Address Natural Resource Management Issues

Developing a comprehensive farm plan that integrates natural resource management with business objectives and production plans is crucial for ensuring the long-term sustainability and productivity of agricultural operations. This process involves several key steps aimed at assessing, prioritizing, and implementing strategies to manage natural resources effectively while supporting the economic viability of the farm.

Begin by conducting a thorough assessment of the property to identify areas requiring improvement in natural resource management. This may involve auditing water sources, soil quality, biodiversity, and existing land use practices. Based on this assessment, develop property improvement plans or action plans that outline specific measures to enhance natural resource management. These plans should prioritize actions based on their potential impact, feasibility, and cost-effectiveness.

Identify areas of the farm that have been degraded due to erosion, soil compaction, overgrazing, or other factors. Develop targeted strategies to rehabilitate degraded land, such as implementing erosion control measures, restoring soil fertility through organic amendments, and establishing native vegetation corridors to promote biodiversity. Incorporate revegetation initiatives, such as planting native species and creating habitat structures, to enhance biodiversity and ecosystem resilience.

Assess the farm's water resources, including surface water sources, groundwater aquifers, and rainfall patterns. Develop water supply and management plans that optimize water use efficiency, minimize waste, and mitigate risks of water scarcity or contamination. Implement strategies for sustainable irrigation practices, water conservation measures, and infrastructure upgrades to enhance water storage and distribution systems.

Develop vegetation management plans that prioritize the conservation and restoration of native vegetation communities. Identify areas for revegetation efforts, such as riparian zones, buffer strips, and marginal lands, to improve habitat quality and ecological connectivity. Incorporate strategies for controlling invasive species, such as targeted herbicide application, mechanical removal, and ecological restoration techniques, to prevent biodiversity loss and maintain ecosystem integrity.

Develop wildlife management strategies to mitigate conflicts between wildlife conservation and agricultural production objectives. Identify key wildlife species and their habitat requirements, and implement measures to protect sensitive habitats and minimize negative interactions with farm activities. Incorporate wildlife-friendly farming practices, such as providing nesting sites, creating wildlife corridors, and implementing predator control measures where necessary.

Develop integrated weed and pest animal control plans that combine biological, mechanical, cultural, and chemical control methods to effectively manage invasive species and pest populations. Prioritize control measures based on the level of infestation, potential economic impact, and environmental risks. Implement monitoring programs to assess the effectiveness of control efforts and adjust management strategies as needed.

Assess the farm's vulnerability to wildfires and develop fire risk management plans to minimize the risk of fire damage to property, infrastructure, and natural resources. Implement fire prevention measures, such as creating firebreaks, reducing fuel loads, and maintaining firefighting equipment and access roads. Develop emergency response plans and evacuation procedures to ensure the safety of farm personnel and livestock in the event of a wildfire.

Ensure that natural resource management strategies are integrated into the broader business and production plans for the farm. Align resource management objectives with business goals, financial budgets, and operational timelines to ensure coherence and consistency across all aspects of farm management. Monitor and evaluate the implementation of natural resource management strategies regularly, and adjust business plans as needed to reflect changing environmental conditions, market dynamics, and stakeholder priorities.

By systematically developing and implementing property improvement plans that address key aspects of natural resource management, farmers can enhance the sustainability, resilience, and profitability of their agricultural operations while safeguarding the health and integrity of the surrounding ecosystem.

A soil health management plan (SHMP) is a comprehensive strategy developed by farmers or land managers to sustainably manage and improve the health and productivity of their soil. It outlines specific actions and practices aimed at maintaining or enhancing soil quality, fertility, structure, and biological activity.

Key components of a soil health management plan typically include:

- Soil Inventory: Identification and characterization of different soil types across the farm, including soil texture, depth, drainage, and other relevant properties.

- Assessment and Monitoring: Regular monitoring of soil health indicators through methods such as soil and plant sampling, laboratory analyses, and visual inspections to track changes over time.

- Planning and Management: Development of management strategies and practices tailored to address soil health issues identified through assessment and monitoring. This may involve decisions related to crop selection, rotation, tillage, nutrient management, erosion control, and more.

- Implementation: Execution of the planned management practices, ensuring proper implementation and adherence to the strategies outlined in the SHMP.

- Evaluation and Adjustment: Continuous evaluation of the effectiveness of management practices and adjustment of the plan as needed based on feedback, changing conditions, and new information.

A soil health management plan is required when:
- Starting a New Farm Operation: When establishing a new farm or agricultural enterprise, it's crucial to develop a SHMP to ensure sustainable soil management practices from the outset.

- Expanding or Diversifying Operations: Any significant changes or expansions in farming activities, such as introducing new crops or livestock, may necessitate the development of a SHMP to address the potential impact on soil health.

- Addressing Soil Degradation: In cases where soil degradation, erosion, nutrient depletion, compaction, or other soil health issues are identified, a SHMP is essential to implement corrective measures and restore soil health.

- Complying with Regulations: In some jurisdictions, agricultural regulations or certification programs may require farmers to develop and implement soil management plans as part of environmental conservation efforts or to meet specific standards.

Overall, a soil health management plan is essential for promoting sustainable agricultural practices, optimizing productivity, and ensuring the long-term health and resilience of agricultural soils.

A Soil Health Management Plan (SHMP) can be tailored to suit the knowledge, skills, and time available to the farm manager, ranging from simple to more complex approaches. The key components of a SHMP typically include soil inventory, monitoring, planning, and management.

Soil inventory involves identifying major soil differences across the farm, providing detailed soil profile descriptions, recording critical differences between soils, ranking them from best to worst based on specific purposes, and noting any special considerations for management.

The importance of having a SHMP lies in recognizing soil as a finite resource on the farm, where its inherent qualities, along with land class, seasonal temperatures, and water availability, influence agricultural decisions and productivity. By explicitly integrating soil health management into the overall planning process alongside economic considerations, machinery operations, seed selection, and labour management, the SHMP ensures a holistic approach to farming success.

Assessment and monitoring are crucial aspects of the SHMP, focusing on understanding the impact of farming practices on soil health. This includes soil and plant sampling, laboratory analyses, systematic paddock walks, and evaluation of crop and animal production, quality, and health. The frequency and methods of monitoring depend on various factors such as land use, management practices, and critical issues like erosion.

Planning and management translate knowledge of soil properties and conditions into actionable strategies to maintain or improve soil health while optimizing agricultural production. Decisions regarding crop selection, rotations, stock management, soil amendments, tillage practices, and traffic management are all informed by the SHMP. Special attention is given to minimizing soil disturbance, optimizing nutrient management, and controlling traffic across soil to ensure sustainable soil health and productivity.

A successful SHMP involves ongoing evaluation and adjustment based on feedback and changing circumstances. It should be integrated with the overall farm business plan, utilizing available soil information, adopting appropriate technologies, and seeking expert advice when needed. Key questions related to the farm business, soil health, machinery, and stock management help guide the development and implementation of the SHMP, ensuring its relevance and effectiveness over time.

A soil health management plan (SHMP) is a specific component of a broader farm plan, focusing specifically on strategies and practices related to soil management. While a farm plan encompasses various aspects of agricultural operations, including production goals, marketing strategies, financial management, and infrastructure development, a SHMP specifically addresses soil health and sustainability.

The relationship between a SHMP and a farm plan is that the SHMP is integrated into the overall farm plan as one of its essential elements. Here's how they relate:

1. Integration: A SHMP is integrated into the broader framework of the farm plan, aligning soil management strategies with overall farm goals and objectives. It ensures that soil health considerations are incorporated into decision-making processes related to crop selection, rotation, nutrient management, and land use.

2. Complementary Goals: While the farm plan may encompass goals related to profitability, production efficiency, and market expansion, the SHMP contributes to these goals by ensuring the long-term productivity and sustainability of the land. Healthy soils support higher yields, reduce input costs, and enhance the resilience of the farm ecosystem.

3. Coordination of Activities: The implementation of a SHMP involves coordination with other farm activities and practices outlined in the farm plan. For example, soil conservation practices may be integrated with water management strategies, pest control measures, and crop rotation schedules outlined in the broader farm plan.

4. Monitoring and Evaluation: Both the farm plan and the SHMP involve monitoring and evaluation processes to assess progress towards established goals and objectives. While the farm plan evaluates overall farm performance, the SHMP monitors soil health indicators and the effectiveness of soil management practices.

5. Adaptation and Adjustment: As conditions change and new information becomes available, both the farm plan and the SHMP require adaptation and adjustment. Changes in market conditions, environmental regulations, or soil health assessments may necessitate updates to both plans to ensure continued success and sustainability.

While a farm plan encompasses various aspects of agricultural operations, a SHMP specifically focuses on soil management strategies and practices. The relationship between the two lies in their integration, complementary goals, coordination of activities, monitoring and evaluation processes, and the need for adaptation and adjustment over time. Together, they contribute to the overall success and sustainability of the farm enterprise.

Documenting the farm plan is a crucial step in ensuring clarity, accountability, and progress tracking. While the extent of documentation may vary, writing down goals, objectives, and strategies adds rigor to the planning process. Research indicates that setting written goals can significantly enhance personal satisfaction and overall well-being. Additionally, written plans facilitate collaboration and input from stakeholders like staff, suppliers, and advisers, fostering a sense of involvement and achievement among all involved parties.

A written plan serves as a reference point for accountability and progress tracking. It helps to delineate short-term, medium-term, and long-term objectives, providing clarity and direction for the farm's future. By documenting priorities and core values, farmers can better balance conflicting objectives and make informed decisions, especially in areas where expertise may be lacking.

Moreover, a formalized plan is beneficial when seeking financing or external investments, providing potential investors with a clear understanding of the farm's direction and goals. Regularly revisiting the written plan allows for reflection on growth and success over time.

Tools like Tool 1.4 offer a structured format for creating a written business plan, offering guidance on the essential information to include during the planning process.

Once the plan is established, the focus shifts to implementation and monitoring progress. Clear communication of roles and responsibilities among family and staff members is vital for successful execution. Establishing ground rules, setting expectations, and holding regular business meetings ensure everyone is aligned and contributing to the farm's goals.

Implementing any changes should be integrated into the farm's annual operating plan, aiming for a swift transition while maintaining financial stability. Monitoring progress involves regular assessments of physical resources, livestock performance, and financial outcomes. This ongoing evaluation allows for adjustments to be made based on real-time data and ensures that the farm remains on track towards its objectives.

Monitoring key performance indicators (KPIs) provides valuable insights into the farm's performance and helps identify areas for improvement. Benchmarks and comparisons with industry standards aid in evaluating success and guiding future decisions.

Managing risks associated with change is essential for mitigating potential setbacks. Strategies for risk management include scheduling investments based on the highest rate of return, ensuring accurate analysis and predictions, and revising strategies as needed based on feedback and changing circumstances.

Regular reviews, both short-term and long-term, allow farmers to assess progress, make necessary adjustments, and ensure that the farm remains aligned with its goals and objectives. By monitoring both physical and financial indicators, farmers can make informed decisions to optimize performance and drive the farm towards success.

As a basic example of a whole farm plan, the following provides a whole farm plan for the hypothetical agribusiness with vertical farming expansion.

Whole Farm Plan for Hypothetical Agribusiness with Vertical Farming Expansion
Introduction:

The following comprehensive farm plan outlines strategies for the hypothetical agribusiness to integrate vertical farming into its existing operations. Leveraging strengths, addressing weaknesses, and capitalizing on opportunities, this plan aims to ensure sustainable growth and competitiveness in the agricultural sector.

Vision and Mission:

Vision: To become a leading provider of high-quality specialty greens through innovative vertical farming practices.

Mission: To cultivate a diverse range of premium greens using sustainable methods, meeting the evolving demands of consumers while maximizing profitability and environmental stewardship.

Goals and Objectives:

a. Establish vertical farming facilities for the production of arugula, butterhead lettuce, red leaf lettuce, and romaine lettuce.

b. Secure partnerships with local restaurants, supermarkets, and food service providers for distribution channels.

c. Achieve certification for organic and sustainable farming practices.

d. Maintain profitability and financial stability through efficient resource management.

e. Enhance brand recognition and market presence through strategic marketing initiatives.

Operational Plan:

a. Vertical Farming Infrastructure: Invest in advanced technology and infrastructure to support vertical farming operations, including hydroponic systems, LED lighting, and climate control systems. Establish dedicated facilities with sufficient space and utilities for optimal crop production.

b. Crop Selection and Cultivation: Conduct market research to identify high-demand varieties of specialty greens. Implement crop rotation and succession planting techniques to maximize yield and minimize disease risk. Monitor plant health and growth parameters regularly to ensure optimal growing conditions.

c. Sustainability Practices: Implement sustainable farming practices such as water recycling, energy efficiency measures, and integrated pest management to minimize environmental impact and conserve natural resources. Obtain organic certification and adhere to strict quality standards to meet consumer expectations.

d. Supply Chain Management: Establish partnerships with reliable suppliers for seeds, nutrients, and equipment. Maintain inventory levels and logistics management systems to ensure uninterrupted production and timely delivery of products to customers.

e. Workforce Development: Provide comprehensive training programs for employees to acquire the necessary skills and knowledge for vertical farming operations. Foster a culture of innovation, teamwork, and continuous improvement to enhance productivity and efficiency.

Marketing and Sales Strategy:

a. Branding and Packaging: Develop a distinctive brand identity and packaging design that communicates the agribusiness's commitment to quality and sustainability. Emphasize the freshness, flavour, and nutritional value of specialty greens to attract consumers.

b. Market Segmentation: Identify target consumer segments, including health-conscious individuals, food enthusiasts, and upscale restaurants. Tailor marketing messages and promotional activities to resonate with each segment's preferences and lifestyle choices.

c. Distribution Channels: Establish partnerships with local restaurants, supermarkets, and specialty food stores to showcase and sell vertical farming products. Explore online sales platforms and farmers' markets to reach a wider audience and increase market penetration.

d. Promotional Campaigns: Launch promotional campaigns highlighting the unique benefits of locally grown specialty greens, such as freshness, flavour, and environmental sustainability. Utilize social media, influencer partnerships, and experiential marketing events to engage with consumers and build brand awareness.

Financial Plan:

a. Budgeting and Forecasting: Develop detailed financial projections for vertical farming operations, including startup costs, ongoing expenses, and revenue forecasts. Allocate resources efficiently to optimize profitability and return on investment.

b. Financing Options: Explore financing options such as loans, grants, and investment partnerships to fund vertical farming expansion projects. Negotiate favourable terms and interest rates to minimize financial risk and maximize capital availability.

c. Cost Management: Implement cost-saving measures such as bulk purchasing, energy efficiency upgrades, and waste reduction strategies to reduce operating expenses and improve profit margins.

d. Performance Monitoring: Track key performance indicators such as crop yield, production costs, and sales revenue to monitor the financial health and viability of vertical farming operations. Adjust budgetary allocations and business strategies as needed to achieve financial goals and objectives.

Conclusion:

By implementing the strategies outlined in this comprehensive farm plan, the hypothetical agribusiness can successfully integrate vertical farming into its existing operations, capitalize on emerging market opportunities, and establish itself as a leader in the production of high-quality specialty greens. With a focus on sustainability, innovation, and market responsiveness, the agribusiness can achieve long-term growth, profitability, and resilience in the dynamic agricultural industry.

Reviewing Farm Plan

In the dynamic environment of agriculture, reviewing and revising business and production plans is essential to adapt to changing circumstances and optimize farm performance. From the perspective of developing a farm plan that integrates natural resource management with business objectives and production plans, this process involves continuous assessment, adjustment, and optimization to ensure alignment between resource management strategies and overall farm goals.

Regularly review business and production plans to assess their effectiveness in achieving desired outcomes and identify any emerging challenges or opportunities. Consider factors such as market trends, weather patterns, regulatory changes, and technological advancements that may impact farm operations. Evaluate the performance of existing strategies and identify areas for improvement or adjustment to better align with evolving circumstances and stakeholder priorities.

When reviewing business and production plans, it's crucial to follow a structured approach to ensure comprehensive assessment and optimization. The process typically involves several key steps:

- Assessment of Performance: Start by evaluating the performance of critical indicators like crop yields, livestock productivity, financial performance, and resource utilization. This evaluation helps gauge the effectiveness of current strategies and pinpoint areas requiring improvement.

- Identification of Challenges and Opportunities: Identify emerging challenges or opportunities that could impact farm operations, such as shifts in market demand, input costs, environmental regulations, or climatic conditions. Assessing the potential impact of these factors on business objectives and production plans is essential for informed decision-making.

- Stakeholder Consultation: Engage with a diverse range of stakeholders, including farm owners, managers, employees, suppliers, customers, and community members. Gathering feedback, insights, and perspectives from stakeholders helps gain a comprehensive understanding of current farm practices and future directions. Incorporating stakeholder input into the revision process ensures alignment with broader goals and priorities.

- Goal Setting and Prioritization: Define clear, measurable goals and objectives for the farm based on stakeholder input, market trends, and resource availability. Prioritize goals based on their significance, feasibility, and potential impact on farm sustainability and profitability. This step lays the foundation for focused action and strategic decision-making.

- Development of Action Plans: Develop detailed action plans that outline specific strategies, tasks, timelines, and responsibilities for achieving identified goals and objectives. Breaking down larger goals into smaller, actionable steps facilitates implementation and monitoring, ensuring progress towards desired outcomes.

- Monitoring and Evaluation: Implement robust monitoring and evaluation mechanisms to track progress towards goals, measure performance indicators, and identify any deviations from planned outcomes. Regularly reviewing and analysing data enables assessment of the effectiveness of implemented strategies and informs decisions about future adjustments or refinements. This iterative process of monitoring and evaluation is critical for continuous improvement and optimization of farm operations.

Incorporating digital finance and farm management tools into the farm planning process offers a streamlined approach to data collection, analysis, and decision-making. These tools provide real-time insights into various aspects of farm operations, including financial performance, resource utilization, production efficiency, and risk management, empowering farmers to make informed and proactive decisions.

Digital finance and farm management tools bring several benefits to agricultural enterprises:

- Data Integration: These tools integrate financial, operational, and environmental data from diverse sources such as accounting software, precision agriculture technologies, environmental monitoring systems, and market databases. This integration offers a comprehensive understanding of farm performance and resource dynamics.

- Decision Support: Advanced analytics, modelling, and forecasting capabilities help assess different scenarios, evaluate potential outcomes, and identify optimal strategies for achieving business objectives while minimizing risks and maximizing returns.

- Automation and Efficiency: Routine tasks, processes, and workflows can be automated to reduce administrative burden, streamline operations, and enhance overall efficiency. Digital tools facilitate record-keeping, inventory management, budgeting, and financial planning, saving time and resources.

- Collaboration and Communication: These tools facilitate collaboration and communication among farm stakeholders by providing access to centralized data repositories, interactive dashboards, and collaborative platforms. Real-time sharing of information, insights, and feedback fosters transparency, accountability, and alignment among team members.

By leveraging digital finance and farm management tools, farmers can enhance their capacity to review, revise, and optimize business and production plans in response to changing circumstances and emerging opportunities. These tools enable data-driven decision-making, improve operational efficiency, and support the achievement of long-term sustainability and profitability in agricultural enterprises.

Chapter Seventeen

Planning and Monitor Production Processes

Determining Production Process Requirements

Businesses engaged in agricultural production encompass a diverse range of operations, such as breeding facilities, farms specializing in both crops and select animals, as well as greenhouses and nurseries focused on cultivating products intended for retail sale (Chait, 2020). Additionally, ranches and tree and sod farms are included within this category, provided their products are sold directly to consumers at retail and are not replanted elsewhere by the grower (Chait, 2020).

Agricultural production processes encompass a series of activities crucial for the cultivation, harvesting, processing, and distribution of agricultural products. These processes vary depending on the specific crop or livestock being produced and the methods employed by farmers. Among the common agricultural production processes are land preparation, planting or sowing, irrigation, fertilization, pest and disease management, crop maintenance, harvesting, post-harvest handling, livestock management, slaughtering and processing, and storage and distribution.

Land preparation marks the initial phase, involving clearing and tilling the land to make it suitable for planting. Mechanized equipment such as tractors and plows is often employed for this purpose. Following land preparation, seeds or seedlings are planted into the soil, either manually or using mechanical planters, depending on the scale of the operation. In areas with insufficient rainfall, irrigation systems such as sprinklers, drip irrigation, or furrow irrigation may be employed to ensure adequate moisture levels.

Fertilization is essential to promote plant growth, with farmers applying fertilizers to supplement essential nutrients like nitrogen, phosphorus, and potassium. Managing pests, diseases, and weeds is crucial to prevent damage and maintain optimal yields. Integrated pest management techniques, combining biological controls, cultural practices, and chemical pesticides as needed, are commonly employed.

Throughout the growing season, crop maintenance involves tasks like cultivation, weeding, and monitoring for signs of stress or nutrient deficiencies to ensure healthy growth and maximum yields. Once crops reach maturity, they are harvested using various methods, such as hand harvesting or mechanized harvesting with specialized equipment like combines or harvesters.

Post-harvest handling processes follow harvesting, including cleaning, sorting, grading, and packaging, to maintain product quality and freshness. Livestock management involves caring for the health and well-being of animals through feeding, watering, sheltering, and veterinary care. This encompasses breeding, birthing, and raising young animals.

In livestock farming, animals may undergo slaughtering and processing to produce meat products for consumption. Specialized facilities and equipment are employed to ensure food safety and quality. Finally, agricultural products are stored in facilities designed to preserve their freshness and quality before being distributed to markets, retailers, or consumers through various channels like wholesalers, farmers' markets, or direct sales.

Overall, agricultural production processes play a vital role in efficiently and sustainably producing food, fibre, and other agricultural products to meet global consumer needs.

The agricultural production process initiates with the acquisition or breeding of qualifying animals and the preparation of soil for crop planting (Minnesota Department of Revenue, 2024). This commencement marks the onset of a series of activities aimed at cultivating, harvesting, and managing agricultural products. The process culminates when livestock or crops, whether packaged or unpackaged, are integrated into finished goods inventory, or when grain reaches a sellable stage or can be commingled. It is important to note that if a product is not included in finished goods, the process concludes upon completion of the last step preceding product shipment, excluding loading or preparation for shipping, which are not considered part of the production process.

Agricultural production encompasses various activities that yield products intended for retail sale. This includes agriculture, which entails cultivating soil, planting, nurturing, and harvesting crops, as well as rearing, feeding, and overseeing animals. Additionally, aquaculture involves raising private aquatic animals such as fish, while floriculture focuses on cultivating flowering plants. Horticulture encompasses the growth of fruits, vegetables, and plants, alongside activities like maple syrup harvesting and silviculture, which involves growing and nurturing forest trees (Minnesota Department of Revenue, 2024).

Businesses engaged in agricultural production span a range of operations. These include breeding facilities, farms dedicated to crops and certain animals, greenhouses, and nurseries where products are cultivated for retail sale. Ranches, tree farms, and sod farms are also included if their products are sold at retail rather than installed by the grower (Minnesota Department of Revenue, 2024).

Conversely, agricultural production excludes certain activities. This comprises storing or preserving raw materials before the production process commences, as well as storing, preserving, handling, or transporting finished goods. Additionally, activities such as storing or processing agricultural products at co-ops, grain elevators, dairies, or meat packers fall outside the scope of agricultural production. Lastly, raising animals for personal use is not considered part of agricultural production (Minnesota Department of Revenue, 2024).

To effectively plan and monitor production processes in agricultural operations, it is important to start by accessing comprehensive information about the products being grown or manufactured and the market in which they will be sold. This involves conducting market research to understand consumer demand, market trends, and competitor offerings. By gathering data on consumer preferences, pricing dynamics, and distribution channels, farmers can make informed decisions about which products to produce and how to position them in the market.

Once the market landscape is understood, the next step is to confirm the characteristics of the land under production and assess the quality of associated infrastructure. This includes evaluating factors such as soil quality, topography, drainage patterns, and access to water sources. Additionally, assessing the condition of infrastructure such as irrigation systems, buildings, and equipment is essential to ensure that production processes can be carried out efficiently and effectively.

Analysing recent and historical data from organizational records is another critical aspect of production planning. By reviewing data on past yields, input usage, labour productivity, and financial performance, farmers can identify patterns, trends, and areas for improvement. This data-driven approach enables informed decision-making and helps set realistic production targets based on past performance and future projections.

To identify potential improvements or innovations, farmers should actively seek out opportunities to enhance production processes and optimize resource utilization. This may involve adopting new technologies, implementing best practices, or exploring alternative farming methods to increase efficiency, reduce costs, and minimize environmental impact. By staying abreast of advancements in agricultural science and technology, farmers can innovate their production systems and stay competitive in the market.

Once the production processes required to achieve targeted production levels are identified, it's essential to compare them with those that currently exist on the farm. This involves evaluating existing workflows, equipment, and practices to determine their effectiveness and suitability for meeting production targets. By identifying gaps or areas where improvements are needed, farmers can develop strategies to enhance production efficiency and productivity.

Finally, it's important to ensure that production planning information is integrated into other organizational planning processes. This includes aligning production targets with financial budgets, resource allocation plans, and marketing strategies. By coordinating production planning with other functional areas of the business, such as finance, marketing, and supply chain management, farmers can ensure a cohesive approach to achieving overall business objectives and maximizing profitability.

Conducting market research is a crucial step for farmers aiming to understand consumer demand, market trends, and competitor offerings within the agricultural industry. Market research involves gathering comprehensive data on various aspects of the market to inform decision-making processes. One key aspect is understanding consumer preferences, which entails collecting information on what consumers are looking for in agricultural products, including quality, price sensitivity, and preferred varieties. For example, farmers may conduct surveys, interviews, or focus groups to gain insights into consumer preferences for specific crops or livestock products.

Moreover, analysing pricing dynamics is essential to determine the appropriate pricing strategy for agricultural products. Farmers need to consider factors such as production costs, market demand, and competitive pricing when setting prices for their products. By researching market prices for similar products and understanding how pricing affects consumer behaviour, farmers can establish competitive pricing strategies that maximize profitability while remaining attractive to consumers.

Furthermore, understanding distribution channels is critical for ensuring that agricultural products reach consumers efficiently and effectively. Farmers need to identify the most appropriate distribution channels for their products, whether through direct sales to consumers, wholesale markets, farmers' markets, supermarkets, or online platforms. By assessing the advantages and disadvantages of different distribution channels and understanding the preferences of target consumers, farmers can develop effective distribution strategies that optimize sales and reach the desired market segments.

Conducting market research provides farmers with valuable insights into consumer behaviour, market dynamics, and competitor strategies, enabling them to make informed decisions about which products to produce and how to position them in the market. By gathering data on consumer preferences, pricing dynamics, and distribution channels, farmers can develop tailored marketing strategies that resonate with target consumers and drive sales growth in the competitive agricultural marketplace.

Confirming the characteristics of the land under production and assessing the quality of associated infrastructure are crucial steps in agricultural planning and production management. Soil quality is a fundamental consideration, as it directly impacts crop growth and productivity. Farmers may conduct soil tests to analyse key parameters such as pH levels, nutrient content, and soil texture. By understanding the soil composition, farmers can make informed decisions about crop selection, fertilization practices, and soil management techniques to optimize agricultural production.

Topography and drainage patterns also play significant roles in land suitability for agriculture. Farmers need to assess factors such as slope, elevation, and natural drainage to identify potential areas of waterlogging or erosion. By mapping out the topography of the land and analysing drainage patterns, farmers can implement appropriate land management practices such as contour ploughing, terracing, or installing drainage systems to mitigate soil erosion and waterlogging issues.

Access to water sources is another critical aspect of land assessment, particularly in regions where irrigation is necessary for crop production. Farmers need to evaluate the availability and reliability of water sources such as rivers, streams, ponds, or groundwater aquifers. Additionally, assessing the condition and capacity of irrigation systems is essential to ensure efficient water distribution to crops. This may involve inspecting pipelines, pumps, and irrigation infrastructure to identify any leaks, blockages, or inefficiencies that could affect water delivery to fields.

Furthermore, assessing the condition of associated infrastructure such as buildings, barns, and equipment is essential for operational efficiency and productivity. Farmers need to inspect existing infrastructure to identify any maintenance or repair needs and ensure that facilities are in good working condition. For example, evaluating the functionality of irrigation pumps, checking the integrity of storage facilities, and inspecting the safety of farm equipment are essential tasks to ensure smooth production processes.

Analysing recent and historical data from organizational records is an essential component of production planning in agriculture. This process involves systematically reviewing data on various aspects of farm operations, including past yields, input usage, labour productivity, and financial performance. By delving into this wealth of information, farmers can gain valuable insights into the performance of their operations over time, identify patterns and trends, and pinpoint areas for improvement.

For example, by examining historical yield data for different crops or livestock, farmers can assess how production has fluctuated seasonally or in response to changes in management practices or environmental conditions. Understanding these trends can inform decisions about crop rotation, breeding programs, or adjustments to livestock management strategies to optimize productivity and profitability.

Similarly, analysing input usage data such as fertilizer and pesticide applications allows farmers to evaluate the effectiveness of their input management practices and identify opportunities to optimize resource use and reduce costs. For instance, if data analysis reveals that certain inputs are being over-applied or applied inefficiently, farmers can adjust their application rates or timing to minimize waste and environmental impact while maintaining crop yields.

Labor productivity is another key metric that farmers can analyse to improve operational efficiency. By examining data on labour hours, tasks performed, and output levels, farmers can assess the productivity of their workforce and identify potential bottlenecks or inefficiencies in production processes. This insight can help farmers streamline workflows, allocate resources more effectively, and implement training or skill development programs to enhance workforce performance.

Financial performance data, including income, expenses, and profit margins, provide critical insights into the overall economic health of the farm operation. By analysing financial records, farmers can evaluate the profitability of different enterprises, assess the impact of input costs or market fluctuations on profitability, and identify opportunities to optimize revenue streams or reduce expenses. For example, if financial analysis reveals that certain crops or livestock enterprises are consistently more profitable than others, farmers may consider reallocating resources or adjusting production plans to capitalize on these opportunities.

Identifying potential improvements or innovations in agricultural production processes is crucial for farmers to enhance efficiency, reduce costs, and remain competitive in the ever-evolving agricultural landscape. To achieve this, farmers should actively seek out opportunities to optimize their production processes and resource utilization. This proactive approach may involve adopting new technologies, implementing best practices, or exploring alternative farming methods that align with their operational goals and sustainability objectives.

For example, advancements in precision agriculture technologies, such as GPS-guided machinery and drones, offer farmers the ability to optimize inputs like water, fertilizers, and pesticides by precisely targeting application rates based on real-time data and field variability. By incorporating these technologies into their operations, farmers can reduce input wastage, minimize environmental impact, and improve overall productivity.

Furthermore, farmers can explore innovative farming techniques such as vertical farming, hydroponics, or aquaponics to maximize land use efficiency and overcome limitations associated with traditional farming methods. These alternative production systems allow farmers to cultivate crops in controlled environments, independent of soil quality or climatic conditions, thereby enabling year-round production, higher yields, and greater resource efficiency.

Additionally, adopting regenerative agricultural practices, such as cover cropping, crop rotation, and agroforestry, can enhance soil health, biodiversity, and resilience to climate change while reducing reliance on synthetic inputs and mitigating environmental degradation. By integrating these practices into their production systems, farmers can improve soil structure, water retention, and nutrient cycling, leading to more sustainable and resilient farming operations.

Moreover, collaboration with agricultural researchers, extension services, and industry partners can provide farmers with access to cutting-edge research, knowledge, and expertise to drive innovation and continuous improvement in their production processes. By participating in field trials, demonstration projects, and technology transfer initiatives, farmers can evaluate new technologies and practices in real-world settings and assess their potential impact on their operations.

Overall, identifying potential improvements or innovations in agricultural production requires a proactive and open-minded approach to embracing new technologies, practices, and methodologies. By leveraging advancements in agricultural science and technology and adopting sustainable farming practices, farmers can optimize resource utilization, enhance productivity, and ensure the long-term viability and resilience of their operations in a rapidly changing agricultural environment.

Once the production processes required to achieve targeted production levels are identified, the next crucial step is to compare them with the existing processes already in place on the farm. This comparison involves a comprehensive evaluation of current workflows, equipment, and practices to determine their effectiveness and suitability for meeting the desired production targets. Farmers need to closely examine each aspect of their existing production system to identify any gaps or deficiencies that may hinder their ability to reach the desired production levels.

For example, if the identified production processes necessitate the use of modern irrigation techniques to optimize water usage, farmers would assess their current irrigation systems to determine if they are capable of meeting the required efficiency standards. This evaluation may reveal outdated equipment or inefficient practices that need to be upgraded or replaced to align with the new production requirements.

Similarly, if the targeted production levels call for the adoption of precision agriculture technologies for precise input application, farmers would assess the compatibility of their existing machinery and equipment with these advanced technologies. They would also evaluate the level of training and expertise among their workforce to ensure they can effectively operate and manage the new systems.

Moreover, farmers need to consider the integration of new production processes with existing ones to ensure seamless operations and minimize disruptions. This requires careful planning and coordination to identify potential bottlenecks or conflicts between different production activities and implement strategies to optimize workflow efficiency.

Integrating production planning information into other organizational processes is crucial for ensuring coherence and alignment across different functional areas of the agricultural business. This integration involves harmonizing production targets with financial budgets, resource allocation plans, and marketing strategies to create a unified approach towards achieving overall business objectives.

For instance, suppose a farm sets a production target to increase the yield of a particular crop by 20% in the upcoming season. In that case, this information needs to be communicated to the finance department to ensure that adequate funds are allocated for purchasing additional inputs such as seeds, fertilizers, and machinery required to achieve the target. The finance team will then adjust the budget accordingly to accommodate the increased production expenses while maintaining profitability.

Similarly, integrating production planning with marketing strategies is essential for ensuring that the produced goods align with consumer demand and market trends. For example, if market research indicates a growing demand for organic produce, farmers may adjust their production plans to prioritize organic farming practices and allocate resources accordingly. This alignment between production and marketing ensures that the farm produces goods that are in high demand, thereby enhancing market competitiveness and revenue generation.

Moreover, integrating production planning information into supply chain management processes helps optimize inventory levels, streamline distribution channels, and minimize wastage. By coordinating production schedules with procurement and logistics operations, farmers can ensure that harvested crops are efficiently transported to markets or processing facilities without delays or disruptions. This seamless coordination enhances operational efficiency and reduces the risk of inventory stockouts or overages, thereby maximizing resource utilization and minimizing costs.

To exemplify the process, returning to a previously utilised hypothetical agribusiness operation, determining production process requirements for vertical farming of arugula, butterhead lettuce, red leaf lettuce, and romaine lettuce involves several steps:

1. Market Research: The agribusiness manager should conduct thorough market research to understand consumer demand for these specific crops. This includes analysing trends in urban areas where vertical farming is popular, as well as identifying potential distribution channels such as local farmers' markets, restaurants, and grocery stores.

2. Crop Selection and Varieties: Assess which varieties of arugula, butterhead lettuce, red leaf lettuce, and romaine lettuce are well-suited for vertical farming. Consider factors such as growth characteristics, yield potential, and market demand for each variety.

3. Technology and Infrastructure: Evaluate the technology and infrastructure required for vertical farming, including vertical farming systems, grow lights, irrigation systems, climate control mechanisms, and automation tools. Determine the optimal setup based on crop requirements and production targets.

4. Growing Medium and Nutrient Management: Decide on the growing medium to be used in the vertical farming system, such

as hydroponics or aeroponics. Develop a nutrient management plan to ensure that the crops receive the necessary nutrients for healthy growth.

5. Lighting Requirements: Determine the lighting requirements for each crop stage, including seed germination, vegetative growth, and flowering. Choose appropriate grow lights and establish lighting schedules to optimize plant growth and energy efficiency.

6. Temperature and Humidity Control: Establish temperature and humidity control parameters to create an ideal growing environment for the crops. Install climate control systems such as HVAC units, fans, and ventilation systems to maintain optimal conditions throughout the vertical farming facility.

7. Watering and Irrigation: Implement a precise watering and irrigation system to deliver water and nutrients to the crops efficiently. Consider using drip irrigation, misting systems, or nutrient film technique (NFT) to ensure uniform moisture distribution and minimize water wastage.

8. Pest and Disease Management: Develop an integrated pest management (IPM) plan to prevent and manage pests and diseases in the vertical farming system. Incorporate biological controls, cultural practices, and organic pesticides to minimize chemical inputs and maintain crop health.

9. Harvesting and Post-Harvest Handling: Determine the harvesting techniques and post-harvest handling procedures for each crop to maintain freshness and quality. Establish protocols for washing, packing, and storing the harvested produce before distribution to market.

10. Quality Assurance and Food Safety: Implement quality assurance measures and food safety protocols to ensure that the vertical farming operation complies with regulatory standards and meets consumer expectations. Conduct regular inspections and audits to monitor product quality and safety throughout the production process.

By systematically evaluating these factors, the agribusiness manager can determine the production process requirements for vertical farming of arugula, butterhead lettuce, red leaf lettuce, and romaine lettuce and develop a comprehensive plan to optimize crop production and maximize profitability.

Planning Production and Monitoring Systems

Establishing targets for the enterprise, as well as for each product, crop, herd, or flock, is a critical step in planning and monitoring production processes. This begins by gathering information from organizational management and strategic plans. The targets should align with the overall goals and objectives of the agribusiness, considering factors such as market demand, resource availability, and financial constraints. For example, if the strategic plan aims to increase market share in a specific segment, production targets can be set accordingly to meet the projected demand for that product.

Once targets are established, the next step is to schedule production processes, taking organizational factors into consideration. This involves coordinating activities such as planting, harvesting, breeding, feeding, and processing to ensure that production goals are met efficiently and effectively. Factors such as labour availability, equipment capacity, and seasonal variations must be considered when developing production schedules. For instance, scheduling planting activities based on weather forecasts and soil conditions can optimize crop yields and minimize production risks.

In addition to production targets and schedules, it's essential to establish waste management controls to minimize waste and dispose of waste sustainably. This includes implementing practices to reduce, reuse, and recycle waste generated during production processes.

For example, composting crop residues or animal bedding can help convert waste into valuable organic fertilizer, while proper disposal of hazardous materials prevents environmental contamination.

To monitor and evaluate production processes effectively, it's crucial to establish monitoring points and performance indicators using target, environmental management, and scheduling information. This involves identifying key metrics such as yield per hectare, feed conversion ratio, water usage efficiency, and labour productivity to track performance against established targets. Regular monitoring allows for timely adjustments and corrective actions to optimize production efficiency and resource utilization.

Moreover, establishing appropriate biosecurity and quality assurance requirements is essential to safeguard the health and integrity of livestock and crops. This includes implementing measures to prevent the introduction and spread of diseases, pests, and contaminants on the farm. Biosecurity protocols may involve practices such as quarantine procedures, vaccination programs, and sanitation practices to maintain a healthy production environment. Similarly, quality assurance measures ensure that products meet regulatory standards and consumer expectations for safety, freshness, and consistency.

Furthermore, establishing risk management strategies to control hazards is critical to minimize production disruptions and losses. This involves identifying potential risks such as adverse weather events, market fluctuations, supply chain disruptions, and regulatory changes, and developing mitigation plans to address them. For example, diversifying crop varieties or implementing crop insurance can help mitigate the impact of crop failures due to weather-related risks.

Finally, documenting the production plan, including targets and monitoring requirements, is essential for accountability, transparency, and continuous improvement. A well-documented production plan serves as a roadmap for guiding production activities, communicating expectations to stakeholders, and evaluating performance over time. It should include detailed information on production targets, schedules, waste management practices, monitoring protocols, biosecurity measures, quality assurance standards, risk management strategies, and any other relevant information necessary for effective production planning and management.

Setting production targets commences with collecting pertinent information from organizational management and strategic plans. These targets must harmonize with the overarching goals and objectives of the agribusiness, taking into account various factors such as market demand, resource availability, and financial constraints. For instance, suppose the strategic plan outlines an objective to augment market share in a particular segment. In that case, production targets can be established to align with the anticipated demand for that product. By closely aligning production targets with strategic objectives, agribusinesses ensure that their production efforts are directed towards meeting the evolving needs and aspirations of their target market while leveraging available resources optimally. This strategic alignment facilitates a coherent approach to production planning and enhances the likelihood of achieving desired outcomes within specified timeframes.

Once production targets have been established, the subsequent step is to schedule production processes, a crucial aspect of ensuring the efficient and effective achievement of those targets. This entails coordinating various activities such as planting, harvesting, breeding, feeding, and processing to align with the predetermined goals. When developing production schedules, it is essential to take into account organizational factors that can influence the execution of these activities. One of the key considerations is labour availability, as the availability of skilled workers to carry out tasks at specific times can significantly impact the overall productivity of the operation. Additionally, assessing equipment capacity is vital to ensure that machinery and tools are utilized optimally and that production processes are not hindered by equipment shortages or breakdowns.

Seasonal variations play a significant role in agricultural production and must be carefully considered when scheduling production processes. Different crops and livestock may have specific growing or breeding seasons, and production schedules need to accommodate these natural cycles. For example, planting activities should be scheduled based on weather forecasts and soil conditions to optimize crop yields and minimize the risk of losses due to adverse weather events or poor soil conditions. Similarly, breeding activities for livestock may be timed to coincide with periods of optimal fertility or favourable environmental conditions for offspring survival.

By aligning production schedules with labour availability, equipment capacity, and seasonal variations, agribusinesses can maximize efficiency and productivity while minimizing the risks associated with production disruptions. Effective scheduling ensures that tasks are carried out in a timely manner, resources are utilized efficiently, and production targets are met according to plan. Moreover, it allows for better coordination and communication among workers, suppliers, and other stakeholders involved in the production process, facilitating smoother operations and improved overall performance of the agricultural enterprise.

In conjunction with setting production targets and schedules, implementing robust waste management controls is crucial to minimizing waste and ensuring sustainable disposal practices. This entails adopting strategies aimed at reducing, reusing, and recycling waste generated throughout production processes. For instance, agricultural operations can employ practices like composting crop residues or animal bedding to transform waste into valuable organic fertilizer, thereby closing the loop on waste disposal while simultaneously enhancing soil health. Moreover, proper disposal protocols for hazardous materials are imperative to prevent environmental contamination and safeguard public health. By establishing comprehensive waste management controls, agribusinesses not only mitigate the environmental impact of their operations but also harness opportunities to derive value from waste materials, contributing to overall sustainability and resource optimization efforts.

In order to ensure the efficiency and effectiveness of production processes, it is imperative to establish comprehensive monitoring points and performance indicators that draw from target objectives, environmental management considerations, and scheduling information. This involves identifying key metrics, such as yield per hectare, feed conversion ratio, water usage efficiency, and labour productivity, to systematically track and evaluate performance against predetermined targets. For example, in crop production, monitoring yield per hectare allows farmers to assess the effectiveness of cultivation techniques and make informed decisions regarding seed selection, irrigation practices, and fertilization strategies.

Similarly, in livestock operations, tracking feed conversion ratios enables producers to optimize feed formulations and management practices to maximize animal growth rates while minimizing feed costs. By regularly monitoring these performance indicators, agribusiness managers can proactively identify areas for improvement, implement timely adjustments, and take corrective actions to enhance production efficiency and resource utilization, ultimately driving towards the achievement of production goals and organizational objectives.

Biosecurity measures are designed to prevent the introduction and spread of diseases, pests, and contaminants that could potentially jeopardize the well-being of animals or crops. For instance, in a poultry farm, biosecurity protocols may include restricting access to the premises, implementing strict sanitation practices, and monitoring for signs of infectious diseases among the flock. Similarly, in crop production, biosecurity measures may involve controlling access to fields to prevent the introduction of invasive pests or pathogens, as well as ensuring the cleanliness of equipment and vehicles to minimize the risk of cross-contamination.

In addition to biosecurity, establishing robust quality assurance requirements is essential to ensure that agricultural products meet regulatory standards and consumer expectations for safety, freshness, and consistency. Quality assurance practices may encompass a range of activities, including regular testing and analysis of soil and water samples to assess contamination levels, as well as monitoring product quality throughout the production process. For example, in a dairy farm, quality assurance measures may involve conducting regular milk tests for bacterial contamination and somatic cell counts to ensure product safety and integrity. Similarly, in vegetable production, quality assurance protocols may include visual inspections of crops for signs of disease or pest damage, as well as implementing proper post-harvest handling practices to maintain product freshness and extend shelf life. By establishing and adhering to stringent biosecurity and quality assurance requirements, agribusinesses can mitigate risks, uphold product integrity, and maintain consumer trust and confidence in their products.

Implementing effective risk management strategies is crucial for agribusinesses to minimize production disruptions and mitigate potential losses. This process begins with identifying various hazards and risks that could impact agricultural operations, including adverse weather conditions, market fluctuations, supply chain disruptions, and regulatory changes. For instance, a sudden drought or heavy rainfall could adversely affect crop yields, while changes in consumer preferences or trade policies could impact market demand and prices. Once these risks are identified, agribusiness managers can develop proactive mitigation plans to address them and reduce their impact on production.

One example of a risk management strategy is diversification, which involves spreading risk across different crops or enterprises. By planting a variety of crops with different maturity times and market demand profiles, farmers can reduce their vulnerability to adverse weather conditions or market fluctuations affecting a single crop. Additionally, implementing crop insurance can provide financial protection against losses due to weather-related events such as droughts, floods, or hailstorms. Crop insurance policies compensate farmers for yield losses or revenue declines, helping them recover from production setbacks and maintain financial stability.

Another risk management approach involves strengthening supply chain resilience to minimize disruptions and ensure the timely delivery of inputs and products. This may include establishing relationships with multiple suppliers for critical inputs such as seeds, fertilizers, and equipment, as well as maintaining adequate inventory levels to buffer against potential shortages or delays. Additionally, agribusinesses can explore alternative transportation routes and distribution channels to mitigate the impact of logistical challenges such as road closures or port congestion.

Furthermore, staying informed about regulatory changes and compliance requirements is essential for managing regulatory risks effectively. Agribusiness managers should regularly monitor developments in agricultural policies, environmental regulations, and food safety standards to ensure compliance and minimize the risk of fines, penalties, or legal disputes. By proactively addressing regulatory requirements and adopting best practices, agribusinesses can mitigate the risk of non-compliance and maintain their reputation and credibility in the marketplace.

Documenting the production plan is a critical step in ensuring the clarity, consistency, and accountability of agricultural operations. A comprehensive production plan serves as a foundational document that outlines key objectives, strategies, and procedures for achieving production targets and managing resources effectively. By documenting production targets and monitoring requirements, agribusiness managers can establish clear benchmarks for performance evaluation and continuous improvement.

For example, a production plan for a crop farm may include specific targets for yield per hectare, planting and harvesting schedules, irrigation and fertilization protocols, and pest management strategies. Each target is accompanied by monitoring requirements, such as regular field inspections, soil testing, and yield assessments, to track progress and identify areas for optimization.

Moreover, documenting waste management controls, biosecurity measures, and quality assurance standards ensures compliance with regulatory requirements and industry best practices. For instance, a livestock production plan may outline procedures for manure management, vaccination schedules, and hygiene protocols to minimize the risk of disease transmission and maintain animal health and welfare.

Additionally, risk management strategies should be clearly documented to ensure preparedness for potential hazards and emergencies. This may include contingency plans for adverse weather events, market disruptions, or equipment failures, as well as procedures for implementing corrective actions and communicating with stakeholders in the event of an incident.

By documenting the production plan in a structured and accessible format, agribusiness managers can facilitate communication and collaboration among team members, suppliers, customers, and regulatory authorities. This promotes transparency and alignment of efforts towards common goals, fostering a culture of accountability and continuous improvement within the organization.

Furthermore, regular review and updates of the production plan are essential to reflect changes in market conditions, technological advancements, regulatory requirements, and internal capabilities. Agribusiness managers should periodically evaluate the effectiveness of production processes, monitor performance against targets, and solicit feedback from stakeholders to identify opportunities for optimization and refinement.

For the hypothetical agribusiness using vertical farming for the production of arugula, butterhead lettuce, red leaf lettuce, and romaine lettuce, a production plan could include, for example:

Production Plan for Vertical Farming of Arugula, Butterhead Lettuce, Red Leaf Lettuce, and Romaine Lettuce

1. *Production Targets:*

   ○ *Arugula: 500 lbs (227 kg) per week*

   ○ *Butterhead Lettuce: 800 heads per week*

   ○ *Red Leaf Lettuce: 600 heads per week*

   ○ *Romaine Lettuce: 700 heads per week*

2. *Planting Schedule:*

  ○ *Arugula: Start planting seeds every two weeks for continuous harvest.*

  ○ *Butterhead Lettuce: Start planting seedlings every week for staggered maturity.*

  ○ *Red Leaf Lettuce: Start planting seedlings every week for staggered maturity.*

  ○ *Romaine Lettuce: Start planting seedlings every week for staggered maturity.*

3. *Irrigation and Fertilization:*

  ○ *Implement automated drip irrigation system to ensure consistent moisture levels.*

  ○ *Apply hydroponic nutrient solution according to crop growth stages.*

  ○ *Monitor pH and EC levels regularly to maintain optimal nutrient uptake.*

4. *Lighting and Climate Control:*

  ○ *Install LED grow lights with adjustable spectrum and intensity for optimal plant growth.*

  ○ *Maintain temperature and humidity levels within the ideal range for each crop.*

  ○ *Use ventilation and CO_2 supplementation to promote photosynthesis and growth.*

5. *Pest and Disease Management:*

  ○ *Implement integrated pest management (IPM) practices to monitor and control pests.*

  ○ *Use biological controls, such as predatory insects, to manage pest populations.*

  ○ *Maintain a clean and sanitized growing environment to prevent disease outbreaks.*

6. *Harvesting and Packaging:*

  ○ *Harvest arugula leaves when they reach desired size and flavour profile.*

  ○ *Harvest lettuce heads when they reach full maturity with tight leaf formation.*

  ○ *Package harvested produce in food-grade containers or bags for freshness and quality.*

7. *Waste Management:*

  ○ *Compost crop residues and organic waste to minimize environmental impact.*

  ○ *Recycle nutrient solution and irrigation water to conserve resources.*

  ○ *Dispose of non-recyclable waste in accordance with local regulations.*

8. *Quality Assurance:*

  ○ *Conduct regular quality checks on harvested produce for size, colour, texture, and flavour.*

    ◦ *Implement food safety practices to prevent contamination during handling and packaging.*

    ◦ *Maintain records of production activities, inputs used, and quality control measures.*

  9. *Monitoring and Evaluation:*

    ◦ *Monitor crop growth and development using sensors and monitoring systems.*

    ◦ *Track production yields, labour efficiency, and resource utilization for performance evaluation.*

    ◦ *Analyse data to identify trends, patterns, and areas for improvement in production processes.*

  10. *Risk Management:*

    ◦ *Develop contingency plans for power outages, equipment failures, and environmental emergencies.*

    ◦ *Maintain insurance coverage for property damage, crop loss, and liability risks.*

    ◦ *Stay informed about market trends, regulatory changes, and emerging risks to adjust production strategies accordingly.*

This production plan provides a framework for efficiently managing vertical farming operations for arugula, butterhead lettuce, red leaf lettuce, and romaine lettuce. By setting clear targets, implementing best practices, and monitoring performance, the agribusiness can optimize production efficiency, ensure product quality, and achieve business objectives effectively.

Monitoring and Evaluating the Effectiveness of Production Processes

To effectively plan and monitor production processes in agriculture, it's crucial to ensure that performance indicators, targets, and specifications are met. This involves regularly checking the progress of production against predetermined benchmarks and making amendments to the process where necessary. For example, if production targets for a specific crop are not being met due to unexpected weather conditions or pest infestations, adjustments may need to be made to planting schedules or pest management strategies to mitigate losses and maintain productivity levels. By closely monitoring performance indicators and promptly addressing any deviations from targets, farmers can optimize production efficiency and maximize yields.

Evaluation of the effectiveness of production processes at key points is essential for identifying areas of improvement and making necessary adjustments. This involves conducting periodic reviews of production methods, equipment performance, and resource utilization to assess their impact on overall productivity and profitability. For instance, if certain farming practices are found to be inefficient or resource-intensive, modifications can be made to streamline workflows, upgrade equipment, or adopt alternative techniques to enhance efficiency and reduce costs. Regular evaluation and adjustment of production processes ensure that resources are used effectively and production goals are consistently met.

Monitoring and assessing environmental impacts and workplace health and safety hazards throughout the production cycle are critical for ensuring sustainable and responsible agricultural practices. This includes conducting risk assessments to identify potential environmental risks such as soil erosion, water pollution, or habitat destruction, as well as occupational hazards such as exposure to pesticides or machinery accidents. Farmers must implement appropriate measures to mitigate these risks, such as erosion control measures, water conservation practices, and safety training for workers. By prioritizing environmental stewardship and workplace safety, farmers can minimize negative impacts on the environment and protect the health and well-being of farm workers and surrounding communities.

Modifying the production process by shifting priorities and resources when necessary is essential for adapting to changing conditions and market demands. This may involve reallocating resources, adjusting production schedules, or diversifying crop or livestock portfolios in response to market trends or emerging opportunities. For example, if consumer demand for a particular crop suddenly increases, farmers may need to allocate more land, labour, and resources to meet this demand, even if it means temporarily reducing production of other crops. Flexibility and agility in responding to market dynamics allow farmers to capitalize on opportunities and optimize returns on investment.

analysing data, observations, and documentation from the production process against the plan is critical for identifying trends, patterns, and areas for improvement. This involves systematically reviewing performance data, field observations, and production records to identify factors influencing productivity, resource efficiency, and profitability. By analysing production data alongside the initial production plan, farmers can identify discrepancies, inefficiencies, or bottlenecks in the production process and take corrective actions to address them. This iterative approach to production planning and monitoring enables continuous improvement and optimization of production processes over time.

In terms of the hypothetical example, to effectively monitor the production plan for vertical farming of arugula, butterhead lettuce, red leaf lettuce, and romaine lettuce, it's crucial to establish a comprehensive monitoring process aligned with the principles of effective production planning in agriculture. This involves regularly checking the progress of production against predetermined targets and specifications to ensure that performance indicators are met and making necessary adjustments to the process where required. For instance, if the production targets for arugula are not being met due to unexpected weather conditions affecting growth rates, adjustments may need to be made to irrigation schedules or lighting intensity to mitigate losses and maintain productivity levels.

Evaluation of the effectiveness of production processes at key points is essential for identifying areas of improvement and making necessary adjustments. This involves conducting periodic reviews of production methods, equipment performance, and resource utilization to assess their impact on overall productivity and profitability. For example, if the effectiveness of the automated drip irrigation system in maintaining consistent moisture levels for arugula is found to be suboptimal, modifications may be required to improve system efficiency and reduce water wastage.

Monitoring and assessing environmental impacts and workplace health and safety hazards throughout the production cycle are critical for ensuring sustainable and responsible agricultural practices. This includes conducting risk assessments to identify potential environmental risks such as soil erosion, water pollution, or habitat destruction, as well as occupational hazards such as exposure to pesticides or machinery accidents. By regularly monitoring environmental conditions and implementing appropriate measures to mitigate risks, farmers can minimize negative impacts on the environment and ensure the health and safety of farm workers and surrounding communities.

Modifying the production process by shifting priorities and resources when necessary is essential for adapting to changing conditions and market demands. This may involve reallocating resources, adjusting production schedules, or diversifying crop portfolios in response to market trends or emerging opportunities. For instance, if consumer demand for butterhead lettuce suddenly increases, farmers may need to allocate more space and resources to this crop to meet market demand while adjusting production targets for other crops accordingly.

Analysing data, observations, and documentation from the production process against the plan is critical for identifying trends, patterns, and areas for improvement. This involves systematically reviewing performance data, field observations, and production records to identify factors influencing productivity, resource efficiency, and profitability. By analysing production data alongside the initial production plan, farmers can identify discrepancies, inefficiencies, or bottlenecks in the production process and take corrective actions to address them. This iterative approach to production planning and monitoring enables continuous improvement and optimization of production processes over time, ensuring the long-term success and sustainability of vertical farming operations.

Recording and Managing Information

In the realm of planning and monitoring production processes, one crucial aspect is preparing recommendations for future plans based on the analysis of production data. This involves meticulously analysing the data collected during the production process, including performance metrics, yield data, resource utilization, and any other relevant factors. By scrutinizing this data, agricultural managers can identify trends, patterns, and areas for improvement, which serve as the foundation for developing recommendations to enhance future production plans. For example, if the analysis reveals inefficiencies in resource allocation or production bottlenecks, recommendations may include implementing new technologies, adjusting planting schedules, or improving operational workflows to optimize productivity and profitability.

Another essential component is preparing a comprehensive report that documents the implementation of the production plan. This report serves as a formal record of the production activities carried out, the progress made towards achieving production targets, and any deviations or challenges encountered during the process. By documenting the plan's implementation, agricultural managers can track the effectiveness of strategies deployed, evaluate the success of interventions, and identify lessons learned for future planning efforts. Additionally, the report provides stakeholders with transparency and accountability regarding production activities, ensuring clear communication and alignment with organizational goals and objectives.

Maintaining meticulous records and documentation as outlined in the production plan is imperative for effective planning and monitoring of production processes. This involves systematically recording data related to production targets, resource inputs, operational activities, quality control measures, and any other relevant information. By maintaining detailed records, agricultural managers can track progress, monitor performance, and facilitate informed decision-making throughout the production cycle. Moreover, comprehensive documentation serves as a valuable reference for future planning efforts, audits, and regulatory compliance, ensuring continuity and consistency in production management practices.

Lastly, it's essential to check that the required information is available, accessible, meaningful, and useful for effective planning and monitoring of production processes. This involves ensuring that the data collected is accurate, reliable, and relevant to the production objectives and performance metrics established in the production plan. Additionally, information accessibility is vital to ensure that stakeholders can easily access and interpret the data needed to make informed decisions and take appropriate actions. By verifying the availability and relevance of information, agricultural managers can ensure that production planning efforts are well-informed, data-driven, and conducive to achieving organizational goals and objectives effectively.

In the context of the production plan for vertical farming by means of an example, preparing recommendations for future plans based on the analysis of production data is vital for continuous improvement and optimization. For instance, if the analysis of production data reveals that the yield of arugula is consistently below the target of 500 lbs (227 kg) per week due to inadequate lighting conditions, a recommendation may include investing in additional LED grow lights or adjusting their spectrum and intensity to better support arugula growth. Similarly, if the analysis indicates that the irrigation system is not delivering consistent moisture levels to the crops, recommendations could involve upgrading the irrigation system or adjusting the timing and frequency of irrigation cycles to improve water efficiency and crop health.

Furthermore, a comprehensive report documenting the implementation of the production plan is essential for transparency, accountability, and organizational learning. For example, the report may highlight deviations from production targets, such as fluctuations in yield or quality issues, along with the corresponding actions taken to address these challenges. By documenting successes, failures, and lessons learned, agricultural managers can gain valuable insights into the effectiveness of production strategies and identify opportunities for refinement in future plans. Additionally, the report serves as a communication tool for stakeholders, providing them with a clear understanding of the progress made and the challenges encountered during production.

Maintaining meticulous records and documentation as outlined in the production plan ensures that relevant data is captured, organized, and readily accessible for analysis and decision-making. For instance, records of crop yields, resource usage, pest management activities, and quality control checks provide valuable insights into the performance of production processes and the factors influencing productivity and profitability. By maintaining detailed records, agricultural managers can track trends over time, identify areas for improvement, and make data-driven decisions to optimize production efficiency and effectiveness.

Lastly, ensuring that the required information is available, accessible, meaningful, and useful for planning and monitoring production processes is crucial for success. For example, data on crop yields, resource utilization, and environmental conditions should be accurately collected and stored in a centralized database or management system that is easily accessible to relevant stakeholders. Additionally, information should be regularly reviewed, analysed, and interpreted to extract meaningful insights and guide decision-making. By prioritizing data integrity, accessibility, and relevance, agricultural managers can enhance the effectiveness of production planning and monitoring efforts and ultimately achieve the desired business objectives.

Chapter Eighteen

Selecting and Using Agricultural Technology

Evaluating the Need for Agricultural Technology

T he landscape of modern farms and agricultural practices has evolved significantly compared to a few decades ago, largely due to technological advancements. These innovations encompass a range of tools including sensors, devices, machinery, and information technology. Contemporary agriculture integrates sophisticated technologies like robots, temperature and moisture sensors, aerial imagery, and GPS systems (USDA, 2024b). By leveraging these advanced tools and embracing precision agriculture and robotics, businesses in the agricultural sector can achieve greater profitability, efficiency, safety, and environmental sustainability.

The significance of agricultural technology lies in its ability to revolutionize traditional farming practices. Rather than uniformly applying water, fertilizers, and pesticides across entire fields, farmers now have the capability to precisely target specific areas or individual plants with the minimum required resources. This approach offers several advantages, including increased crop productivity, reduced resource usage such as water, fertilizer, and pesticides, thereby helping to maintain food prices, and minimized impact on natural ecosystems. Moreover, it curtails the runoff of chemicals into rivers and groundwater, consequently enhancing worker safety (USDA, 2024b).

Robotic technologies play a pivotal role in enhancing the monitoring and management of natural resources such as air and water quality (USDA, 2024b). These advancements empower producers with greater control over various stages of plant and animal production, processing, distribution, and storage. This control translates into enhanced efficiencies, leading to lower prices for consumers, while simultaneously ensuring safer growing conditions and food products. Furthermore, the adoption of robotic technologies contributes to reducing the environmental and ecological footprint of agricultural operations, aligning with the imperative of sustainability in modern farming practices (USDA, 2024b).

In the realm of selecting and applying modern agricultural technology in production systems, the process begins with identifying and evaluating organizational tasks and processes that could benefit from technological applications. This involves conducting a thorough assessment of existing farm activities, such as planting, irrigation, pest management, harvesting, and record-keeping, to identify areas where technological interventions can streamline operations and enhance efficiency. For example, tasks that involve repetitive manual labour or require precise monitoring and control, such as irrigation scheduling or fertilizer application, are prime candidates for automation or sensor-based technologies.

Once potential areas for technological adoption are identified, the next step is to assess opportunities and limitations for operational improvements that may result from adopting specific technological applications. This involves evaluating the potential benefits, such as increased productivity, reduced labour costs, improved resource efficiency, and enhanced data accuracy, as well as potential challenges or constraints, such as compatibility with existing infrastructure, training requirements, and upfront investment costs. For instance, while precision agriculture technologies can offer significant benefits in terms of optimizing inputs and maximizing yields, they may require substantial upfront investments in equipment and training.

After assessing the opportunities and limitations, agricultural managers need to evaluate equipment, tools, and resource requirements and select the most appropriate options. This involves researching and comparing available technological solutions, considering factors such as functionality, reliability, ease of use, scalability, and compatibility with existing systems. For example, when choosing an irrigation management system, farmers may evaluate options based on features such as soil moisture sensors, weather data integration, remote monitoring capabilities, and support services.

In addition to evaluating equipment and tools, it's crucial to assess the cost-benefit of using technology to ensure that investments align with the organization's goals and financial resources. This involves conducting a comprehensive cost-benefit analysis that considers not only upfront costs but also ongoing operational expenses, potential savings or revenue gains, and intangible benefits such as improved data accuracy or decision-making capabilities. For instance, while adopting precision farming technologies may require significant upfront investments, the long-term benefits in terms of increased yields, reduced inputs, and improved sustainability may outweigh the initial costs.

Furthermore, seeking independent technical advice and sources of information is essential to ensure informed decision-making and successful implementation of technological solutions. This may involve consulting with agricultural experts, technology vendors, industry associations, and research institutions to gain insights into the latest developments, best practices, and case studies related to agricultural technology adoption. Leveraging external expertise can help agricultural managers navigate complex technical considerations, identify optimal solutions, and mitigate risks associated with technology implementation.

Finally, developing a plan to incorporate the use of technology to improve operational efficiency, productivity, and sustainability is critical for successful adoption and integration into production systems. This plan should outline specific objectives, timelines, resource allocations, training requirements, and performance metrics to guide the implementation process and measure outcomes. It should also incorporate strategies for monitoring and evaluating the effectiveness of technological applications over time, allowing for continuous improvement and optimization of production systems. By following a systematic approach and leveraging appropriate technology, agricultural organizations can enhance their competitiveness, resilience, and sustainability in an increasingly complex and dynamic agricultural landscape.

The application of agricultural technology involves utilizing various techniques to manage the growth and harvesting of animal and vegetable products (Stewart, 2024). This includes soil preparation, which encompasses mechanical processes aimed at achieving the proper physical condition for planting and enhancing agricultural operations through tilling and fertilizing. Tilling, for instance, involves manipulating soil using mechanical tools to alter its structure, eliminate weeds, and manage crop residues. The objective is to create an optimal environment for seeds and roots, ensuring efficient water intake, storage, and transmission, as well as facilitating aeration of the subsoil.

Different soil textures require tailored approaches to seedbed preparation, with some soils necessitating adjustments to aggregate size for optimal cultivation conditions. For example, coarse-textured soils may benefit from tilling to increase aggregate size, while fine-textured soils may require clod breakage to facilitate seedling growth. However, timing is crucial, as tillage under inappropriate moisture conditions can result in less favourable soil structure (Stewart, 2024).

Primary tillage equipment, such as moldboard, disk, rotary, chisel, and subsoil ploughs, is utilized to break and loosen soil to various depths, depending on the specific agricultural requirements. Each type of plough serves distinct purposes and is selected based on factors like soil type, moisture content, and desired furrow dimensions. For instance, moldboard plough are well-suited for turning under crop residues and vary in design to accommodate different soil conditions (Stewart, 2024).

Secondary tillage further refines the seedbed through increased soil pulverization, weed destruction, and residue management using implements like harrows, rollers, and pulverisers. These tools are employed after primary tillage operations to enhance soil structure and moisture retention while promoting favourable conditions for seed germination and crop growth (Stewart, 2024).

Minimum tillage practices aim to minimize soil disturbance to preserve soil structure and reduce erosion. Techniques like mulch tillage leave crop residues on the soil surface, providing protective cover and promoting moisture retention. However, successful implementation requires careful consideration of factors such as rainfall patterns, soil type, and cropping systems to optimize agricultural productivity and sustainability (Stewart, 2024).

The selection and application of agricultural technology involve a comprehensive understanding of soil dynamics, crop requirements, and environmental factors to enhance production efficiency, conserve resources, and promote sustainable agricultural practices.

Soil fertility refers to a soil's ability to provide plants with necessary compounds in appropriate quantities and balance to support their growth under favourable conditions, including light, moisture, temperature, and soil structure. When soil fertility is lacking, natural or synthetic materials can be added to supply essential plant nutrients, known as fertilizers, though typically excluding lime or gypsum. Fertilizer grade, denoted by figures like 10–20–10, signifies the percentage of key nutrients present, such as nitrogen, phosphoric oxide, and potash, yet plants often require more nutrients than these (Stewart, 2024).

Plants require at least 16 elements for their growth, with key ones including carbon, hydrogen, oxygen, nitrogen, phosphorus, sulphur, potassium, calcium, and magnesium. While carbon and hydrogen dioxide are sourced from the atmosphere, other nutrients are absorbed from the soil. Although sodium, iodine, cobalt, silicon, and aluminium are present in plants, they are deemed nonessential. Despite the abundance of nutrients in soils, much of it remains bound in forms not readily available to plants for optimal growth, necessitating the assessment of available nutrient supply versus total quantity.

Soil solid content is broadly categorized as organic and inorganic. Organic materials range from fresh plant tissue to stable humus formed through biological decay, serving as a potential source of nitrogen, phosphorus, and sulphur. Conversely, the mineral fraction, mainly comprising sand, silt, and clay, originates from rocks and plays a crucial role in nutrient supply to plants. However, the slow decomposition of minerals in soil often leads to inadequate nutrient availability for optimal plant growth (Stewart, 2024).

Determining a crop's nutrient requirements is crucial in fertilizer technology. While some crops may display visible signs of nutrient deficiency, others may not, necessitating detailed examination of plants and soil conditions in the field. Diagnostic tests, plant tissue analysis, and soil tests aid in identifying nutrient deficiencies and determining appropriate fertilizer types and quantities. Field experiments and soil-testing laboratories are commonly used to estimate nutrient availability, while advanced techniques like remote sensing hold promise for future estimations.

The economic viability of fertilizers depends on balancing the cost of nutrient inputs against potential yield increases. Farmers aim to maximize profits by determining the most cost-effective fertilizer rate that optimizes crop production. However, the law of diminishing returns underscores the need for strategic fertilizer application to achieve optimal yield increases (Stewart, 2024).

Farm manure, a significant source of organic matter and plant nutrients, has been traditionally valued in agriculture. While manure management requires careful handling to maximize its benefits, including nutrient preservation and proper timing of application, its role has somewhat diminished due to the availability of commercial fertilizers. Nevertheless, manure remains valuable for its soil-enhancing properties and potential erosion prevention. Green Manuring and Other Organic Amendments

Green manuring, the practice of ploughing under green crops for soil improvement, has gained prominence in humid regions. Despite its benefits, such as nitrogen addition, erosion reduction, and soil fertility enhancement, challenges like crop failure, increased pests and diseases, and soil moisture depletion persist. Additionally, other organic amendments like compost, peat, and sludge contribute to soil structure and water retention, albeit with varying nutrient contents and application methods.

Liming, aimed at reducing soil acidity, is crucial in areas where rainfall leaches calcium and magnesium, creating an acidic environment. Ground limestone, basic slag, marl, chalk, and oyster shells are common liming materials used to neutralize soil acidity and supply essential plant nutrients. Proper application methods ensure uniform distribution and efficient nutrient utilization by crops (Stewart, 2024).

Fertilizers can be applied in solid, liquid, or gaseous forms, depending on various factors such as crop type, soil conditions, and equipment availability. Techniques like solid-manure spreading, granulated or pelleted fertilizer application, and liquid fertilizer spraying offer flexibility and efficiency in nutrient delivery, ensuring optimal crop growth and yield (Stewart, 2024).

Future developments in fertilizer technology are expected to focus on high-nutrient mixtures and materials, improved nitrogen delivery methods, and the utilization of micronutrients tailored to specific geographical areas. Innovations like biodegradable seeding tapes and advanced fertilizing machinery promise to enhance nutrient application precision and efficiency, contributing to sustainable agriculture practices (Stewart, 2024).

The arrangement and succession of crops cultivated on a specific area of land over time constitute the cropping system. It can involve a regular rotation of diverse crops or the continuous cultivation of a single crop year after year on the same plot of land.

Historically, agricultural studies have demonstrated the benefits of crop rotations, especially those incorporating leguminous crops. Such rotations generally sustain soil fertility, improve soil structure, and mitigate erosion. Legumes like alfalfa, sweet clover, red clover, and Ladino clover are known for their ability to enhance soil nitrogen levels. However, certain legumes, such as soybeans, do not contribute nitrogen to the soil since it is stored in the harvested seed protein. Incorporating the top growth of legumes into the soil helps replenish nitrogen levels. While grain yields tend to increase when rotated with legumes, determining the extent of this improvement attributable to nitrogen addition versus improved soil structure or pest and disease reduction can be challenging (Stewart, 2024).

The effectiveness of crop rotation depends on whether the crops compete or complement each other. Complementary relationships exist when one crop or soil-management practice benefits the growth of another by providing necessary nutrients or favourable conditions. Grasses and legumes, for instance, can complement grains or row crops by supplying nitrogen, controlling erosion and pests, and enhancing soil structure, resulting in increased productivity. Conversely, continuous cultivation of deep-rooted legumes may deplete soil moisture, necessitating alternate practices to improve subsequent forage yields.

Profit considerations, both short-term and long-term, also influence cropping systems, particularly concerning soil erosion. Excessive soil loss, detrimental to both agricultural viability and environmental conservation, prompts farmers to minimize erosion-promoting crop rotations. Soil erosion is least severe in continuous sod fields and most severe in fields with continuous row crops. Rotating row crops with sod over time reduces the erosive vulnerability of row crops. Certain row crops requiring frequent cultivation, like peanuts and potatoes, pose higher erosion risks compared to others like corn and sugarcane, which leave more residue and require less tillage. Small grains generally contribute less to erosion than row crops, while grasses or grass-legume mixtures are less erosive than pure legume stands such as alfalfa (Stewart, 2024).

The practice of cultivating the same crop annually on a given plot of land, known as monoculture, has historically faced challenges, particularly in humid regions where non-leguminous crops deplete soil nitrogen, resulting in reduced yields. However, the availability of low-cost nitrogen fertilizers has sparked renewed interest in monoculture. The advantages and disadvantages of monoculture must be carefully weighed against those of rotational systems to maintain productivity while meeting crop diversity and soil conservation goals.

Monoculture offers several potential benefits, including the ability to match each crop with the soil type best suited to it, optimize soil fertility levels for specific crops, avoid regular reseeding for perennial forage crops, and provide greater flexibility in adapting to changing crop demands and business needs.

However, successful monoculture requires rigorous management practices, such as meeting all non-legume crop nitrogen needs through purchased fertilizers or manure, addressing soil erosion issues, and managing soil structure problems associated with continuous cultivation of inter-tilled crops. Dependence on chemical insecticides, disease-resistant plant varieties, and other control methods further complicates monoculture management, as crop rotation's natural pest and disease control benefits are absent (Stewart, 2024).

Selecting cropping systems that sustain productivity, minimize soil erosion, and align with market demand and operational objectives is a complex task. The growing utilization of systems analysis is expected to facilitate informed decision-making among the various cropping options available.

Crops are susceptible to various threats such as pests, diseases, competition, and environmental factors. Insects, plant pathogens, nematodes, rodents, weeds, and air pollutants are among the adversaries that can diminish crop yields and jeopardize stored crops.

For instance, insects have the potential to decimate crops rapidly, posing a persistent challenge to farmers and scientists alike. Despite ongoing efforts, complete success in insect control remains elusive, with control measures often inadvertently affecting beneficial organisms like honey bees and natural predators of pests.

Numerous insect species are considered pests, with several hundred known for their particularly destructive impact on crops and livestock feed. These pests not only consume food and forage but also serve as vectors for plant and animal diseases.

Insecticides are widely used due to their effectiveness, affordability, and relative safety when handled properly. However, their benefits can be offset by adverse effects such as toxicity to beneficial organisms and environmental persistence. Additionally, some insect species develop resistance to certain insecticides, necessitating ongoing research for safer alternatives and the promotion of nonchemical pest control methods.

A variety of organophosphate and carbamate insecticides offer alternatives with fewer residual issues. For instance, chemicals like malathion and carbaryl are effective against pests while minimizing risks to non-target organisms. Nevertheless, their application requires careful handling due to potential hazards to applicators and the environment (Stewart, 2024).

Technological advancements have led to the development of ultralow-volume techniques, enabling precise application of minimal amounts of insecticides for cost efficiency and environmental protection. Granular formulations also offer advantages in reducing application rates and minimizing adverse effects on beneficial organisms.

Mechanical and cultural controls, such as light traps and reflective mulches, offer alternatives to chemical insecticides. These methods aim to disrupt insect behaviour or create unfavourable conditions for pest development. Additionally, biological control agents, including parasites, predators, and microbial pathogens, can help regulate pest populations, although their effectiveness depends on various ecological factors.

Integrated control approaches, which combine multiple control methods, show promise in reducing reliance on chemical insecticides while effectively managing pest populations. Strategies like combining resistant crop varieties with targeted insecticide applications or employing biological control agents alongside cultural practices represent holistic approaches to pest management.

Plant diseases and nematodes pose significant threats to crop production, often necessitating a combination of chemical, biological, and cultural control measures. Early detection, quarantine, and timely application of control measures are critical in mitigating disease outbreaks and minimizing crop losses.

Weed management is essential for maximizing crop yields and reducing production costs. While traditional methods like cultivation and crop rotation remain important, herbicides have revolutionized weed control since World War II. Herbicides, available in various formulations and application methods, offer effective and economical weed management solutions, albeit with environmental considerations and the need for careful application (Stewart, 2024).

Debates surrounding the use of chemical pesticides and herbicides continue, reflecting concerns over environmental impacts and the need for sustainable agricultural practices. Balancing the benefits of chemical control with environmental stewardship requires ongoing research, regulatory oversight, and adoption of integrated pest management approaches to ensure food and fibre security while safeguarding ecological health (Stewart, 2024).

The equipment used for harvesting is typically categorized based on the type of crop being harvested. Reapers are employed for cutting cereal grains, while threshers are used to separate the seeds from the plant. Modern combines have integrated functions, performing cutting, threshing, and cleaning of grain in a single operation. Mechanical corn pickers are utilized for corn harvesting, snapping the ears from the stalks and collecting only the grain and cobs. Cotton harvesters of the stripper type strip the entire plant of both open and unopened bolls, ideally late in the season after frost has killed off the vegetative growth. Hay and forage machines encompass various equipment such as mowers, crushers, windrowers, field choppers, balers, and machines for compacting hay into wafers or pellets.

Silage, a succulent and fermented form of storage for grass, legumes, corn, and other crops, involves cutting the crops to allow tight packing in silos, leading to anaerobic fermentation and preventing mold formation. Silage crops are typically cut in the field using a forage harvester that cuts and chops the crop immediately or picks up and chops a previously cut and raked windrow.

Root crops are harvested using diggers and digger-pickers, often resulting in the extraction of clods, stones, and vines alongside the crop. While some machines incorporate manual sorting of extraneous material by workers, this task is increasingly being mechanized. Modern sugar-beet harvesters lift the entire root from the ground, remove soil, and deposit it into bins or wagons, sometimes after the removal of beet tops for cattle feed. Peanuts are lifted with their vines and allowed to dry before the pods are removed.

Tobacco-harvesting methods vary depending on the type of tobacco and its intended use. Flue-cured tobacco is harvested using machines that carry workers who cut the leaves and place them on conveyor belts, where they are tied mechanically or by hand. Burley tobacco has traditionally been harvested by workers using machete-type knives, with the stalks then hung by hand in tobacco barns for curing. Efforts are underway to mechanize the harvesting and curing of burley tobacco, although harvesting small aromatic tobacco leaves, grown in shade, remains predominantly manual (Stewart, 2024).

Harvesting of tree crops can be done by hand or with mechanical shakers, while vegetable crops like asparagus, lettuce, and cabbage are still predominantly harvested by hand, although mechanization is increasing due to labour shortages and costs.

Machinery plays a crucial role in preparing crops for transportation, storage, market sale, and livestock feed. Rapid advancements in crop-processing machines have been driven by factors such as new crops, increased yields, diversified cropping practices, and evolving techniques.

A common method of crop drying involves spreading the crop, usually grain, on floors or mats and frequently stirring it while exposed to sunlight. While widespread in underdeveloped countries, these systems are slow and weather-dependent. Forced-air-drying systems offer more flexibility, allowing farmers to select grain varieties and harvest times more freely. These systems, relatively simple to operate, have gained popularity, especially in tropical regions, with the option of adding heat to increase air temperatures during drying. In dryeration, wet corn is placed in a dryer where it loses moisture, then transferred to a cooling bin to stabilize, reducing kernel damage and increasing dryer output.

High moisture in stored hay can lead to rapid deterioration and even spontaneous combustion. Hay, initially containing over 70% moisture, is typically wilted and dried to about 40% moisture before being further dried to a safe storage condition of around 15% moisture by blowing air through it, sometimes with supplemental heat (Stewart, 2024).

Feed-processing mills, known as feed grinders, are primarily used to mill cereals for livestock feed, aiding digestion. Modern mills allow farmers to grind grain and mix in other ingredients as desired.

Other types of crop-processing machinery include machines for seed separation, grading machinery to classify seeds, fruit graders and separators, and cotton gins, which separate cotton seeds from fibres (Stewart, 2024).

Hydroponics refers to the cultivation of plants without soil, a technique that has garnered significant attention in recent years. Originating from laboratory practices, hydroponics involves growing plants with their roots immersed in a water solution containing essential minerals or rooted in a moistened sand medium. While similar in principle to soilless culture, hydroponics is typically larger in scale. In a typical hydroponics setup, plants are supported on a bed of peat, wood fibre, or similar material, with their roots dipping into the solution below. Aeration of the solution is ensured. Alternatively, plants may be rooted in sand or gravel contained in a shallow tank, with the solution pumped at intervals by automatic control. Hydroponic techniques are employed both outdoors and in greenhouses on a small scale.

Plants obtain carbon, oxygen, and hydrogen from atmospheric gases or soil water, while other necessary elements are absorbed as mineral salts from the soil. In hydroponic systems, nitrogen, phosphorus, sulphur, potassium, calcium, and magnesium are provided as dissolved salts in various solutions tailored to meet plant requirements. Some crops grown using hydroponic methods can achieve yields comparable to those grown in fertile soils. However, wide-scale hydroponic crop production is economically viable only for certain intensive agricultural practices or under specific conditions. Nonetheless, hydroponic techniques have proven beneficial in regions lacking soil fertility, such as certain coral islands in the Pacific.

A greenhouse is a structure with transparent or translucent roofs and sides, allowing sufficient solar radiation for photosynthesis and providing a controlled environment for crop growth independent of external climate conditions. Ranging from small hobby structures to large commercial units, greenhouses vary in complexity. Hot beds, smaller than conventional greenhouses, use fermenting organic matter to generate heat, facilitating early seedling growth.

Greenhouse construction typically involves a lightweight but sturdy frame supporting glass panes, though plastic film or fiberglass panels are also used. Temperature regulation within greenhouses can be challenging due to external fluctuations. Heating systems, such as hot water, steam, or electric furnaces, are controlled by thermostats. Ventilation helps regulate temperature, and additional cooling may be necessary, especially in warm climates. Evaporative coolers can efficiently cool the structure and increase humidity. Carbon dioxide supplementation may be used to enhance photosynthetic efficiency.

Commercial greenhouse operators cultivate vegetables or ornamental plants, requiring meticulous management of environmental factors. This includes temperature regulation, ventilation, sunlight exposure, soil moisture, fertilization, and pollination. Maintenance tasks during the off-season include cleaning, fumigation, soil restructuring, and equipment checks. Disease management is crucial in greenhouse farming and often involves the use of chemicals. Despite advancements, mechanization of greenhouse operations lags behind traditional agriculture practices.

In terms of emerging technologies, the digitization of the agriculture industry is expected to unlock new value, with input providers and agritech companies playing significant roles in the emerging data ecosystem. For instance, fertilizer distributors now offer software to analyse field data for optimal fertilizer application, while equipment manufacturers are developing precision controls utilizing satellite imagery and vehicle-to-vehicle connections. Advanced connectivity also presents opportunities for telcos and LPWAN providers to establish essential connectivity infrastructure in rural areas (Goedde et al., 2020).

Agritech companies offer innovative products utilizing technology and data to enhance decision-making and increase yields. Subscription models and risk-reduction pricing strategies are likely to expedite product adoption among farmers. However, widespread adoption depends on rural areas gaining access to high-speed broadband networks.

Investment in connectivity infrastructure can occur through telco-driven, provider-driven, or farmer-driven deployment models (Goedde et al., 2020). Collaboration among industry stakeholders is crucial for success, with partnerships between agriculture players and connectivity providers offering significant advantages. Additionally, the public sector can support broadband network development through subsidies or tax breaks (Goedde et al., 2020).

Various use cases demonstrate the potential benefits of connectivity in agriculture, including crop monitoring, livestock monitoring, building and equipment management, farming by drone, and autonomous farming machinery (Goedde et al., 2020). These technologies have the potential to generate significant value through improved resource use, disease prevention, cost savings, and enhanced productivity.

Use Case 1: Crop Monitoring

Connectivity offers numerous avenues to enhance the observation and care of crops. By integrating weather data, irrigation systems, nutrient management, and other factors, farmers can improve resource utilization and increase yields by accurately identifying and predicting deficiencies. For example, sensors deployed to monitor soil conditions can communicate via LPWAN, enabling sprinklers to adjust water and nutrient application accordingly. Additionally, these sensors can provide imagery from remote areas of fields, aiding farmers in making timely decisions and receiving early warnings about issues like disease or pests. Smart monitoring can optimize the harvesting window by enabling farmers to monitor crop quality characteristics such as sugar content and fruit colour, thereby maximizing revenue potential.

However, current IoT networks often lack the capacity to support imagery transfer between devices or to monitor large fields accurately due to bandwidth and connection-density limitations. The emergence of Narrowband Internet of Things (NB-IoT) and 5G technology promises to address these challenges, potentially unlocking significant value ranging from $130 billion to $175 billion by 2030.

Use Case 2: Livestock Monitoring

In large-scale livestock management, preventing disease outbreaks and identifying animals in distress are paramount. Utilizing chips and body sensors that measure temperature, pulse, and blood pressure, among other indicators, can facilitate early detection of illnesses, preventing herd infection and enhancing food quality. Technologies such as ear-tag systems from providers like Smartbow and electronic tracing solutions from companies like Allflex are already employed to monitor various aspects of livestock health and location, contributing to comprehensive disease management strategies.

Moreover, environmental sensors can trigger automated adjustments in ventilation or heating within barns, improving living conditions for animals and addressing consumer concerns about animal welfare. Better monitoring of animal health and growth conditions has the potential to generate substantial value, estimated at $70 billion to $90 billion by 2030.

Use Case 3: Building and Equipment Management

Chips and sensors designed to monitor levels of silos and warehouses can automate inventory management processes, reducing costs for farmers. These tools also contribute to improving the shelf life of inputs and reducing post-harvest losses by monitoring and optimizing storage conditions. Monitoring conditions and usage of buildings and equipment can lead to decreased energy consumption, while computer vision and predictive maintenance systems attached to equipment can lower repair costs and extend machinery life. Such solutions have the potential to achieve significant cost savings, projected to reach $40 billion to $60 billion by 2030.

Use Case 4: Farming by Drone

Drones have been utilized in agriculture for decades, initially for tasks like crop spraying. The latest generation of drones can survey crops and herds over large areas efficiently, serving as relays for real-time data transmission and analysis. Additionally, drones equipped with computer vision technology can analyse field conditions and deliver precise interventions such as fertilizers, nutrients, and pesticides where needed most. They can also assist in planting seeds in remote locations, reducing equipment and labour costs. The adoption of drones in agriculture has the potential to generate substantial value, estimated at between $85 billion and $115 billion by 2030.

Figure 60: A drone intended for agricultural use. Agridrones Solutions Israel, CC BY-SA 4.0, via Wikimedia Commons.

Use Case 5: Autonomous Farming Machinery

Advancements in GPS technology, coupled with computer vision and sensors, are driving the deployment of smart and autonomous farm machinery. These machines can operate multiple equipment simultaneously without human intervention, freeing up time and resources for farmers. Autonomous machinery offers higher efficiency and precision in field operations compared to human-operated counterparts, leading to fuel savings and increased yields. Improved connectivity further enhances the autonomy of machinery, potentially creating an additional value of $50 billion to $60 billion by 2030.

Agritech advancements can play a crucial role in promoting sustainable agriculture by offering innovative solutions that address environmental concerns, enhance resource efficiency, and improve agricultural practices. Some key agritech advancements applicable to sustainable agriculture include:

1. Precision Agriculture: Precision agriculture technologies utilize sensors, GPS, drones, and data analytics to optimize resource management, including water, fertilizer, and pesticides. By precisely targeting inputs based on real-time data and field variability,

precision agriculture minimizes waste, reduces chemical usage, and improves crop yields, thereby promoting sustainability.

2. Vertical Farming and Indoor Agriculture: Vertical farming and indoor agriculture utilize controlled environments and advanced hydroponic or aeroponic systems to grow crops vertically, often in urban settings. These technologies reduce the need for land and water, minimize pesticide use, and enable year-round production, thereby promoting sustainable food production in densely populated areas.

3. Biological Crop Protection: Agritech companies are developing biological alternatives to chemical pesticides, such as biopesticides, pheromones, and beneficial insects. These biological crop protection methods reduce chemical residues in soil and water, minimize harm to non-target organisms, and support ecosystem health, contributing to sustainable pest management.

4. Soil Health Monitoring and Management: Soil health monitoring technologies, including sensors, imaging, and data analytics, enable farmers to assess soil quality, moisture levels, and nutrient content in real time. By implementing precision soil management practices, such as variable-rate fertilization and conservation tillage, farmers can enhance soil fertility, reduce erosion, and mitigate nutrient runoff, promoting sustainable soil management.

5. Climate-Smart Agriculture: Agritech advancements focused on climate-smart agriculture aim to help farmers adapt to and mitigate climate change impacts. These technologies include drought-resistant crop varieties, weather forecasting tools, and carbon sequestration practices. By improving resilience to extreme weather events and reducing greenhouse gas emissions, climate-smart agriculture contributes to sustainable farming practices.

6. Agroforestry and Permaculture: Agritech innovations support agroforestry and permaculture practices by providing tools for integrated land management, agroecological design, and biodiversity conservation. Agroforestry systems, which combine trees with crops and livestock, enhance soil fertility, provide habitat for wildlife, and diversify farm income, promoting sustainable land use and ecosystem resilience.

7. Blockchain and Traceability Systems: Blockchain technology and traceability systems enable transparent and traceable supply chains, facilitating the verification of sustainable farming practices, fair labour conditions, and product authenticity. By enhancing trust and accountability in food production and distribution, blockchain solutions support sustainable agriculture initiatives and consumer preferences.

Overall, agritech advancements offer a wide range of tools and solutions to support sustainable agriculture practices, from precision farming and soil health management to climate adaptation and supply chain transparency. By harnessing these technologies, farmers can improve productivity, reduce environmental impacts, and build resilient and sustainable food systems for the future.

Implementing Technology to Manage Production

In modern agriculture, the integration of advanced technologies has become imperative for efficient farm management, production optimization, and sustainable practices. To record, analyse, and manage production data effectively, farmers can leverage various technological tools and platforms. For instance, farm management software allows farmers to digitally record and organize farm activities, such as planting schedules, irrigation routines, pesticide applications, and harvest dates. These software solutions often include features for data analysis, enabling farmers to track trends, identify patterns, and make informed decisions about crop management strategies.

Furthermore, the use of sensor technology plays a crucial role in data collection and analysis on the farm. Sensors can be deployed throughout the field to monitor soil moisture levels, temperature variations, and crop health indicators in real time. This continuous data collection provides valuable insights into the performance of crops and soil conditions, allowing farmers to adjust their management practices accordingly. Additionally, drones equipped with multispectral imaging cameras can capture aerial data, generating detailed maps of crop health and identifying areas of concern for further investigation.

In developing treatment strategies or input requirements, technology offers sophisticated tools for precision agriculture. By integrating data from soil sensors, weather forecasts, and historical yield records, farmers can develop tailored treatment plans for each field or crop variety. For example, based on soil nutrient levels and weather predictions, farmers can determine the optimal timing and dosage of fertilizer applications to maximize crop productivity while minimizing environmental impact. Similarly, pest and disease management strategies can be customized using data analytics to identify potential threats and implement targeted interventions, reducing reliance on broad-spectrum pesticides.

To effectively collect, store, and analyse data across land-based businesses, farmers can adopt comprehensive farm management platforms that centralize data from various sources. These platforms enable seamless integration of data from field operations, equipment sensors, financial records, and supply chain logistics. By consolidating information into a unified database, farmers gain a holistic view of their operations, facilitating informed decision-making and strategic planning. Moreover, cloud-based storage solutions ensure data accessibility and security, allowing stakeholders to access relevant information anytime, anywhere.

In terms of research, reporting, and communication, technology provides farmers with access to a wealth of information resources and collaborative tools. Online databases, scientific journals, and agronomic research publications offer valuable insights into emerging trends, best practices, and innovative technologies in agriculture. Farmers can leverage these resources to stay updated on industry developments and incorporate new knowledge into their production management strategies. Additionally, digital reporting and communication platforms enable farmers to share data, insights, and performance metrics with stakeholders, fostering transparency and accountability throughout the supply chain.

Finally, integrating technology to improve operational efficiency, production, profitability, and sustainability requires a holistic approach that encompasses both hardware and software solutions. From automated irrigation systems and precision planting equipment to advanced analytics software and predictive modelling tools, farmers have access to a wide range of technologies that can streamline workflows, optimize resource utilization, and enhance overall farm performance. By embracing technology-driven innovations, farmers can achieve higher yields, lower input costs, and greater environmental stewardship, ultimately ensuring the long-term viability and sustainability of their agricultural operations.

As specific examples, indoor vertical farming represents a revolutionary approach to agriculture that offers numerous benefits, including increased crop yields, overcoming limited land area, and reducing farming's environmental impact by shortening the supply chain (Ku & Serna, 2023). This innovative concept involves growing produce stacked vertically in controlled environments, significantly reducing the land space required compared to traditional farming methods. Moreover, vertical farming often utilizes hydroponic or aeroponic systems, eliminating the need for soil and conserving water resources. Vertical farms can use up to 70% less water than traditional farms, making them highly efficient and sustainable alternatives.

In addition to water conservation, indoor vertical farming enables precise control over environmental variables such as light, humidity, and water, resulting in year-round production with reliable harvests. Furthermore, labour costs are reduced through the integration of automation technologies, including robots for harvesting, planting, and logistics. By streamlining operations and minimizing manual labour, vertical farms can address labour shortages in the agriculture industry while improving efficiency and profitability (Ku & Serna, 2023).

Another area of technological advancement in agriculture is farm automation, also known as smart farming. This technology automates various aspects of crop or livestock production cycles, enhancing efficiency and reducing labour requirements. AgTech companies are developing robotics innovations such as drones, autonomous tractors, robotic harvesters, and automatic watering and seeding robots to streamline farm operations. While farm automation is still relatively new, traditional agriculture companies are increasingly adopting these technologies to optimize their processes and improve productivity (Ku & Serna, 2023).

Livestock farming is also benefiting from technological advancements, with developments in livestock management technologies improving productivity, animal welfare, and operational efficiency. The concept of the "connected cow" involves equipping herds with sensors to monitor health and increase productivity (Ku & Serna, 2023). Additionally, advancements in animal genomics enable producers to make informed breeding decisions, optimizing the profitability and yields of livestock herds.

Furthermore, modern greenhouses are incorporating advanced technologies such as LED lights and automated control systems to create optimal growing environments. These high-tech greenhouses are increasingly capital-infused and strategically located near urban hubs to meet the growing demand for local food. Precision agriculture technology is another area of innovation, allowing farmers to maximize yields by controlling variables such as moisture levels, soil conditions, and microclimates. Precision agriculture companies are leveraging technology to provide more accurate techniques for planting and growing crops, ultimately increasing efficiency and managing costs (Ku & Serna, 2023).

Blockchain technology is also transforming the agriculture industry by enhancing food traceability and transparency. By creating a decentralized ledger for tracking food products from farm to table, blockchain ensures verified products and practices, addressing issues such as food fraud, safety recalls, and supply chain inefficiencies. Additionally, artificial intelligence is playing a vital role in agriculture by monitoring plant health, soil conditions, and environmental variables. Through advanced algorithms and remote sensors, AI enables farmers to make data-driven decisions and achieve better harvests, ultimately enhancing productivity and sustainability in agriculture (Ku & Serna, 2023).

The evolution of livestock technologies is reshaping the landscape of farming, empowering farmers with data-driven insights to revolutionize farm management, enhance animal welfare, and drive productivity. Here's a glimpse into some of the transformative innovations redefining livestock farming:

Automated Dairy Installations: These installations automate the milking process without human intervention, while sophisticated milk sensors monitor milk quality in real-time, ensuring optimal production standards.

Automated Cleaning Systems: Utilizing automated cleaning systems, farmers can efficiently remove waste, creating cleaner and disease-free environments for livestock, thereby promoting their health and well-being.

Armenta's Non-Antibiotic Treatment: Armenta introduces a revolutionary non-antibiotic treatment for bovine mastitis, leveraging acoustic pulse technology (APT) to combat this prevalent disease, which incurs significant annual losses.

Automated Feeder Systems: These systems provide animals with precisely tailored feeding mixtures, addressing their specific nutritional needs and optimizing feed utilization for enhanced health and productivity.

Faromatics: Faromatics integrates robotics, artificial intelligence (AI), and big data analytics to bolster animal welfare and farm productivity, ushering in a new era of sustainable livestock management.

Further, traditional scarecrows are being replaced by high-tech laser scarecrows equipped with motion sensors, designed to deter birds and rodents from damaging crops. Emitting green laser light, these devices startle birds, significantly reducing crop damages by up to 70% to 90%.

Farm automation also integrates agricultural machinery, computer systems, and data management to streamline operations and minimize human error. This technology, encompassing automated harvesters, drones, autonomous tractors, and seeding/weeding systems, optimizes resource utilization and enhances productivity, allowing farmers to focus on strategic decision-making.

Real-Time Kinematic (RTK) Technology enables precise field mapping and vehicle navigation, facilitating the implementation of controlled traffic systems for soil health and productivity. By providing centimetre-level accuracy, RTK technology ensures optimal positioning of farm machinery, minimizing soil compaction and maximizing output.

Minichromosome technology offers a breakthrough solution to enhance crop traits without genetic modification, enabling the development of pest-resistant and drought-tolerant crops with minimal environmental impact. This innovation reduces reliance on pesticides and fertilizers while enhancing crop nutritional content.

Farm management software solutions provide farmers with real-time data and insights to optimize farm operations, from procurement and supply chain management to financial planning and processing. Streamlining workflows and fostering collaboration, these platforms empower farmers to make informed decisions and drive efficiency.

Innovative water management technologies, such as micro drip irrigation systems like N-Drip, minimize water wastage and improve crop quality by delivering water directly to plant roots. By reducing water usage by up to 50%, these technologies offer sustainable solutions to address water scarcity and enhance agricultural productivity.

Implementing drone technology on a farm offers numerous opportunities to enhance agricultural operations and improve productivity. One of the key applications of drones in agriculture is crop monitoring. By attaching sensors to drones, farmers can collect valuable data on crop health, soil conditions, moisture levels, and pest or weed infestations. This data allows farmers to make informed decisions about irrigation, fertilization, and pest management, ultimately leading to improved crop health and higher yields. By regularly monitoring crops with drones, farmers can detect issues early and take proactive measures to address them, reducing the risk of crop loss and optimizing production.

Furthermore, the use of drones in agriculture can save farmers valuable time and resources. Instead of manually inspecting crops or walking through fields, farmers can deploy drones to quickly and efficiently survey large areas of land. This not only saves time but also reduces labour costs and minimizes the need for manual intervention in crop management tasks. Additionally, drones can cover terrain that is difficult or dangerous for humans to access, such as steep slopes or remote areas of the farm, allowing farmers to gather data from areas that would otherwise be inaccessible.

In addition to crop monitoring, drone technology is also being utilized in the livestock industry. Drones can be used to check water points, monitor herds, and assist with livestock movements. This enables farmers to efficiently manage their livestock operations, ensuring that animals have access to water and are healthy and accounted for. By using drones for livestock management tasks, farmers can save time and fuel that would otherwise be spent on manual inspections or driving through pastures. This allows farmers to allocate their resources more effectively and focus on other aspects of farm management.

Drones exemplify a groundbreaking technology that empowers farmers to envision the possibilities of the future while mitigating present-day constraints. Their integration into agriculture offers a multitude of advantages, streamlining labour, reducing expenses, and optimizing resource allocation within the industry. They can be specifically utilised for (UAV Training Australia, 2024):

1. Imaging Capabilities: Drones, as aerial devices, effortlessly navigate agricultural landscapes, whether confined spaces or expansive fields, capturing high-resolution images. These images can be utilized independently or processed into intricate three-dimensional maps, all without disturbing the farmers, their crops, or the surrounding ecosystem. This facilitates better seed planting strategies, soil analysis, and real-time monitoring of crop growth, enabling prompt intervention for issues like pest infestations, yield fluctuations, adverse weather conditions, and other agricultural challenges.

2. Data Collection and Monitoring: The visual data and real-time coverage provided by drones empower farmers to inspect irrigation systems, monitor pesticide and fertilizer applications, and collect plant and environmental data. This facilitates informed decision-making regarding production output, resource management, and overall crop health. Additionally, drones enable precise crop surveys, furnishing invaluable yield data for future seasons. This continuous data refinement enhances operational efficiency, enabling farmers to optimize yield while minimizing resource wastage over time.

3. Pesticide and Fertilizer Deployment: Drones offer a safer, more cost-effective alternative to traditional crop-dusting methods. Equipped with large reservoirs for fertilizers, herbicides, or pesticides, drones mitigate risks associated with manned aircraft, such as pilot safety and potential ground accidents. Furthermore, drones facilitate precise chemical application, reducing environmental impact by minimizing chemical usage and ensuring targeted treatment, thus promoting sustainability in agricultural practices.

4. Irrigation Management: Utilizing sensors, drones can pinpoint areas of a field experiencing water stress or excess moisture, enabling precise irrigation adjustments. Additionally, drones provide real-time assessments of crop health, temperature, and density, facilitating adaptive irrigation strategies throughout the crop growth cycle. Early detection of irrigation issues allows for optimized drainage, prevention of water pooling, and protection of delicate crops from unnecessary damage.

5. Planting Innovation: Some companies are exploring drone planting technologies, where drones distribute seeds and nutrients directly into the soil, eliminating the need for manual labour or conventional machinery. This efficient method not only accelerates the planting process but also reduces associated costs, enhancing overall productivity.

In summary, drones revolutionize farming practices by enhancing efficiency and enabling proactive problem-solving. As the demand for drones grows alongside their proven efficacy in agriculture, this marks a pivotal moment for drone operation and development worldwide.

Overall, the implementation of drone technology in agriculture offers significant benefits for farmers. By leveraging drones for crop monitoring and livestock management, farmers can improve efficiency, reduce costs, and optimize production. As drone technology continues to advance, its potential applications in agriculture are likely to expand, offering even more opportunities for farmers to enhance their operations and achieve sustainable agricultural practices.

Implementing livestock GPS tracking software on a farm involves several steps to ensure effective utilization and integration into existing operations. Firstly, selecting the appropriate GPS tracking software is crucial. Farmers should research and evaluate different software options available in the market, considering factors such as features, compatibility with existing systems, ease of use, and cost-effectiveness. It's essential to choose software that offers functionalities specifically tailored to livestock management, such as real-time tracking, health monitoring, and data analytics.

Once the software is selected, the next step is installation and setup. This involves installing GPS tracking devices on individual animals, typically in the form of ear tags or collars equipped with GPS receivers. Farmers need to ensure proper placement of these devices to ensure accurate tracking without causing discomfort or hindering the animals' movements. Additionally, setting up the software interface on computers or mobile devices and configuring it to sync with the GPS devices is essential.

Training farm personnel is another crucial aspect of implementation. Farmers and farmworkers need to be trained on how to use the GPS tracking software effectively, including how to interpret tracking data, monitor animal health indicators, and troubleshoot any technical issues that may arise. Training sessions can be conducted by the software provider or through online tutorials and resources.

Integration with existing farm management systems is also essential for seamless operation. Farmers should ensure that the GPS tracking software can integrate with other farm management tools and databases, such as inventory management systems, electronic health records, and financial software. This integration enables comprehensive data analysis and decision-making across different aspects of farm operations.

Regular monitoring and maintenance of the GPS tracking system are necessary to ensure optimal performance. Farmers should routinely check the functionality of tracking devices, replace batteries as needed, and troubleshoot any connectivity issues. It's also essential to update the software regularly to access new features, security patches, and performance improvements.

Lastly, farmers should continuously evaluate the effectiveness of the GPS tracking system in meeting their goals and objectives. This involves analysing tracking data, monitoring animal health and productivity indicators, and soliciting feedback from farm personnel. Based on this evaluation, adjustments and refinements can be made to optimize the use of GPS tracking technology and maximize its benefits for livestock management.

Implementing soil moisture sensors on a farm involves several steps to ensure accurate monitoring and effective water management. Farmers need to select the appropriate soil moisture sensors based on factors such as soil type, crop type, and climate conditions. There are various types of sensors available, including capacitance-based, tensiometric, and volumetric sensors, each with its own advantages and limitations. Farmers should consider the specific needs of their farm and choose sensors that provide reliable and accurate measurements for their soil conditions.

Once the sensors are selected, the next step is installation. Farmers need to properly install the sensors in representative locations across their fields to ensure comprehensive coverage. This involves burying the sensors at appropriate depths in the soil according to the rooting depth of the crops being grown. It's essential to follow manufacturer guidelines for installation to ensure accurate readings and minimize interference from external factors.

After installation, the sensors need to be connected to a data logging system or telemetry system for data collection and analysis. This system can be a standalone device installed on the farm or a cloud-based platform that allows remote access to real-time data. Farmers should ensure that the data logging system is compatible with the sensors and provides features such as data storage, visualization, and alerts for threshold levels.

Once the sensors are installed and connected, farmers need to calibrate the sensors to ensure accurate measurements. Calibration involves comparing sensor readings to known moisture levels in the soil and adjusting the sensor settings accordingly. This step is crucial for obtaining reliable data that can be used for decision-making.

With the sensors in place and calibrated, farmers can begin monitoring soil moisture levels in their fields. Regular data collection and analysis allow farmers to track changes in soil moisture over time and make informed decisions about irrigation scheduling. By knowing exactly how much moisture is present in the soil, farmers can optimize their irrigation practices, applying water only when and where it's needed to maximize crop growth and minimize water waste.

In addition to monitoring soil moisture levels, farmers should also integrate sensor data with other relevant information such as weather forecasts, crop water requirements, and irrigation system performance. This holistic approach to water management enables farmers to develop effective irrigation strategies that consider various factors impacting soil moisture levels and crop health.

Continuous monitoring and adjustment are key to successful implementation of soil moisture sensors on a farm. Farmers should regularly review sensor data, analyse trends, and make necessary adjustments to irrigation schedules and practices based on changing conditions. By leveraging soil moisture sensors as part of their irrigation management toolkit, farmers can improve water efficiency, optimize crop yields, and promote sustainable agriculture practices.

To implement remote monitoring systems on a farm, farmers need to assess their specific needs and determine which areas of their operation would benefit most from remote monitoring. In the case of monitoring livestock water points, farmers should identify the locations of water troughs, dams, tanks, or bores that require regular inspection and maintenance.

Once the monitoring points are identified, farmers can select appropriate sensors and monitoring devices that are compatible with their infrastructure and requirements. These sensors should be capable of measuring relevant parameters such as water level, temperature, and flow rate, depending on the specific needs of the farm.

Next, farmers need to install the sensors and monitoring devices at the designated locations. This may involve mounting sensors on water troughs or tanks, installing monitoring equipment near dams or bores, and ensuring that all components are securely fixed and properly calibrated.

After installation, farmers need to set up the remote monitoring system to collect data and transmit it to a centralized platform for analysis and visualization. This may involve configuring wireless communication networks, such as Wi-Fi or cellular networks, to transmit data from the sensors to a cloud-based data storage system.

Once the monitoring system is operational, farmers can access real-time data and insights about the status of their livestock water points remotely using a computer, tablet, or smartphone. This allows farmers to conveniently monitor water levels and other parameters from anywhere, reducing the need for manual inspections and saving time and effort.

Furthermore, remote monitoring systems can be configured to send alerts and notifications to farmers via email, text message, or mobile app when predefined thresholds or anomalies are detected. This enables farmers to respond promptly to issues such as low water levels, equipment malfunction, or leaks, minimizing downtime and potential losses.

Implementing remote monitoring systems on a farm involves careful planning, selection of appropriate sensors and equipment, installation, configuration, and ongoing management. By leveraging remote monitoring technology, farmers can streamline operations, improve efficiency, and enhance productivity while reducing manual labour and ensuring timely response to critical events.

Implementing remote security cameras on a farm to mitigate the risk of theft involves requires farmers to assess their security needs and identify areas of vulnerability on their farm, such as access points, equipment storage areas, and livestock enclosures. Once potential security threats are identified, farmers can determine the locations where remote security cameras should be installed to provide optimal coverage.

Next, farmers should select suitable remote security cameras that meet their requirements in terms of functionality, resolution, and connectivity. There are various types of remote security cameras available on the market, including wireless cameras that operate on batteries, making them easy to install and move around as needed.

After acquiring the cameras, farmers need to install them in strategic locations across the farm to maximize surveillance coverage. This may involve mounting cameras on buildings, fences, poles, or other structures to provide clear visibility of the targeted areas. It's essential to position the cameras in such a way that they capture relevant activity and provide adequate coverage of vulnerable areas.

Once the cameras are installed, farmers need to set up the monitoring system to ensure continuous surveillance and remote access to camera feeds. This typically involves connecting the cameras to a central monitoring station or a cloud-based platform where farmers can access live or recorded footage from their computer, smartphone, or tablet.

Additionally, farmers may choose to integrate motion sensors or other detection devices with the security cameras to trigger alerts when unusual activity is detected. This enables farmers to receive real-time notifications via email, text message, or mobile app whenever the cameras detect motion or unauthorized access, allowing them to respond promptly to potential security threats.

Regular maintenance and monitoring of remote security cameras are essential to ensure their proper functioning and effectiveness in deterring theft and vandalism. Farmers should periodically check camera batteries, clean lenses, and adjust camera angles as needed to maintain optimal performance.

Implementing remote security cameras on a farm involves assessing security needs, selecting suitable cameras, installing them in strategic locations, setting up the monitoring system, and integrating additional detection devices as necessary. By leveraging remote security camera technology, farmers can enhance farm security, deter theft and vandalism, and protect valuable assets more effectively.

Evaluate Use of Technology

To effectively select and apply modern agricultural technology in production systems, it's crucial to develop and review strategies that align with operational goals while ensuring cost-effectiveness. Initially, farmers should conduct a comprehensive assessment of their farm's specific needs and objectives. This involves identifying areas where technology can optimize efficiency, enhance productivity, and reduce operational costs. By prioritizing these objectives, farmers can narrow down the selection of technologies that offer the most significant benefits to their production systems. Regular reviews of these strategies are essential to adapt to changing market conditions, technological advancements, and evolving operational goals.

Identifying areas where technology can optimize efficiency, enhance productivity, and reduce operational costs involves a systematic approach to evaluating current practices and pinpointing areas that could benefit from technological intervention. The process begins with conducting a comprehensive assessment of the farm's operations. This entails evaluating each stage of the production process, from planting or raising crops to harvesting and distribution, with a keen eye on tasks that are labour-intensive, time-consuming, or prone to errors.

Following the assessment, the next step is to analyse the operational challenges that hinder efficiency and productivity. This may encompass issues such as manual data collection and recording, inefficient resource allocation, inconsistent quality control, or reliance on outdated equipment and techniques. By identifying these challenges, farmers can better understand the areas where technology can make a significant impact.

Once the operational challenges have been identified, the farmer can proceed to research available agricultural technologies that address these issues. This includes exploring options such as farm management software, precision agriculture tools, automation systems, sensor technologies, and equipment upgrades. It's essential to consider technologies that offer features such as real-time monitoring, data analytics, automation, and remote access.

With a range of technologies to choose from, the next step is to prioritize technology adoption based on its potential to optimize efficiency, enhance productivity, and reduce operational costs. Farmers should focus on technologies that offer tangible benefits in terms of labour savings, resource optimization, yield improvement, or risk mitigation. Assessing the return on investment (ROI) for each technology under consideration is crucial at this stage, comparing the expected cost savings and productivity gains against the initial investment and ongoing operational expenses.

Assessing the return on investment (ROI) for each technology under consideration is a critical step in the process of modernizing agricultural operations. This assessment involves a thorough examination of the potential benefits and costs associated with adopting a particular technology. By carefully comparing the expected cost savings and productivity gains against the initial investment and ongoing operational expenses, farmers can make informed decisions about which technologies offer the best value for their farm.

First and foremost, it's essential to identify and quantify the potential benefits that a technology is expected to deliver. This may include factors such as labour savings, increased efficiency, improved crop yields, reduced waste, or enhanced data insights. By clearly defining these benefits, farmers can develop realistic expectations about the impact that the technology will have on their operations.

On the flip side, it's equally important to assess the costs associated with implementing and maintaining the technology. This includes not only the upfront investment required to purchase the technology but also ongoing expenses such as maintenance, training, and support. Additionally, farmers should consider any potential risks or uncertainties that could affect the cost-effectiveness of the technology over time.

Once the expected benefits and costs have been identified, farmers can calculate the potential return on investment using a variety of financial metrics. One common method is to calculate the payback period, which represents the amount of time it will take for the cumulative cost savings generated by the technology to equal or exceed the initial investment. A shorter payback period indicates a higher ROI and greater financial viability.

Another important metric to consider is the net present value (NPV), which takes into account the time value of money by discounting future cash flows back to their present value. A positive NPV indicates that the technology is expected to generate more value than it costs to implement, while a negative NPV suggests the opposite.

In addition to these quantitative measures, it's also valuable to consider qualitative factors that may influence the overall return on investment. This could include intangible benefits such as improved farm management practices, enhanced decision-making capabilities, or increased competitiveness in the market.

Ultimately, the goal of assessing the return on investment for each technology under consideration is to ensure that farmers are making sound financial decisions that align with their operational goals and objectives. By carefully weighing the potential benefits and costs of adopting new technologies, farmers can identify opportunities to improve efficiency, productivity, and profitability on their farms.

Before fully committing to implementation, it's advisable to pilot test selected technologies on a smaller scale to evaluate their effectiveness in real-world farm conditions. Gathering feedback from farmworkers and stakeholders during this phase can help identify any challenges or areas for improvement. Continuous monitoring of the performance of adopted technologies is essential, collecting data on key performance metrics such as yield, labour efficiency, input usage, and cost savings. This data can then be used to identify areas for further optimization and adjustment.

Finally, fostering a culture of continuous improvement is vital for ensuring ongoing success. By regularly reviewing and updating technology adoption strategies based on evolving operational goals, technological advancements, and market trends, farmers can stay ahead of the curve. Remaining informed about emerging technologies and opportunities for innovation in the agricultural sector is key to driving continuous improvement and maximizing overall farm performance and profitability.

Assessing the need for additional training and support is paramount to maximize the effective use of technology on the farm. Investing in training programs for farmworkers ensures that they possess the necessary skills and knowledge to operate new technologies efficiently. Moreover, establishing support channels, such as technical assistance hotlines or on-site troubleshooting services, can address any challenges or questions that arise during technology implementation. Continuous assessment of training needs and providing ongoing support will empower farmworkers to leverage technology effectively, contributing to improved productivity and operational outcomes.

Developing strategies to address barriers to the effective use of technology is essential for overcoming challenges that may impede its adoption and utilization. Common barriers include limited access to financing, inadequate infrastructure, and resistance to change among farmworkers. Farmers can mitigate these barriers by seeking financial assistance through grants or loans tailored to technology adoption, investing in infrastructure upgrades to support technology implementation, and fostering a culture of innovation and openness to change through effective communication and leadership.

Evaluating the impact of technology use on production levels, input costs, and cash flow budget is critical to measure the return on investment and assess the overall effectiveness of technology adoption. Farmers should establish key performance indicators (KPIs) to track the impact of technology on various aspects of production, such as yield, labour efficiency, input usage, and profitability. Regular monitoring and analysis of these KPIs enable farmers to identify areas of improvement, optimize resource allocation, and make informed decisions to enhance operational efficiency and financial sustainability.

Developing strategies for monitoring, evaluating, and incorporating future developments in technology ensures that farms remain adaptive and responsive to emerging trends and innovations in the agricultural sector. This involves staying informed about advancements in agricultural technology through industry publications, research studies, and participation in relevant workshops or conferences. By actively monitoring technological developments, farmers can identify opportunities to integrate new technologies that offer potential benefits to their production systems. Additionally, establishing a process for evaluating and piloting new technologies allows farmers to assess their feasibility and performance before full-scale implementation, ensuring strategic adoption and optimization of resources.

Chapter Nineteen

Monitoring and Managing Soils for Production

Research Information About Soil

Researching and analysing soil type, characteristics, and properties is a foundational step in soil management for agricultural purposes. Farmers and land managers must thoroughly understand the soil on their property to make informed decisions about crop selection, fertilization, irrigation, and other management practices. This process typically begins with gathering existing soil information from sources such as soil surveys, geological maps, and historical records. These resources provide valuable insights into the general soil types present on the land, as well as their basic characteristics and properties.

The quality of agricultural productivity heavily relies on the condition of farmland, making soil testing a crucial tool in identifying potential problems affecting crop growth. Field suitability analysis, facilitated by soil testing, aids in selecting appropriate crops and determining suitable land use for farming activities.

Regular soil testing is not only vital for growers but also provides valuable insights to various stakeholders in the agribusiness sector, including agricultural cooperatives, crop insurers, banks, input suppliers, and commodity traders. However, it's important to note that soil testing conducted by designated laboratories only provides a snapshot of current field properties. Therefore, when analysing soil test results to make decisions regarding field improvement, it's advisable to supplement lab reports with historical data obtained from satellite imagery analytics.

Once the initial research is complete, it's important to identify particular soil properties that are relevant to the specific workplace or agricultural operation. This may include factors such as soil texture, structure, pH levels, nutrient content, organic matter content, and drainage characteristics. By focusing on these key properties, farmers can prioritize their soil management efforts and target areas that require further investigation or intervention.

Next, farmers must determine acceptable soil physical and chemical parameters for the specified crop or crops they intend to grow. This involves consulting published data, research studies, and agronomic recommendations to identify optimal soil conditions for crop growth and productivity. These parameters may vary depending on the crop's nutrient requirements, moisture tolerance, and other environmental preferences.

With acceptable soil parameters established, the next step is to identify relevant properties of soils and areas with homogeneous soil characteristics for testing. This helps ensure that soil samples are representative of the entire production area and provide accurate information for decision-making. Farmers may use techniques such as grid sampling or zone sampling to target specific areas of interest within their fields.

Once suitable sampling locations have been identified, farmers must take soil samples from the land under production using appropriate sampling tools and techniques. Soil samples should be collected at the appropriate depth and location to capture variations in soil properties across the landscape. Care must be taken to avoid contaminating the samples with foreign materials and to collect a sufficient quantity of soil for analysis.

After sampling, farmers must carefully collect, package, label, and dispatch the soil samples according to the requirements of the testing agency or laboratory. This typically involves placing the samples in clean, airtight containers, labelling them with accurate identification information, and submitting them to the testing facility in a timely manner. Proper sample handling and documentation are essential to ensure accurate and reliable test results that can inform soil management decisions effectively.

Soil testing involves the analysis of farmland for multiple parameters such as chemical content, toxicity, pH level, salinity, presence of earth-dwelling biota, and more. These tests also provide information on chemical contamination, humic or organic content, electric conductivity, cation exchange capacity, and other physical and chemical properties (Cherlinka, 2022).

The type of soil analysis conducted depends on the specific components or properties of the soil that may impact crop development positively or negatively. Some of the commonly performed soil tests include those assessing mineral content, pH level, soil moisture, salinity, pesticides and chemical contamination, structure, and texture.

Soil nutrient testing provides valuable information on the nutrient content of the soil, enabling precise fertilization practices to support plant needs within precision agriculture implementations. This type of testing primarily focuses on determining the levels of essential nutrients such as nitrogen (N), phosphorus (P), and potassium (K), along with secondary nutrients like calcium (Ca), sulphur (S), and magnesium (Mg). Extended tests may also include minor elements like iron (Fe), manganese (Mn), boron (B), molybdenum (Mo), among others (Cherlinka, 2022).

Maintaining proper soil pH is crucial for plant productivity, as excessively high or low pH levels can adversely affect crop growth. Soil pH testing involves measuring the concentration of hydrogen ions in the soil, with pH values ranging from 0 to 14. Neutral pH is indicated by a value of 7, while lower values indicate acidity and higher values indicate alkalinity. Soil pH tests help determine the need for corrective measures such as lime application to adjust pH levels accordingly.

Soil salinity testing is essential for understanding the suitability of lands for agricultural activities, as excessive salinity can lead to osmotic stresses that hinder water absorption by plants. This type of testing typically involves analysing the total soluble salts (TSS) through evaporation or measuring the electric conductivity (EC) of soil extracts. Soil salinity tests can be performed in the field or a designated laboratory (Cherlinka, 2022).

Soil testing for pesticides and chemical contaminants is crucial for assessing potential risks to crop health and yield quality. Pesticides help control harmful organisms that can damage crops, but excessive use or improper application can lead to contamination of soil and groundwater, posing risks to human health and the environment. Soil testing helps identify the presence of chemical contaminants and allows for informed decision-making regarding crop treatments and management practices.

In addition to chemical content, soil testing also involves analysing the physical properties of the soil, including texture, structure, and moisture. Soil texture refers to the proportions of clay, sand, and silt in the soil, which determine its ability to retain nutrients and moisture. Soil structure describes the arrangement of soil particles and pore spaces, which influence water and air movement in the soil. Understanding soil texture and structure helps in planning irrigation, fertigation, and crop selection based on soil type suitability (Cherlinka, 2022).

Soil moisture testing is essential for assessing water availability for plant growth, as adequate moisture is crucial for proper plant development. Soil moisture tests measure the amount of water present in the soil, with high-temperature evaporation methods being a common technique. Monitoring soil moisture levels throughout the growing season helps farmers make informed decisions regarding irrigation scheduling and water management practices to optimize crop yields.

Soil testing is a critical component of modern agricultural practices, providing valuable information on soil properties and conditions that influence crop growth and productivity. By conducting comprehensive soil tests and interpreting the results accurately, farmers can make informed decisions regarding soil management practices, nutrient applications, and crop selection, ultimately leading to improved farm profitability and sustainability (Cherlinka, 2022).

When considering how to conduct a soil test, several options are available (Cherlinka, 2022):

1. DIY with Special Kits: You can choose to perform the soil test yourself using specialized kits designed for this purpose.

2. Professional Laboratory Analysis: Alternatively, you can send soil samples to a state or private laboratory for professional analysis. Some local labs may offer free soil test procedures.

3. Hire a Soil Testing Company: Another option is to hire a company that handles the entire soil testing process, from sampling to providing recommendations for field improvement.

The entire soil testing procedure typically takes several weeks to complete.

Accurate soil sampling is crucial for obtaining reliable soil test results. There are two primary soil sampling methods (Cherlinka, 2022).

Grid Sampling: This method involves dividing the field into equally spaced grids, typically ranging from one to five acres in size. The smaller the grid, the more accurate the results, but it requires more subsamples. For example, a one-acre grid cell requires a minimum of five subsamples, while up to ten subsamples may be needed for grids of five acres. Grid soil sampling provides spatial information on the field but does not account for ground variability.

It is recommended to sample soil every two or three years in the same seasonal period. For fields under intensive cultivation, sampling should occur annually.

Zone Soil Sampling: In this method, the field is divided into zones based on crop maps or soil type. Each zone exhibits similar conditions, and the number and size of zones depend on their variability. Typically, zones range from two to ten acres in size. Larger zones require more soil test probes, with at least five subsamples per zone and two subsamples per acre.

While agronomic soil testing provides valuable information on field health, combining data from multiple sources can offer even better insights. The Zoning feature on EOSDA Crop Monitoring platform, for instance, utilizes satellite-based productivity maps to delineate different zones within a field more accurately. This assists in making informed decisions regarding necessary chemical inputs based not only on vegetation status but also on specific properties of each zone.

Upon receiving a laboratory report, it typically presents the results for specified parameters and the concentrations of chemical elements found in the soil samples. Understanding the nutrient norms and adjusting their concentrations accordingly is crucial.

For instance, nitrogen concentrations required vary depending on the soil type, with the highest levels needed for clay fields (Cherlinka, 2022):

- Sandy soils (25 – 50 mg-N/kg)

- Loam soils (50 – 75 mg-N/kg)

- Clay soils (75 – 125 mg-N/kg)

Soil testing provides essential insights into the nutrient levels present in the soil, critical for maintaining healthy crop growth and maximizing agricultural productivity. Understanding the recommended ranges for different chemical elements is fundamental for effective soil management strategies. These include (Cherlinka, 2022):

- Phosphorus (P): Phosphorus is crucial for plant growth, particularly in facilitating energy transfer and root development. The normal range for phosphorus levels in the soil typically falls between 25 to 35 parts per million (ppm). Ensuring phosphorus levels within this range promotes robust plant growth and development.

- Sulphur (S): Sulphur plays a vital role in enzyme activation and protein synthesis within plants. The recommended soil sulphur levels typically range from 7 to 15 ppm. Adequate sulphur content in the soil is essential for optimizing crop yield and quality.

- Zinc (Zn): Zinc is an essential micronutrient required for various physiological processes in plants, including enzyme activity and hormone regulation. The optimal range for soil zinc levels is generally between 1 to 3 ppm. Maintaining adequate zinc levels ensures proper plant growth and development.

- Iron (Fe): Iron is essential for chlorophyll formation and photosynthesis, crucial processes for plant growth and vitality. The recommended range for soil iron levels typically ranges from 10 to 20 ppm. Adequate iron availability in the soil is necessary for healthy leaf development and overall plant vigour.

- Manganese (Mn): Manganese plays a vital role in photosynthesis, nitrogen metabolism, and enzyme activation in plants. The normal range for soil manganese levels usually falls between 8 to 11 ppm. Maintaining adequate manganese levels is essential for optimal plant growth and reproductive development.

- Copper (Cu): Copper is essential for various enzymatic reactions in plants, including photosynthesis and respiration. The optimal range for soil copper levels is typically between 0.8 to 1 ppm. Ensuring adequate copper availability in the soil is crucial for overall plant health and productivity.

- Potassium (K): Potassium is vital for water uptake, nutrient transport, and enzyme activation in plants. The recommended range for soil potassium levels typically ranges from 165 to 220 ppm. Adequate potassium content in the soil is essential for maintaining proper plant growth, stress tolerance, and disease resistance.

- Calcium (Ca): Calcium plays a crucial role in cell wall structure, membrane integrity, and nutrient uptake in plants. The optimal soil calcium levels are considered to be 1400 ppm or higher. Maintaining adequate calcium levels is essential for promoting strong root development and overall plant health.

- Magnesium (Mg): Magnesium is essential for chlorophyll formation, enzyme activation, and photosynthesis in plants. The recommended range for soil magnesium levels is generally 100 ppm or higher. Adequate magnesium availability in the soil is necessary for optimal plant growth and nutrient uptake.

- Sodium (Na): Sodium, while not considered an essential nutrient for most plants, can impact soil structure and plant health in high concentrations. The recommended range for soil sodium levels typically falls between 80 to 120 ppm. Maintaining sodium levels within this range helps prevent adverse effects on soil structure and plant growth.

In addition to nutrient content, a soil test report may encompass other parameters:

- Soil Class: This indicates the humic matter content and is categorized into mineral, mineral-organic, and organic classes. Fields rich in minerals (MIN) have lower humic concentrations, while organic-rich soils (ORG) contain the highest levels of humus. Mineral-organic fields (M-O) have medium humic content.

- Organic Matter (OM): This indicates the organic matter content in the samples. Organic matter norms for row crops, excluding sandy soils, typically range from 2.5% or higher.

- Cation Exchange Capacity (CEC): This measures the soil's ability to retain positively charged ions (cations). Average values for CEC range from 13 to 25, but can vary from 1 to 40.

Soil test results from a laboratory may include recommendations for amelioration to enhance yields, such as adjustments in fertilizer application or lime treatments to reduce soil acidity. However, it's important to note that a laboratory report offers insights solely into the current state of the field. Therefore, historical data on field conditions must also be considered when interpreting soil test results for nutrient supplies. This includes factors such as crop types, crop rotation practices, tillage depth (conventional, conservation, or no-till), soil type, and past fertilization or fertigation events.

Interpreting soil test results comprehensively requires consideration of various aspects, as each contributes to the overall soil health and nutrient balance. Adjustments to nutrient content should also account for the upcoming crop season, as each plant type has specific requirements for healthy development (Cherlinka, 2022).

Apply Information from Soil Testing to Production or Management Plan

From the perspective of someone involved in researching soil information, applying soil testing data to production or management plans, developing and monitoring soil amendment practices, as well as reviewing and documenting soil management plans, several key steps can be outlined to effectively carry out these tasks.

The first step involves classifying soil types from the sample area according to established standards for soil classification. This process entails analysing various physical and chemical properties of the soil samples collected from different locations within the area of interest. Parameters such as texture, structure, pH level, organic matter content, and nutrient composition are evaluated to determine the soil type classification. This classification provides a foundational understanding of the soil characteristics, which is essential for making informed decisions regarding crop selection, land use planning, and soil management practices.

Once the soil samples have been analysed, the next step is to compare the results from soil testing with the identified soil physical and chemical parameters required for the specified crop. This comparison involves assessing whether the soil meets the optimal conditions necessary for the successful growth and development of the target crop. Parameters such as nutrient levels, pH suitability, moisture retention capacity, and drainage characteristics are evaluated to determine if any adjustments or amendments are needed to optimize soil conditions for crop production.

In addition to comparing soil testing results with identified parameters for the specified crop, it is essential to compare these findings with existing soil records and production history. Historical data on previous crop yields, soil amendments applied, and management practices implemented can provide valuable insights into the long-term trends and patterns observed in soil health and fertility. By comparing current soil testing results with historical data, trends and changes in soil properties over time can be identified, allowing for more informed decision-making regarding soil management practices.

After conducting the necessary comparisons between soil testing results, identified parameters, and historical data, it is crucial to review and record the comparative results. Documenting these findings in detail ensures that all relevant information is properly documented for future reference and analysis. This documentation also facilitates communication and collaboration among stakeholders involved in soil management and agricultural decision-making processes.

Based on the findings from the comparative analysis, a production or management plan is developed to address any identified deficiencies or areas of improvement in soil health and fertility. This plan outlines specific soil amendment practices, nutrient management strategies, irrigation plans, and other interventions aimed at optimizing soil conditions for crop production. Additionally, all testing results, comparative analyses, and management decisions are recorded and documented as part of the soil management plan to track progress over time and inform future decision-making processes.

By following these steps, individuals responsible for researching soil information and implementing soil management practices can effectively classify soil types, compare soil testing results, develop comprehensive production or management plans, and maintain detailed records to ensure informed decision-making and sustainable soil stewardship practices. A sample production plan follows:

Hypothetical Sample Production Plan: Soil Health and Fertility Optimization
Objective:
To enhance soil health and fertility for optimal crop production and yield.

1. Soil Amendment Practices:

- *Conduct soil pH adjustment based on soil testing results to ensure optimal pH levels for crop growth.*

- *Apply organic matter amendments such as compost or manure to improve soil structure and nutrient retention.*

- *Incorporate lime or gypsum as needed to address calcium or magnesium deficiencies identified in soil testing.*

- *Implement cover cropping and crop rotation practices to replenish soil nutrients and reduce soil erosion.*

2. Nutrient Management Strategies:

- *Apply fertilizers based on soil test recommendations to meet crop nutrient requirements while minimizing excess nutrient runoff.*

- *Utilize precision agriculture techniques such as variable rate application to tailor nutrient inputs to specific soil zones and crop needs.*

- *Monitor soil nutrient levels throughout the growing season and adjust fertilizer applications accordingly to maintain optimal nutrient balance.*

3. Irrigation Plans:

- *Implement drip irrigation systems to deliver water directly to plant roots, minimizing water waste and soil erosion.*

- *Schedule irrigation based on soil moisture monitoring data to ensure timely and adequate water supply to crops.*

- *Use mulching techniques to conserve soil moisture and reduce evaporation losses, especially during dry periods.*

4. Other Interventions:

- *Introduce soil biological amendments such as beneficial microbes or mycorrhizal fungi to enhance nutrient uptake and improve soil structure.*

- *Implement conservation tillage practices to minimize soil disturbance and preserve soil organic matter.*

- *Monitor and manage soil compaction through timely cultivation or use of cover crops to improve soil aeration and root penetration.*

5. Monitoring and Documentation:

- *Regularly monitor soil health indicators such as pH, nutrient levels, and soil structure through periodic soil testing.*

- *Document all soil management activities, including soil amendments, nutrient applications, and irrigation schedules, in a comprehensive soil management plan.*

- *Continuously track progress and changes in soil health over time, comparing current results with baseline data to assess the effectiveness of soil management practices.*

- *Use monitoring data and documentation to inform future decision-making processes and make adjustments to the production plan as needed.*

By implementing this production plan, we aim to optimize soil conditions for crop production, improve overall soil health and fertility, and ultimately achieve higher yields and sustainable agricultural practices.

Note that several assumptions were made to create the Hypothetical Sample Production Plan: Soil Health and Fertility Optimization:

1. Soil Testing Results: It is assumed that soil testing has been conducted and that the results accurately reflect the current soil conditions, including nutrient levels, pH, and other relevant parameters.

2. Crop Specificity: The plan assumes that specific crops to be grown on the farm have been identified, and soil management practices are tailored to meet the nutritional requirements and growth conditions of these crops.

3. Access to Resources: It is assumed that the farm has access to necessary resources such as organic matter amendments, fertilizers, irrigation systems, and equipment required to implement the proposed soil management practices.

4. Technical Knowledge: The plan assumes that there is sufficient technical knowledge and expertise available to interpret soil testing results, develop appropriate soil amendment practices, and implement effective nutrient management strategies.

5. Environmental Considerations: Environmental factors such as climate, weather patterns, and local regulations are considered in the plan to ensure that soil management practices are environmentally sustainable and comply with relevant guidelines.

6. Monitoring and Documentation: The plan assumes that there is a system in place for regular monitoring of soil health indicators and documentation of soil management activities to track progress over time and inform future decision-making processes.

7. Flexibility: It is assumed that the production plan is flexible and can be adjusted as needed based on changing soil conditions, crop requirements, or external factors that may impact soil health and fertility.

Develop Soil Amendment Practices

Identifying soil properties capable of being ameliorated is the foundational step in optimizing soil health and fertility. This process involves a thorough understanding of the soil's physical and chemical properties, including nutrient levels, pH, texture, structure, and organic matter content. By analysing soil testing information and considering production goals, farmers can identify specific soil properties that need improvement to support optimal crop growth. For example, if soil testing reveals low levels of essential nutrients like phosphorus or potassium, or if the soil pH is too acidic or alkaline, these are indicators of soil properties that require amelioration.

Soil amelioration refers to the process of improving the physical, chemical, and biological properties of soil to enhance its fertility, structure, and overall health. It involves various practices aimed at remedying soil deficiencies, restoring balance, and creating optimal conditions for plant growth. Soil amelioration can address a range of issues, including nutrient deficiencies, soil compaction, poor drainage, soil acidity or alkalinity, erosion, and degradation.

There are several methods of soil amelioration, including:

- Adding organic matter: Incorporating organic materials such as compost, manure, crop residues, or biochar into the soil can improve its structure, water retention, nutrient content, and microbial activity. Organic matter acts as a source of nutrients and energy for soil organisms, promoting soil health and fertility.

- Adjusting soil pH: Soil pH influences nutrient availability and microbial activity. Amending soil pH by adding materials like lime to raise pH or sulphur to lower pH can optimize nutrient uptake by plants and create favourable conditions for microbial processes.

- Nutrient management: Applying fertilizers or soil amendments containing essential nutrients such as nitrogen, phosphorus, potassium, calcium, and magnesium can correct nutrient deficiencies and promote healthy plant growth. Nutrient management practices aim to provide crops with the right balance of nutrients for optimal performance.

- Soil conservation practices: Implementing soil conservation measures such as crop rotation, cover cropping, reduced tillage, contour farming, terracing, and mulching can help prevent soil erosion, improve soil structure, and enhance water infiltration and retention.

- Drainage improvement: Installing drainage systems or using practices like subsoiling or deep ripping can alleviate soil compaction, enhance water drainage, and reduce waterlogging, thereby improving aeration and root development.

- Remediation of soil contaminants: Treating soils contaminated with pollutants or heavy metals through techniques such as phytoremediation, bioremediation, or soil washing can detoxify the soil and restore its suitability for agricultural use.

Overall, soil amelioration is essential for maintaining soil fertility, productivity, and sustainability in agricultural systems. By implementing appropriate soil management practices, farmers can optimize soil conditions and support healthy plant growth, leading to improved crop yields and environmental stewardship.

Once soil properties requiring improvement are identified, the next step is to identify relevant soil ameliorants or cultural practices that can address these deficiencies. Soil ameliorants may include organic materials such as compost, manure, or biochar to improve soil structure and fertility. Cultural practices such as crop rotation, cover cropping, reduced tillage, or mulching can also help enhance soil health by promoting nutrient cycling, improving soil structure, and reducing erosion. Incorporating these ameliorants and practices into the production plan is essential for implementing targeted soil management strategies.

Costing soil ameliorating activities is crucial for budgeting and financial planning. Farmers need to estimate the expenses associated with purchasing soil ameliorants, implementing cultural practices, and any additional labour or equipment required for soil improvement activities. This cost analysis should consider factors such as the quantity and quality of soil ameliorants needed, the cost of application or implementation, and any potential savings or benefits resulting from improved soil health, such as increased crop yields or reduced input costs.

Once the cost of soil ameliorating activities has been determined, farmers can quantify and purchase the necessary soil ameliorants. This may involve sourcing organic materials from local suppliers, calculating the quantity of amendments needed based on soil testing recommendations and field size, and scheduling the application or implementation of cultural practices at appropriate times in the production cycle. By quantifying and procuring soil ameliorants in advance, farmers can ensure timely implementation of soil improvement measures as part of their overall soil management plan.

Integrating soil amelioration into the Hypothetical Sample Production Plan: Soil Health and Fertility Optimization, nets:

Hypothetical Sample Production Plan: Soil Health and Fertility Optimization
Objective:
The objective of this production plan is to enhance soil health and fertility to ensure optimal crop production and yield.

   1. *Soil Amendment Practices:*

To address soil deficiencies and improve soil structure:

- *Adjust soil pH based on soil testing results to ensure optimal levels for crop growth.*

- *Apply organic matter amendments such as compost or manure to enhance soil structure and nutrient retention.*

- *Incorporate lime or gypsum as needed to rectify calcium or magnesium deficiencies identified in soil testing.*

- *Implement cover cropping and crop rotation practices to replenish soil nutrients and mitigate soil erosion.*

To optimize nutrient availability and minimize nutrient runoff:

- *Apply fertilizers according to soil test recommendations to meet crop nutrient requirements.*

- *Utilize precision agriculture techniques like variable rate application to tailor nutrient inputs to specific soil zones and crop needs.*

- *Monitor soil nutrient levels throughout the growing season and adjust fertilizer applications as necessary to maintain optimal nutrient balance.*

To efficiently manage water resources and maintain soil moisture:

- *Implement drip irrigation systems to deliver water directly to plant roots, reducing water waste and soil erosion.*

- *Schedule irrigation based on soil moisture monitoring data to ensure timely and adequate water supply to crops.*

- *Utilize mulching techniques to conserve soil moisture and minimize evaporation losses, particularly during dry periods.*

To further enhance soil health and fertility:

- *Introduce soil biological amendments such as beneficial microbes or mycorrhizal fungi to promote nutrient uptake and improve soil structure.*

- *Implement conservation tillage practices to minimize soil disturbance and preserve soil organic matter.*

- *Monitor and manage soil compaction through timely cultivation or cover crop utilization to enhance soil aeration and root penetration.*

To track soil health indicators and management activities:

- *Regularly monitor soil pH, nutrient levels, and soil structure through periodic soil testing.*

- *Document all soil management activities, including soil amendments, nutrient applications, and irrigation schedules, in a comprehensive soil management plan.*

- *Continuously evaluate progress and changes in soil health over time, comparing current results with baseline data to assess the effectiveness of soil management practices.*

- *Utilize monitoring data and documentation to inform future decision-making processes and make adjustments to the production plan as necessary.*

By implementing this comprehensive production plan, we aim to optimize soil conditions for crop production, improve overall soil health and fertility, and ultimately achieve higher yields and sustainable agricultural practices.

Monitor Soil Amendment Practices

Sampling areas for ameliorant activities for testing across a representative sampling area involves a systematic approach to selecting locations that accurately represent the variability of soil conditions within the production or management area. To begin, researchers or soil scientists must identify key factors influencing soil health and fertility, such as soil texture, pH levels, nutrient content, and organic matter. Using this information, they strategically divide the area into manageable zones based on similarities in soil properties, topography, and historical land use practices.

Once the zones are delineated, researchers proceed to select specific sampling sites within each zone to collect soil samples for testing. It's crucial to ensure that the sampling sites are distributed evenly across the area and adequately represent the variation in soil properties within each zone. This may involve employing a grid or random sampling approach to ensure comprehensive coverage. Careful attention is paid to factors such as soil depth, proximity to water sources, and potential sources of contamination, which could affect the accuracy of the test results.

After collecting soil samples from the identified sampling sites, researchers proceed with laboratory analysis to measure various soil parameters, including nutrient levels, pH, organic matter content, and soil texture. The results obtained from the soil testing provide valuable insights into the current state of soil health and fertility within each zone, serving as a baseline for assessing the effectiveness of soil amelioration activities.

Analysing the results to measure the performance of soil amelioration activities involves comparing the pre- and post-amendment soil test results to evaluate the impact of the intervention. Researchers assess changes in soil properties such as nutrient availability, pH levels, and organic matter content to determine the effectiveness of the ameliorant applied. Positive changes, such as increased nutrient levels or improved soil structure, indicate successful soil amelioration, while negative trends may necessitate further adjustments to the management plan.

Recording a monitoring program and including it in the production plan is essential for tracking the progress of soil amelioration activities over time. Researchers establish a systematic monitoring schedule to regularly assess soil health indicators and evaluate the long-term effects of soil management practices. This may involve conducting periodic soil testing, observing changes in plant growth and yield, and documenting any observed improvements or challenges. By integrating the monitoring program into the production plan, stakeholders can make informed decisions based on real-time data and adapt management strategies as needed to achieve their soil health and fertility goals.

Review and Document the Soil Management Plan

Recording the analysis of the sampling methods and amending them as required is a crucial step in ensuring the accuracy and reliability of soil information gathered for agricultural purposes. Initially, researchers meticulously document the details of the sampling methods employed, including the sampling locations, depths, and techniques used for collecting soil samples. They also record any challenges encountered during the sampling process, such as variations in soil texture or difficulties accessing certain areas. This comprehensive documentation serves as a reference point for future assessments and helps ensure consistency and transparency in data collection.

Following the collection of soil samples, researchers proceed with laboratory analysis to assess various soil properties, such as nutrient levels, pH, texture, and organic matter content. Once the results are obtained, they meticulously review the findings to identify any discrepancies or inconsistencies that may indicate errors in the sampling or analysis process. If necessary, researchers may need to amend the sampling methods to address any identified shortcomings or improve the accuracy of future soil assessments. This may involve refining sampling protocols, adjusting sample collection techniques, or revising sampling intervals to better capture variations in soil conditions.

Reviewing the monitoring program, amending it for effectiveness, and recording the adjustments made is essential for ensuring the ongoing success of soil management practices. Researchers regularly evaluate the effectiveness of the monitoring program in capturing relevant soil health indicators and providing timely insights into soil conditions. They assess factors such as the frequency of monitoring, the selection of monitoring parameters, and the reliability of monitoring tools and techniques. Based on this review, researchers may identify areas for improvement or refinement in the monitoring program and implement necessary amendments to enhance its effectiveness. These amendments are carefully documented to track changes over time and facilitate continuous improvement in soil management practices.

Documenting the production plan incorporating reviews and reports is essential for maintaining a comprehensive record of soil management activities and facilitating informed decision-making. Researchers document all aspects of the production plan, including soil testing results, soil amendment practices, monitoring program reviews, and any adjustments made based on the findings. This documentation provides a clear record of the steps taken to optimize soil health and fertility, enabling stakeholders to track progress, identify trends, and make data-driven decisions to support sustainable agricultural practices. By documenting reviews and reports within the production plan, researchers ensure transparency, accountability, and traceability in soil management efforts, ultimately contributing to long-term soil health and productivity.

As a final example, the principles outlined in this chapter can be applied to precision farming. Precision farming involves leveraging technology and data-driven approaches to optimize agricultural practices, including soil management. Here's how the steps outlined can be applied to precision farming:

1. Research and Analyse Soil Type, Characteristics, and Properties: Utilize remote sensing technologies, soil surveys, and GIS mapping tools to gather detailed information about soil type, texture, pH, nutrient levels, and other relevant characteristics across the farm. Analyse this data to understand spatial variability and trends in soil properties.

2. Identify Particular Soil Properties Relevant to Precision Farming: Focus on soil properties that have a significant impact on crop growth and yield, such as nutrient availability, water holding capacity, compaction, and pH levels. Research existing literature and consult with agronomists to identify key soil parameters relevant to precision farming practices.

3. Determine Acceptable Soil Parameters for Specified Crops: Refer to published data, crop nutrient requirements, and soil fertility guidelines to determine target soil parameters for specific crops. These parameters may include optimal pH levels, nutrient concentrations, and soil moisture content required for maximum crop productivity.

4. Identify Relevant Soil Properties and Areas for Testing: Prioritize areas within the farm where soil properties exhibit significant variability or where historical yield data suggests potential productivity gaps. Use precision soil sampling techniques to collect representative soil samples from these areas for laboratory analysis.

5. Take Soil Samples and Dispatch According to Requirements: Employ high-resolution soil sampling techniques, such as grid or zone sampling, to collect soil samples at precise locations within the identified areas of interest. Package, label, and dispatch soil samples following established protocols to ensure accurate and reliable analysis by testing agencies.

6. Classify Soil Types and Compare Testing Results: Classify soil types based on standardized classification systems and compare soil testing results with identified soil parameters for specified crops. Use advanced data analysis tools to assess spatial variability and identify areas requiring targeted soil amelioration practices.

7. Develop Production or Management Plan: Develop a comprehensive production or management plan tailored to address specific soil health and fertility issues identified through soil testing and analysis. Incorporate precision farming techniques, such as variable rate application of fertilizers and soil amendments, into the plan to optimize resource utilization and maximize crop yields.

8. Identify Soil Properties Capable of Being Ameliorated: Identify soil properties that can be improved through targeted soil amelioration practices, such as pH adjustment, nutrient supplementation, and soil conditioning. Prioritize interventions based on their potential to address identified soil deficiencies and improve overall soil health.

9. Identify Relevant Soil Ameliorants and Cost Activities: Research and select appropriate soil ameliorants, including fertilizers, lime, gypsum, and organic amendments, based on soil testing results and crop nutrient requirements. Cost out soil ameliorating activities and incorporate them into the production plan budget.

10. Quantify and Purchase Soil Ameliorants: Calculate the quantities of soil ameliorants required based on soil test recommendations and field-specific conditions. Source high-quality soil ameliorants from reputable suppliers and ensure timely delivery and application according to the production plan schedule.

11. Monitor Soil Amendment Practices: Implement precision monitoring techniques, such as soil sensors and remote sensing technologies, to track the effectiveness of soil amelioration practices in real-time. Monitor key soil parameters, such as pH, nutrient levels, and soil moisture content, to assess the impact of soil amendments on soil health and fertility.

12. Sample Areas for Testing Across Representative Sampling Area: Continuously sample soil across the farm to monitor changes in soil properties over time and assess the effectiveness of soil amelioration activities. Ensure that sampling locations are representative of the farm's variability and target areas with known soil health issues.

13. Analyse Results to Measure Performance of Soil Amelioration Activities: Analyse soil testing results and compare them with baseline data to evaluate the performance of soil amelioration activities. Assess changes in soil properties, crop response, and yield outcomes to determine the effectiveness of implemented soil management practices.

14. Record Monitoring Program and Include in the Production Plan: Document the monitoring program, including sampling protocols, data collection methods, and analysis procedures, and integrate it into the production plan. Maintain detailed records of monitoring activities, including dates, locations, and results, to track progress and inform future decision-making.

15. Review and Document the Soil Management Plan: Regularly review and update the soil management plan based on ongoing monitoring data, research findings, and feedback from stakeholders. Document all revisions and amendments to the plan to ensure continuity and accountability in soil management practices.

16. Record Analysis of Sampling Methods and Amend as Required: Record detailed analyses of sampling methods used and any adjustments made based on performance evaluations or feedback from soil testing results. Continuously refine and improve sampling techniques to enhance the accuracy and reliability of soil data collected.

17. Review Monitoring Program, Amend for Effectiveness, and Record: Review the effectiveness of the monitoring program periodically and make amendments as necessary to improve its accuracy and relevance. Document all changes made to the monitoring program and record their impact on soil management practices and outcomes.

18. Document Production Plan Incorporating Reviews and Reports: Document all aspects of the production plan, including soil testing results, soil amendment practices, monitoring program reviews, and any adjustments made based on performance evaluations. Incorporate feedback from reviews and reports into future planning and decision-making processes to support continuous improvement in soil management practices.

By following these steps, precision farmers can effectively research soil information, apply soil testing information to production or management plans, develop and monitor soil amendment practices, review and document soil management plans, and ultimately optimize soil health and fertility for sustainable crop production.

Chapter Twenty

Monitoring and Reviewing Business Performance

Identifying Performance Requirements

M onitoring and reviewing agricultural business performance is imperative for securing long-term sustainability and profitability. To achieve these objectives, several key steps must be taken.

Firstly, developing realistic performance indicators is essential. These indicators should be aligned with the business objectives and available resources, encompassing quantifiable metrics such as crop yield, production costs, revenue generated, profitability margins, resource utilization efficiency, and customer satisfaction ratings. By establishing clear and achievable targets within feasible timeframes, farmers can effectively track progress and make informed decisions to drive performance improvement.

Identifying and minimizing inhibiting factors is equally crucial. This involves identifying potential obstacles such as adverse weather conditions, pest infestations, soil degradation, supply chain disruptions, regulatory changes, or market price fluctuations. Regular assessments and risk analyses enable farmers to anticipate challenges and implement mitigation strategies promptly, thereby safeguarding business operations and maintaining resilience in the face of adversity.

Furthermore, staying informed about market conditions is paramount. Farmers should gather relevant data and make justifiable assumptions based on industry trends, consumer preferences, and economic forecasts. Monitoring factors such as supply and demand dynamics, competitor activities, input costs, trade policies, and consumer behaviour patterns is essential. By utilizing market intelligence tools and industry reports, agricultural businesses can assess market opportunities and risks, identify emerging trends, and make informed strategic decisions to optimize profitability and competitive advantage.

Additionally, assessing the capacity to promote sustainability is vital for long-term success. Farmers should evaluate factors such as resource efficiency, environmental impact, social responsibility, and resilience to climate change. Implementing sustainable farming practices such as conservation tillage, crop rotation, integrated pest management, and water conservation measures is crucial. Collaboration with stakeholders such as government agencies, NGOs, and industry associations can provide access to resources, funding, and expertise to support sustainability initiatives. By integrating sustainability into business operations, agricultural enterprises can enhance long-term viability, mitigate risks, and meet the evolving expectations of consumers and stakeholders.

Monitoring and reviewing agricultural business performance entail developing realistic performance indicators, identifying and minimizing inhibiting factors, assessing market conditions, and promoting sustainability within enterprise procedures. By adopting a proactive and data-driven approach, agricultural businesses can enhance resilience, optimize decision-making, and achieve long-term success in a dynamic and competitive market environment.

In the agricultural sector, the need for metrics and key performance indicators (KPIs) is essential for effective business management. Despite this, KPIs are often overlooked in agriculture, despite the sector being equally affected by market dynamics as any other industry. It's crucial for farms and related businesses to track their performance using metrics to ensure profitability and sustainability.

Over the years, advancements in science and technology have propelled agricultural practices forward. However, this progress comes with its own set of challenges. A study conducted in 2021 reveals that food and agriculture activities contribute significantly to global greenhouse gas emissions, ranging from 25% to 35% (Vintila, 2022).

Distinguishing modern agricultural systems from organic farming systems, the former heavily relies on agrochemicals, which contribute to pollution and pose risks to underground water supplies. In contrast, organic systems utilize nutrients primarily from biological sources like manure, compost, and cover crops, benefiting both plant growth and soil microorganisms.

Technology has also left its mark on agriculture, with the use of key performance indicators (KPIs) emerging as a method to assess its impact. While modern machinery aims to enhance efficiency, it often leads to increased energy consumption and emissions. By leveraging KPIs, farmers can gauge both the positive and negative environmental effects of their practices, allowing for more informed decision-making.

The adoption of KPIs in the agricultural sector serves multiple purposes, including increasing productivity, managing daily operations, and facilitating informed business decisions. These metrics offer insights into various aspects of farm management, from crop yields to resource utilization and environmental impact.

Specific KPIs relevant to the agricultural sector include farm size, percentage of irrigated land, pesticide consumption, hiring costs, on-farm trials, cultivated land area, and harvesting expenses (Vintila, 2022). These indicators aid in decision-making processes and provide a comprehensive overview of farm operations over time.

Moreover, different types of farms require specific KPIs tailored to their unique characteristics and objectives. For instance, dairy farms may focus on metrics such as milk yield per cow, dairy calf mortality rate, and daily cow replacement costs, among others.

In terms of productivity and costs, farmers can monitor unit production time, energy costs per unit of production, energy usage, input waste, and daily production rates. By tracking these metrics, farmers can optimize their operations, reduce expenses, and mitigate environmental impacts while maximizing productivity and revenue.

Profit delineates the remaining income subsequent to settling all expenses and is determined by deducting costs from gross farm income. Gross farm income, sometimes termed as total revenue or gross farm receipts, is typically computed by multiplying the price by the volume of production sold (Alexander, 2018).

Net profit is the gross farm income minus all expenses linked with production and managing the business. Expenses encompass fixed and variable operating costs, allocations for livestock, plant and machinery replacement (often in the form of depreciation), finance costs (such as interest and lease costs), management allowances (if a salary is not withdrawn), and taxation.

While many individuals concentrate on gross margins, significant costs not included in the gross margin must be factored in when making comparisons across enterprises, scrutinizing profitability, and establishing price targets.

Analysis across different profit points (such as gross margin versus operating surplus versus EBIT versus EBT) will vary depending on the objective. For instance, a farm manager might scrutinize the earnings before interest and tax (EBIT) level when contrasting enterprises, evaluate the net profit after tax level when contemplating farm expansion, or examine the gross margin when comparing crops with similar fixed costs. Typically, analysis is conducted at the EBIT profit level to compare farm business performance, as taxation and finance costs fluctuate depending on the business structure and financing arrangement (Alexander, 2018).

Profit fluctuates across farm businesses and years, influenced by alterations in price, production, and costs. Farm business managers should comprehend how each of these crucial profit drivers impacts their business profit. This comprehension facilitates crucial decisions, particularly regarding returns on investment of time and money. Recognizing a farm's profit drivers enables managers to assess the risk and resilience of their business and make expenditure decisions within and between seasons.

The capacity of farmers to adeptly manage each of these drivers determines a farm's profitability under various conditions. Management proficiency often serves as the distinguishing factor between top and bottom performers operating in similar environments.

In most businesses, adjustments to price exert the most significant impact on profit, marginally ahead of alterations in production volume and costs. This is primarily because any price increment directly translates into profit net of any sales commissions or levies. In our case study illustration, a 1% rise in price augmented pre-tax profit by nearly 4% (Alexander, 2018).

Market prices of most agricultural commodities are largely dictated by global and local demand and supply fundamentals. Agricultural producers do not establish the prices they receive. Prices for many agricultural commodities are highly erratic and challenging to forecast. They are influenced by the unpredictability of global weather conditions and their subsequent impact on global and local production (i.e., supply). Prices are further affected by fluctuations in the exchange rate.

To optimize profitability, farm business managers must grasp their cost structures, discern breakeven production volumes required to cover all expenses, and implement strategies to control spending when production targets may be at risk.

Costs come in two forms: fixed and variable. Fixed costs are those incurred regardless of production volume or season, encompassing permanent wages, overheads, depreciation, and finance charges. Variable costs fluctuate with production volume or season, including fuel, repairs, fertilizers, livestock purchases, and more.

Given the unpredictable nature of seasonal conditions, farm businesses must adopt strategies to navigate their cost structures, allowing them to seize production opportunities in favourable seasons while minimizing expenses during poor ones.

A farm with a high proportion of fixed costs faces greater risk in volatile production areas than one with lower fixed costs. Hence, managers must be mindful of this when considering additional land or machinery, as interest and capital allowance expenses will persist regardless of future seasons' favourability.

Farm managers should assess the return on investment (ROI) for every option when contemplating an investment decision. ROI is calculated as profit divided by the cost of investment.

When evaluating an investment in new technology or upgrades, it's crucial to first calculate the expected cost savings or increased production value to determine the justified expenditure or payback period (Alexander, 2018).

Grain enterprises are increasingly capital-intensive due to larger, more advanced machinery. It's essential to align the funds invested in machinery with the farm's size to avoid overcapitalization. Always factor in a replacement allowance for machinery when assessing enterprise profitability.

Different crops require specific management approaches to optimize returns on inputs. Implementing precision farming techniques, regular equipment servicing, and adhering to label instructions for treatments are critical to minimizing costs and maximizing yields.

Precision feeding, preg-scanning ewes, and employing contractors for specialized tasks can reduce costs and enhance productivity in sheep enterprises.

Post-farm gate expenses, including freight, commissions, and levies, can significantly impact profitability. Exploring alternative supply chain paths and negotiating efficient logistics arrangements can help lower these costs.

Efficient time management and labour utilization are crucial for maximizing profitability. Planning ahead, investing in staff training, and employing technology to enhance efficiency can all contribute to cost reduction.

While tax management strategies can influence cash flow, profitability should remain the primary driver of investment decisions. Strategies such as timing income and expenditure, utilizing farm management deposits, and exploring government rebate schemes can help optimize tax outcomes while ensuring profitability remains paramount (Alexander, 2018).

Evaluating Enterprise Performance

Monitoring and reviewing agricultural business performance is essential for ensuring its sustainability and profitability. To achieve this, several key steps must be taken.

Firstly, gathering and analysing data relating to enterprise production and financial performance is crucial. This involves collecting data on crop yields, input usage, labour costs, revenue, and expenses. By analysing historical and current performance data, farmers can identify trends, patterns, and areas for improvement. This data provides valuable insights into the efficiency and effectiveness of production processes, helping farmers make informed decisions to optimize performance.

To facilitate the gathering and analysis of data, farmers can utilize a variety of tools and technologies. Farm management software, for example, allows farmers to input and track data related to production and finances in a centralized system. This makes it easier to organize and analyse data, as well as generate reports and forecasts.

Additionally, farmers can use specialized sensors and monitoring devices to collect real-time data on various aspects of their operations, such as soil moisture levels, weather conditions, and equipment performance. This data can then be integrated into their overall data analysis process to provide a more comprehensive understanding of their operations.

Reviewing and analysing operational structures is also important. Farmers must assess the suitability of their organizational processes to enterprise objectives. This involves evaluating factors such as workflow efficiency, resource allocation, communication channels, and decision-making processes. By identifying strengths and weaknesses in operational structures, farmers can streamline processes, improve productivity, and enhance overall business performance.

Furthermore, evaluating enterprise strengths and weaknesses against market conditions is essential. Farmers need to assess how well their business is positioned to compete in the market. This involves analysing factors such as product quality, pricing strategies, marketing efforts, and customer satisfaction levels. By understanding market dynamics and their own competitive advantages and disadvantages, farmers can capitalize on opportunities and mitigate risks to ensure long-term success.

Monitoring the impact of natural conditions on enterprise performance is another critical aspect. Weather events, pest infestations, and other natural factors can significantly affect agricultural productivity and profitability. Farmers must closely monitor these conditions and implement appropriate risk management strategies to minimize losses and maximize yields.

Assessing the sustainability of resource use is also imperative. Farmers must evaluate their use of land, water, energy, and other resources to ensure they are being utilized efficiently and responsibly. This involves implementing sustainable farming practices such as soil conservation, water management, and biodiversity conservation. By promoting resource sustainability, farmers can protect the environment, reduce costs, and enhance long-term viability.

Finally, monitoring performance against enterprise objectives is essential for identifying variations and opportunities for future development. Farmers must regularly assess their progress towards achieving business goals and objectives. This involves comparing actual performance against targets, analysing variances, and adjusting strategies as needed. By continuously monitoring performance and making data-driven decisions, farmers can adapt to changing market conditions, capitalize on opportunities, and drive business growth.

Farm managers often lack real-time data and rely on yearly accounting feedback, making it challenging to assess performance accurately. Therefore, attention to data in farming operations is paramount. The following highlights some key metrics that agricultural professionals should monitor closely:

1. Asset Turnover Ratio: Measures the effectiveness of assets in generating revenue, indicating better resource management with a higher ratio.

2. Yield of Stock: Evaluates resource utilization, such as land and facilities, in agricultural operations, considering factors like weather and farming practices.

3. Wages to Revenue: Assesses workforce productivity by comparing income to compensation spent.

4. Overall Feed and Water Use: Monitors resource efficiency by reducing water and feed inputs without compromising productivity.

5. Debt Over Assets Ratios: Indicates operational risk exposure by measuring borrowed assets against total assets.

6. Working Capital: Evaluates financial flexibility by comparing available funds to operating costs, aiming for a healthy ratio to avoid financial constraints.

7. Goods Pricing by Region: Tracks market prices for commodities to make informed pricing decisions.

8. Human Capital: Considers employment prospects and skills development to ensure a skilled workforce and future sustainability.

9. Commodity Pricing: Monitors commodity prices to make informed production decisions and avoid overproduction.

By monitoring these metrics, agricultural businesses can make informed decisions, optimize resource utilization, and ensure long-term success in a dynamic market environment. Following are examples of how to calculate and use each of these metrics:

1. Asset Turnover Ratio:

   ○ Calculation: Asset Turnover Ratio = Total Revenue / Average Total Assets

   ○ Example: If a farm generates $500,000 in revenue and has an average total asset value of $250,000, then the asset turnover ratio would be 2 ($500,000 / $250,000).

   ○ Use: A higher asset turnover ratio indicates that assets are being used more effectively to generate revenue, suggesting better resource management. It can help identify areas where assets may be underutilized or where improvements can be made to increase revenue generation.

2. Yield of Stock:

   ○ Calculation: Yield of Stock = Total Output / Total Input

   ○ Example: If a farm produces 10,000 bushels of corn using 100 acres of land, then the yield of stock would be 100 bushels per acre.

   ○ Use: This metric evaluates how efficiently resources such as land and facilities are utilized in agricultural operations. It helps farmers identify areas where productivity can be improved through better resource management practices.

3. Wages to Revenue:

   ○ Calculation: Wages to Revenue Ratio = Total Wages / Total Revenue

   ○ Example: If a farm spends $50,000 on wages and generates $500,000 in revenue, then the wages to revenue ratio would be 0.1 ($50,000 / $500,000).

   ○ Use: This metric assesses the productivity of the workforce by comparing the amount spent on wages to the revenue generated. A lower ratio indicates more efficient use of labour resources.

4. Overall Feed and Water Use:

   ○ Calculation: Monitor the amount of feed and water used per unit of output, such as per pound of meat or per bushel of grain.

   ○ Example: If a farm produces 1,000 pounds of beef and uses 2,000 pounds of feed and 10,000 gallons of water, then the overall feed and water use per pound of beef would be 2 pounds of feed and 10 gallons of water.

   ○ Use: Monitoring feed and water use efficiency helps farmers reduce input costs while maintaining productivity levels, leading to improved profitability and sustainability.

5. Debt Over Assets Ratios:

- ○ Calculation: Debt Over Assets Ratio = Total Debt / Total Assets

- ○ Example: If a farm has $100,000 in debt and $500,000 in assets, then the debt over assets ratio would be 0.2 ($100,000 / $500,000).

- ○ Use: This ratio indicates the proportion of assets that are financed by debt, providing insight into the farm's financial leverage and risk exposure. A lower ratio suggests lower financial risk.

6. Working Capital:

- ○ Calculation: Working Capital = Current Assets - Current Liabilities

- ○ Example: If a farm has $100,000 in current assets and $50,000 in current liabilities, then the working capital would be $50,000 ($100,000 - $50,000).

- ○ Use: Working capital measures the farm's ability to cover short-term operating costs. A higher ratio indicates greater financial flexibility and resilience to unexpected expenses.

7. Goods Pricing by Region:

- ○ Observation: Track market prices for agricultural commodities in different regions through market reports, online databases, or industry publications.

- ○ Use: By monitoring commodity prices, farmers can make informed pricing decisions to maximize revenue and profitability. This information helps them adjust production levels and marketing strategies accordingly.

8. Human Capital:

- ○ Observation: Assess the skills, training, and experience of the farm's workforce, as well as recruitment and retention efforts.

- ○ Use: Human capital metrics help farmers evaluate their workforce's capabilities and identify areas for skills development and training. Investing in human capital improves productivity, innovation, and long-term sustainability.

9. Commodity Pricing:

- ○ Observation: Monitor market prices for agricultural commodities through commodity exchanges, market reports, and industry publications.

- ○ Use: By tracking commodity prices, farmers can make informed decisions about production levels, timing of sales, and allocation of resources. This helps them avoid overproduction and maximize profitability in volatile markets.

Financial analysis techniques play a crucial role in helping farmers assess the financial health and performance of their operations. Here's an explanation of each technique from a farming perspective:

1. Benchmarking: Benchmarking involves comparing various aspects of farm performance, such as yields, costs, and profitability, against industry standards or similar farms. By benchmarking, farmers can identify areas where their operations excel or lag behind, allowing them to set realistic goals for improvement.

2. Cost benefit analysis: Cost benefit analysis evaluates the costs and benefits associated with a particular decision or investment on the farm. Farmers use this technique to assess the financial feasibility of implementing new practices, purchasing equipment, or undertaking projects by comparing the expected costs against the anticipated benefits.

3. 'What if?' analyses: 'What if?' analyses involve exploring different scenarios or hypothetical situations to understand their potential impact on farm finances. Farmers can use this technique to assess the consequences of various decisions or external factors, such as changes in commodity prices, weather conditions, or input costs, helping them make informed decisions and develop risk management strategies.

4. Time series and trend analysis: Time series and trend analysis involve examining historical financial data over time to identify patterns, trends, and changes in farm performance. Farmers use this technique to detect long-term trends, seasonal variations, and cycles in revenues, expenses, and profitability, providing valuable insights for strategic planning and decision-making.

5. Expenditure and revenue ratios: Expenditure and revenue ratios compare different components of farm expenses and income to assess their proportionate relationship. Common ratios include the operating expense ratio, debt-to-income ratio, and gross profit margin. These ratios help farmers evaluate the efficiency of resource utilization, profitability, and financial sustainability.

6. Break-even analysis: Break-even analysis determines the level of production or sales required for a farm to cover its total costs and achieve a zero-profit scenario. Farmers use this technique to set pricing strategies, evaluate investment decisions, and assess the financial viability of new ventures or enterprises.

7. Accounting standards: Accounting standards provide guidelines and principles for recording, reporting, and disclosing financial transactions and information in a consistent and transparent manner. Farmers adhere to accounting standards to ensure accurate financial reporting, compliance with regulatory requirements, and informed decision-making.

8. Cash flow schedules: Cash flow schedules forecast the inflows and outflows of cash over a specific period, typically monthly or annually. Farmers use cash flow schedules to monitor liquidity, plan for upcoming expenses, and ensure sufficient funds are available to meet financial obligations, such as loan repayments, operating expenses, and investments.

These financial analysis techniques enable farmers to assess their financial performance, identify opportunities for improvement, mitigate risks, and make informed decisions to achieve their financial goals and ensure the long-term viability and sustainability of their farm operations.

The following provides examples of how each financial analysis technique can be calculated and used from a farming perspective:

1. Benchmarking: Example: A farmer wants to benchmark their farm's yield of corn per acre against the industry average. They collect data on their yield for the past three years and compare it to the average yield of similar-sized farms in their region. If their yield is consistently lower, they may investigate factors such as soil quality, irrigation methods, or crop management practices to identify areas for improvement.

2. Cost benefit analysis: Example: A farmer is considering investing in new irrigation equipment to improve water efficiency on their farm. They calculate the total cost of purchasing and installing the equipment, including any ongoing maintenance expenses. Then, they estimate the potential savings in water usage and increased crop yields that the equipment could generate over its expected lifespan. If the projected benefits outweigh the costs, the farmer may proceed with the investment.

3. 'What if?' analyses: Example: A farmer wants to assess the potential impact of a drought on their farm's profitability. They create a 'what if?' scenario where they simulate a 20% reduction in crop yields due to drought conditions. By adjusting their revenue and expense projections accordingly, the farmer can evaluate how the drought scenario would affect their cash flow, break-even point, and overall financial health, allowing them to develop contingency plans.

4. Time series and trend analysis: Example: A farmer analyses their farm's monthly milk production over the past five years to identify patterns and trends. They observe that milk production tends to increase during the spring months and decrease in the fall, reflecting seasonal variations in feed quality and animal health. By understanding these trends, the farmer can better

anticipate fluctuations in income and adjust their management practices accordingly.

5. Expenditure and revenue ratios: Example: A farmer calculates their operating expense ratio by dividing total operating expenses (e.g., seed, fertilizer, labour) by total revenue. If the ratio is high compared to industry benchmarks, it may indicate inefficiencies in resource management or cost overruns that need to be addressed to improve profitability.

6. Break-even analysis: Example: A farmer wants to determine the minimum number of bushels of wheat they need to sell to cover their total production costs. They calculate their fixed costs (e.g., land rent, equipment depreciation) and variable costs (e.g., seed, fertilizer) per acre, then divide the total costs by the expected price per bushel of wheat to find the break-even point. Knowing this threshold helps the farmer set realistic sales targets and pricing strategies.

7. Accounting standards: Example: A farmer follows generally accepted accounting principles (GAAP) to record their financial transactions, prepare financial statements (e.g., income statement, balance sheet), and comply with tax regulations. By adhering to accounting standards, the farmer ensures accurate and transparent reporting of their farm's financial performance, which is essential for decision-making and regulatory compliance.

8. Cash flow schedules: Example: A farmer creates a monthly cash flow schedule to track the inflows and outflows of cash on their farm. They list all sources of income (e.g., crop sales, government subsidies) and expenses (e.g., loan payments, equipment purchases) expected for each month. By comparing projected cash flows to available funds, the farmer can anticipate any shortfalls or surpluses and plan accordingly to maintain liquidity and meet financial obligations.

Review Business Performance

Monitoring and reviewing agricultural business performance involves a comprehensive assessment of various aspects of the operation to identify opportunities for improvement. Firstly, reviewing business operations is crucial to identifying opportunities for enhancing performance. This entails evaluating all aspects of the production process, including planting, cultivation, harvesting, and post-harvest handling. Farmers need to assess factors such as workflow efficiency, resource utilization, equipment maintenance, and labour management. By scrutinizing each step of the operation, farmers can identify bottlenecks, inefficiencies, and areas where processes can be optimized for better performance.

In addition to reviewing operational aspects, it's essential to assess the financial performance of the agricultural business. This involves analysing income, expenses, profitability, cash flow, and financial ratios. Farmers need to identify areas where costs can be reduced, revenue can be increased, and overall financial management can be improved. By conducting a thorough review of financial performance, farmers can make informed decisions to enhance financial stability and profitability.

Furthermore, reviewing business marketing performance is crucial for identifying opportunities to reach target markets more effectively. Farmers need to assess their marketing strategies, including product positioning, pricing, distribution channels, and promotional efforts. By analysing market trends, consumer preferences, and competitive landscape, farmers can identify areas where marketing efforts can be strengthened to attract more customers and increase sales.

Additionally, reviewing business risk management performance is essential for identifying opportunities to mitigate risks and safeguard business operations. Farmers need to evaluate their risk management strategies, including insurance coverage, diversification, and contingency planning. By identifying potential risks such as weather-related disasters, pest outbreaks, or market fluctuations, farmers can implement proactive measures to minimize the impact of these risks on business performance.

Finally, documenting opportunities for improvement is critical to ensure that identified issues are addressed systematically. Farmers should maintain detailed records of their findings, including specific areas for improvement, recommended actions, and timelines for implementation. By documenting opportunities for improvement, farmers can track progress over time, prioritize actions, and hold themselves accountable for achieving performance goals.

In summary, monitoring and reviewing agricultural business performance involve a thorough assessment of operations, financials, marketing strategies, risk management practices, and documentation of opportunities for improvement. By systematically analysing each aspect of the business, farmers can identify areas where performance can be enhanced and take targeted actions to achieve their business objectives.

References

Acosta-Silva, Y. J., Torres-Pacheco, I., Matsumoto, Y., Toledano-Ayala, M., Soto-Zarazúa, G. M., Zelaya-Ángel, O., & Méndez-López, A. (2019). Applications of Solar and Wind Renewable Energy in Agriculture: A Review. *Science Progress*.

Adamchak, R. (2023). organic farming. In *Encyclopedia Britannica*.

Adão, T., Hruška, J., Pádua, L., Bessa, J., Peres, E., Morais, R., & Sousa, J. J. (2017). Hyperspectral Imaging: A Review on UAV-Based Sensors, Data Processing and Applications for Agriculture and Forestry. *Remote Sensing*.

Afroj, M., Kazal, M. M. H., & Rahman, M. (2016). Precision Agriculture in the World and Its Prospect in Bangladesh. *Research in Agriculture Livestock and Fisheries*.

Agbaje, K. A. A., Martin, R. A., & Williams, D. L. (2001). Impact of Sustainable Agriculture on Secondary School Agricultural Education Teachers and Programs in the North Central Region. *Journal of Agricultural Education*.

Agriculture Victoria. (2017). *Shelterbelt design*. Retrieved 8/2/2024 from

Agriculture Victoria. (2024a). *Water quality for farm water supplies*. Retrieved 7/2/2024 from

Agriculture Victoria. (2024b). *What is soil?* Retrieved 8/2/2024 from

Agüera, J., Carballido, J., Gil, J., Gliever, C., & Pérez-Ruiz, M. (2013). Design of a Soil Cutting Resistance Sensor for Application in Site-Specific Tillage. *Sensors*.

Ahmadi, A., Mosammam, H. M., & Mirzaei, N. (2017). Land Suitability Evaluation to Determine the Appropriate Areas of Development: A Case Study of Hormuz Island. *Open Journal of Ecology*.

Aksüt, G., & Eren, T. (2023). Evaluation of the Reasons for Health and Safety Risks in Agriculture by Integration of ANP and PROMETHEE Methods

Akter, M., Islam, M. N., Afrin, H., Shammi, S. A., Begum, F., & Haque, S. (2016). Comparative Profitability Analysis of IPM and Non-Ipm Technology on Vegetable Cultivation in Selected Areas of Kishoreganj District in Bangladesh. *Progressive Agriculture*.

Al-Subaiee, S., Yoder, E. P., & Thomson, J. (2005). Extension Agents' Perceptions of Sustainable Agriculture in the Riyadh Region of Saudi Arabia. *Journal of International Agricultural and Extension Education*.

Alexander, T. (2018). *Generating more profit from your farm business*. D. o. P. I. a. R. Development.

Amare, G., & Gacheno, D. (2021). Indigenous Knowledge for Climate Smart Agriculture—A Review. *International Journal of Food Science and Agriculture*.

Amran, M. A., Palaniveloo, K., Fauzi, R., Satar, N. M., Mohidin, T. B. M., Mohan, G., Razak, S. A., Arunasalam, M., Nagappan, T., & Seelan, J. S. S. (2021). Value-Added Metabolites From Agricultural Waste and Application of Green Extraction Techniques. *Sustainability*.

Anaba, L. A., Banadda, N., Kiggundu, N., Wanyama, J., Engel, B., & Moriasi, D. N. (2017). Application of SWAT to Assess the Effects of Land Use Change in the Murchison Bay Catchment in Uganda. *Computational Water Energy and Environmental Engineering*.

Arjoo, N., Kumar, V., & Shreya. (2022). Climate-Smart Agriculture : Need of the Hour.

Ashworth, A. J., Lindsay, K., Popp, M. P., & Owens, P. R. (2018). Economic and Environmental Impact Assessment of Tractor Guidance Technology. *Agricultural & Environmental Letters*.

Aslan, M. F., Durdu, A., Sabancı, K., Ropelewska, E., & Gültekin, S. (2022). A Comprehensive Survey of the Recent Studies With UAV for Precision Agriculture in Open Fields and Greenhouses. *Applied Sciences*.

Assefa, F., Elias, E., Soromessa, T., & Ayele, G. T. (2020). Effect of Changes in Land-Use Management Practices on Soil Physicochemical Properties in Kabe Watershed, Ethiopia. *Air Soil and Water Research*.

Aznar-Sánchez, J. A., Velasco-Muñoz, J. F., López-Felices, B., & Torres, F. d. M. (2020). Barriers and Facilitators for Adopting Sustainable Soil Management Practices in Mediterranean Olive Groves. *Agronomy*.

Baumhardt, R. L. (2015). Crop Choices and Rotation Principles.

Bhattacharyya, P. N., Goswami, M., & Bhattacharyya, L. H. (2016). Perspective of Beneficial Microbes in Agriculture Under Changing Climatic Scenario: A Review. *Journal of Phytology*.

Biró, K., Csete, M. S., & Németh, B. (2021). Climate-Smart Agriculture: Sleeping Beauty of the Hungarian Agribusiness. *Sustainability*.

Blake, G., Sandler, H. A., Coli, W. M., Pober, D. M., & Coggins, C. (2007). An Assessment of Grower Perceptions and Factors Influencing Adoption of IPM in Commercial Cranberry Production. *Renewable Agriculture and Food Systems*.

Blount, J. D., Horns, J. J., Kittelberger, K. D., Neate-Clegg, M. H. C., & Şekercioğlu, Ç. H. (2021). Avian Use of Agricultural Areas as Migration Stopover Sites: A Review of Crop Management Practices and Ecological Correlates. *Frontiers in Ecology and Evolution*.

Bottarelli, M., & Gallero, F. J. G. (2020). Energy Analysis of a Dual-Source Heat Pump Coupled With Phase Change Materials. *Energies*.

Brata, S., Tanasa, C., Stoian, V., Stoian, D., Dan, D., Păcurar, C., & Brata, S. (2019). Measured and Calculated Energy Saving on Ventilation of a Residential Building Equipped With Ground-Air Heat Exchanger. *E3s Web of Conferences*.

Brenya, R., Akomea-Frimpong, I., Ofosu, D., & Adeabah, D. (2022). Barriers to Sustainable Agribusiness: A Systematic Review and Conceptual Framework. *Journal of Agribusiness in Developing and Emerging Economies*.

Brodt, S., Goodell, P. B., Krebill-Prather, R., & Vargas, R. (2007). California Cotton Growers Utilize Integrated Pest Management. *California Agriculture*.

Brodt, S., Six, J., Feenstra, G., Ingels, C., & Campbell, D. (2011). Sustainable Agriculture. *Nature, 3*(10).

Buchori, D., Rizali, A., Larasati, A., Hidayat, P., Ngo, H. T., & Gemmil-Herren, B. (2019). Natural Habitat Fragments Obscured the Distance Effect on Maintaining the Diversity of Insect Pollinators and Crop Productivity in Tropical Agricultural Landscapes. *Heliyon*.

Bufebo, B., & Elias, E. (2020). Effects of Land Use/Land Cover Changes on Selected Soil Physical and Chemical Properties in Shenkolla Watershed, South Central Ethiopia. *Advances in Agriculture*.

Buick, R. (1997). Precision Agriculture: An Integration of Information Technologies With Farming. *Proceedings of the New Zealand Plant Protection Conference*.

Bulkis, S., Rahmadanih, R., & Nasruddin, A. (2020). Rice Farmers' Adoption and Economic Benefits of Integrated Pest Management in South Sulawesi Province, Indonesia. *Journal of Agricultural Extension*.

Caush, P. (2024, 12/2/2024). How To Write Your Company's Sustainability Policy.

Chait, J. (2020, 14/2/2024). What Is Agricultural Production?

Chamber of Commerce of Metropolitan Montreal. (2023a, 12/2/2024). How to write an environmental policy.

Chamber of Commerce of Metropolitan Montreal. (2023b). Prepare a business plan for growth.

Chang, E. H., Wang, C.-H., Chen, C. L., & Chung, R. S. (2014). Effects of Long-Term Treatments of Different Organic Fertilizers Complemented With Chemical N Fertilizer on the Chemical and Biological Properties of Soils. *Soil Science & Plant Nutrition*.

Chel, A., & Kaushik, G. (2011). Renewable energy for sustainable agriculture. *Agronomy for Sustainable Development, 31*, 91-118.

Chen, M., Shang, S., & Li, W. (2020). Integrated Modeling Approach for Sustainable Land-Water-Food Nexus Management. *Agriculture*.

Cherlinka, V. (2022, 14/2/2024). Soil Testing: How To Take Samples And Read Results.

Comi, M., Becot, F., & Bendixsen, C. (2023). Automation, Climate Change, and the Future of Farm Work: Cross-Disciplinary Lessons for Studying Dynamic Changes in Agricultural Health and Safety. *International journal of environmental research and public health*.

Congedo, P. M., Lorusso, C., Giorgi, M. G. D., & Laforgia, D. (2014). Computational Fluid Dynamic Modeling of Horizontal Air-Ground Heat Exchangers (HAGHE) for HVAC Systems. *Energies*.

Cruickshank, C. (2022, 8/2/2024). Rural Agricultural Health and Safety.

Dai, L., Li, H., Tan, F., Zhu, N., He, M., & Hu, G. (2016). Biochar: A Potential Route for Recycling of Phosphorus in Agricultural Residues. *GCB Bioenergy*.

Danbaki, C. A., Onyemachi, N. C., Gado, D. S. M., Mohammed, G., Agbenu, D., & Ikegwuiro, P. U. (2020). Precision Agriculture Technology: A Literature Review. *Asian Journal of Advanced Research and Reports*.

Dawid, I., & Workalemahu, S. (2021). Review on Windbreaks Agroforestry as a Climate Smart Agriculture Practices. *American Journal of Agriculture and Forestry*.

Dawson, J., Turner, C., Pileng, O., Farmer, A., McGary, C., Walsh, C., Tamblyn, A., & Yosi, C. K. (2011). Bird Communities of the Lower Waria Valley, Morobe Province, Papua New Guinea: A Comparison Between Habitat Types. *Tropical Conservation Science*.

Degieter, M., Steur, H. D., Tran, D., Gellynck, X., & Schouteten, J. J. (2023). Farmers' Acceptance of Robotics and Unmanned Aerial Vehicles: A Systematic Review. *Agronomy Journal*.

Díaz-Pérez, J. C., & Batal, K. D. (2002). Colored Plastic Film Mulches Affect Tomato Growth and Yield via Changes in Root-Zone Temperature. *Journal of the American Society for Horticultural Science*.

Diekötter, T., & Crist, T. O. (2013). Quantifying Habitat-specific Contributions to Insect Diversity in Agricultural Mosaic Landscapes. *Insect Conservation and Diversity*.

Digi Clip. (2024). *Essential Farm Safety Practices: Protecting Yourself and Your Workers*. Retrieved 11/2/2024 from

Dokin, B. D., & Aletdinova, A. (2021). Automation of Crop Production in the Siberian Region With the Development of Precision Farming Technologies. *E3s Web of Conferences*.

Duan, S. X., Wibowo, S., & Chong, J. (2021). A Multicriteria Analysis Approach for Evaluating the Performance of Agriculture Decision Support Systems for Sustainable Agribusiness. *Mathematics*.

Dubey, A. (2024). sustainable agriculture. In *Encyclopedia Britannica*: Encyclopedia Britannica.

Dunn, P. K., Powierski, A., & Hill, R. (2006). Statistical Evaluation of Data From Tractor Guidance Systems. *Precision Agriculture*.

Edwards, J., Santos-Medellín, C., Nguyen, B. A. T., Kilmer, J., Liechty, Z., Veliz, E., Ni, J., Phillips, G. C., & Sundaresan, V. (2019). Soil Domestication by Rice Cultivation Results in Plant-Soil Feedback Through Shifts in Soil Microbiota. *Genome Biology*.

Eldridge, B. M., Manzoni, L. R., Graham, C. A., Rodgers, B., Farmer, J. R., & Dodd, A. N. (2020). Getting to the Roots of Aeroponic Indoor Farming. *New Phytologist*.

Fallah-Alipour, S., Boshrabadi, H. M., Mehrjerdi, M. R. Z., & Hayati, D. (2018). A Framework for Empirical Assessment of Agricultural Sustainability: The Case of Iran. *Sustainability*.

Farooq, A., Farooq, N., Akbar, H., Hassan, Z. U., & Gheewala, S. H. (2023). A Critical Review of Climate Change Impact at a Global Scale on Cereal Crop Production. *Agronomy*.

Filho, A. A. R., Adams, C., & Murrieta, R. S. S. (2013). The Impacts of Shifting Cultivation on Tropical Forest Soil: A Review. *Boletim Do Museu Paraense Emílio Goeldi Ciências Humanas*.

Foereid, B. (2019). Nutrients Recovered From Organic Residues as Fertilizers: Challenges to Management and Research Methods. *World Journal of Agriculture and Soil Science*.

Food and Agriculture Organisation if the United Nations. (2022). *Agroforestry* Retrieved 7/2/2024 from

Fujiwara, K., Aoyama, C., Takano, M., & Shinohara, M. (2012). Suppression of Ralstonia Solanacearum Bacterial Wilt Disease by an Organic Hydroponic System. *Journal of General Plant Pathology*.

Gairhe, J. J., Adhikari, M. P., Ghimire, D., Khatri-Chhetri, A., & Panday, D. (2021). Intervention of Climate-Smart Practices in Wheat Under Rice-Wheat Cropping System in Nepal. *Climate*.

Gajda, A. M., Czyż, E. A., & Dexter, A. R. (2016). Effects of Long-Term Use of Different Farming Systems on Some Physical, Chemical and Microbiological Parameters of Soil Quality. *International Agrophysics*.

Garrison, A. J., Miller, A. D., Ryan, M. R., Roxburgh, S. H., & Shea, K. (2014). Stacked Crop Rotations Exploit Weed-Weed Competition for Sustainable Weed Management. *Weed Science*.

Gashaw, T., Worqlul, A. W., Dile, Y. T., Addisu, S., Bantider, A., & Zeleke, G. (2020). Evaluating Potential Impacts of Land Management Practices on Soil Erosion in the Gilgel Abay Watershed, Upper Blue Nile Basin. *Heliyon*.

Gaudin, A. C. M., Westra, S. V., Loucks, C., Janovicek, K., Martin, R. C., & Deen, W. M. (2013). Improving Resilience of Northern Field Crop Systems Using Inter-Seeded Red Clover: A Review. *Agronomy*.

Genito, D., Gburek, W. J., & Sharpley, A. N. (2002). Response of Stream Macroinvertebrates to Agricultural Land Cover in a Small Watershed. *Journal of Freshwater Ecology*.

Gilbert, A. D., & Sovacool, B. K. (2016). Looking the Wrong Way: Bias, Renewable Electricity, and Energy Modelling in the United States. *Energy*.

Gnanavelrajah, N., Shrestha, R. P., Schmidt-Vogt, D., & Samarakoon, L. (2007). Carbon Stock Assessment and Soil Carbon Management in Agricultural Land-uses in Thailand. *Land Degradation and Development*.

GoCodes. (2024). *How To Read Construction Plans Like a Pro*. Retrieved 8/2/2024 from

Goedde, L., Katz, J., Ménard, A., & Revellat, J. (2020). Agriculture's connected future: How technology can yield new growth. *McKinsey and Company*.

Gold, M. A. (2023). agroforestry. In *Encyclopedia Britannica*.

Gorjian, S., Fakhraei, O., Gorjian, A., Sharafkhani, A., & Aziznejad, A. (2022). Sustainable Food and Agriculture: Employment of Renewable Energy Technologies. *Current robotics reports*, *3*(3), 153–163.

Government of British Columbia. (2022). *Reducing agricultural greenhouse gases*. Retrieved 10/2/2024 from

Guerry, A. D., Polasky, S., Lubchenco, J., Chaplin-Kramer, R., Daily, G. C., Griffin, R., Ruckelshaus, M., Bateman, I. J., Duraiappah, A. K., Elmqvist, T., Feldman, M. W., Folke, C., Hoekstra, J., Kareiva, P., Keeler, B. L., Li, S., McKenzie, E., Ouyang, Z., Reyers, B., . . . Vira, B. (2015). Natural Capital and Ecosystem Services Informing Decisions: From Promise to Practice. *Proceedings of the National Academy of Sciences*.

Gusev, A. S., Betin, O., Skvortsov, E. A., Ziablitckaia, N. V., Vashukevich, N. V., & Malkova, Y. V. (2020). The Analysis of Factors and Motivative Aspects Promoting and Hindering the Implementation of Precision Farming Technologies. *Wseas Transactions on Environment and Development*.

Hải, N. T. T., Dung, T. A., & Quy, N. T. (2023). Application of Electrochemically Activated Solution to Control Pathogens of Hydroponic Solution. *Vietnam Journal of Science and Technology*.

Halder, J., Sardana, H. R., Pandey, M. K., Nagendran, K., & Bhat, M. N. (2020). Synthesis and Validation IPM Technology and Its Economic Analysis for Bottle Gourd (Lagenaria Siceraria). *The Indian Journal of Agricultural Sciences*.

Hashim, M. H., Adam, A. H. M., Dawoods, M. A., & Fangama, I. M. (2018). Towards Implementing the Integrated Technology of Precision Agriculture in Sudan. *Journal of Agronomy Research*.

He, D.-C., Ma, Y., Li, Z.-Z., Zhong, C.-S., Cheng, Z., & Zhan, J. (2021). Crop Rotation Enhances Agricultural Sustainability: From an Empirical Evaluation of Eco-Economic Benefits in Rice Production. *Agriculture*.

He, Y., Xu, M., Qi, Y., Dong, Y., He, X., Li, J., Liu, X., & Sun, L. (2017). Differential Responses of Soil Microbial Community to Four-Decade Long Grazing and Cultivation in a Semi-Arid Grassland. *Sustainability*.

Holzschuh, A., Dormann, C. F., Tscharntke, T., & Steffan-Dewenter, I. (2012). Mass-Flowering Crops Enhance Wild Bee Abundance. *Oecologia*.

Hrabalova, A., & Zander, K. (2006). Organic Beef Farming in the Czech Republic: Structure, Development and Economic Performance. *Agricultural Economics (Zemědělská Ekonomika)*.

Ifeanyi-obi, C. C., & Henri-Ukoha, A. (2022). Strengthening Climate Change Adaptive Capacity of Rural Women Crop Farmers Through Reduced Social Exclusion in Nigeria. *European Journal of Agriculture and Food Sciences*.

Isgin, T., Özel, R., Bilgiç, A., Florkowski, W. J., & Sevinç, M. R. (2020). DEA Performance Measurements in Cotton Production of Harran Plain, Turkey: A Single and Double Bootstrap Truncated Regression Approaches. *Agriculture*.

Jadidi, S., Badihi, H., & Zhang, Y. (2021). Fault-Tolerant Cooperative Control of Large-Scale Wind Farms and Wind Farm Clusters. *Energies*.

Janati, M. E., Akkal-Corfini, N., Bouaziz, A., Oukarroum, A., Robin, P., Sabri, A., Chikhaoui, M. T., & Thomas, Z. (2021). Benefits of Circular Agriculture for Cropping Systems and Soil Fertility in Oases. *Sustainability*.

Janker, J., Mann, S., & Rist, S. (2018). What Is Sustainable Agriculture? Critical Analysis of the International Political Discourse. *Sustainability*.

Jebli, M. B., & Youssef, S. B. (2017). The Role of Renewable Energy and Agriculture in Reducing CO 2 Emissions: Evidence for North Africa Countries. *Ecological Indicators*.

Jena, P. K. (2021). Nexus Between Climate Change and Agricultural Production in Odisha, India: An ARDL Approach. *International Journal of Environment Agriculture and Biotechnology*.

Jiang, W., Jacobson, M. G., & Langholtz, M. (2018). A Sustainability Framework for Assessing Studies About Marginal Lands for Planting Perennial Energy Crops. *Biofuels Bioproducts and Biorefining*.

Jones, D. L., Cross, P., Withers, P. J. A., DeLuca, T. H., Robinson, D. A., Quilliam, R. S., Harris, I. M., Chadwick, D. R., & Edwards-Jones, G. (2013). REVIEW: Nutrient Stripping: The Global Disparity Between Food Security and Soil Nutrient Stocks. *Journal of Applied Ecology*.

Joshi, G. R., & Bhandari, R. (2023). Determinants of Intensity of Adoption of Climate Change Adaptation Practices in the Agriculture Sector in Nepal. *International Journal of Environment and Climate Change*.

Kabanda, T. (2015). Land Capability Evaluation for Crop Production Using Remote Sensing, Gis and Geostatistics in Rietfontein, North West Province of South Africa. *Geo Uerj*.

Kandul, N. P., Liu, J., C, H. M. S., Wu, S. L., Marshall, J. M., & Akbari, O. S. (2019). Transforming Insect Population Control With Precision Guided Sterile Males With Demonstration In flies. *Nature Communications*.

Kassa, G. (2021). Agroforestry a Pathway to Climate Smart Agribusiness: Lessons to Smallholder Farmers.

Kekkonen, H., Ojanen, H., Haakana, M., Latukka, A., & Regina, K. (2019). Mapping of Cultivated Organic Soils for Targeting Greenhouse Gas Mitigation. *Carbon Management*.

Khajuria, A., & Ravindranath, N. H. (2012). Climate Change in Context of Indian Agricultural Sector. *Journal of Earth Science & Climatic Change*.

Khanal, A., Regmi, P. P., Kc, G. B., Kc, D. B., & Dahal, K. C. (2021). Cost Effective Strategy to Disseminate IPM Technology: A Case of Banke and Surkhet District of Nepal. *International Journal of Social Sciences and Management*.

Kimura, M., & Rodriguez–Amaya, D. B. (2003). Carotenoid Composition of Hydroponic Leafy Vegetables. *Journal of Agricultural and Food Chemistry*.

Koleshko, V. M., Gulay, A., Полынкова, E. B., Gulay, V. A., & Varabei, Y. (2012). Intelligent Systems in Technology of Precision Agriculture and Biosafety.

Ku, L., & Serna, I. (2023, 14/2/2024). New Agriculture Technology in Modern Farming.

Kumar, S., Singh, M., Yadav, K., & Singh, P. (2021). Opportunities and Constraints in Hydroponic Crop Production Systems: A Review. *Environment Conservation Journal*.

Lee, J., & Eom, A.-H. (2009). Effect of Organic Farming on Spore Diversity of Arbuscular Mycorrhizal Fungi and Glomalin in Soil. *Mycobiology*.

Levavasseur, F., & Houot, S. (2022). Predicting the Short- and Long-term Effects of Recycling Organic Wastes in Cropping Systems With the PROLEG Tool. *Soil Use and Management*.

Liebig, M. A., Archer, D. W., Dobrowolski, J., Duiker, S. W., Franzluebbers, A. J., Hendrickson, J., Mitchell, R., Mohamed, A. I., Russell, J. R., & Strickland, T. C. (2017). Aligning Land Use With Land Potential: The Role of Integrated Agriculture. *Agricultural & Environmental Letters*.

Liu, S., Zhang, P., Marley, B., & Liu, W. (2019). The Factors Affecting Farmers' Soybean Planting Behavior in Heilongjiang Province, China. *Agriculture*.

Liu, Y., Yang, Z., Zhu, C., Zhang, B., & Li, H. (2023). The Eco-Agricultural Industrial Chain: The Meaning, Content and Practices. *International journal of environmental research and public health*.

Liu, Y., Yu, Q., Zhou, Q., Wang, C., Bellingrath-Kimura, S. D., & Wu, W. (2022). Mapping the Complex Crop Rotation Systems in Southern China Considering Cropping Intensity, Crop Diversity, and Their Seasonal Dynamics. *Ieee Journal of Selected Topics in Applied Earth Observations and Remote Sensing*.

Liu, Z., Xiong, W., & Cao, X. (2012). Design of Precision Fertilization Management Information System on GPS and GIS Technologies.

Liuzzo, L., & Freni, G. (2018). Implications of land use change on river flow in South West England. *EPiC Series in Engineering*, 3, 1240-1247.

Lu, B., Dao, P. D., Liu, J., He, Y., & Shang, J. (2020). Recent Advances of Hyperspectral Imaging Technology and Applications in Agriculture. *Remote Sensing*.

Luo, G., Feng, Y., Zhang, B., & Cheng, W. (2010). Sustainable Land-Use Patterns for Arid Lands: A Case Study in the Northern Slope Areas of the Tianshan Mountains. *Journal of Geographical Sciences*.

Lv, Y., Zhang, C., Ma, J.-L., Wang, Y., Gao, L., & Li, P. (2019). Sustainability Assessment of Smallholder Farmland Systems: Healthy Farmland System Assessment Framework. *Sustainability*.

Ma, W., Ma, C., Su, Y., & Nie, Z. (2017). Organic Farming. *China Agricultural Economic Review*.

Maho, A., Skenderasi, B., & Cara, M. (2019). Changes in Potato Cultivation Technology in Korça Region as Adaptation to Climate Change. *Italian Journal of Agronomy*.

Makaju, S., & Kurunju, K. (2021). A Review on Use of Agrochemical in Agriculture and Need of Organic Farming in Nepal. *Archives of Agriculture and Environmental Science*.

Manogaran, G., Shakeel, P. M., Fouad, H., Nam, Y., Baskar, S., Chilamkurti, N., & Sundarasekar, R. (2019). Wearable IoT Smart-Log Patch: An Edge Computing-Based Bayesian Deep Learning Network System for Multi Access Physical Monitoring System. *Sensors*.

Margaret, G., Becky, A., Billy, M., Baptist, T., Patrick, N., & Ernest, M. (2022). Prospects for soilless farming in Africa: A review on the aids of plant growth-promoting rhizobacteria inoculants in hydroponics.

Maurya, R. P., Agnihotri, M., Tiwari, S., & Yadav, L. B. (2017). Validation of Integrated Pest Management Module Against Insect Pests of Pigeonpea, Cajanus Cajan in Tarai Region of Uttarakhand. *Journal of Applied and Natural Science*.

McGee, J. A., & Alvarez, C. (2016). Sustaining Without Changing: The Metabolic Rift of Certified Organic Farming. *Sustainability*.

Melstrom, R. T. (2020). The Effect of Land Use Restrictions Protecting Endangered Species on Agricultural Land Values. *American Journal of Agricultural Economics*.

Mengesha, A. K., Mansberger, R., Damyanovic, D., & Stoeglehner, G. (2019). Impact of Land Certification on Sustainable Land Use Practices: Case of Gozamin District, Ethiopia. *Sustainability*.

Messina, C. D., Eeuwijk, F. v., Tang, T., Truong, S. K., McCormick, R. F., Technow, F., Powell, O., Mayor, L., Gutterson, N., Jones, J. W., Hammer, G., & Cooper, M. (2022). Crop Improvement for Circular Bioeconomy Systems. *Journal of the Asabe*.

Metson, G. S., & Bennett, E. M. (2015). Phosphorus Cycling in Montreal's Food and Urban Agriculture Systems. *Plos One*.

Metson, G. S., Powers, S. M., Hale, R. L., Sayles, J. S., Öberg, G., MacDonald, G. K., Kuwayama, Y., Springer, N. P., Weatherley, A., Hondula, K. L., Jones, K., Chowdhury, R. B., Beusen, A., & Bouwman, A. F. (2017). Socio-Environmental Consideration of Phosphorus Flows in the Urban Sanitation Chain of Contrasting Cities. *Regional Environmental Change*.

Minnesota Department of Revenue. (2024). *Defining Agricultural Production*. Retrieved 14/2/2024 from

Mitchell, S., Weersink, A., & Erickson, B. (2018). Adoption of Precision Agriculture Technologies in Ontario Crop Production. *Canadian Journal of Plant Science.*

Mohapatra, S., Mohapatra, S., Han, H., Ariza-Montes, A., & López-Martín, M. d. C. (2022). Climate Change and Vulnerability of Agribusiness: Assessment of Climate Change Impact on Agricultural Productivity. *Frontiers in psychology.*

Mok, H. F., Williamson, V. G., Grove, J., Burry, K., Barker, F., & Hamilton, A. J. (2013). Strawberry Fields Forever? Urban Agriculture in Developed Countries: A Review. *Agronomy for Sustainable Development.*

Moreno-Espíndola, I. P., Ferrara-Guerrero, M. J., Luna-Guido, M., Ramírez-Villanueva, D. A., León-Lorenzana, A. S. d., Gómez-Acata, S., González-Terreros, E., Ramírez-Barajas, B., Navarro-Noya, Y. E., Sánchez-Rodríguez, L. M., Fuentes-Ponce, M., Macedas-Jímenez, J. U., & Dendooven, L. (2018). The Bacterial Community Structure and Microbial Activity in a Traditional Organic Milpa Farming System Under Different Soil Moisture Conditions. *Frontiers in Microbiology.*

Mostafaeipour, A., Goudarzi, H., Khanmohammadi, M., Jahangiri, M., Sedaghat, A., Norouzianpour, H., Chowdhury, S., Techato, K., Issakhov, A., Almutairi, K., & Dehshiri, S. J. H. (2021). Techno-economic Analysis and Energy Performance of a Geothermal Earth-to-air Heat Exchanger (EAHE) System in Residential Buildings: A Case Study. *Energy Science & Engineering.*

Mostefaoui, Z., & Amara, S. (2019). Renewable energy analysis in the agriculture–greenhouse farms: A case study in the mediterranean region (sidi bel abbes, algeria). *Environmental Progress & Sustainable Energy, 38*(3), e13029.

Mugera, A., & Langemeier, M. R. (2011). Does Farm Size and Specialization Matter for Productive Efficiency? Results From Kansas. *Journal of Agricultural and Applied Economics.*

Mwalukasa, N. (2013). Agricultural Information Sources Used for Climate Change Adaptation in Tanzania. *Library Review.*

Nie, J., & Yang, B. (2021). A Detailed Study on GPS and GIS Enabled Agricultural Equipment Field Position Monitoring System for Smart Farming. *Scalable Computing Practice and Experience.*

Niu, W., Shi, J., Xu, Z., Wang, T., Zhang, H., & Su, X. (2022). Evaluating the Sustainable Land Use in Ecologically Fragile Regions: A Case Study of the Yellow River Basin in China. *International journal of environmental research and public health.*

Nowak, B., Nesme, T., David, C., & Pellerin, S. (2013). To What Extent Does Organic Farming Rely on Nutrient Inflows From Conventional Farming? *Environmental Research Letters.*

NSW EPA. (2024). *Integrated pest management.* Retrieved 7/2/2024 from

Ntakirutimana, L., Li, F., Huang, X., Wang, S., & Yin, C. (2019). Green Manure Planting Incentive Measures of Local Authorities and Farmers' Perceptions of the Utilization of Rotation Fallow for Sustainable Agriculture in Guangxi, China. *Sustainability.*

Ogunbameru, B. O., Mustapha, S., & Idrisa, Y. L. (2013). Capacity Building for Climate Change Adaptation: Modules for Agricultural Extension Curriculum Development. *Russian Journal of Agricultural and Socio-Economic Sciences.*

Oostendorp, R., Asseldonk, M. v., Gathiaka, J., Mulwa, R., Radeny, M. A. O., Recha, J. W. M., Wattel, C. J., & Wesenbeeck, C. F. A. v. (2019). Inclusive Agribusiness Under Climate Change: A Brief Review of the Role of Finance. *Current Opinion in Environmental Sustainability.*

Paul, C. J. M., Nehring, R. F., Banker, D. E., & Somwaru, A. (2004). Scale Economies and Efficiency in U.S. Agriculture: Are Traditional Farms History? *Journal of Productivity Analysis.*

Paul McAlister Architects. (2024). *Embodied Energy of Materials - 76.* Retrieved 8/2/2024 from

Pawlewicz, A., Brodzińska, K., Zvirbule, A., & Popluga, D. (2020). Trends in the Development of Organic Farming in Poland and Latvia Compared to the EU. *Rural Sustainability Research.*

Piñeiro, V., Arias, J., Dürr, J., Elverdin, P., Ibáñez, A. M., Kinengyere, A. A., Opazo, C., Owoo, N. S., Page, J., Prager, S. D., & Torero, M. (2020). A Scoping Review on Incentives for Adoption of Sustainable Agricultural Practices and Their Outcomes. *Nature Sustainability.*

Preißel, S., Reckling, M., Schläfke, N., & Zander, P. (2015). Magnitude and Farm-Economic Value of Grain Legume Pre-Crop Benefits in Europe: A Review. *Field Crops Research.*

Purvis, E. E. N., Meehan, M. L., & Lindo, Z. (2019). Agricultural Field Margins Provide Food and Nesting Resources to Bumble Bees (<i>Bombus</I> Spp., Hymenoptera: Apidae) in Southwestern Ontario, Canada. *Insect Conservation and Diversity*.

Rao, N. H. (2018). Big Data and Climate Smart Agriculture-Status and Implications for Agricultural Research and Innovation in India. *Proceedings of the Indian National Science Academy*.

Rauw, W. M., Gomez-Raya, L., Star, L., Øverland, M., Delezie, E., Grīviņš, M., Hamann, K. T., Pietropaoli, M., Klaassen, M. T., Klemetsdal, G., Gil, M. G., Torres, O., Dvergedal, H., & Formato, G. (2022). Sustainable Development in Circular Agriculture: An Illustrative Bee↺legume↺poultry Example. *Sustainable Development*.

Ray, D. K., West, P., Clark, M., Gerber, J., Prishchepov, A. V., & Chatterjee, S. (2019). Climate Change Has Likely Already Affected Global Food Production. *Plos One*.

Rejesus, R. M., & Jones, M. (2020). Perspective: Enhancing Economic Evaluations and Impacts of Integrated Pest Management Farmer Field Schools (<scp>IPM-FFS</Scp>) in Low-income Countries. *Pest Management Science*.

Rokicki, T., Ratajczak, M., Bórawski, P., Bełdycka-Bórawska, A., Gradziuk, B., Gradziuk, P., & Siedlecka, A. (2021). Energy Self-Subsistence of Agriculture in EU Countries. *Energies*.

Sarcinelli, O., Romeiro, A. R., Pereira, L. B., & Tosto, S. G. (2022). Land Use Capability and the Sustainable Scale: An Overview of Agriculture in São Paulo State, Brazil. *Natural Resources*.

Schimmelpfennig, D., & Lowenberg-DeBoer, J. (2020). Farm Types and Precision Agriculture Adoption: Crops, Regions, Soil Variability, and Farm Size. *SSRN Electronic Journal*.

Schöll, E. M., Eschberger-Friedl, A., Schai-Braun, S. C., & Frey-Roos, F. (2023). Habitat Preferences and Similarities of Grey Partridges and Common Pheasants in Agricultural Landscapes Under Organic and Conventional Farming. *European Journal of Wildlife Research*.

Segarra, J., Buchaillot, M. L., Araus, J. L., & Kefauver, S. C. (2020). Remote Sensing for Precision Agriculture: Sentinel-2 Improved Features and Applications. *Agronomy*.

Seyoum, B. (2016). Assessment of Soil Fertility Status of Vertisols Under Selected Three Land Uses in Girar Jarso District of North Shoa Zone, Oromia National Regional State, Ethiopia. *Environmental Systems Research*.

Shafi, U., Mumtaz, R., García-Nieto, J., Hassan, S. A., Zaidi, S. A. R., & Iqbal, N. (2019). Precision Agriculture Techniques and Practices: From Considerations to Applications. *Sensors*.

Sharma, S., & Srushtideep, A. (2022). Precision Agriculture and Its Future. *International Journal of Plant & Soil Science*.

Shrestha, G. (2015). Soil Properties and Soil Management Practices in Commercial Organic and Conventional Vegetable Farms in Kathmandu Valley. *Nepal Journal of Science and Technology*.

Simončič, A., Stopar, M., Bolta, Š. V., Bavčar, D., Leskovšek, R., & Česnik, H. B. (2015). Integrated Pest Management of "Golden Delicious" Apples. *Food Additives and Contaminants Part B*.

Skotnicka-Siepsiak, A. (2020). Operation of a Tube GAHE in Northeastern Poland in Spring and Summer—A Comparison of Real-World Data With Mathematically Modeled Data. *Energies*.

Sobczak, W., & Sobczak, A. (2022). Farmers' Attitudes Towards Renewable Energy Sources. *Annals of the Polish Association of Agricultural and Agribusiness Economists*.

Stephens, M., Hazard, K., Moser, D. K., Cox, D. R., Rose, R., & Alkon, A. (2017). An Integrated Pest Management Intervention Improves Knowledge, Pest Control, and Practices in Family Child Care Homes. *International journal of environmental research and public health*.

Stevens, R. J. A. M. (2023). Understanding Wind Farm Power Densities. *Journal of Fluid Mechanics*.

Stevenson, F. C., & Kessel, C. v. (1996). The Nitrogen and Non-Nitrogen Rotation Benefits of Pea to Succeeding Crops. *Canadian Journal of Plant Science*.

Stewart, R. E. (2024). agricultural technology. In *Encyclopedia Britannica*.

Stratton, M. (2022, 7/2/2024). What Is Community Supported Agriculture (CSA)?

Subedi, N., & Poudel, S. (2020). Effects of Climate Change on Agriculture and Its Mitigation Through Climate Smart Agriculture Practices in Nepal. *Tropical Agrobiodiversity*.

Sun, C., Zhou, J., Ma, Y., Xu, Y., Pan, B., & Zhang, Z. (2022). A Review of Remote Sensing for Potato Traits Characterization in Precision Agriculture. *Frontiers in Plant Science*.

Suwanmaneepong, S., Kerdsriserm, C., Iyapunya, K., & Wongtragoon, U. (2020). Farmers' Adoption of Organic Rice Production in Chachoengsao Province, Thailand. *Journal of Agricultural Extension*.

Tadesa, E. (2020). Review on Climate Change Adaptation Strategies in Ethiopia. *International Journal of Energy and Environmental Science*.

Talukder, B., Blay-Palmer, A., Hipel, K. W., & vanLoon, G. W. (2017). Elimination Method of Multi-Criteria Decision Analysis (MCDA): A Simple Methodological Approach for Assessing Agricultural Sustainability. *Sustainability*.

Tanigawa, S. (2017). *Fact Sheet | Biogas: Converting Waste to Energy*. Environmental and Energy Study Institute. Retrieved 10/2/2024 from

Thảo, N. T. T., Tung, T. V., Thao, N. T. P., Hải, L. T., Braunegg, S., Braunegg, G., & Schnitzer, H. (2020). Energy Efficiency in an Integrated Agro-Ecosystem Within an Acidic Soil Area of the Mekong Delta, Vietnam.

The Constructor. (2024). *What is Embodied Energy of Building Materials?* Retrieved 8/2/2024 from

Tiffin, R., & Balcombe, K. (2011). The Determinants of Technology Adoption by UK Farmers Using Bayesian Model Averaging: The Cases of Organic Production and Computer Usage. *Australian Journal of Agricultural and Resource Economics*.

Turyasingura, B., & Chavula, P. (2022). Climate-Smart Agricultural Extension Service Innovation Approaches in Uganda: Review Paper. *International Journal of Food Science and Agriculture*.

UAV Training Australia. (2024, 14/2/2024). How Drones Are Used in Agriculture.

Union of Concerned Scientists. (2008, 9/2/2024). Renewable Energy and Agriculture: A Natural Fit.

USDA. (2024a). Retrieved 12/2/2024 from

USDA. (2024b). *Agriculture Technology*. Retrieved 14/2/2024 from

Vadlamudi, S. (2020). Internet of Things (IoT) in Agriculture: The Idea of Making the Fields Talk. *Engineering International*.

Vaish, S., Garg, N., & Ahmad, I. (2020). Microbial Basis of Organic Farming Systems With Special Reference to Biodynamic Preparations. *The Indian Journal of Agricultural Sciences*.

Valdez-Juarez, S. O., Krebs, E., Drake, A., & Green, D. J. (2019). Assessing the Effect of Seasonal Agriculture on the Condition and Winter Survival of a Migratory Songbird in Mexico. *Conservation Science and Practice*.

Vintila, A. (2022). KPIs in Agriculture: Sustainable Practices for Farmers. *Performance Magazine*.

Viola, E., & Mendes, V. (2022). Agriculture 4.0 and Climate Change in Brazil. *Ambiente & Sociedade*.

Widjonarko, W., & Maryono, M. (2022). Sustainable Land Use Model in Garang Watershed. *Iop Conference Series Earth and Environmental Science*.

Williams, D. L., & Wise, K. (1997). Perceptions of Iowa Secondary School Agricultural Education Teachers and Students Regarding Sustainable Agriculture. *Journal of Agricultural Education*.

Wolff, A., Paul, J.-P., Martin, J. L., & Bretagnolle, V. (2001). The Benefits of Extensive Agriculture to Birds: The Case of the Little Bustard. *Journal of Applied Ecology*.

Wolińska, A., Górniak, D., Zielenkiewicz, U., Goryluk-Salmonowicz, A., Kuźniar, A., Stępniewska, Z., & Błaszczyk, M. (2017). Microbial Biodiversity in Arable Soils Is Affected by Agricultural Practices. *International Agrophysics*.

Wood, J., Wong, C., & Paturi, S. (2021). Addressing Food Security in Constrained Urban Environments.

Xu, J., Wang, J., Wang, H., & Li, C. (2022). Evolution Trend and Promotion Potential of Environmental Efficiency of Dairy Farming in China From the Perspective of "Club Convergence". *Frontiers in Environmental Science*.

Xuan, X., Liu, B., & Zhang, F. (2021). Climate Change and Adaptive Management: Case Study in Agriculture, Forestry and Pastoral Areas. *Land*.

Yakubiv, V., Hryhoruk, I., Maksymiv, Y., & Popadynets, N. (2019). Strategic Analysis of the Potential of Bioenergy: Outlook for Ukraine.

Yang, C. (2018). High Resolution Satellite Imaging Sensors for Precision Agriculture. *Frontiers of Agricultural Science and Engineering*.

Yu, T., Mahe, L. P., Li, Y., Xue, W., Deng, X., & Zhang, D. (2022). Benefits of Crop Rotation on Climate Resilience and Its Prospects in China. *Agronomy*.

Yu, X., Xin, L., Li, X., Tan, M., & Wang, Y. (2019). Exploring a Moderate Operation Scale in China's Grain Production: A Perspective on the Costs of Machinery Services. *Sustainability*.

Zhang, M., Arendshorst, M. G., & Stevens, R. J. A. M. (2018). Large Eddy Simulations of the Effect of Vertical Staggering in Large Wind Farms. *Wind Energy*.

Zhao, C., Liu, B., Piao, S., Wang, X., Lobell, D. B., Huang, Y., Huang, M., Yao, Y., Bassu, S., Ciais, P., Durand, J. L., Elliott, J., Ewert, F., Janssens, I. A., Li, T., Lin, E., Liu, Q., Martre, P., Müller, C., . . . Asseng, S. (2017). Temperature Increase Reduces Global Yields of Major Crops in Four Independent Estimates. *Proceedings of the National Academy of Sciences*.

Zhao, F. (1991). Impact and Implications of Price Policy and Land Degradation on Agricultural Growth in Developing Countries. *Agricultural Economics*.

Zhou, D., Rongqun, Z., Liu, L., & Gao, L. (2009). Evaluation of the Sustainable Land Use Status of the North China Plain. *International Journal of Sustainable Development & World Ecology*.

Безносов, Г. А., Semin, A. N., Skvortsov, E. A., & Volkova, S. A. (2019). The Economic Essence of the Category of Precision Agriculture.

Index

www.ingramcontent.com/pod-product-compliance
Lightning Source LLC
Chambersburg PA
CBHW051749200326
41597CB00025B/4494